Tearing Mode Dynamics in Tokamak Plasmas

Online at: https://doi.org/10.1088/978-0-7503-5367-0

IOP Series in Plasma Physics

Series Editor

About the series

The IOP Plasma Physics ebook series aims at comprehensive coverage of the physics and applications of natural and laboratory plasmas, across all temperature regimes. Books in the series range from graduate and upper-level undergraduate textbooks, to research monographs and reviews.

The conceptual areas of plasma physics addressed in the series include:

- Equilibrium, stability, and control.
- Waves: fundamental properties, emission, and absorption.
- Nonlinear phenomena and turbulence.
- Transport theory and phenomenology.
- Laser–plasma interactions.
- Non-thermal and suprathermal particle populations.
- Beams and non-neutral plasmas.
- High energy density physics.
- Plasma–solid interactions, dusty, complex, and non-ideal plasmas.
- Diagnostic measurements and techniques for data analysis.

The fields of application include:

- Nuclear fusion through magnetic and inertial confinement.
- Solar–terrestrial and astrophysical plasma environments and phenomena.
- Advanced radiation sources.
- Materials processing and functionalisation.
- Propulsion, combustion, and bulk materials management.
- Interaction of plasma with living matter and liquids.
- Biological, medical, and environmental systems.
- Low-temperature plasmas, glow discharges, and vacuum arcs.
- Plasma chemistry and reaction mechanisms.
- Plasma production by novel means.

A full list of titles published in this series can be found here: https://iopscience.iop.org/bookListInfo/iop-plasma-physics-series.

Tearing Mode Dynamics in Tokamak Plasmas

Richard Fitzpatrick
Institute for Fusion Studies, University of Texas at Austin, TX, USA

IOP Publishing, Bristol, UK

ISBN 978-0-7503-5367-0 (ebook)
ISBN 978-0-7503-5365-6 (print)
ISBN 978-0-7503-5368-7 (myPrint)
ISBN 978-0-7503-5366-3 (mobi)

DOI 10.1088/978-0-7503-5367-0

Version: 20230601

IOP ebooks

British Library Cataloguing-in-Publication Data: A catalogue record for this book is available from the British Library.

Published by IOP Publishing, wholly owned by The Institute of Physics, London

IOP Publishing, No.2 The Distillery, Glassfields, Avon Street, Bristol, BS2 0GR, UK

US Office: IOP Publishing, Inc., 190 North Independence Mall West, Suite 601, Philadelphia, PA 19106, USA

Contents

Preface

The development of humankind's ultimate energy source, *nuclear fusion*, has proceeded slowly but surely over the course of the last 60 years. Moreover, the perceived need for such an energy source has never been more acute than it is at present. Of all of the plasma confinement schemes that have been attempted over the years, *magnetic confinement*, by which a thermonuclear plasma equilibrium is contained by a strong magnetic field, seems to be the most practical. Moreover, by far and away the most successful magnetic confinement device is the *tokamak*.

A tokamak is a device whose purpose is to confine a thermonuclear plasma on a set of *axisymmetric*, nested, toroidal magnetic flux-surfaces generated by a combination of electrical currents flowing in external field coils and currents induced within the plasma itself by transformer action. Confinement is possible because, although heat and charged particles stream along magnetic field lines very rapidly, they can only diffuse across magnetic flux surfaces comparatively slowly.

Unlike most naturally occurring plasmas (e.g. the solar wind), tokamak plasmas are extremely quiescent. (Of course, this is by design.) Tokamak plasma discharges usually last tens of millions of Alfvén times. (The Alfvén time is the typical timescale on which Alfvén waves traverse the plasma and also on which ideal magnetohydrodynamic (MHD) instabilities grow; it is of the order of a tenth of a microsecond in conventional tokamak plasmas.)

Tokamak plasmas are sometimes terminated by violent events known as *disruptions*. One major class of disruption is caused by the plasma discharge crossing an ideal MHD stability boundary. However, such disruptions are easy to avoid, because the locations of the stability boundaries in operational space can be calculated very accurately.

The overwhelming majority of disruptions that are not caused by crossing ideal stability boundaries are associated with *tearing modes*. Tearing modes are slow-growing, macroscopic instabilities in tokamak plasmas that tear and reconnect magnetic field lines at various resonant surfaces in the plasma to produce radially localized magnetic island chains. Tearing modes are driven by radial current and pressure gradients within the plasma and can be unstable even when the plasma is ideally stable. Tearing modes degrade plasma confinement because heat and particles can flow very rapidly from one (radial) side of a magnetic island chain to another by streaming along magnetic field lines, rather than by having to slowly diffuse across magnetic flux surfaces. Tearing modes in tokamak plasmas generally saturate at fairly low amplitudes (such that the associated magnetic island chains have radial extents that are a few percent of the plasma's minor radius), and can persist over a large fraction of the lifetime of the plasma discharge.

Tearing modes in tokamak plasmas usually rotate rapidly (at many kilo-radians per second) as a consequence of plasma flows induced by the radial density and temperature gradients in the plasma. However, tearing modes that grow to comparatively large amplitudes tend to *slow down* due to eddy currents induced in the vacuum vessel surrounding the plasma and eventually *lock* (i.e. become stationary in the laboratory frame) to static imperfections in the externally generated magnetic field known as *error fields*. Such tearing modes often trigger disruptions. In fact, there is a very clear correlation between the occurrence of so-called *locked modes* and disruptions.

Tearing modes are generally driven to instability by radial current and pressure gradients within tokamak plasmas. However, there exists a particularly virulent class of tearing modes, known as *neoclassical tearing modes*, that is driven by the loss of the neoclassical bootstrap current inside the separatrix of a magnetic island chain consequent on the flattening of the plasma pressure profile within the separatrix.

Tearing modes in tokamak plasmas are very poorly described by conventional single-fluid resistive MHD because of the relatively low collisionality of such plasmas combined with the significantly different drift velocities of the various plasma species. Tearing modes are also not always well described by linear analysis, which becomes invalid as soon as the radial widths of the magnetic island chains at the various resonant surfaces exceed the (very narrow) linear layer widths.

The aim of this book is to outline a realistic, comprehensive, self-consistent, analytic theory of tearing-mode dynamics in tokamak plasmas. The theory in question models the plasma as a multi-component fluid (a kinetic approach would be infeasible) and makes extensive use of asymptotic matching methods.

Chapter 1 estimates the typical plasma parameters needed to achieve thermonuclear fusion in a conventional tokamak and goes on to give a general overview of tearing modes in tokamaks. Chapter 2 outlines the fundamental fluid theory that underpins the analysis of tearing mode dynamics in tokamak plasmas. This task is complicated by the low collisionality of tokamak plasmas, which requires a so-called *neoclassical* closure of the parallel (to the magnetic field) dynamics, as well as by the presence of small-scale plasma turbulence, which necessitates a phenomenological closure of the perpendicular dynamics. Chapter 3 introduces an approximation that forms the basis of much of the analysis in this book, in which a tokamak plasma is treated as a periodic cylinder. The chapter also introduces the fundamental asymptotic matching method that underpins all tearing-mode theory. In chapter 4, a reduced drift-MHD model is extracted from the fundamental fluid equations derived in chapter 2 by first neglecting all specifically neoclassical terms in the equations and then removing from the remaining equations the irrelevant physics of compressible Alfvén waves. Chapter 5 uses the reduced drift-MHD model of chapter 4 to determine all linear response regimes of a resonant layer interacting with a rotating magnetic perturbation. In chapter 6, the linear response theory of chapter 5 is employed to estimate the linear growth rates, rotation frequencies, and resonant-layer thicknesses of tearing modes in tokamak plasmas. Chapter 7 uses the linear response theory of chapter 5 to determine the critical error-field amplitude above which such a field is able to introduce a locked magnetic island chain into a tokamak plasma. Chapter 8 employs the reduced drift-MHD model of chapter 5 to determine the nonlinear response of a magnetic island chain to a rotating magnetic perturbation. In chapter 9, the nonlinear response theory of chapter 8 is used to analyze the growth, saturation, and rotation of nonlinear tearing modes in tokamak plasmas. Chapter 10 employs the nonlinear response theory of chapter 8 to investigate the braking of a magnetic island chain's rotation when it interacts electromagnetically with a resistive vacuum vessel. In chapter 11, a neoclassical reduced drift-MHD model is extracted from the fundamental fluid equations derived in chapter 2 by first retaining all specifically neoclassical terms in the equations and then removing from the equations the irrelevant physics of compressible-Alfvén waves. Chapter 12 uses the neoclassical reduced drift-MHD model of chapter 11 to investigate the physics of neoclassical tearing modes. In chapter 13, the neoclassical reduced drift-MHD model of chapter 11 is employed to analyze the locking of a rotating magnetic island chain to a static error field. Finally, chapter 14 generalizes the analysis of the book to take the true toroidal geometry of tokamak plasmas into account.

The author would like to express his gratitude to the teachers, colleagues, and students with whom he has interacted over the years in his quest to gain a more complete understanding of tearing mode dynamics in tokamak plasmas. These include R J Hastie, J W Connor, J B Taylor, T C Hender, A W Morris, G M Fishpool, C G Gimblett, H R Wilson, C M Roach, A Thyagaraja, and D A Gates at the UKAEA Culham Laboratory; J A Wesson, P G Carolan, and M F F Nave at the Joint European Torus (JET); T H Jensen, R J La Haye, J T Scoville, M S Chu, and C Paz-Solden at General Atomics; F L Waelbroeck, R Hazeltine, P G Watson, F Militello, E P Yu, E Rossi, A J Cole, R L White, R Carrera, W L Rowan, R D Bengston, E R Solano, P H Edmonds, H Gasquet, G Cima, and A J Wooten at the University of Texas at Austin; J D Callen, C C Hegna, B E Chapman, D Craig, and S C Prager at the University of Wisconsin-Madison; M E Mauel, G A Navratil, A M Garofalo, S A Sabbagh, and D A Maurer at Columbia University; J P Freidberg, I H Hutchinson, R S Granetz, S M Wolfe, and A Hubbard at the Massachusetts Institute of Technology (MIT); A H Glasser and J M Finn at Los Alamos National Laboratory; K M McGuire, J Bialek, M Okabayashi, H R Strauss, A O Nelson, A H Reiman, J-K Park, S Kim, A Bhattacharjee, D P Brennan, N C Logan, Q M Hu, and R Nazikian at Princeton Plasma Physics Laboratory; A I Smolyakov at the University of Saskatchewan; S C Guo, D F Escande, P Zanca, and P Martin at the Consorzio RFX, Padua; F Porcelli and D Grasso at the Politecnico di Torino; J Lee at the Korea Institute of Fusion Energy, Daejeon; K H Finken at Jülich; and H Zohm and S Günter at the Max-Planck-Institut für Plasmaphysik, Garching.

Author biography

Richard Fitzpatrick

Richard Fitzpatrick earned a Master's degree in physics from the University of Cambridge and a DPhil in astronomy from the University of Sussex. After working in the UKAEA's Culham Laboratory for seven years, he became a professor of physics at the University of Texas at Austin, where he has been a faculty member since 1994. He has published over 100 papers in scientific journals, most of which are concerned with various aspects of the theory of tearing mode dynamics in tokamak plasmas. He is a member of the Royal Astronomical Society and a fellow of the American Physical Society. He is the author of *Maxwell's equations and the Principles of Electromagnetism* (2008), *An Introduction to Celestial Mechanics* (2012), *Oscillations and Waves: An Introduction* (2013), *Plasma Physics: An Introduction* (2014), *Quantum Mechanics* (2015), *Theoretical Fluid Mechanics* (2017), *Oscillations and Waves: An Introduction*, 2nd edn (2019), *Thermodynamics and Statistical Mechanics* (2020), *Newtonian Dynamics: An Introduction* (2022), and *Plasma Physics: An Introduction*, 2nd edn (2022).

IOP Publishing

Tearing Mode Dynamics in Tokamak Plasmas

Richard Fitzpatrick

Chapter 1

Introduction

1.1 Introduction

The aims of this chapter are, first, to estimate the typical plasma parameters needed to achieve thermonuclear fusion in a conventional magnetic confinement device such as a tokamak, and second, to give a general overview of tearing modes in tokamaks.

1.2 Thermonuclear fusion

Existing nuclear reactors derive energy from the fission of heavy nuclei (i.e., ^{235}U and ^{239}Pu). On the other hand, a nuclear fusion reactor would derive energy from the fusion of light nuclei (i.e. ^{2}H and ^{3}H). The main advantages of nuclear fusion reactors are, first, that they do not generate heat-producing, highly radioactive, long-lived fission products, and, second, that the principal fuel for such reactors (namely, deuterium) is abundantly present on the Earth. In fact, deuterium has a natural abundance in the Earth's oceans of about one atom in 6420 of hydrogen.

In order for two light nuclei to fuse together, they must collide with sufficient violence that one of the nuclei is able to quantum mechanically tunnel through the other's repulsive Coulomb barrier [1]. In so-called *thermonuclear* fusion, both species of nuclei are confined within a plasma, and have *thermal* (i.e. Maxwellian) velocity distribution functions characterized by a common temperature. The collisions that give rise to nuclear fusion reactions are associated with the random motions of the nuclei within the plasma, and have mean energies (in the center of mass frame) that are directly proportional to the plasma temperature. Thus, the higher the temperature, the more violent the collisions. It follows that if the temperature is made sufficiently high, nuclear fusion will occur. However, a high-temperature plasma is, necessarily, also a high-pressure plasma and thus requires confinement to prevent it from dispersing. The thermonuclear plasma at the center of the Sun is confined by gravity. Unfortunately, gravity is too weak a force to confine a terrestrial thermonuclear plasma of realistic size. Instead, such plasmas can be confined by a strong magnetic field. In principle, magnetic confinement is possible if the gyroradii

doi:10.1088/978-0-7503-5367-0ch1

of the charged particles that make up the plasma are much smaller than the plasma dimensions [2].

1.3 Nuclear fusion reactions

The nuclear fusion reaction that is of principal interest for the purposes of energy production in a magnetically confined thermonuclear plasma is

$$\mathrm{D} + \mathrm{T} \rightarrow \mathrm{n} + \alpha. \tag{1.1}$$

Here, D denotes a deuterium nucleus (^2H), T denotes a tritium nucleus (^3H), n denotes a neutron, and α denotes an alpha particle (^4He). At achievable mean plasma temperatures (i.e. less than about 10 keV), a D–T fusion reaction has a cross section that is approximately 100 times greater than that of a D–D fusion reaction (or a T–T fusion reaction) [1]. (There is, unfortunately, no H–H fusion reaction.) For this reason, D–T fusion is considered to be more practical than D–D fusion, despite the fact that there is no natural source of tritium on the Earth. In fact, it is envisaged that D–T fusion reactors will breed the requisite tritium within a blanket that surrounds the plasma via nuclear reactions such as

$$^6Li + \mathrm{n} \rightarrow \alpha + \mathrm{T}. \tag{1.2}$$

Note that ^6Li makes up 7.6% of terrestrial lithium, which makes up about 0.002% of the Earth's crust.

A D–T fusion reaction liberates

$$\Delta\mathcal{E}_{\mathrm{DT}} = 17.59\ \mathrm{MeV} \tag{1.3}$$

of kinetic energy that is subsequently carried off by the fusion products [1]. In a thermonuclear plasma, the momenta of the fusion products exceed those of the fusion reagents by many orders of magnitude, so the fusion reaction effectively takes place in the center of mass frame. Consequently, momentum and energy conservation require each product to carry off the fraction of the liberated energy that is inversely proportional to its mass (given that the products are moving non-relativistically) [3]. The masses of a neutron and an alpha particle are $m_{\mathrm{n}} = 1.675 \times 10^{-27}$ kg and $m_\alpha = 6.645 \times 10^{-27}$ kg, respectively. Hence, the kinetic energies of the neutron and the alpha particle generated by a D–T fusion reaction are

$$\mathcal{E}_{\mathrm{n}} = \left(\frac{m_\alpha}{m_{\mathrm{n}} + m_\alpha}\right)\Delta\mathcal{E}_{\mathrm{DT}} = 14.05\ \mathrm{MeV}, \tag{1.4}$$

$$\mathcal{E}_\alpha = \left(\frac{m_{\mathrm{n}}}{m_{\mathrm{n}} + m_\alpha}\right)\Delta\mathcal{E}_{\mathrm{DT}} = 3.541\ \mathrm{MeV}, \tag{1.5}$$

respectively. In a magnetic confinement device, the alpha particles generated by D–T fusion reactions are confined by the magnetic field that permeates the plasma and subsequently slow down and heat the deuterium and tritium nuclei, thereby maintaining the nuclear fusion reactions. On the other hand, the neutrons generated by D–T fusion reactions exit the plasma, and are absorbed by the surrounding blanket. The energy absorbed by the blanket is extracted via a conventional heat exchanger and used to generate electrical power. As has already been mentioned, the neutrons also breed tritium within the blanket, thereby replacing the tritium burned up in the fusion reactions.

Consider a thermonuclear plasma consisting principally of electrons, deuterium ions (that are fully stripped of electrons), and (fully stripped) tritium ions. All three species are assumed to have Maxwellian velocity distribution functions characterized by a common temperature, T. The rate of D–T fusion reactions occurring per unit volume within the plasma is [4]

$$f = n_{\rm D}\, n_{\rm T}\, \langle\sigma\, v\rangle_{\rm DT}(T), \tag{1.6}$$

where $n_{\rm D}$ is the deuteron number density, $n_{\rm T}$ the triton number density, σ the cross section for D–T fusion reactions, v the relative velocity of the reacting species, and $\langle\ \rangle$ denotes an average over the Maxwellian distributions of the reacting species. In the range of temperatures from 1–100 keV, the D–T fusion reactivity, $\langle\sigma\, v\rangle_{\rm DT}(T)$, is accurately fitted by the following formula [1, 5]

$$\langle\sigma\, v\rangle_{\rm DT}(T) = C_0\, \zeta^{\,-5/6}\, \xi^{\,2}\, \exp(-3\, \zeta^{\,1/3}\, \xi), \tag{1.7}$$

where

$$\xi = \frac{C_1}{T^{1/3}}, \tag{1.8}$$

$$\zeta = 1 - \frac{C_2\, T + C_4\, T^2 + C_6\, T^3}{1 + C_3\, T + C_5\, T^2 + C_7\, T^3}. \tag{1.9}$$

Here, T is measured in units of keV. The parameters C_1–C_7 are specified in table 1.1. Figure 1.1 plots $\langle\sigma\, v\rangle_{\rm DT}(T)$ for a realistic range of plasma temperatures. It can be seen that $\langle\sigma\, v\rangle_{\rm DT}(T)$ is a rapidly increasing function of increasing temperature, and that $\langle\sigma\, v\rangle_{\rm DT} \simeq 10^{-22}$ m^3 s^{-1} when $T \simeq 10$ keV.

Table 1.1. Parameters for formulae (1.7)–(1.9).

C_0 (m^3 s^{-1})	6.4341×10^{-20}
C_1 (keV$^{1/3}$)	6.6610
C_2 (keV^{-1})	1.5136×10^{-2}
C_3 (keV^{-1})	7.5189×10^{-2}
C_4 (keV^{-2})	4.6064×10^{-3}
C_5 (keV^{-2})	1.3500×10^{-2}
C_6 (keV^{-3})	-1.0675×10^{-4}
C_7 (keV^{-3})	1.366×10^{-5}

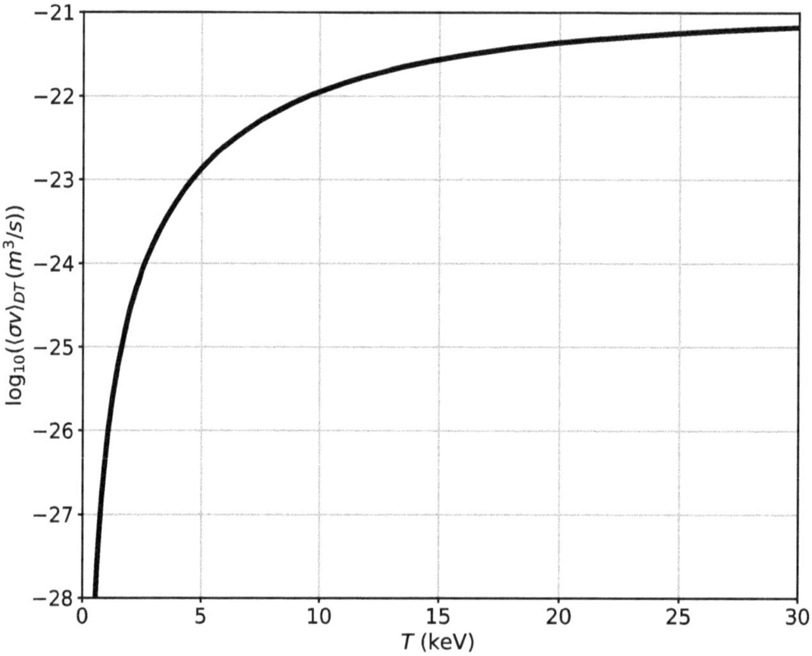

Figure 1.1. D–T fusion reactivity plotted as a function of the assumed common temperature of the fusion reagents.

1.4 The Lawson criterion

Consider a thermonuclear plasma that consists of an optimal 50%–50% mixture of deuterium and tritium ions as well as electrons. Suppose that there are no impurity ions or helium ash particles (i.e. thermalized alpha particles) present in the plasma. *Quasi-neutrality* [2] demands that

$$n_{\rm D} = n_{\rm T} = \frac{n_{\rm e}}{2}, \tag{1.10}$$

where $n_{\rm e}$ is the number density of electrons. Suppose that the two ions species have the same temperature, $T_{\rm e}$ (measured in energy units), as the electrons. The total thermal energy density of the plasma is thus [2]

$$W \equiv \frac{3}{2}\, n_{\rm D}\, T_{\rm e} + \frac{3}{2}\, n_{\rm T}\, T_{\rm e} + \frac{3}{2}\, n_{\rm e}\, T = 3\, n_{\rm e}\, T_{\rm e}. \tag{1.11}$$

The rate of nuclear fusion reactions per unit volume is (see equation (1.6))

$$f \equiv n_{\rm D}\, n_{\rm T}\, \langle \sigma\, v\rangle_{\rm DT}(T_{\rm e}) = \frac{n_{\rm e}^{\,2}}{4}\, \langle \sigma\, v\rangle_{\rm DT}(T_{\rm e}). \tag{1.12}$$

In order to achieve a self-sustaining nuclear fusion reaction in a thermonuclear plasma, the fusion heating power per unit volume, $f\, \mathcal{E}_\alpha$, must exceed the energy loss rate per unit volume, $P_{\rm loss}$. (Recall that the alpha particles produced by nuclear fusion reactions heat the plasma, whereas the neutrons exit the plasma without heating it.) Thus, we require

$$f\,\mathcal{E}_\alpha \geqslant P_{\rm loss}. \tag{1.13}$$

Let us write

$$P_{\rm loss} = \frac{W}{\tau_{\rm E}}. \tag{1.14}$$

Here, the *energy confinement time*, $\tau_{\rm E}$, is a measure of how long the plasma's thermal energy is confined within the plasma before escaping. Note that, at this stage, we are making no statement about the nature of the energy loss mechanism. In fact, equation (1.14) can be thought of as the definition of $\tau_{\rm E}$. The previous four equations can be combined to give

$$n_{\rm e}\,\tau_{\rm E} \geqslant F_{\rm lawson}(T_{\rm e}), \tag{1.15}$$

where

$$F_{\rm lawson}(T_{\rm e}) = \frac{12\,T_{\rm e}}{\langle\sigma\,v\rangle_{\rm DT}(T_{\rm e})\,\mathcal{E}_\alpha}. \tag{1.16}$$

Figure 1.2 plots $F_{\rm lawson}$ as a function of the electron temperature, $T_{\rm e}$. It can be seen that $F_{\rm lawson}(T_{\rm e})$ attains a minimum value of 1.49×10^{20} s m^{-3} when $T_{\rm e} = 25.67$ keV.

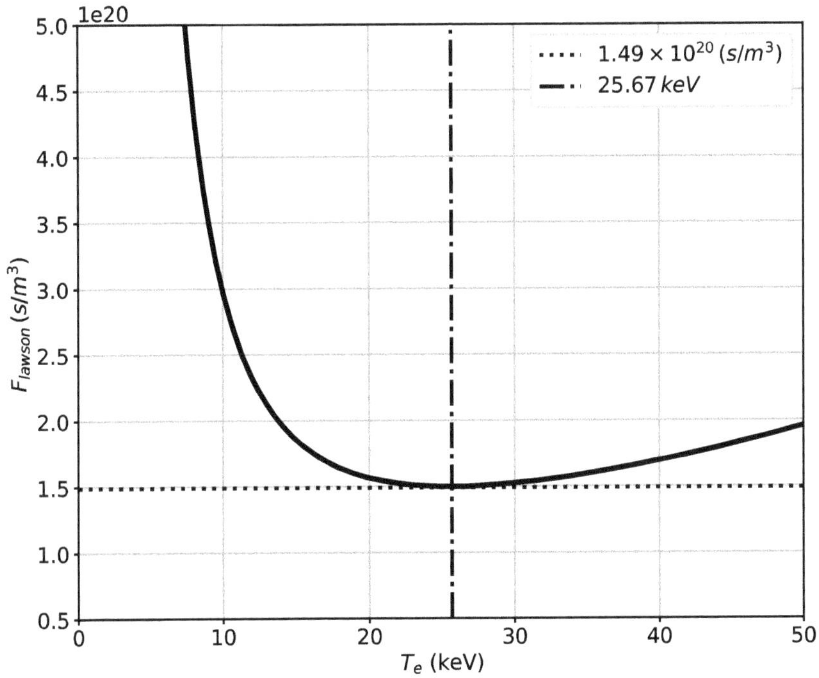

Figure 1.2. The Lawson function, $F_{\rm lawson}$, versus the electron temperature, $T_{\rm e}$.

Thus, we conclude that a self-sustaining nuclear fusion reaction is only possible in a thermonuclear plasma if

$$n_e\,\tau_E \geqslant 1.49 \times 10^{20}\ \mathrm{s\,m^{-3}}. \tag{1.17}$$

This criterion is known as the *Lawson criterion* [6].

In conventional magnetic confinement devices, n_e and T_e can be varied over a wide range of values. However, the maximum value of the plasma pressure, which is proportional to $n_e\,T_e$, is fixed by plasma stability considerations [7, 8] (see section 1.10). According to equations (1.15) and (1.16), the criterion for a self-sustaining nuclear fusion reaction can be recast in the form

$$n_e\,T_e\,\tau_E \geqslant F_{\rm triple}(T_e), \tag{1.18}$$

where

$$F_{\rm triple}(T_e) = \frac{12\,T_e^{\,2}}{\langle\sigma\,v\rangle_{\rm DT}(T_e)\,\mathcal{E}_\alpha}. \tag{1.19}$$

Given that the maximum value of the product $n_e\,T_e$ is fixed, it follows that fusion reactivity is maximized at the temperature that minimizes the function $F_{\rm triple}(T_e)$. As illustrated in figure 1.3, $F_{\rm triple}(T_e)$ attains a minimum value of 2.76×10^{21} keV s m^{-3} when $T_e = 13.54$ keV. Thus, a more useful form of the Lawson criterion is

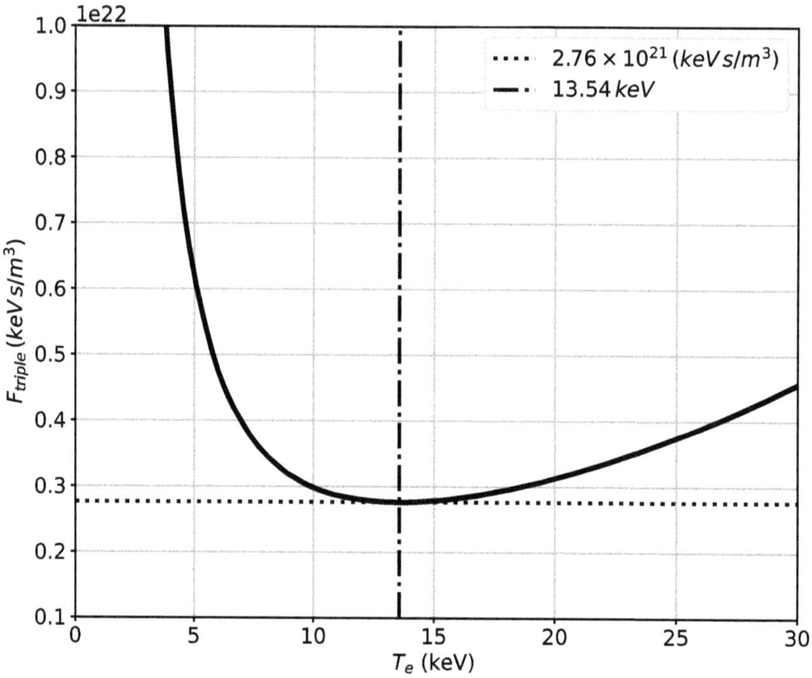

Figure 1.3. The triple product function, $F_{\rm triple}$, versus the electron temperature, T_e.

$$n_e\, T_e\, \tau_E \geqslant 2.76 \times 10^{21}\ \text{keV s m}^{-3}. \tag{1.20}$$

Here, $n_e\, T_e\, \tau_E$ is known as the *fusion triple product* and is the conventional figure of merit for thermonuclear fusion reactions in magnetic confinement devices [8].

1.5 Fusion plasma parameters

We are now in a position to estimate the characteristic properties of a thermonuclear plasma trapped in a magnetic confinement device. As we have seen, the critical fusion triple product required to achieve nuclear fusion is minimized when $T_e \simeq 13.5$ keV. Nevertheless, the two burning plasma experiments that are currently under construction, namely the ITER tokamak [9] and the SPARC tokamak [10], aim to operate at a somewhat lower volume-averaged plasma temperature of $T_e \simeq 7$ keV. Note that $F_{\text{triple}}(T_e = 7\ \text{keV}) = 3.97 \times 10^{21}\ \text{keV s m}^{-3}$. Thus, confinement devices operating at this lower plasma temperature have to satisfy the slightly more stringent Lawson criterion

$$n_e\, T_e\, \tau_E \geqslant 3.97 \times 10^{21}\ \text{keV s m}^{-3}. \tag{1.21}$$

The total plasma pressure is written as

$$p \equiv n_D\, T_e + n_T\, T_e + n_e\, T_e = 2\, n_e\, T_e. \tag{1.22}$$

Let B be the strength of the magnetic field that confines the plasma. It is helpful to define the dimensionless parameter [2]

$$\beta \equiv \frac{2\,\mu_0\, p}{B^2} = 8.05 \times 10^{-2}\, \frac{n_e(10^{20}\ \text{m}^{-3})\ T_e\ (\text{keV})}{B^2\ (\text{T})}, \tag{1.23}$$

which measures the ratio of the plasma's thermal energy density to the energy density of the magnetic field. It turns out that plasma stability considerations ensure that conventional magnetic confinement devices, such as tokamaks, cannot safely operate at beta values that exceed a few percent [11] (see section 1.10).

We can write

$$\tau_E = \frac{a^2}{\chi_\perp}, \tag{1.24}$$

where a is the *minor radius* of the plasma (i.e. the shortest distance from the plasma core to the plasma boundary) and $\chi_\perp$ is the mean rate at which energy diffuses through the plasma. In practice, energy diffuses out of a magnetic confinement device through the action of small-scale turbulence driven by density and temperature gradients within the plasma [8, 12]. Moreover, the typical energy diffusion rate is about 1 $\text{m}^2\ \text{s}^{-1}$ [8].

The previous four equations can be combined to give

$$B\, a \geqslant 17.9 \left[\frac{\chi_\perp}{\beta(\%)}\right]^{1/2}\ \text{T m}. \tag{1.25}$$

Thus, we deduce that in order to obtain a self-sustaining nuclear fusion reaction in a magnetic confinement device characterized by $\beta \simeq 0.02$ (see equation (1.81)) and $\chi_\perp \simeq 1\ \mathrm{m^2\,s^{-1}}$, the product of the magnetic field strength and the plasma's minor radius must exceed about 13 tesla-meters. Conventional superconducting magnet technology limits practical magnetic field strengths in magnetic confinement devices to approximately $B \simeq 5$ T. Thus, to achieve nuclear fusion, such devices must have minor radii of about $a \simeq 2.5$ m [9]. High-temperature superconducting magnet technology allows the practically achievable magnetic field strength to be increased to approximately $B \simeq 12$ T. Thus, to achieve nuclear fusion, high-field devices need only have minor radii of about $a \simeq 1.1$ m, which implies a reduction in the plasma volume by a factor of approximately 12 [10]. This is significant because the cost of a fusion experiment scales roughly with the volume of the experiment.

The rules of thumb which state that $\beta \simeq 0.02$ and $\chi_\perp \simeq 1\ \mathrm{m^2\,s^{-1}}$, combined with equations (1.23)–(1.25), allow us to estimate the characteristic properties of thermonuclear plasmas trapped in both low-field magnetic confinement devices and high-field magnetic confinement devices, as given in table 1.2. Note that these estimates agree well with the much more carefully worked out estimates given in references [9] and [10]. Roughly speaking, thermonuclear fusion requires a deuterium–tritium plasma with a minor radius of at least 1 m and an electron number density of at least 10^{20} particles per cubic meter to be heated to a mean temperature of about 7 keV, and the heat must be subsequently confined within the plasma for at least 1 s. The requisite strength of the confining magnetic field is at least 5 tesla.

Of course, in order for magnetic confinement to work properly, the gyroradii of all of the charged particle species must be significantly less than the minor radius of the plasma. Obviously, the most stringent requirement is that the gyroradii of alpha particle fusion products, ρ_α, must be less than the minor radius. We can write

Table 1.2. The characteristic properties of thermonuclear plasmas trapped in low-field magnetic confinement devices and high-field magnetic confinement devices. Here, B is the magnetic field strength, T_e the electron temperature, n_e the electron number density, R_0 the plasma major radius, a the plasma minor radius, τ_E the energy confinement time, β the ratio of the thermal energy density to the magnetic energy density, and ρ_α the gyroradius of alpha particle fusion products.

	Low field	High field
B (T)	5.0	12.0
T_e (keV)	7.0	7.0
n_e ($10^{20}\ \mathrm{m^{-3}}$)	0.89	5.1
R_0 (m)	7.6	3.2
a (m)	2.5	1.1
τ_E (s)	6.3	1.1
β	0.02	0.02
ρ_α/a	4.3×10^{-2}	4.3×10^{-2}

$$\rho_\alpha = \frac{\sqrt{2\,\mathcal{E}_\alpha\,m_\alpha}}{e\,B}, \tag{1.26}$$

where e is the magnitude of the electron charge [2]. Table 1.2 gives ρ_α/a estimates for both low-field magnetic confinement devices and high-field magnetic confinement devices. In both cases, ρ_α/a is significantly less than unity, indicating that the magnetic field is, in principle, strong enough to confine all of the charged particles within the plasma (see also table 2.1).

1.6 Particle balance

Let us now examine particle balance in a thermonuclear plasma trapped in a magnetic confinement device. The number densities of the deuterons, tritons, and helium ash particles in the plasma evolve over time according to the following idealized equations:

$$\frac{dn_{\rm D}}{dt} = -f - \frac{n_{\rm D}}{\tau_{\rm p}} + \frac{n_{\rm D}}{\tau_{\rm f}}, \tag{1.27}$$

$$\frac{dn_{\rm T}}{dt} = -f - \frac{n_{\rm T}}{\tau_{\rm p}} + \frac{n_{\rm T}}{\tau_{\rm f}}, \tag{1.28}$$

$$\frac{dn_\alpha}{dt} = f - \frac{n_\alpha}{\tau_{\rm p}}. \tag{1.29}$$

Here, f is the rate of nuclear fusion reactions per unit volume (see equation (1.6)). Moreover, $\tau_{\rm p}$ is the *particle confinement time*, which is defined as the average time that a particle is confined within the plasma. For the sake of simplicity, the deuterons, tritons, and helium ash particles are assumed to have the same particle confinement times. Finally, $\tau_{\rm f}$ is the *fueling time*, which is the average time in which the fusion reagents are replenished in the plasma (e.g. via the injection of frozen deuterium/tritium pellets) [13]. We are assuming that deuterons and tritons are replenished at the same rate. Of course, there is no helium ash particle fueling (other than that provided by nuclear fusion reactions).

Suppose that the plasma has attained a steady state (i.e. $d/dt = 0$). Let us assume that $n_{\rm T} = n_{\rm D}$ (i.e. the deuterium–tritium mix is optimal for nuclear fusion). The *burn fraction*,

$$f_{\rm b} = \frac{f\,\tau_{\rm f}}{n_{\rm D}}, \tag{1.30}$$

is defined as the fraction of fuel ions injected into the plasma that undergo fusion reactions, rather than escaping. Equations (1.27) and (1.29) imply that

$$\tau_{\rm f} = (1 - f_{\rm b})\,\tau_{\rm p}, \tag{1.31}$$

$$\frac{n_\alpha}{n_{\rm D}} = \frac{f_{\rm b}}{1 - f_{\rm b}}. \tag{1.32}$$

Quasi-neutrality [2] demands that

$$n_{\rm D} + n_{\rm T} + 2\, n_\alpha + Z_{\rm I}\, n_{\rm I} = n_{\rm e}. \tag{1.33}$$

Here, we are supposing that the plasma contains a fixed fraction of impurity ions of number density $n_{\rm I}$ and mean charge number $Z_{\rm I} > 2$. Usually, such ions originate from the interaction between the plasma and the surrounding solid surfaces. Typical impurity ion species are beryllium, boron, carbon, and tungsten. The *effective charge number* of the plasma ions is defined as [8]

$$Z_{\rm eff} = \frac{n_{\rm D} + n_{\rm T} + 4\, n_\alpha + Z_{\rm I}^{\,2}\, n_{\rm I}}{n_{\rm e}}. \tag{1.34}$$

The previous six equations can be combined with equation (1.6) to give

$$\frac{n_{\rm D}}{n_{\rm e}} = \frac{1}{2}\,\frac{(1 - f_{\rm b})\,(Z_{\rm I} - Z_{\rm eff})}{Z_{\rm I} - (1 + f_{\rm b})}, \tag{1.35}$$

$$\frac{n_\alpha}{n_{\rm e}} = \frac{1}{2}\,\frac{f_{\rm b}\,(Z_{\rm I} - Z_{\rm eff})}{Z_{\rm I} - (1 + f_{\rm b})}, \tag{1.36}$$

$$\frac{f_{\rm b}\,[Z_{\rm I} - (1 + f_{\rm b})]}{(1 - f_{\rm b})^2} = \frac{(Z_{\rm I} - Z_{\rm eff})\,\langle\sigma v\rangle_{\rm DT}\, n_{\rm e}\, a^2}{2\, D_\perp}. \tag{1.37}$$

Here, we have written

$$\tau_{\rm p} = \frac{a^2}{D_\perp}, \tag{1.38}$$

where $D_\perp$ is the mean rate at which particles diffuse through the plasma. In practice, particles diffuse out of a magnetic confinement device through the action of small-scale turbulence driven by density and temperature gradients within the plasma [8, 12]. Moreover, the typical particle diffusion rate is about $0.2\ {\rm m^2\,s^{-1}}$ [8].

Assuming that $D_\perp \simeq 0.2\ {\rm m^2\,s^{-1}}$, $Z_{\rm eff} = 1.5$ [9, 10], and $Z_{\rm I} = 6$ (which corresponds to fully stripped carbon impurity ions), making use of the data given in table 1.2, and employing equations (1.31) and (1.34)–(1.38), we can derive the particle balance data given in table 1.3. (Incidentally, equation (1.37) is most easily solved for $f_{\rm b}$ via iteration.) Note that, for both low-field confinement devices and high-field confinement devices, the burn fraction is only about 5%. In other words, only 5% of the fuel ions injected into the plasma undergo nuclear fusion reactions; the remainder escape. Unfortunately, this is necessary, otherwise there would be an unacceptable buildup of helium ash within the plasma (see equation (1.32)). In fact, we estimate that the helium ash makes up about 3% of the ion content of the plasma. The presence of helium ash and impurity ions in the plasma dilutes the concentration of fuel ions

Table 1.3. Additional characteristic properties of thermonuclear plasmas trapped in low-field magnetic confinement devices and high-field magnetic confinement devices. Here, B is the magnetic field strength, τ_p the particle confinement time, τ_f the fueling time, f_b the burn fraction, n_e the electron number density, n_D the deuteron number density, n_α the alpha particle number density, n_I the impurity ion number density, and τ_s the alpha particle slowing-down time.

	Low field	High field
B (T)	5.0	12.0
τ_p (s)	31	6.1
τ_f (s)	30	5.7
f_b	4.8×10^{-2}	5.3×10^{-2}
n_D/n_e	0.433	0.431
n_α/n_e	2.2×10^{-2}	2.4×10^{-2}
n_I/n_e	1.5×10^{-2}	1.5×10^{-2}
τ_s/τ_p	4.1×10^{-3}	3.7×10^{-3}

with respect to that which would be present if the helium ash and impurities were absent by about 14%. This, in turn, reduces the nuclear fusion reaction rate in the plasma by about 25%. We conclude that the helium ash and impurity ion contents of the plasma must be kept as low as possible. Incidentally heavy impurity ions, such as tungsten, have such a high mean charge number in a thermonuclear plasma (about 60, in the case of tungsten) that their relative concentration must be significantly less than that given in table 1.3 (about $n_I/n_e \simeq 1.2 \times 10^{-4}$, in the case of tungsten) in order to avoid unacceptable fuel ion dilution.

If thermonuclear fusion is to work properly, the time required for alpha particle fusion products to slow down due to collisions with thermal particles and thereby heat the plasma must be much less than the particle confinement time. (Here, we are making the simplifying assumption that thermonuclear alpha particles have the same confinement time as the thermalized plasma species.) Otherwise, alpha particles would be able to escape from the plasma without giving up all of their fusion energy. The slowing-down time for alpha particles is approximately given by [2, 14, 15]

$$\tau_s = \frac{3\,\pi^{3/2}\,\epsilon_0{}^2\,m_\alpha\,T_e^{3/2}}{\sqrt{8}\,e^4\,n_e\,m_e{}^{1/2}\,\ln\Lambda}, \tag{1.39}$$

where $\ln\Lambda \simeq 16$ is the Coulomb logarithm [2] and m_e the electron mass. As shown in table 1.3, the ratio τ_s/τ_p is indeed very much less than unity in thermonuclear magnetic confinement devices.

1.7 Energy balance

Let us now perform the energy balance calculation outlined in section 1.4 more exactly. In the presence of helium ash and impurities, expression (1.11) for the total thermal energy density of the plasma generalizes to give

$$W = \frac{3}{2}\left(1 + \frac{2\,n_\mathrm{D}}{n_\mathrm{e}} + \frac{n_\alpha}{n_\mathrm{e}} + \frac{n_\mathrm{I}}{n_\mathrm{e}}\right) n_\mathrm{e}\, T_\mathrm{e}. \tag{1.40}$$

As before, the energy loss rate per unit volume due to the turbulent transport of thermal energy out of the plasma is given by (see equation (1.14))

$$P_\mathrm{loss} = \frac{W}{\tau_\mathrm{E}}, \tag{1.41}$$

where τ_E is the energy confinement time. However, fusion plasmas also lose energy due to the emission of electromagnetic radiation (because they are optically thin). In fact, the principal radiation loss mechanism is *electron–ion bremsstrahlung*. The energy loss rate per unit volume due to electron–ion bremsstrahlung is

$$P_\mathrm{rad} = C\, Z_\mathrm{eff}\, n_\mathrm{e}^{\,2}\, T_\mathrm{e}^{\,1/2}, \tag{1.42}$$

where $C = 4.07 \times 10^{-29}\ \mathrm{J}^{1/2}\ \mathrm{s}^{-1}\ \mathrm{m}^{-3}$ [4]. Impurity ions that are not fully stripped of electrons can also radiate via line emission [16]. For light impurities, such as carbon, which are fully stripped in the plasma core, such radiation is concentrated at the edge of the plasma and is relatively unimportant. On the other hand, heavy impurities, such as tungsten, are not fully stripped in the plasma core and therefore emit line radiation throughout the plasma. The energy loss due to line radiation from heavy impurities is typically similar in magnitude to the energy loss due to bremsstrahlung [10, 16]. As before, the heating power per unit volume due to nuclear fusion reactions is

$$f\,\mathcal{E}_\alpha = \left(\frac{n_\mathrm{D}}{n_\mathrm{e}}\right)^2 n_\mathrm{e}^{\,2}\,\langle\sigma\, v\rangle_\mathrm{DT}(T_\mathrm{e})\,\mathcal{E}_\alpha. \tag{1.43}$$

(See equation (1.6).) The criterion that must be satisfied to obtain a self-sustaining nuclear fusion reaction is

$$f\,\mathcal{E}_\alpha \geqslant P_\mathrm{loss} + P_\mathrm{rad}. \tag{1.44}$$

(See equation (1.13).) The previous equations can be rearranged to give

$$n_\mathrm{e}\, T_\mathrm{e}\, \tau_\mathrm{E} \geqslant F_\mathrm{triple}, \tag{1.45}$$

where

$$F_\mathrm{triple} = \left(\frac{n_\mathrm{e}}{n_\mathrm{D}}\right)^2 \frac{\left[\frac{3}{2}\left(1 + \frac{2\,n_\mathrm{D}}{n_\mathrm{e}} + \frac{n_\alpha}{n_\mathrm{e}} + \frac{n_\mathrm{I}}{n_\mathrm{e}}\right) T_\mathrm{e}^{\,2} + C\, Z_\mathrm{eff}\, n_\mathrm{e}\, T_\mathrm{e}^{\,3/2}\,\tau_\mathrm{E}\right]}{\langle\sigma v\rangle_\mathrm{DT}(T_\mathrm{e})\,\mathcal{E}_\alpha}. \tag{1.46}$$

Finally, the energy per unit time per unit plasma volume that ends up being absorbed in the blanket via neutron emission is

$$P_n = f\,\mathcal{E}_\mathrm{n}. \tag{1.47}$$

Hence, the fractions of the nuclear fusion energy generated within the plasma that end up being absorbed in the blanket, lost as heat conducted to the plasma-facing components, and radiated away, are

$$f_{\rm n} = \frac{P_{\rm n}}{P_{\rm n} + P_{\rm loss} + P_{\rm rad}}, \tag{1.48}$$

$$f_{\rm loss} = \frac{P_{\rm loss}}{P_{\rm n} + P_{\rm loss} + P_{\rm rad}}, \tag{1.49}$$

$$f_{\rm rad} = \frac{P_{\rm rad}}{P_{\rm n} + P_{\rm loss} + P_{\rm rad}}, \tag{1.50}$$

respectively.

The information in tables 1.2 and 1.3 can be used in combination with equations (1.46) and (1.48)–(1.50) to produce the estimates that are given in table 1.4. It can be seen that the dilution of the fusion reagent ions due to the presence of helium ash and impurities increases the critical fusion triple product needed for self-sustained nuclear fusion by about 30%. Radiation losses increase the critical fusion triple product by at least (because we have neglected impurity line radiation) a further 40%. Thus, the table highlights the importance of keeping the helium ash and impurity content of the plasma as small as possible. Note, however, that, even in a pure D–T plasma, there is an irreducible radiation loss due to bremsstrahlung between the electrons and the fueling ions that increases the critical triple product over that specified in equation (1.21) by about 25%. Finally, about 70% of the fusion

Table 1.4. Additional characteristic properties of thermonuclear plasmas trapped in low-field magnetic confinement devices and high-field magnetic confinement devices. Here, $F_{\rm triple\,1}$ is the critical fusion triple product needed for self-sustained nuclear fusion in the absence of both ion dilution and radiation losses, $F_{\rm triple\,2}$ the critical fusion triple product in the presence of ion dilution and the absence of radiation losses, $F_{\rm triple\,3}$ the critical fusion triple product in the presence of both ion dilution and radiation losses, $f_{\rm n}$ the fraction of the nuclear fusion energy generated within the plasma that is absorbed by the blanket, $f_{\rm loss}$ the fraction of the fusion energy that is lost as heat conducted to the plasma-facing components, and $f_{\rm rad}$ the fraction of the fusion energy that is radiated away.

	Low field	High field
B (T)	5.0	12.0
$F_{\rm triple\,1}$ (10^{21} keV s m^{-3})	3.97	3.97
$F_{\rm triple\,2}$ (10^{21} keV s m^{-3})	5.04	5.09
$F_{\rm triple\,3}$ (10^{21} keV s m^{-3})	6.84	7.09
$f_{\rm n}$	0.69	0.71
$f_{\rm loss}$	0.23	0.21
$f_{\rm rad}$	0.08	0.08

energy created within the plasma is absorbed by the blanket, about 22% is lost as heat conducted to the plasma-facing components, and about 8% is radiated away.

1.8 Linear pinches

Let us examine a particularly simple magnetic confinement device. Suppose that the plasma is contained in a cylindrical vacuum vessel of radius a and finite length, capped by conducting end plates (see figure 1.4). Let r, θ, z be a conventional cylindrical coordinate system whose axis corresponds to that of the vessel. Suppose that a uniform axial current, I_p, is driven through the plasma by electrically biasing the end plates. In other words, suppose that the electric current density within the plasma is

$$\mathbf{j} = \frac{I_p}{\pi\, a^2}\,\mathbf{e}_z. \tag{1.51}$$

The axial current generates a 'poloidal' magnetic field that circulates around the axis of the cylinder:

$$\mathbf{B} = \frac{\mu_0\, I_p\, r}{2\pi\, a^2}\,\mathbf{e}_\theta. \tag{1.52}$$

Here, $\mathbf{e}_z = \nabla z/|\nabla z|$ and $\mathbf{e}_\theta = \nabla\theta/|\nabla\theta|$. In combination with the poloidal magnetic field, the axial current generates an inward radial force that compresses the plasma. This is known as the *pinch effect* [17], and the type of magnetic confinement device described here is called a *linear pinch* (or sometimes a Z pinch) [7]. Under equilibrium, we expect that [2]

$$-\nabla p + \mathbf{j}\times\mathbf{B} = 0, \tag{1.53}$$

where $p(r)$ is the plasma pressure. Thus, we obtain

$$\frac{dp}{dr} = -j_z\, B_\theta = -\frac{\mu_0\, I_p^{\,2}\, r}{2\pi^2\, a^4}, \tag{1.54}$$

which can be integrated to give

$$p = \frac{\mu_0\, I_p^{\,2}}{4\pi^2\, a^4}(a^2 - r^2). \tag{1.55}$$

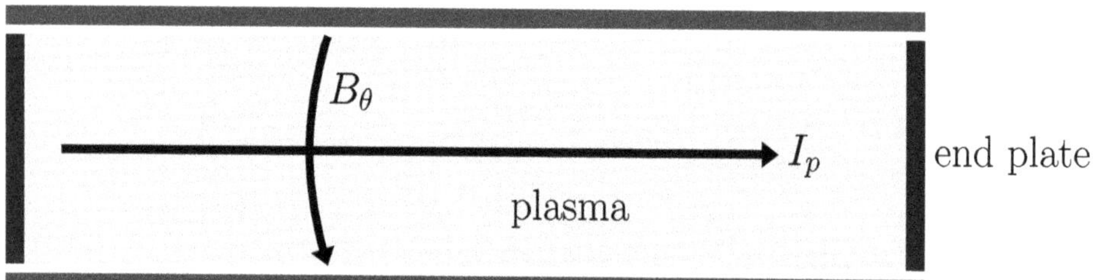

Figure 1.4. Schematic diagram of a linear pinch.

The volume-averaged pressure is

$$\langle p\rangle = \frac{1}{\pi\,a^2}\int_0^a 2\pi\,r\,p(r)\,dr = \frac{\mu_0\,I_p^{\,2}}{8\pi^2\,a^4}. \tag{1.56}$$

Let us make the identification (see equation (1.22))

$$\langle p\rangle = 2\,n_e\,T_e, \tag{1.57}$$

where n_e and T_e are the volume-averaged electron number density and the electron temperature within the plasma, respectively.

The previous two equations yield

$$I_p = 4\pi\,a\left(\frac{n_e\,T_e}{\mu_0}\right)^{1/2}. \tag{1.58}$$

Table 1.5 uses the information in table 1.2 to estimate the critical plasma current needed to achieve self-sustaining nuclear fusion. It can be seen that, in both low-field confinement devices and high-field confinement devices, the critical current is about 9 MA. The critical poloidal magnetic field strength needed to achieve nuclear fusion is

$$B_\theta = \frac{\mu_0\,I_p}{2\pi\,a}. \tag{1.59}$$

Table 1.5. Further characteristic properties of thermonuclear plasmas trapped in low-field magnetic confinement devices and high-field magnetic confinement devices. Here, B is the magnetic field strength, I_p the plasma current, B_θ the poloidal magnetic field strength, $T_{e\,\mathrm{ohm}}$ the plasma temperature due to ohmic heating alone, P_{loss} the energy loss rate per unit volume, τ_A the Alfvén time, τ_R the resistive time, S the Lundquist number, δ the maximum thickness of the wall material that can be ablated into the plasma, l_e the electron mean free path between collisions, $\tau_{E\,\parallel}$ the parallel energy confinement time, $U_\parallel$ the electron energy flux into the end plates of a linear pinch, and $U_\perp$ the energy flux into the curved wall of a linear pinch.

	Low field	High field
B (T)	5.0	12.0
I_p (MA)	8.9	8.9
B_θ (T)	0.71	1.7
$T_{e\,\mathrm{ohm}}$ (keV)	0.48	0.48
P_{loss} (MW m^{-3})	4.7×10^{-2}	1.5
τ_A (s)	1.7×10^{-6}	7.2×10^{-7}
τ_R (s)	1.9×10^{3}	3.3×10^{2}
S	1.1×10^{9}	4.6×10^{8}
δ (m)	1.5×10^{-11}	3.7×10^{-11}
l_e (m)	4.7×10^{3}	8.2×10^{2}
$\tau_{E\,\parallel}$ (s)	2.0×10^{-7}	2.0×10^{-7}
$U_\parallel$ (TW m^{-2})	7.4	43
$U_\perp$ (MW m^{-2})	5.9×10^{-2}	0.81

Note from table 1.5 that the critical poloidal field strength is much less than the total field strength. This is the case because another, much larger component of the magnetic field (in this case, an axial component) is needed to stabilize the plasma [7] (see section 1.10).

As well as generating a poloidal magnetic field, the axial current that passes through the plasma heats it ohmically. The ohmic heating rate per unit volume is

$$P_{\rm ohm} = \frac{j_z^{\,2}}{\sigma} = \left(\frac{I_{\rm p}}{\pi\,a^2}\right)^2\frac{1}{\sigma}, \tag{1.60}$$

where

$$\sigma = \frac{6\sqrt{2}\,\pi^{3/2}\,\epsilon_0^{\,2}\,T_{\rm e}^{\,3/2}}{Z_{\rm eff}\,e^2\,m_{\rm e}^{1/2}\,\ln\Lambda} \tag{1.61}$$

is the electrical conductivity of the plasma [2] (see section 2.6). Let us investigate whether ohmic heating alone is capable of generating a high enough plasma temperature for nuclear fusion. A steady state is achieved when the ohmic heating rate balances the energy loss rate:

$$P_{\rm ohm} = P_{\rm loss} \tag{1.62}$$

Here,

$$P_{\rm loss} = \frac{3\,n_{\rm e}\,T_{\rm e}\,\chi_\perp}{a^2} = \frac{3\,\mu_0\,I_{\rm p}^{\,2}\,\chi_\perp}{16\,\pi^2\,a^4}, \tag{1.63}$$

where use has been made of equations (1.11), (1.14), (1.24), and (1.58). $P_{\rm loss}$ is estimated in table 1.5. The previous four equations suggest that the plasma temperature achieved by ohmic heating alone is

$$T_{\rm e\,ohm} = \left(\frac{8\,Z_{\rm eff}\,e^2\,m_{\rm e}^{\,1/2}\,\ln\Lambda}{9\sqrt{2}\,\pi^{3/2}\,\epsilon_0^{\,2}\,\mu_0\,\chi_\perp}\right)^{2/3}. \tag{1.64}$$

According to table 1.5, if $Z_{\rm eff} = 1.5$ and $\chi_\perp \simeq 1\,{\rm m^2\,s^{-1}}$ (see sections 1.5 and 1.6), $T_{\rm e\,ohm}$ is only about 0.48 keV. However, this is the volume-averaged electron temperature. If we assume that the temperature profile is peaked in the same manner as the pressure profile in equation (1.55), then we conclude that the central electron temperature due to ohmic heating alone is about 0.96 keV. This is still well below the plasma temperature (i.e. 7 keV) needed for self-sustaining nuclear fusion. Thus, we deduce that, in order to achieve thermonuclear fusion in a magnetic confinement device, the plasma must be subject to additional heating. This so-called *auxiliary heating* is usually provided by high-energy neutral particles or radio-frequency electromagnetic waves injected into the plasma [8]. If $P_{\rm aux}$ is the auxiliary heating rate per unit volume, then we can write

$$P_{\rm ohm} + P_{\rm aux} = P_{\rm loss}. \tag{1.65}$$

The previous five equations yield

$$T_e = \frac{T_{e\,ohm}}{(1 - P_{aux}/P_{loss})^{2/3}}. \tag{1.66}$$

Hence, we conclude that plasma temperatures sufficient for nuclear fusion can be achieved by means of auxiliary heating, but this requires $P_{aux} \to P_{loss}$.

There are two serious problems with linear pinch magnetic confinement devices. The first problem is that the plasma is ideally unstable. A so-called *ideal* plasma instability is one that does not change the topology of the magnetic field (i.e. it does not require the reconnection of magnetic field lines). For the case of a linear pinch, the relevant instabilities are the 'sausage' mode and the 'kink' mode [7]. These instabilities are global in nature, cause distortions in the shape of the plasma that are consistent with their names, and lead to the complete disruption of the plasma discharge [4]. Ideal instabilities arise because the force balance criterion described in equation (1.53) breaks down. Consequently, the electromagnetic pinch force is balanced by plasma inertia, rather than by plasma pressure. In other words,

$$\rho\,\frac{\partial \mathbf{V}}{\partial t} \simeq \mathbf{j} \times \mathbf{B}, \tag{1.67}$$

where ρ is the plasma mass density and $\mathbf{V}$ is the plasma (i.e. ion) velocity. We can make the following estimates: $B \simeq B_\theta$, where B_θ is specified in equation (1.59); $j \simeq 2\,B_\theta/(\mu_0\,a)$; $\rho = (1/2)\,n_e\,(m_D + m_T)$, where $m_D = 3.344 \times 10^{-27}$ kg and $m_T = 5.007 \times 10^{-27}$ kg are the deuteron and triton masses, respectively; $t \simeq \tau_A$, where τ_A is the typical timescale on which an ideal mode grows; and $V \simeq a/\tau_A$, which is appropriate to a global instability. These estimates lead to the following estimate for the characteristic timescale on which an ideal instability grows:

$$\tau_A = \frac{a\sqrt{\mu_0\,n_e\,(m_D + m_T)}}{2\,B_\theta}. \tag{1.68}$$

Here, τ_A is known as the *Alfvén time* [7]. According to table 1.5, the Alfvén time is of order a microsecond in a linear pinch. We conclude that ideal instabilities would disrupt a plasma confined in a linear pinch on a timescale that is about a million times shorter than the energy confinement time needed to achieve nuclear fusion (see table 1.2). Clearly, it is imperative that plasmas in magnetic confinement devices be rendered stable when exposed to ideal instabilities. This goal can be achieved in a linear pinch by adding an axial magnetic field, $B_z\,\mathbf{e}_z$, whose strength greatly exceeds that of the poloidal field [7].

The second problem with a linear pinch is associated with the need to keep the impurity content of the plasma within acceptable limits. Suppose, for the sake of example, that the inner surface of the vacuum vessel is lined with graphite tiles, as is the case in many magnetic confinement devices. The mass density of graphite is $\rho_g = 2.2 \times 10^3$ kg m^{-3}. Moreover, the mass of a carbon atom is $m_c = 1.994 \times 10^{-26}$ kg. Thus, the number density of carbon atoms within the tiles is $n_c = \rho_g/m_c = 1.1 \times 10^{29}$ m^{-3}. Note that this number density exceeds that of the particles in the plasma by nine orders of magnitude (see table 1.2). We need to keep the impurity content of the plasma such that $Z_{\rm eff} \simeq 1.5$, otherwise the fuel ion dilution and radiation losses become unacceptable. According to table 1.3, this implies that the number density of carbon atoms within the plasma, $n_{\rm I}$, cannot exceed 1.5% of the electron number density. Suppose that, as a consequence of plasma–wall interactions, a layer of carbon atoms of thickness δ is ablated into the plasma. It can easily be demonstrated that

$$n_{\rm I} = \frac{2\,\delta\,n_{\rm c}}{a}, \tag{1.69}$$

which yields

$$\delta = \frac{n_{\rm I}\,a}{2\,n_{\rm c}}. \tag{1.70}$$

It can be seen from table 1.5 that $\delta \sim 10^{-11}$ m, which is about 10 times smaller than the interatomic spacing of carbon atoms in graphite. Thus, we conclude that in order to avoid unacceptable fuel ion dilution and radiation losses, plasma–wall interactions must result in less than an atomic monolayer of the plasma-facing components ending up in the plasma. Obviously, this is a very stringent criterion. If the inner surface of the vacuum vessel is lined with tungsten instead of carbon, the criterion becomes even more stringent. In fact, only about 1 tungsten atom per 1000 making up the first atomic monolayer of the plasma-facing components can end up in the plasma.

In a linear pinch, the interaction of the plasma with the curved surface of the vacuum vessel is moderated by the small gyroradii of charged particles within the plasma, which ensure that particles cannot freely stream to the surface, but instead have to slowly diffuse across magnetic field lines in order to reach it (see section 2.6). However, in the presence of an axial magnetic field, charged particles can reach the end plates by moving along magnetic field lines. This process is not moderated by the small particle gyroradii. In fact, the only possible moderation mechanism is collisions. The mean free path between collisions of an electron moving parallel to a magnetic field line is [2]

$$l_{\rm e} = v_{\rm t\,e}\,\tau_{\rm e}, \tag{1.71}$$

where $v_{\rm t\,e} = (2\,T_{\rm e}/m_{\rm e})^{1/2}$ is the electron thermal velocity, and

$$\tau_{\rm e} = \frac{6\sqrt{2}\,\pi^{3/2}\,\epsilon_0^{\,2}\,m_{\rm e}^{\,1/2}\,T_{\rm e}^{3/2}}{Z_{\rm eff}\,e^4\,n_{\rm e}\,\ln\Lambda} \tag{1.72}$$

is the electron–ion collision time [2] (see section 2.4). According to table 1.5, the electron mean free path between collisions is of the order of a kilometer. The ion mean free path between collisions is of similar magnitude. (See table 2.1. Note that $l_{\rm e}$ is calculated with $Z_{\rm eff} = 1$ in this table.) Given that a kilometer is very much longer than the conceivable length of a practical linear pinch, we conclude that collisions do not moderate the flow of charged particles along magnetic field lines to the end plates at all. In other words, electron and ion thermal energy effectively flows along magnetic field lines to the end plates at the appropriate thermal velocity. Suppose that the pinch has a length of $L = 10\,\mathrm{m}$. Electron thermal energy that flows along magnetic field lines to the end plates is effectively only confined within the plasma for a time

$$\tau_{E\,\parallel} = \frac{L}{v_{t\,e}}. \tag{1.73}$$

As shown in table 1.5, this parallel energy confinement time is less than a microsecond, which means that it is at least a million times too small for the purposes of nuclear fusion (see table 1.2). The flux of electron thermal energy into the end plates is

$$U_\parallel = \frac{3\,n_e\,T_e\,L}{2\,\tau_{E\,\parallel}}. \tag{1.74}$$

As shown in table 1.5, this flux is of the order of a terawatt per meter squared. Obviously, it is impossible to believe that ablation at a solid surface subject to such an enormous energy flux could be limited to a monolayer or less of atoms. By contrast, the flux of plasma thermal energy into the curved wall is

$$U_\perp = \frac{3\,n_e\,T_e\,a}{2\,\tau_E} = \frac{3\,n_e\,T_e\,\chi_\perp}{2\,a}. \tag{1.75}$$

As shown in table 1.5, this flux is much more manageable, being less than a megawatt per meter squared.

1.9 Toroidal pinches

An obvious way of dealing with the problem of the end plates is to bend a linear pinch into a circle to produce a toroidal plasma. See figure 1.5. This procedure allows us to carry over most of the results that we previously obtained for a linear pinch while obviating the need for the end plates. In a so-called *toroidal pinch*, the plasma is confined on a series of axisymmetric, nested, toroidal magnetic flux surfaces [7]. As such, the plasma thermal energy is not able to reach the toroidal

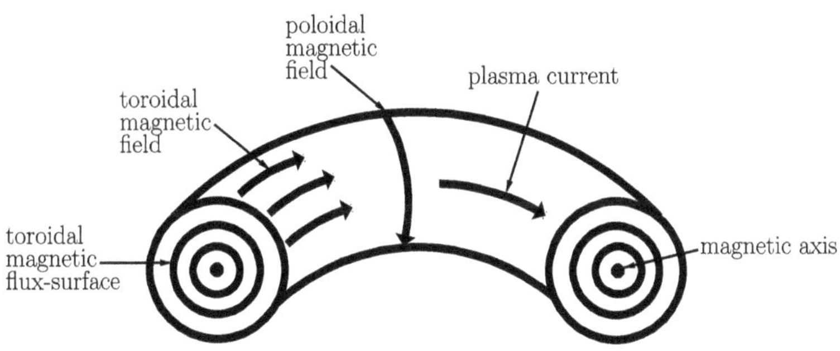

Figure 1.5. Schematic diagram of a toroidal pinch.

vacuum vessel surrounding the plasma by freely streaming along magnetic field lines —which, as we have seen, is a very rapid process—but must instead diffuse across the magnetic flux surfaces, which is a comparatively slow process. The magnetic field of a toroidal pinch consists of a poloidal component that circulates about the magnetic axis and a toroidal component that runs parallel to the axis. See figure 1.5. The poloidal magnetic field is generated by a toroidal current that is induced in the plasma via transformer action. In essence, the plasma forms the single-turn secondary winding of a transformer circuit. The toroidal magnetic field is generated by currents flowing in magnetic field coils that surround the plasma [8]. The poloidal magnetic field is responsible for confining the plasma via the pinch effect (when it is combined with the toroidal plasma current). The toroidal magnetic field is needed to stabilize the plasma against ideal kink modes but does not contribute greatly to confinement (other than by reducing the gyroradii of charged particles within the plasma).

Let R_0 and a be the major and minor radii of the plasma torus, respectively. Consider an idealized magnetic flux surface of circular poloidal cross section. Let $r < a$ be the minor radius of this surface, and let θ and φ be the poloidal and toroidal angles, respectively. Furthermore, let $B_\theta(r)$ and B_φ be the mean poloidal and toroidal magnetic field strengths, respectively, on the surface. The *safety factor*,

$$q(r) = \frac{r\,B_\varphi}{R_0\,B_\theta}, \tag{1.76}$$

is the mean number of toroidal circuits of the plasma that a magnetic field line within the flux surface completes for every poloidal circuit [8]. A more exact definition of this important quantity is given in equation (2.128). Incidentally, throughout this book, we shall make the conventional assumptions that the r, θ, φ coordinate system is right-handed and that $B_\theta > 0$ and $B_\varphi > 0$. This implies that $I_{\rm p} > 0$ (in other words, the toroidal plasma current runs in the $+\varphi$ direction), and $q > 0$.

1.10 Tokamaks

A conventional *tokamak* is a type of toroidal pinch characterized by a monotonically increasing $q(r)$ profile with a central value that is slightly less than unity (typically, $q(0) \simeq 0.8$) [8].

Let $q_a = q(a)$ be the safety factor at the edge of the plasma. One criterion that must be met if a tokamak plasma is to remain stable to ideal kink modes is [7]

$$q_a > 1. \tag{1.77}$$

This criterion is known as the *Kruskal–Shafranov criterion* [18, 19] and sets an upper limit on the toroidal plasma current at a fixed toroidal magnetic field strength. In practice, a tokamak can only safely operate when $q_a \gtrsim 3$ [8]. A conventional tokamak has an *inverse aspect ratio* of [9, 10]

$$\epsilon_a \equiv \frac{a}{R_0} \simeq \frac{1}{3}. \tag{1.78}$$

(see table 1.2). It follows that

$$\frac{B_\varphi}{B_\theta(a)} = \frac{q_a}{\epsilon_a} \simeq 9. \tag{1.79}$$

In other words, the strength of the toroidal magnetic field which is needed to stabilize the plasma against ideal kink modes exceeds that of the poloidal magnetic field (which actually confines the plasma) by almost an order of magnitude. Obviously, this is a rather steep price to pay for plasma stability (given that most of the expense of a tokamak experiment is associated with generating the toroidal magnetic field).

A second criterion that must be met if a tokamak plasma is to remain stable to ideal kink modes is [7]

$$\beta(\%) < \beta_N \, \frac{I_{\rm p}\ ({\rm MA})}{a({\rm m})\ B_\varphi\ ({\rm T})}, \tag{1.80}$$

where $\beta_N = 2.8$. This criterion is known as the *Troyon limit* [11]; it sets an upper limit on the plasma pressure at a fixed toroidal magnetic field strength (see equation (1.23)). The Troyon limit can also be expressed in the form

$$\beta(\%) < 5\, \beta_N \, \frac{\epsilon_a}{q_a} \simeq 1.6, \tag{1.81}$$

where use has been made of equations (1.58) and (1.79). Thus, we can see that the maximum beta value at which a conventional tokamak plasma can safely be operated is indeed limited to about 0.02 (see section 1.5).

1.11 Tearing modes

Consider a magnetic perturbation of a tokamak plasma equilibrium that has m periods in the poloidal direction and n periods in the toroidal direction. Here, m is termed the *poloidal mode number*, while n is termed the *toroidal mode number*. Such a perturbation resonates with the plasma (i.e. satisfies $\mathbf{k}\cdot\mathbf{B} = 0$, where $\mathbf{k}$ is the wavenumber of the perturbation, and $\mathbf{B}$ is the equilibrium magnetic field) at the magnetic flux surface of minor radius $r_{\rm s}$, at which [20]

$$q(r_{\rm s}) = \frac{m}{n}. \tag{1.82}$$

Such a surface is termed a *rational* magnetic flux surface, for obvious reasons. The well-known *flux-freezing constraint* [2] of ideal magnetohydrodynamics forbids any change in the topology of magnetic field lines due to the perturbation [7]. In particular, the constraint requires the perturbed radial magnetic field at the rational surface to be zero.

An ideal kink mode is an instability of a tokamak plasma that remains unstable even when the perturbed radial magnetic field is constrained to be zero at the associated rational surface (assuming that the surface lies within the plasma).

As we have seen, such instabilities grow on the very short Alfvén timescale. Fortunately, ideal kink modes are relatively easy to avoid. In fact, ideal kink modes can be avoided by not allowing the edge safety factor to fall significantly below three and by not allowing the plasma beta to approach the Troyon limit. It is also necessary to prevent the plasma density from exceeding the so-called *Greenwald limit* [21] that is associated with the radiative collapse of the plasma current profile.

A *tearing mode* [22] is an instability of a tokamak plasma that is driven by the same free energy source as an ideal kink mode (namely, radial current and pressure gradients within the plasma [7]), but changes the topology of the magnetic field by tearing and reconnecting magnetic field lines at the rational surface. This implies that the mode is not subject to the constraint that the perturbed radial magnetic field at the rational surface must be zero. Consequently, a tearing mode can be unstable even when the corresponding ideal kink mode (i.e. one that possesses the same poloidal and toroidal mode numbers) is stable [20].

Magnetic reconnection is made possible by a plasma's finite electrical resistivity [2]. Magnetic flux diffuses through a tokamak plasma due to the action of resistivity in the characteristic *resistive time*

$$\tau_R = \mu_0\, a^2\, \sigma, \tag{1.83}$$

where the plasma's electrical conductivity is specified in equation (1.61) [2]. As shown in table 1.5, the resistive time in a tokamak fusion reactor exceeds the Alfvén time by many orders of magnitude. To be more exact, the ratio of the two timescales,

$$S = \frac{\tau_R}{\tau_A}, \tag{1.84}$$

is known as the *Lundquist number* [2] and is of the order of 10^9 in a tokamak fusion reactor.

It is not surprising that a tearing mode grows and saturates on a timescale that is related to the resistive time [23]. In fact, the saturation time is typically a few percent of the resistive time [24]. This implies that a tearing mode is an *extremely slowly growing* instability. Nevertheless, the pulse duration of a tokamak plasma is generally sufficiently long for tearing instabilities to develop fully. As illustrated in figure 1.6, a tearing mode changes the topology of the magnetic field in the vicinity of the rational magnetic flux surface to produce a *magnetic island chain* with m periods in the poloidal

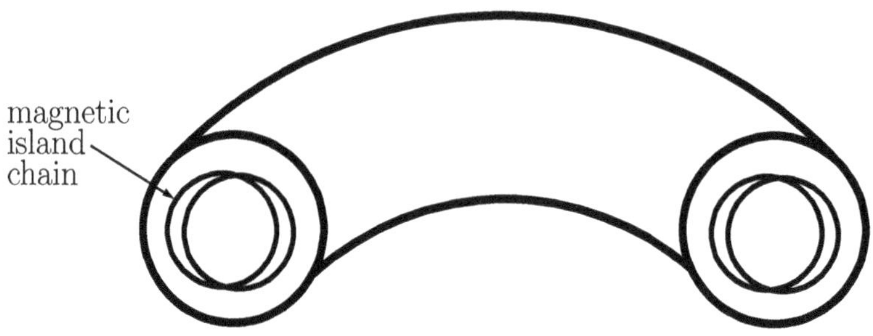

Figure 1.6. Schematic diagram of an $m = 2/n = 1$ magnetic island chain in a tokamak plasma.

direction and n periods in the toroidal direction. The chain typically has a radial width (defined as the width of the magnetic separatrix that separates reconnected from unreconnected magnetic field lines) that is a few percent of the plasma's minor radius. The presence of the island chain is significant because heat and particles are able to get from one (radial) side of the magnetic separatrix to the other by rapidly streaming along magnetic field lines, rather than by having to slowly diffuse across magnetic flux surfaces. Consequently, the plasma pressure profile is flattened within the magnetic separatrix [25], giving rise to a degradation of the energy and particle confinement properties of the plasma [26]. Incidentally, the flattening of the electron temperature within the separatrix of a magnetic island allows the island structure to be imaged by an electron cyclotron emission diagnostic [27, 28].

1.12 Tearing mode rotation

Under normal circumstances, a tearing mode in a tokamak plasma has a nonzero real frequency in the laboratory frame; this frequency is determined by the equilibrium plasma flow at the rational surface. In essence, a magnetic island chain is a helical pattern in the magnetic field generated by a helical current perturbation that is localized in the vicinity of the rational surface [23]. Given that plasma current is predominately carried by electrons, it is natural to suppose that a magnetic island chain (as well as the tearing mode perturbation away from the rational surface) is convected by the electron fluid in the immediate vicinity of the rational surface. This is indeed the case in the so-called linear regime [29], in which the radial thickness of the island chain is less than the (very thin) characteristic linear layer width [22]. Of course, as a consequence of diamagnetic flows (see section 2.11), if the island chain is convected by the electron fluid at the rational surface, then it propagates with respect to the local ion fluid. However, this is not a problem because a linear layer is sufficiently thin that the magnetic field can diffuse through the plasma very rapidly, which implies that the ion fluid is not tied to the magnetic structure of the island chain. The situation is very different in the nonlinear regime, in which the radial thickness of the island chain exceeds the linear layer width. The region inside the magnetic separatrix of a nonlinear magnetic island chain is governed by a combination of flux freezing and perturbed force balance [23]. This implies that both the electron and the ion fluids are trapped inside the separatrix and are therefore forced to corotate with the island chain. There is no such constraint outside the separatrix, so the electron and ion fluids flow at different speeds in this region as a consequence of diamagnetism. It follows that one or other of the electron and the ion fluid rotation profiles must exhibit a strong gradient across the separatrix. The island propagation velocity is determined by whichever of the two fluids is most resistant to the formation of such a gradient. Of course, it is the ion fluid which is more resistant because of its much greater viscosity [30–32] (see section 2.6). Hence, a nonlinear magnetic island chain is convected by the ion fluid in the vicinity of the resonant surface, because this choice of propagation speed minimizes the ion fluid velocity gradient across the separatrix. Interestingly, although early measurements of tearing mode rotation in tokamak plasmas suggested that tearing modes corotate with the electron fluid at the rational surface [33, 34], later measurements obtained the result that tearing modes corotate with the ion fluid [35, 36]. The difference between these two sets of results may be due to the fact that the early measurements were made in comparatively cold plasmas with comparatively wide linear layer widths (the layer width scales roughly as $T_e^{-3/5}$ [2, 22]).

Rotating tearing modes in tokamaks are often observed to slow down as they grow in amplitude; they finally stop rotating or 'lock' with a definite poloidal and toroidal phase when the mode exceeds a critical amplitude [37, 38]. So-called *locked modes* are of concern because they are strongly correlated with sudden disruptions of the plasma discharge [39]. The slowing down of the rotation of a growing tearing mode is associated with eddy currents excited in the resistive vacuum vessel surrounding the plasma; such currents generate a magnetic perturbation that exerts an electromagnetic torque at the rational surface that acts to brake the local plasma rotation; in turn, this brakes the rotation of the tearing mode [40, 41] (see chapter 10). The locking of the tearing mode is associated with a so-called *error field*, which is an accidentally produced, static, helical magnetic perturbation generated by misalignments of magnetic field coils and uncompensated coil feeds. An error field with the same helicity as a tearing mode exerts an electromagnetic locking torque at the mode's rational surface that acts to arrest the rotation of the mode [40] (see chapter 13).

1.13 Error field penetration

Error fields can drive magnetic reconnection, resulting in the formation of locked magnetic island chains in intrinsically tearing-stable plasmas. As has already been mentioned, locked-mode formation is strongly correlated with plasma disruptions. Error-field-driven reconnection is strongly suppressed by plasma rotation at the associated rational surface. However, when the error field amplitude rises above a certain critical value, the rotation at the rational surface is suddenly arrested, and error-field-driven reconnection proceeds unhindered. This phenomenon is known as *error-field penetration* [40, 42]. The scenario just outlined has been observed in a number of tokamak experiments [43–51] (see chapter 7).

1.14 Neoclassical tearing modes

Transformer action is not the only source of toroidal plasma current in a tokamak plasma. In fact, there is an additional, noninductive component of the toroidal plasma current, known as the *bootstrap current* [52–54], which is driven by radial gradients in the plasma pressure (see section 2.20). The flattening of the pressure profile inside the magnetic separatrix of a magnetic island chain gives rise to a helical hole in the bootstrap current that has a destabilizing effect on the chain [55]. A tearing mode that is driven into instability by this mechanism, rather than by the usual free energy sources for a tearing mode (i.e. global current and pressure gradients), is known as a *neoclassical tearing mode* (see chapter 12). Neoclassical tearing modes were originally identified experimentally in the TFTR tokamak [56] and are regarded as the main obstacle to obtaining β values in tokamak plasmas that are adequate for the achievement of thermonuclear fusion [57–59]. However, a magnetic island chain can only locally flatten the plasma pressure when its radial width exceeds a certain threshold value that depends on the local ratio of the parallel and perpendicular energy diffusivities [25]. This observation leads to the conclusion that neoclassical tearing modes are actually metastable. In other words, some sort of seed perturbation must be applied to the relevant rational magnetic flux surface in order to trigger a neoclassical tearing mode. In practice, the seed perturbation usually takes the form of a transient magnetic perturbation that is resonant at the rational surface [60]. Such perturbations arise naturally in tokamak plasmas as a consequence of plasma instabilities such as internal kink modes and edge-localized modes [8]. Neoclassical tearing modes can be stabilized by driving a parallel (to the magnetic field) current in the vicinity of the rational surface by means of electron cyclotron waves injected into the plasma; the idea is to fill in the helical hole in the bootstrap current profile [61–66] (see chapter 12).

1.15 Tearing modes in toroidal plasmas

Figure 14.1 shows the contours of the poloidal magnetic flux for a typical plasma discharge in the KSTAR tokamak [67]. It can be seen that there are a number of differences from the idealized tokamak pictured in figure 1.5. The first main difference is that the flux surfaces in figure 14.1 do not have circular cross sections. In particular, the flux surfaces are highly *vertically elongated.* It turns out that this feature allows the toroidal plasma current driven in a tokamak discharge to be increased without violating the Kruskal–Shafranov criterion (see section 1.10). Hence, all modern tokamaks have strongly shaped, vertically elongated cross sections. The second main difference is that the edge of the plasma is defined by a *last closed magnetic flux surface* that features a magnetic X-point (i.e. a hyperbolic null in the poloidal magnetic field). Plasma that crosses the last closed flux surface is rapidly conducted along magnetic field lines, in a thin *scrape-off layer*, to divertor plates located below the plasma. The purpose of this feature is to mitigate the interaction of the plasma with the plasma-facing components and thereby to help limit the flux of impurities into the plasma [68]. Hence, all modern tokamaks have magnetic X-points. Given that the poloidal magnetic field strength at the X-point is zero, the safety-factor value on the last closed flux surface is infinite (see equation (1.76)). Experimentally, it has been found that the safety-factor value on the magnetic flux surface that encloses 95% of the poloidal magnetic flux enclosed by the last closed flux surface, known as q_{95}, plays an analogous role to the edge safety-factor value, q_a, in a tokamak without a magnetic X-point [8] (see section 14.4). Thus, the rule of thumb for safe operation, $q_a \gtrsim 3$, is replaced by $q_{95} \gtrsim 3$ (see section 1.9).

In the type of highly shaped plasma equilibrium pictured in figure 14.1, tearing modes with the same toroidal mode number but different poloidal mode numbers are coupled together [69, 70]. Somewhat confusingly, this effect is known as *toroidal mode coupling*. Toroidal mode coupling allows magnetic island chains resonant on different rational surfaces within the plasma to interact. The interaction is expected to be mutually destabilizing [70]. However, sheared plasma rotation acts to prevent such interaction (see chapter 14).

References

[1] Atzeni S and Meyer-ter-Vehn J 2009 The Physics of Inertial Fusion: Beam Plasma Interaction, Hydrodynamics *Hot Dense Matter* (Oxford: Oxford University Press)

[2] Fitzpatrick R 2014 *Plasma Physics: An Introduction* (Boca Raton, FL: CRC Press)

[3] Fitzpatrick R 2022 *Newtonian Dynamics: An Introduction* (Boca Raton, FL: CRC Press)

[4] Rose D J and Clark M C Jr 1961 *Plasmas and Controlled Fusion* (Boston, MA: MIT Press)

[5] Bosch H-S and Hale G M 1992 Improved formulas for fusion cross-sections and thermal reactivities *Nucl. Fusion* **32** 611

[6] Lawson J D 1957 Some criteria for a power producing thermonuclear reactor *Proc. Phys. Soc.* B **70** 6

[7] Freidberg J P 1987 *Ideal Magnetohydrodynamics* (New York: Plenum)

[8] Wesson J A 2011 *Tokamaks* 4th edn (Oxford: Oxford University Press)

[9] Shimada M *et al* 2007 Chapter 1: Overview and summary *Progress in the ITER Physics Basis Nucl. Fusion* **41** S1

[10] Creely A J *et al* 2020 Overview of the SPARC tokamak *J. Plasma Phys.* **86** 865860502

[11] Troyon F, Gruber R, Saurenmann H, Semenzato S and Succi S 1984 MHD-limits to plasma confinement *Plasma Phys. Control. Fusion* **26** 209

[12] Doyle E J *et al* 2007 Chapter 2: Plasma confinement and transport *Progress in the ITER Physics Basis Nucl. Fusion* **47** S18 Corrected by 2008 *Nucl. Fusion* **48** 099801.

[13] Combs S K 1993 Pellet injection technology *Rev. Sci. Instrum.* **64** 1679

[14] Stacey W M 2005 *Fusion Plasma Physics* (Weinheim: Wiley)

[15] Stix T H 1972 Heating of toroidal plasmas by neutral injection *Plasma Phys.* **14** 367

[16] Jensen R V, Post D E, Grasberger W H, Tarter C B and Lokke W A 1977 Calculations of impurity radiation and its effects on tokamak experiments *Nucl. Fusion* **17** 1187

[17] Laing E W 1976 *Plasma Physics* (Brighton: Sussex University Press)

[18] Kruskal M D and Schwartzschild M 1954 Some instabilities of a completely ionized plasma *Proc. R. Soc.* A **225** 348

[19] Shafranov V D 1956 About the stability of the plasma column in the presence of a longitudinal magnetic field and the conductive casing *Atomnaya energiya* **1** 38–41 (in Russian)

[20] Wesson J A 1978 Hydrodynamic stability of tokamaks *Nucl. Fusion* **18** 87

[21] Greenwald M, Terry J L, Wolfe S M, Ejima S, Bell M G, Kaye S M and Neilson G H 1988 A new look at density limits in tokamaks *Nucl. Fusion* **28** 2199

[22] Furth H P, Killeen J and Rosenbluth M N 1963 Finite-resistivity instabilities of a sheet pinch *Phys. Fluids* **6** 459

[23] Rutherford P H 1973 Nonlinear growth of the tearing mode *Phys. Fluids* **16** 1906

[24] Hastie R J, Militello F and Porcelli F 2005 Nonlinear saturation of tearing mode islands *Phys. Rev. Lett.* **95** 065001

[25] Fitzpatrick R 1995 Helical temperature perturbations associated with tearing modes in tokamak plasmas *Phys. Plasmas* **2** 825

[26] Chang Z and Callen J D 1990 Global energy confinement degradation due to macroscopic phenomena in tokamaks *Nucl. Fusion* **30** 219

[27] Li E, Liqun L, Ling B, Liu Y, Ti A, Chen K, Shen B and Gao X 2010 Electron cyclotron emission reconstruction image and $m/n = 3/2$ mode in HT-7 tokamak *Rev. Sci. Instrum.* **81** 073506

[28] Nagayama Y, Taylor G, Fredrickson E D, Budny R V, Janos A C, Mansfield D K, McGuire K M and Yamada M 1996 Tomography of (2, 1) and (3, 2) magnetic island structures on tokamak fusion test reactor *Phys. Plasmas* **3** 2631
[29] Ara G, Basu B, Coppi B, Laval G, Rosenbluth M N and Waddell B V 1978 Magnetic reconnection and $m = 1$ oscillations in current carrying plasmas *Ann. Phys.* **112** 443–76
[30] Braginskii S I 1965 Transport processes in a plasma *Reviews of Plasma Physics* 1 (New York: Consultants Bureau) p 205
[31] Fitzpatrick R and Waelbroeck F L 2005 Two-fluid magnetic island dynamics in slab geometry. I. Isolated islands *Phys. Plasmas* **12** 022307
[32] Fitzpatrick R, Watson P G and Waelbroeck F L 2005 Two-fluid magnetic island dynamics in slab geometry: determination of the island phase velocity *Phys. Plasmas* **12** 082510
[33] Hosea J C, Jobes F C, Hickok R L and Dellis A N 1973 Rotation and structure of low-frequency oscillations inside the ST-Tokamak plasma *Phys. Rev. Lett.* **30** 839
[34] Vahala G, Vahala L, Harris J H, Bateman G, Waddell B V, Dunlap J L, Paré V K and Burris R D 1980 Perturbed magnetic-field phase slip for tokamaks *Nucl. Fusion* **20** 17
[35] Buratti P, Alessi E, Baruzzo M, Casolari A, Giovannozzi E, Giroud C, Hawkes N, Menmuir S and Pucella G 2016 Diagnostic application of magnetic islands rotation in JET *Nucl. Fusion* **56** 076004
[36] La Haye R J, Petty C C, Strait E J, Waelbroeck F L and Wilson H R 2003 Propagation of magnetic islands in the $E_r = 0$ frame of co-injected neutral beam driven discharges in the DIII-D tokamak *Phys. Plasmas* **10** 3644
[37] Chapman B E, Fitzpatrick R, Craig D, Martin P and Spizzo G 2004 Observation of tearing mode deceleration and locking due to eddy currents induced in a conducting shell *Phys. Plasmas* **11** 2156
[38] Snipes J A, Campbell D J, Haynes P S, Hender T C, Hugon M, Lomas P J, Lopes Cardozo N J, Nave M F F and Schüller F C 1988 Large amplitude quasi-stationary MHD modes in JET *Nucl. Fusion* **28** 1085
[39] de Vries P C, Johnson M F, Alper B, Buratti P, Hender T C, Koslowski H R, Riccardo V and JET EFDA Contributors 2011 Survey of disruption causes at JET *Nucl. Fusion* **51** 053018
[40] Fitzpatrick R 1993 Interaction of tearing modes with external structures in cylindrical geometry *Nucl. Fusion* **33** 1049
[41] Nave M F F and Wesson J A 1990 Mode Locking in Tokamaks *Nucl. Fusion* **30** 2575
[42] Fitzpatrick R 1998 Bifurcated states of a rotating tokamak plasma in the presence of a static error-field *Phys. Plasmas* **5** 3325
[43] Fishpool G M and Haynes P S 1994 Field error instabilities in JET *Nucl. Fusion* **34** 109
[44] Hender T C *et al* 1992 Effect of resonant magnetic perturbations on COMPASS-C tokamak discharges *Nucl. Fusion* **32** 2091
[45] Howell D F, Hender T C and Cunningham G 2007 Locked mode thresholds on the MAST spherical tokamak *Nucl. Fusion* **47** 1336
[46] Menard J E *et al* 2010 Progress in understanding error-field physics in NSTX spherical torus plasmas *Nucl. Fusion* **50** 045008
[47] Scoville J T, La Haye R J, Kellman A K, Osborne T H, Stambaugh R D, Strait E J and Taylor T S 1991 Locked modes in DIII-D and a method for prevention of the low density mode *Nucl. Fusion* **31** 875
[48] Wang H-H, Sun Y-W, Shi T-H, Zang Q, Liu Y-Q, Yang X, Gu S, He K-Y, Gu X and Qian J-P *et al* 2018 Density scaling of $n = 1$ error field penetration in ohmically heated discharges in EAST *Nucl. Fusion* **58** 056024

[49] Wang N, Rao B, Hu Q, Ding Y, Chen Z, Gao L, Jin W, Yi B, Zeng W and Li Q *et al* 2014 Study of the penetration of resonant magnetic perturbations in J-TEXT *Nucl. Fusion* **54** 064014

[50] Wolf R C *et al* 2005 Effect of the dynamic ergodic divertor in the TEXTOR tokamak on MHD stability *Nucl. Fusion* **45** 1700

[51] Wolfe S M, Hutchinson I H, Granetz R S, Rice J and Hubbard A 2005 Nonaxisymmetric field effects on Alcator C-Mod *Phys. Plasmas* **12** 056110

[52] Bickerton R J, Connor J W and Taylor J B 1971 Diffusion driven plasma currents and bootstrap tokamak *Nat. Phys. Sci.* **229** 110

[53] Zarnstorff M C *et al* 1988 Bootstrap current in TFTR *Phys. Rev. Lett.* **60** 1306

[54] Zarnstorff M C and Prager S C 1984 Experimental observation of neoclassical currents in a plasma *Phys. Rev. Lett.* **53** 454

[55] Carrera R, Hazeltine R D and Kotschenreuther M 1986 Island bootstrap current modification of the nonlinear dynamics of the tearing mode *Phys. Fluids* **29** 899

[56] Chang Z, Callen J D, Fredrickson E D, Budny R V, Hegna C C, McGuire K M, Zarnstorff M C and the TFTR Group 1995 Observation of nonlinear neoclassical pressure-gradient-driven tearing modes in TFTR *Phys. Rev. Lett.* **74** 4663

[57] Buttery R J *et al* 2000 Neoclassical tearing modes *Plasma Phys. Control. Fusion* **42** B61

[58] La Haye R J 2006 Neoclassical tearing modes and their control *Phys. Plasmas* **13** 055501

[59] Sauter O *et al* 1997 Beta limits in long-pulse tokamak discharges *Phys. Plasmas* **4** 1654

[60] Hegna C C, Callen J D and La Haye R J 1999 Dynamics of seed magnetic island formation due to geometrically coupled perturbations *Phys. Plasmas* **6** 130

[61] Hegna C C and Callen J D 1977 On the stabilization of neoclassical magnetohydrodynamic tearing modes using localized current drive or heating *Phys. Plasmas* **4** 2940

[62] La Haye R J, Günter S, Humphreys D A, Lohr J, Luce T C, Maraschek M E, Petty C C, Prater R, Scoville J T and Strait E J 2002 Control of neoclassical tearing modes in DIII-D *Phys. Plasmas* **9** 2051

[63] Nagasaki K, Isayama A, Ide S, and JT-60 Team 2003 Stabilization effect of early ECCD on a neoclassical tearing mode in the JT-60U tokamak *Nucl. Fusion* **43** L7

[64] Reiman A H 1983 Suppression of magnetic islands by RF-driven currents *Phys. Fluids* **26** 1338

[65] Zohm H 1997 Stabilization of neoclassical tearing modes by electron cyclotron current drive *Phys. Plasmas* **4** 3433

[66] Zohm H, Gantenbein G, Gude A, Günter S, Leuterer F, Maraschek M, Meskat J, Suttrop W and Yu Q, ASDEX Upgrade Team and ECRH-Group (AUG) 2001 Neoclassical tearing modes and their stabilization by electron cyclotron current drive in ASDEX upgrade *Phys. Plasmas* **8** 2009

[67] Lee G S *et al* 2001 Design and construction of the KSTAR tokamak *Nucl. Fusion* **41** 1515

[68] Loarte A *et al* 2007 Chapter 4: Power and particle control *Progress in the ITER Physics Basis Nucl. Fusion* **47** S203

[69] Connor J W, Cowley S C, Hastie R J, Hender T C, Hood A and Martin T J 1988 Tearing modes in toroidal geometry *Phys. Fluids* **31** 577

[70] Fitzpatrick R, Hastie R J, Martin T J and Roach C M 1993 Stability of coupled tearing modes in tokamaks *Nucl. Fusion* **33** 1533

Chapter 2

Plasma fluid theory

2.1 Introduction

The aim of this chapter is to outline the fundamental fluid theory that underpins the analysis of tearing-mode dynamics in tokamak plasmas.

2.2 Kinetic theory

Consider an ideal plasma consisting of two species: electrons of mass m_e and electrical charge $e_e = -e$ and ions of mass m_i and electrical charge $e_i = +e$. Note that we are neglecting helium ash particles and impurity ions for the sake of simplicity. Assuming that the plasma contains an optimal 50%–50% mixture of deuterium and tritium ions, it follows that $m_i = (m_D + m_T)/2$, where m_D and m_T are the deuteron and triton masses, respectively.

At the most fundamental level, the dynamics of each plasma species is governed by the *kinetic equation* [1]:

$$\frac{\partial f_s}{\partial t} + \mathbf{v} \cdot \nabla f_s + \frac{e_s}{m_s}(\mathbf{E} + \mathbf{v} \times \mathbf{B}) \cdot \frac{\partial f_s}{\partial \mathbf{v}} = C_s(f_e, f_i). \tag{2.1}$$

Here, $f_s(\mathbf{r}, \mathbf{v}, t)$ is the ensemble-averaged phase-space density of plasma species s (s stands for either e or i) near point $(\mathbf{r}, \mathbf{v})$ at time t, where $\mathbf{r}$ denotes spatial position and $\mathbf{v}$ denotes velocity [2]. Moreover, $\mathbf{E}$ and $\mathbf{B}$ are the ensemble-averaged electric and magnetic field strengths, respectively. Finally, $C_s(f_e, f_i)$ is the *collision operator* for plasma species s [1]. The electron and ion kinetic equations form a complete set when combined with Maxwell's equations.

Equation (2.1) describes plasma dynamics in a six-dimensional space (three spatial dimensions, and three velocity-space dimensions); as such, it is an extremely difficult equation to solve. For dynamics that takes place on timescales that are long compared to the inverse ion gyrofrequency, which certainly applies to tearing-mode dynamics in tokamak plasmas, the equation can be simplified by averaging over the

doi:10.1088/978-0-7503-5367-0ch2

gyromotions of the charged particles [3–5]. The resulting *gyrokinetic equation* is only five-dimensional (three spatial dimensions, and two velocity-space dimensions). The electron and ion gyrokinetic equations have been extensively used to numerically model plasma turbulence in tokamaks [6–8]. It turns out that this is possible because turbulent eddies are localized on toroidal magnetic flux surfaces in flux tubes that run parallel to the magnetic field and whose radial extents are, at most, a few ion gyroradii. Moreover, the eddies attain a quasi-steady state in a matter of a few milliseconds. Hence, despite the high dimensionalities of the gyrokinetic equations, it is practical to solve them in calculations that only simulate a small fraction of the plasma volume over a time interval of a few milliseconds. Tearing modes are *global* plasma instabilities that evolve on timescales of hundreds, if not thousands, of milliseconds. Unfortunately, it is simply not practical to simulate the whole plasma over a time interval of a few hundred to a few thousand milliseconds using the gyrokinetic equations. Hence, a different approach is needed to describe tearing-mode dynamics. In fact, the only practical option is to employ fluid theory.

2.3 Fluid theory

Plasma *fluid equations* are obtained by taking the low-order velocity-space moments of the kinetic equation [1].

The low-order moments of the distribution function, f_s, all have simple physical interpretations. First, we have the particle *number density*,

$$n_s(\mathbf{r}, t) = \int f_s(\mathbf{r}, \mathbf{v}, t)\, d^3\mathbf{v}, \tag{2.2}$$

and the mean *flow velocity*,

$$\mathbf{V}_s(\mathbf{r}, t) = \frac{1}{n_s} \int \mathbf{v}\, f_s(\mathbf{r}, \mathbf{v}, t)\, d^3\mathbf{v}. \tag{2.3}$$

Next, we have the *pressure tensor*,

$$\mathbf{p}_s(\mathbf{r}, t) = \int m_s\, \mathbf{u}_s\mathbf{u}_s f_s(\mathbf{r}, \mathbf{v}, t)\, d^3\mathbf{v}, \tag{2.4}$$

and the *heat flux*,

$$\mathbf{q}_s(\mathbf{r}, t) = \int \frac{1}{2}\, m_s\, u_s^2\, \mathbf{u}_s f_s(\mathbf{r}, \mathbf{v}, t)\, d^3\mathbf{v}. \tag{2.5}$$

Here,

$$\mathbf{u}_s = \mathbf{v} - \mathbf{V}_s. \tag{2.6}$$

The trace of the pressure tensor measures the ordinary (or scalar) pressure,

$$p_s = \frac{1}{3}\, \mathrm{Tr}\, (\mathbf{p}_s). \tag{2.7}$$

The (kinetic) temperature is defined as

$$T_s = \frac{p_s}{n_s}. \tag{2.8}$$

The low-order velocity-space moments of the collision operator also have simple interpretations. The *friction force density* takes the form

$$\mathbf{F}_s = \int m_s\, \mathbf{v}\, C_s(f_e, f_i)\, d^3\mathbf{v}, \tag{2.9}$$

while the *collisional heating rate density* (in the species-s rest frame) is written

$$W_s \equiv \int \frac{1}{2}\, m_s\, u_s^2\, C_s(f_e, f_i)\, d^3\mathbf{v}. \tag{2.10}$$

The zeroth, first, and contracted second velocity-space moments of the kinetic equation, (2.1), yield the following set of fluid equations for species s [1]:

$$\frac{d_s n_s}{dt} + n_s\, \nabla\cdot\mathbf{V}_s = 0, \tag{2.11}$$

$$m_s\, n_s\, \frac{d_s\mathbf{V}_s}{dt} - e_s\, n_s\, (\mathbf{E} + \mathbf{V}_s\times\mathbf{B}) + \nabla p_s + \nabla\cdot\boldsymbol{\pi}_s = \mathbf{F}_s, \tag{2.12}$$

$$\frac{3}{2}\frac{d_s p_s}{dt} + \frac{5}{2}\, p_s\, \nabla\cdot\mathbf{V}_s + \boldsymbol{\pi}_s : \nabla\mathbf{V}_s + \nabla\cdot\mathbf{q}_s = W_s. \tag{2.13}$$

Here,

$$\frac{d_s}{dt} \equiv \frac{\partial}{\partial t} + \mathbf{V}_s\cdot\nabla \tag{2.14}$$

is the well-known *convective derivative*, and we have written

$$\mathbf{p}_s = p_s\, \mathbf{I} + \boldsymbol{\pi}_s, \tag{2.15}$$

where $\mathbf{I}$ is the unit (identity) tensor and $\boldsymbol{\pi}_s$ is the *viscosity tensor*. Obviously, equation (2.11) is a particle conservation equation for species s, equation (2.12) is a momentum conservation equation, and equation (2.13) is an energy conservation equation [1].

2.4 Fundamental quantities

Before proceeding further, it is helpful to define a few fundamental quantities.

Of course, quasi-neutrality demands that [1]

$$n_i \simeq n_e. \tag{2.16}$$

We can estimate typical particle speeds in terms of the so-called *thermal speed* [1],

$$v_{t\,s} = \left(\frac{2\,T_s}{m_s}\right)^{1/2}. \tag{2.17}$$

The typical *gyroradius* of a charged particle gyrating in the magnetic field of a tokamak is given by

$$\rho_s = \frac{v_{ts}}{|\Omega_s|}, \tag{2.18}$$

where

$$\Omega_s = \frac{e_s B}{m_s} \tag{2.19}$$

is the *gyrofrequency* associated with the gyration [1]. (Note that $\Omega_e < 0$, indicating that electrons gyrate around magnetic field lines in the opposite direction to that of ions.)

The electron–ion and ion–ion *collision times* are written

$$\tau_e = \frac{6\sqrt{2}\,\pi^{3/2}\,\epsilon_0{}^2\,\sqrt{m_e}\,T_e^{3/2}}{e^4\,n_e\,\ln\Lambda}, \tag{2.20}$$

$$\tau_i = \frac{12\,\pi^{3/2}\,\epsilon_0{}^2\,\sqrt{m_i}\,T_i^{3/2}}{e^4\,n_e\,\ln\Lambda}, \tag{2.21}$$

respectively [1]. Here, $\ln\Lambda \simeq 16$ is the *Coulomb logarithm* [9]. Note that τ_e is the typical time required for the cumulative effect of electron–ion collisions to deviate the path of an electron through 90°. Likewise, τ_i is the typical time required for the cumulative effect of ion–ion collisions to deviate the path of an ion through 90°.

The electron and ion *collision frequencies* are simply the inverses of the corresponding 90° collision times:

$$\nu_e = \frac{1}{\tau_e}, \tag{2.22}$$

$$\nu_i = \frac{1}{\tau_i}. \tag{2.23}$$

Finally, the *mean free paths* between collisions (i.e. 90° scattering events) for electrons and ions are

$$l_e = v_{te}\,\tau_e, \tag{2.24}$$

$$l_i = v_{ti}\,\tau_i, \tag{2.25}$$

respectively.

Table 2.1 gives estimates for some of the fundamental plasma parameters defined in this section in a low-field tokamak fusion reactor and a high-field tokamak fusion reactor. Here, use has been made of the data shown in table 1.2. For the sake of simplicity, it has also been assumed that $T_i = T_e$.

Table 2.1. Fundamental plasma parameters in a low-field tokamak reactor and a high-field tokamak reactor. Here, B is the toroidal magnetic field strength, R_0 the plasma major radius, a the plasma minor radius, n_e the electron number density, T_e the electron temperature, T_i the ion temperature, Ω_e the electron gyrofrequency, Ω_i the ion gyrofrequency, ρ_e the electron gyroradius, ρ_i the ion gyroradius, τ_e the electron collision time, τ_i the ion collision time, l_e the electron mean free path, and l_i the ion mean free path.

	Low field	High field
B (T)	5.0	12.0
R_0 (m)	7.6	3.2
a (m)	2.5	1.1
n_e (10^{20} m^{-3})	0.89	5.1
T_e (keV)	7.0	7.0
T_i (keV)	7.0	7.0
$\|\Omega_e\|$ (THz)	0.88	2.1
Ω_i (GHz)	0.19	0.46
ρ_e (μm)	56	24
ρ_i (mm)	3.8	1.6
τ_e (ms)	0.14	0.025
τ_i (ms)	13.6	2.36
l_e (km)	7.0	1.2
l_i (km)	10	1.7

2.5 Fluid closure schemes

The fluid equations for the electron and ion species take the respective forms:

$$\frac{d_e n_e}{dt} + n_e \nabla \cdot \mathbf{V}_e = 0, \tag{2.26}$$

$$m_e n_e \frac{d_e \mathbf{V}_e}{dt} + e\, n_e (\mathbf{E} + \mathbf{V}_e \times \mathbf{B}) + \nabla p_e + \nabla \cdot \boldsymbol{\pi}_e = \mathbf{F}_e, \tag{2.27}$$

$$\frac{3}{2}\frac{d_e p_e}{dt} + \frac{5}{2} p_e \nabla \cdot \mathbf{V}_e + \boldsymbol{\pi}_e : \nabla \mathbf{V}_e + \nabla \cdot \mathbf{q}_e = W_e, \tag{2.28}$$

and

$$\frac{d_i n_e}{dt} + n_e \nabla \cdot \mathbf{V}_i = 0, \tag{2.29}$$

$$m_i n_e \frac{d_i \mathbf{V}_i}{dt} - e\, n_e (\mathbf{E} + \mathbf{V}_i \times \mathbf{B}) + \nabla p_i + \nabla \cdot \boldsymbol{\pi}_i = -\mathbf{F}_e, \tag{2.30}$$

$$\frac{3}{2}\frac{d_i p_i}{dt} + \frac{5}{2} p_i \nabla \cdot \mathbf{V}_i + \boldsymbol{\pi}_i : \nabla \mathbf{V}_i + \nabla \cdot \mathbf{q}_i = W_i. \tag{2.31}$$

Here, use has been made of equations (2.11)–(2.13) and (2.16), as well as the fact that $\mathbf{F}_i = -\mathbf{F}_e$ (because collisions conserve momentum) [1]. In their present forms, the electron and ion fluid equations relate interesting fluid quantities, such as the electron number density, n_e, the mean flow velocities, $\mathbf{V}_e$ and $\mathbf{V}_i$, and the scalar pressures, p_e and p_i, to unknown quantities, such as the viscosity tensors, $\boldsymbol{\pi}_e$ and $\boldsymbol{\pi}_i$, the heat fluxes, $\mathbf{q}_e$ and $\mathbf{q}_i$, and the moments of the collision operator, $\mathbf{F}_e$, W_e, and W_i. In order to complete our set of equations, we need to employ some additional information to express the latter quantities in terms of the former; this process is known as *closure*.

There are two basic types of fluid closure scheme. In *truncation schemes*, high-order velocity-space moments of the distribution function are assumed to vanish, or are prescribed in terms of low-order moments [10, 11]. Truncation schemes are relatively straightforward to implement, but the error associated with the closure cannot easily be determined. *Asymptotic schemes*, on the other hand, depend on a rigorous expansion of the kinetic equation in terms of some dimensionless parameter that is small compared to unity [12]. Asymptotic closure schemes have the advantage of providing some estimate of the error involved in the closure. However, the asymptotic approach to closure is mathematically demanding, because it involves working closely with the kinetic equation. In this book, we shall rely on a mixture of truncation and asymptotic closure schemes.

2.6 The classical closure scheme

The so-called *classical* asymptotic closure scheme for the electron and ion fluid equations, (2.26)–(2.31), due to Braginskii [13], is premised on the assumption that both fluids are *highly magnetized*, which means that the electron and ion gyroradii are much smaller than the corresponding mean free paths between collisions. In other words,

$$\frac{\rho_e}{l_e} \ll 1, \tag{2.32}$$

$$\frac{\rho_i}{l_i} \ll 1. \tag{2.33}$$

As is clear from table 2.2, the electron and the ion fluids in tokamak fusion reactors are indeed highly magnetized, which implies that particle, momentum, and energy flows that are perpendicular to magnetic field lines are quite different from those that are parallel to magnetic field lines [1].

For particle, momentum, and energy flows perpendicular to magnetic field lines, the small (compared to unity) dimensionless expansion parameters upon which the classical asymptotic closure scheme is based are ρ_e/a and ρ_i/a. Here, we are assuming that the variation length scales of quantities perpendicular to magnetic field lines are of order the plasma minor radius, a. As is apparent from table 2.2, the expansion parameters ρ_e/a and ρ_i/a are indeed small compared to unity in tokamak fusion reactors.

Table 2.2. Classical fluid closure parameters in a low-field tokamak reactor and a high-field tokamak reactor. Here, B is the toroidal magnetic field strength, ρ_e the electron gyroradius, ρ_i the ion gyroradius, l_e the electron mean free path, l_i the ion mean free path, and L_c the connection length (calculated with $q = 3$).

	Low field	High field
B (T)	5.0	12.0
ρ_e/l_e	8.0×10^{-9}	1.9×10^{-8}
ρ_i/l_i	3.8×10^{-7}	9.2×10^{-7}
ρ_e/a	2.2×10^{-5}	2.2×10^{-5}
ρ_i/a	1.5×10^{-3}	1.5×10^{-3}
l_e/L_c	49	21
l_i/L_c	70	29

For particle, momentum, and energy flows parallel to magnetic field lines, the small (compared to unity) expansion parameters upon which the classical asymptotic closure scheme is based are l_e/L_c and l_i/L_c. Here, we are assuming that the variation length scales of quantities parallel to magnetic field lines are of the order of the *connection length*, $L_c = 2\pi\, q\, R_0$, which is the typical distance a magnetic field line has to travel in order to fully traverse a magnetic flux surface. As is clear from table 2.2, the expansion parameters l_e/L_c and l_i/L_c are actually both large compared to unity in tokamak fusion reactors. It follows that the classical asymptotic closure scheme fails in such reactors (because the confined plasmas are not sufficiently collisional in nature). Nevertheless, in the following, we shall describe the classical asymptotic closure scheme, before attempting to repair it.

According to the classical asymptotic closure scheme [13],

$$\mathbf{F}_e = \mathbf{F}_u + \mathbf{F}_T, \tag{2.34}$$

$$\mathbf{F}_u = n_e\, e\left(\frac{\mathbf{j}_\parallel}{\sigma_\parallel} + \frac{\mathbf{j}_\perp}{\sigma_\perp}\right), \tag{2.35}$$

$$\mathbf{F}_T = -0.71\, n_e\, \nabla_\parallel T_e + \frac{3\, n_e}{2\, \Omega_e\, \tau_e}\, \mathbf{b} \times \nabla_\perp T_e, \tag{2.36}$$

$$W_i = \frac{3\, m_e}{m_i}\frac{n_e\, (T_e - T_i)}{\tau_e}, \tag{2.37}$$

$$W_e = -W_i + \frac{\mathbf{j} \cdot \mathbf{F}_e}{n_e\, e}. \tag{2.38}$$

Here, $\mathbf{b} = \mathbf{B}/B$ is a unit vector parallel to the magnetic field, and

$$\mathbf{j} = n_e\, e\, (\mathbf{V}_i - \mathbf{V}_e) \tag{2.39}$$

is the plasma current density. Moreover, the *parallel electrical conductivity* is given by [13, 14]

$$\sigma_\parallel = 1.96\, \frac{n_e\, e^2\, \tau_e}{m_e}. \tag{2.40}$$

whereas the *perpendicular electrical conductivity* takes the form [13]

$$\sigma_\perp = \frac{n_e\, e^2\, \tau_e}{m_e}. \tag{2.41}$$

Note that $\nabla_\parallel(\cdots) \equiv [\mathbf{b}\cdot\nabla(\cdots)]\,\mathbf{b}$ denotes a gradient parallel to the magnetic field, while $\nabla_\perp \equiv \nabla - \nabla_\parallel$ denotes a gradient perpendicular to the magnetic field. Likewise, $\mathbf{j}_\parallel \equiv (\mathbf{b}\cdot\mathbf{j})\,\mathbf{b}$ represents the component of the plasma current density flowing parallel to the magnetic field, while $\mathbf{j}_\perp \equiv \mathbf{j} - \mathbf{j}_\parallel$ represents the perpendicular component of the plasma current density.

It can be seen that parallel component of the *friction force density*, $\mathbf{F}_u$, is smaller than the perpendicular component by a factor 1.96; this is a consequence of the fact that the collision frequency decreases with increasing velocity ($\nu_e \propto v^{-3}$), causing the distribution of electrons with large parallel velocities to be more distorted from a Maxwellian distribution than that of slower electrons [15]. The *thermal force density*, $\mathbf{F}_T$, is also a consequence of the velocity dependence of the collision frequency [1] (see section 2.23).

The electron and ion heat fluxes are written [13]

$$\mathbf{q}_e = \mathbf{q}_{\parallel\, e} + \mathbf{q}_{\times e} + \mathbf{q}_{\perp\, e} + \mathbf{q}_T, \tag{2.42}$$

$$\mathbf{q}_i = \mathbf{q}_{\parallel\, i} + \mathbf{q}_{\times i} + \mathbf{q}_{\perp\, i}, \tag{2.43}$$

respectively, where

$$\mathbf{q}_{\parallel\, s} = -\kappa_{\parallel\, s}\, \nabla_\parallel T_s, \tag{2.44}$$

$$\mathbf{q}_{\times s} = \kappa_{\times s}\, \mathbf{b}\times\nabla_\perp T_s, \tag{2.45}$$

$$\mathbf{q}_{\perp\, s} = -\kappa_{\perp\, s}\, \nabla_\perp T_s, \tag{2.46}$$

$$\mathbf{q}_T = -0.71\, \frac{T_e}{e}\, \mathbf{j}_\parallel + \frac{3\, T_e}{2\, \Omega_e\, \tau_e\, e}\, \mathbf{b}\times\mathbf{j}_\perp \tag{2.47}$$

are known as the parallel, cross, perpendicular, and thermal heat fluxes, respectively. Here, the *parallel thermal conductivities*, which control the diffusion of heat parallel to magnetic field lines, are given by [13]

$$\kappa_{\parallel\, e} = 3.16\, \frac{n_e\, \tau_e\, T_e}{m_e}, \tag{2.48}$$

$$\kappa_{\parallel\,i} = 3.9\,\frac{n_e\,\tau_i\,T_i}{m_i}. \tag{2.49}$$

Moreover, the *cross thermal conductivities*, which control the (non-diffusive) flow of heat within magnetic flux surfaces, perpendicular to magnetic field lines, are written [13]

$$\kappa_{\times e} = \frac{5\,n_e\,T_e}{2\,m_e\,\Omega_e}, \tag{2.50}$$

$$\kappa_{\times i} = \frac{5\,n_e\,T_i}{2\,m_i\,\Omega_i}. \tag{2.51}$$

(see section 2.11). Finally, the *perpendicular thermal conductivities*, which control the diffusion of heat perpendicular to magnetic flux surfaces, take the forms [13]

$$\kappa_{\perp\,e} = 4.66\,\frac{n_e\,T_e}{m_e\,\Omega_e^{\,2}\,\tau_e}, \tag{2.52}$$

$$\kappa_{\perp\,i} = 2\,\frac{n_e\,T_i}{m_i\,\Omega_i^{\,2}\,\tau_i}. \tag{2.53}$$

Note that $\kappa_{\perp\,s} \sim (\rho_s/l_s)\,\kappa_{\times s}$ and $\kappa_{\times s} \sim (\rho_s/l_s)\,\kappa_{\parallel\,s}$. In other words, in a highly magnetized plasma (i.e. $\rho_s/l_s \ll 1$), the species-s perpendicular thermal conductivity is much less than the cross conductivity, which, in turn, is much less than the parallel conductivity.

According to the previous expressions, the diffusion of heat parallel to magnetic field lines is characterized by the diffusivities

$$\chi_{\parallel\,e} \equiv \frac{\kappa_{\parallel\,e}}{n_e} = 1.58\,\nu_e\,l_e^{\,2}, \tag{2.54}$$

$$\chi_{\parallel\,i} \equiv \frac{\kappa_{\parallel\,i}}{n_e} = 1.95\,\nu_i\,l_i^{\,2}. \tag{2.55}$$

These diffusivities clearly correspond to collision-induced random-walk motions of electrons and ions, parallel to magnetic field lines, with a step frequency of order $\nu_{e,i}$ and a step length of order $l_{e,i}$ [1, 2]. On the other hand, the diffusion of heat perpendicular to magnetic field lines is characterized by the diffusivities

$$\chi_{\perp\,e} \equiv \frac{\kappa_{\perp\,e}}{n_e} = 2.33\,\nu_e\,\rho_e^{\,2}, \tag{2.56}$$

$$\chi_{\perp\,i} \equiv \frac{\kappa_{\perp\,i}}{n_e} = \nu_i\,\rho_i^{\,2}. \tag{2.57}$$

These diffusivities clearly correspond to the collision-induced random-walk motions of electrons and ions, perpendicular to magnetic field lines, with a step frequency of order $\nu_{e,i}$ and a step length of order $\rho_{e,i}$ [1, 2].

In order to describe the viscosity tensor in a highly magnetized plasma, it is helpful to define the species-s *rate-of-strain tensor*:

$$W_{s\,jk} = \frac{\partial V_{s\,j}}{\partial r_k} + \frac{\partial V_{s\,k}}{\partial r_j} - \frac{2}{3}\,\nabla\cdot\mathbf{V}_s\,\delta_{jk}. \tag{2.58}$$

It can easily be demonstrated that this tensor is zero if the species-s fluid translates, or rotates as a rigid body, or if it undergoes isotropic compression. Thus, the rate-of-strain tensor measures the deformation of species-s fluid volume elements.

In a highly magnetized plasma, the viscosity tensor is conveniently described as the sum of three component tensors [13]:

$$\boldsymbol{\pi}_s = \boldsymbol{\pi}_{\parallel\,s} + \boldsymbol{\pi}_{\times s} + \boldsymbol{\pi}_{\perp\,s}, \tag{2.59}$$

where

$$\boldsymbol{\pi}_{\parallel\,s} = -3\,\eta_{\parallel\,s}\left(\mathbf{bb} - \frac{1}{3}\,\mathbf{I}\right)\left(\mathbf{bb} - \frac{1}{3}\,\mathbf{I}\right):\nabla\mathbf{V}_s, \tag{2.60}$$

$$\boldsymbol{\pi}_{\times s} = \frac{\eta_{\times s}}{2}\,(\mathbf{b}\times\mathbf{W}_s\cdot\mathbf{I}_\perp - \mathbf{I}_\perp\cdot\mathbf{W}_s\times\mathbf{b}) + 2\,\eta_{\times s}\,(\mathbf{b}\times\mathbf{W}_s\cdot\mathbf{bb} - \mathbf{bb}\cdot\mathbf{W}_s\times\mathbf{b}), \tag{2.61}$$

$$\begin{aligned}\boldsymbol{\pi}_{\perp\,s} = {}&-\eta_{\perp\,s}\left[\mathbf{I}_\perp\cdot\mathbf{W}_s\cdot\mathbf{I}_\perp + \frac{1}{2}\,\mathbf{I}_\perp\,(\mathbf{b}\cdot\mathbf{W}_s\cdot\mathbf{b})\right]\\ &- 4\,\eta_{\perp\,s}\,(\mathbf{I}_\perp\cdot\mathbf{W}_s\cdot\mathbf{bb} + \mathbf{bb}\cdot\mathbf{W}_s\cdot\mathbf{I}_\perp).\end{aligned} \tag{2.62}$$

Here, $\mathbf{I}$ is the identity tensor, and $\mathbf{I}_\perp = \mathbf{I} - \mathbf{bb}$.

The tensor $\boldsymbol{\pi}_{\parallel\,s}$ describes what is known as *parallel viscosity*; this is a viscosity that controls the diffusion of parallel (to the magnetic field) momentum along magnetic field lines. The parallel viscosity coefficients are given by [13]

$$\eta_{\parallel\,e} = 0.73\,n_e\,\tau_e\,T_e, \tag{2.63}$$

$$\eta_{\parallel\,i} = 0.96\,n_e\,\tau_i\,T_i. \tag{2.64}$$

Moreover, the tensor $\boldsymbol{\pi}_{\times s}$ describes what is known as *gyroviscosity*; this is not really viscosity at all, because the associated viscous stresses are always perpendicular to the velocity, implying that there is no dissipation (i.e. viscous heating) associated with this effect. The gyroviscosity coefficients are given by [13]

$$\eta_{\times e} = \frac{n_e\,T_e}{2\,\Omega_e}, \tag{2.65}$$

$$\eta_{\times i} = \frac{n_e\,T_i}{2\,\Omega_i}. \tag{2.66}$$

Finally, the tensor $\boldsymbol{\pi}_{\perp\,s}$ describes what is known as *perpendicular viscosity*; this is a viscosity that controls the diffusion of perpendicular momentum perpendicular to magnetic field lines. The perpendicular viscosity coefficients take the forms [13]

$$\eta_{\perp\,e} = 0.51\,\frac{n_e\,T_e}{\Omega_e^{\,2}\,\tau_e}, \tag{2.67}$$

$$\eta_{\perp\,i} = \frac{3\,n_e\,T_i}{10\,\Omega_i^{\,2}\,\tau_i}. \tag{2.68}$$

Note that $\eta_{\perp\,s} \sim (\rho_s/l_s)\,\eta_{\times s}$ and $\eta_{\times s} \sim (\rho_s/l_s)\,\eta_{\parallel\,s}$. In other words, in a highly magnetized plasma (i.e. $\rho_s/l_s \ll 1$), the species-s perpendicular viscosity is much less than the gyroviscosity, which, in turn, is much less than the parallel viscosity.

According to the previous expressions, the diffusion of parallel momentum (i.e. parallel to magnetic field lines) is characterized by the diffusivities

$$\Xi_{\parallel\,e} \equiv \frac{\eta_{0\,e}}{n_e\,m_e} = 0.37\,\nu_e\,l_e^{\,2}, \tag{2.69}$$

$$\Xi_{\parallel\,i} \equiv \frac{\eta_{0\,i}}{n_e\,m_i} = 0.48\,\nu_i\,l_i^{\,2}. \tag{2.70}$$

As before, these diffusivities correspond to collision-induced random-walk motions of electrons and ions, parallel to magnetic field lines, with a step frequency of order $\nu_{e,i}$ and a step length of order $l_{e,i}$ [1, 2]. On the other hand, the diffusion of perpendicular momentum perpendicular to magnetic field lines is characterized by the diffusivities

$$\Xi_{\perp\,e} \equiv \frac{\eta_{1\,e}}{n_e\,m_e} = 0.26\,\nu_e\,\rho_e^{\,2}, \tag{2.71}$$

$$\Xi_{\perp\,i} \equiv \frac{\eta_{1\,i}}{n_e\,m_i} = 0.15\,\nu_i\,\rho_i^{\,2}. \tag{2.72}$$

Again, these diffusivities correspond to the collision-induced random-walk motions of electrons and ions, perpendicular to magnetic field lines, with a step frequency of order $\nu_{e,i}$ and a step length of order $\rho_{e,i}$ [1, 2].

Table 2.3 shows estimates for the classical heat and momentum diffusivities in a tokamak fusion reactor. It can be seen that the electron parallel diffusivities are much larger than the ion parallel diffusivities, but that both are extremely large. On the other hand, the ion perpendicular diffusivities are much larger than the electron perpendicular diffusivities, but both are much smaller than the experimentally observed perpendicular diffusivities, which are all approximately 1 $m^2\,s^{-1}$ [16].

2.7 Trapped and passing particles

The motion of a charged particle in the presence of a uniform magnetic field consists of gyration in the plane perpendicular to the local magnetic field line combined with

Table 2.3. Classical diffusivities in a low-field tokamak reactor and a high-field tokamak reactor. Here, B is the toroidal magnetic field strength, $\chi_{\parallel\,e}$ the parallel electron energy diffusivity, $\chi_{\parallel\,i}$ the parallel ion energy diffusivity, $\chi_{\perp\,e}$ the perpendicular electron energy diffusivity, $\chi_{\perp\,i}$ the perpendicular ion energy diffusivity, $\Xi_{\parallel\,e}$ the parallel electron momentum diffusivity, $\Xi_{\parallel\,i}$ the parallel ion momentum diffusivity, $\Xi_{\perp\,e}$ the perpendicular electron momentum diffusivity, and $\Xi_{\perp\,i}$ the perpendicular ion momentum diffusivity.

	Low field	High field
B (T)	5.0	12.0
$\chi_{\parallel\,e}$ ($m^2\,s^{-1}$)	5.5×10^{11}	9.6×10^{10}
$\chi_{\parallel\,i}$ ($m^2\,s^{-1}$)	1.4×10^{10}	2.5×10^{9}
$\chi_{\perp\,e}$ ($m^2\,s^{-1}$)	5.2×10^{-5}	5.2×10^{-5}
$\chi_{\perp\,i}$ ($m^2\,s^{-1}$)	1.1×10^{-3}	1.1×10^{-3}
$\Xi_{\parallel\,e}$ ($m^2\,s^{-1}$)	1.3×10^{11}	2.2×10^{10}
$\Xi_{\parallel\,i}$ ($m^2\,s^{-1}$)	3.5×10^{9}	6.1×10^{8}
$\Xi_{\perp\,e}$ ($m^2\,s^{-1}$)	5.8×10^{-6}	5.8×10^{-6}
$\Xi_{\perp\,i}$ ($m^2\,s^{-1}$)	1.6×10^{-4}	1.6×10^{-4}

a steady drift along the field line [1]. Suppose, however, that the magnetic field is non-uniform, meaning that there are spatial gradients in its direction and strength. As is well known, if the characteristic gradient scale length of the field is much larger than the particle gyroradius, then the motion still consists of gyration in the plane perpendicular to the local field line, combined with drift along the field line (which is no longer straight). However, the *magnetic moment*,

$$\mu_s = \frac{m_s\, v_{\perp\,s}^{2}}{2\,B}, \tag{2.73}$$

of the particle is a conserved quantity during this motion [1, 17]. Here, $\mathbf{v}_{\perp\,s}$ is the particle's perpendicular (to the magnetic field) velocity and B the local magnetic field strength. It is clear from table 2.1 that the gyroradii of both electrons and ions in a tokamak fusion reactor are much smaller than the dimensions of the reactor. Assuming that the characteristic gradient scale length of the reactor's magnetic field is comparable to its dimensions, we conclude that the motions of both ions and electrons in such a reactor consist of rapid (see table 2.1) gyration perpendicular to magnetic field lines, combined with drift along magnetic field lines at a constant magnetic moment. However, these motions are interrupted by occasional collisions (i.e. 90° scattering events).

Consider the gyroaveraged motion of a charged particle around an idealized magnetic flux surface of circular poloidal cross section. Let us set up a right-handed cylindrical coordinate system, R, φ, Z, whose axis corresponds to the symmetry axis of the tokamak. Let R_0 and r be the major and minor radii of the flux surface, respectively. As shown in figure 2.1, we can specify the location of the charged

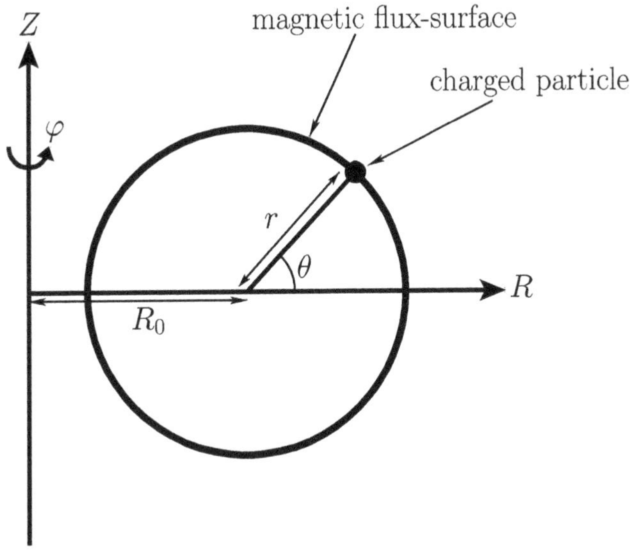

Figure 2.1. Charged particle motion around a magnetic flux surface of circular poloidal cross section.

particle in the poloidal plane in terms of an angular coordinate, θ, which is zero on the outboard midplane (i.e. $Z = 0$, $R > R_0$). In fact, the particle's coordinates in the poloidal plane are

$$R = R_0 + r\cos\theta, \tag{2.74}$$

$$Z = r\sin\theta. \tag{2.75}$$

If we now neglect the influence of the comparatively small poloidal currents flowing in the plasma on the toroidal magnetic field, then we expect the overall magnetic field strength (which is dominated by the toroidal component of the field) to vary as

$$B(\theta) = B_0\,\frac{R_0}{R} = \frac{B_0}{1 + (r/R_0)\cos\theta}. \tag{2.76}$$

Here, B_0 is the toroidal magnetic field strength on the magnetic axis (i.e. $Z = 0$, $R = R_0$). Note that the overall magnetic field strength varies slightly around the flux surface, being larger at smaller major radii and vice versa.

Let us suppose that the parallel (to the magnetic field) electric field strength, $E_\parallel$, is comparatively weak, as is indeed the case in a high-temperature tokamak plasma (otherwise the field would generate an absurdly large parallel current) (see section 2.9). If this is the case, then (neglecting collisions, for the moment) our charged particle drifts around the magnetic flux surface with a constant kinetic energy (recall that a magnetic field cannot do work on the particle). In other words,

$$K_s = \frac{1}{2}\,m_s\,v_{\parallel\,s}^{2} + \frac{1}{2}\,m_s\,v_{\perp\,s}^{2} \tag{2.77}$$

is a constant of the motion. Here, $v_{\parallel\,s}\,\mathbf{b}$ is the particle's parallel velocity. However, the particle's magnetic moment, which is defined in equation (2.73), is also a constant of the motion. Equations (2.73) and (2.77) can be combined to give

$$v_{\parallel\,s}(\theta) = \pm\left(\frac{2}{m_s}\left[K_s - \mu_s\,B(\theta)\right]\right)^{1/2}. \tag{2.78}$$

Hence, we conclude that the particle is excluded from regions of the flux surface in which $K_s < \mu_s\,B(\theta)$ (because $v_{\parallel\,s}$ clearly cannot be imaginary). In fact, if the particle reaches a so-called *bounce point*, characterized by $\theta = \theta_b$, where $K_s = \mu_s\,B(\theta_b)$, then its parallel motion must reverse direction (i.e. the sign in the previous equation must flip).

Let $v_{\parallel\,0\,s}$ and $v_{\perp\,0\,s}$ be the particle's parallel and perpendicular velocities, respectively, at the outermost point of the flux surface (i.e. $\theta = 0$), where the magnetic field is weakest. It follows that μ_s and K_s both take the constant values

$$\mu_s = \frac{m_s\,v_{\perp\,0\,s}^{\,2}}{2\,B(0)}, \tag{2.79}$$

$$K_s = \frac{1}{2}\,m_s\,v_{\parallel\,0\,s}^{\,2} + \frac{1}{2}\,m_s\,v_{\perp\,0\,s}^{\,2}, \tag{2.80}$$

respectively. The previous three equations can be combined to give

$$\frac{v_{\parallel\,s}(\theta)}{|v_{\parallel\,0\,s}|} = \pm\left(1 + \frac{v_{\perp\,0\,s}^{\,2}}{v_{\parallel\,0\,s}^{\,2}}\left[1 - \frac{B(\theta)}{B(0)}\right]\right)^{1/2}. \tag{2.81}$$

Finally, equation (2.76) yields

$$\frac{v_{\parallel\,s}(\theta)}{|v_{\parallel\,0\,s}|} \simeq \pm\left[1 - \frac{v_{\perp\,0\,s}^{\,2}}{v_{\parallel\,0\,s}^{\,2}}\,2\,\epsilon\,\sin^2\!\left(\frac{\theta}{2}\right)\right]^{1/2}, \tag{2.82}$$

where $\epsilon = r/R_0$ is the inverse aspect ratio of the flux surface, and we have made use of the large-aspect-ratio approximation $\epsilon \ll 1$.

The previous equation, combined with the requirement that $v_{\parallel\,s}(\theta)$ not be imaginary, leads to the conclusion that (neglecting collisions) there are two populations of charged particle on a magnetic flux surface. The first population satisfies

$$\tan^{-1}(\sqrt{2\,\epsilon}) < \xi_{0\,s} < \frac{\pi}{2}, \tag{2.83}$$

where

$$\xi_{0\,s} = \tan^{-1}\left(\frac{|v_{\parallel\,0\,s}|}{|v_{\perp\,0\,s}|}\right) \tag{2.84}$$

is known as the *pitch angle* (on the outboard midplane). This first population of particles is called *passing particles* because all such particles have sufficiently large parallel velocities that they can circulate freely around the flux surface. The second population satisfies

$$0 < \xi_{0\,s} < \tan^{-1}(\sqrt{2\,\epsilon}). \tag{2.85}$$

These so-called *trapped particles* have small enough parallel velocities that they are trapped on the outer part of the flux surface, oscillating between bounce points located at $\theta = \pm\theta_b$, where

$$\sin\left(\frac{\theta_b}{2}\right) = \frac{\tan\xi_{0\,s}}{\sqrt{2\,\epsilon}} \tag{2.86}$$

(see figure 2.2). Assuming that the charged particles at the outermost point on the flux surface have a Maxwellian velocity distribution, all possible values of $\xi_{0\,s}$ are equally likely. Thus, we conclude that the fraction of the particles on the flux surface that are trapped is

$$f_t = \sqrt{2\,\epsilon}. \tag{2.87}$$

(A more exact expression for f_t is specified in equation (2.202).) Given that $\epsilon \propto r$, it is clear that the trapped-particle fraction grows as we move from the innermost (i.e. $r = 0$) to the outermost (i.e. $r = a$) magnetic flux surface in the plasma.

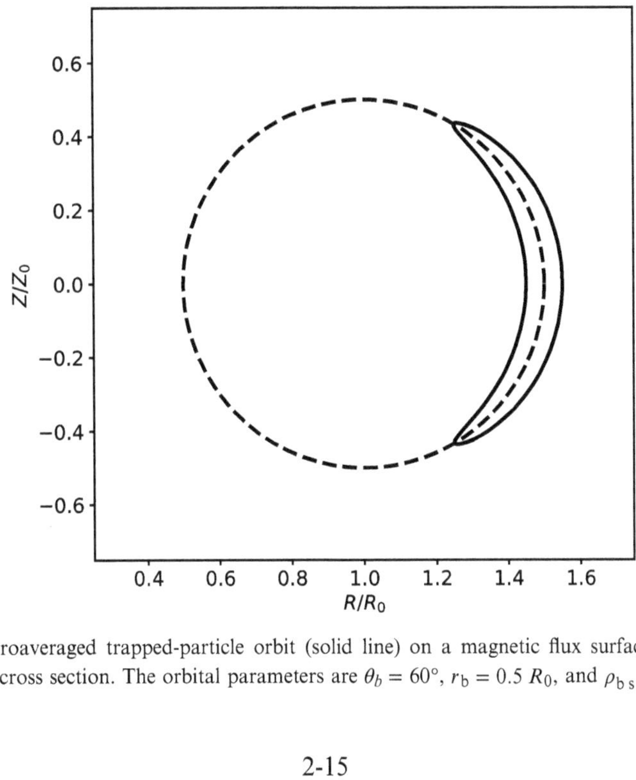

Figure 2.2. A gyroaveraged trapped-particle orbit (solid line) on a magnetic flux surface (dashed line) of circular poloidal cross section. The orbital parameters are $\theta_b = 60°$, $r_b = 0.5\,R_0$, and $\rho_{b\,s} = 0.1\,R_0$.

Consider a trapped particle oscillating between bounce points located at $\theta = \pm\theta_b$. Because the particle is drifting parallel to the magnetic field, its motion is characterized by $d\varphi/d\theta = q$ (see equation (1.76)). Let s represent the path length along a magnetic field line. It follows that $ds \simeq R_0\, d\varphi = q\, R_0\, d\theta$. The particle's parallel equation of motion is

$$\frac{ds}{dt} = v_{\parallel\, s} = \pm|v_{\parallel\, 0\, s}|\left[1 - \frac{\sin^2(\theta/2)}{\sin^2(\theta_b/2)}\right]^{1/2}, \tag{2.88}$$

where use has been made of equations (2.82), (2.84), and (2.86). Hence, we obtain

$$\frac{d\theta}{dt} = \pm\frac{|v_{\parallel\, 0\, s}|}{q\, R_0}\left[1 - \frac{\sin^2(\theta/2)}{\sin^2(\theta_b/2)}\right]^{1/2}. \tag{2.89}$$

Taking the derivative of the previous equation with respect to time, and making use of equations (2.84) and (2.86), we get

$$\frac{d^2\theta}{dt^2} = -\omega_{b\, s}^{\,2} \sin\theta, \tag{2.90}$$

where

$$\omega_{b\, s} = \frac{v_{t\, s}}{q\, R_0}\sqrt{\frac{\epsilon}{2}} \tag{2.91}$$

is known as the *bounce frequency* [16]. Here, we have made the large-aspect-ratio approximation that, on average, $|v_{\perp\, 0\, s}| \simeq v_{t\, s}$ for a trapped particle. Obviously, equation (2.90) has the same form as the equation of motion of a simple pendulum. It follows that a deeply trapped particle, characterized by $\theta_b \ll 1$, executes simple harmonic motion in θ, between bounce points located at $\theta = \pm\theta_b$, at the angular frequency $\omega_{b\, s}$. In this case, the appropriate solution of equation (2.90) is

$$\theta \simeq \theta_b \sin(\omega_{b\, s}\, t). \tag{2.92}$$

A more accurate expression for the bounce frequency, which does not assume that the particle is deeply trapped, is

$$\omega_{b\, s}(\theta_b) = \frac{\omega_{b\, s}}{(2/\pi)\, K(\sin[\theta_b/2])}, \tag{2.93}$$

where $K(k) = \int_0^{\pi/2}(1 - k^2 \sin^2\theta)^{-1/2} d\theta$ is a complete elliptic integral of the first kind [18]. Note that the bounce frequency decreases as the angular motion increases in amplitude, eventually approaching zero logarithmically as $\theta_b \to \pi$.

Let us now take collisions into account. In the absence of collisions, the value of the pitch angle on the outboard midplane, $\xi_{0\, s}$, is a constant of a given particle's motion. Thus, we conclude that if the particle is initially trapped, then it remains trapped at all subsequent times. However, collisions cause the value of $\xi_{0\, s}$ to *diffuse* in velocity space. (Note that Coulomb collisions in a high-temperature plasma are dominated by small-angle scattering events [1], which means that each collision only causes a small change in ξ_{0s}.) It takes a time of order τ_s (where τ_s is the 90° collision time) to change $\xi_{0\,s}$ by order unity. However, it is only necessary to change $\xi_{0\,s}$ by order $\sqrt{\epsilon}$ in order to de-trap a trapped particle (see equation (2.85)). The time required for collisional de-trapping is thus

$$\tau_{d\,s} = (\sqrt{\epsilon})^2\,\tau_s = \epsilon\,\tau_s. \tag{2.94}$$

If $\omega_{b\,s}\,\tau_{d\,s} \ll 1$, then a trapped particle is collisionally de-trapped long before it has time to complete an oscillation between its bounce points. Obviously, in this case, the plasma is sufficiently collisional that it is meaningless to draw a distinction between trapped and passing particles. On the other hand, if $\omega_{b\,s}\,\tau_{d\,s} \gg 1$, then collisions are too infrequent to interfere with the oscillations of a trapped particle between its bounce points.

It is helpful to define the dimensionless *collisionality parameter* [64],

$$\nu_{*\,s} = \frac{q\,R_0\,\nu_s}{\epsilon^{\,3/2}\,v_{t\,s}}. \tag{2.95}$$

Note that $\nu_{*\,s} \simeq (\omega_{b\,s}\,\tau_{d\,s})^{-1}$. It follows that the criterion for species-s trapped particles to exist is $\nu_{*\,s} \ll 1$. Table 2.4 shows that $\nu_{*\,e}$ and $\nu_{*\,i}$ are both much less than unity in a tokamak fusion reactor, implying that populations of trapped electrons as well as trapped ions exist in such reactors. Moreover, it is clear that the trapped-particle

Table 2.4. Neoclassical parameters in a low-field tokamak reactor and a high-field tokamak reactor. Here, B is the toroidal magnetic field strength, f_t the fraction of trapped particles, $\nu_{*\,e}$ the electron collisionality parameter, $\nu_{*\,i}$ the ion collisionality parameter, $\rho_{b\,e}$ the electron banana orbit width, $\rho_{b\,i}$ the ion banana orbit width, δ_e the electron toroidal magnetization parameter, δ_i the ion toroidal magnetization parameter, $\delta_{\theta\,e}$ the electron poloidal magnetization parameter, $\delta_{\theta\,i}$ the ion poloidal magnetization parameter, Δ_e the electron collisionality parameter, and Δ_i the ion collisionality parameter. All quantities are calculated with $q = 3$ and $r = a$ (see table 2.1).

	Low field	High field
B (T)	5.0	12.0
f_t	0.81	0.81
$\nu_{*\,e}$	1.7×10^{-2}	4.0×10^{-2}
$\nu_{*\,i}$	1.2×10^{-2}	2.8×10^{-2}
$\rho_{b\,e}$ (m)	4.1×10^{-4}	1.7×10^{-4}
$\rho_{b\,i}$ (m)	2.8×10^{-2}	1.2×10^{-2}
δ_e	2.2×10^{-5}	2.2×10^{-5}
δ_i	1.5×10^{-3}	1.5×10^{-3}
$\delta_{\theta\,e}$	2.0×10^{-4}	2.0×10^{-4}
$\delta_{\theta\,i}$	1.4×10^{-2}	1.4×10^{-2}
Δ_e	3.2×10^{-3}	7.7×10^{-3}
Δ_i	2.3×10^{-3}	5.5×10^{-3}

fraction, which is the same for both plasma species, is about 80% on the outermost magnetic flux surfaces.

Let us examine the motion of a trapped particle slightly more accurately. In an axisymmetric tokamak plasma, the gyroaveraged motion of the particle takes place at constant toroidal canonical angular momentum:

$$L_{\varphi\,s} \equiv m_s\, R\, v_{\varphi\,s} + e_s\, R\, A_\varphi = \text{constant}, \tag{2.96}$$

where **A** denotes magnetic vector potential. Note that

$$B_\theta = -\frac{dA_\varphi}{dr}. \tag{2.97}$$

Suppose that the particle is trapped on a magnetic flux surface of minor radius r_b. We can write

$$A_\varphi(r) \simeq A_\varphi(r_b) - B_\theta(r_b)\,(r - r_b) \tag{2.98}$$

in the immediate vicinity of the flux surface, where use has been made of equation (2.97). We can now write

$$v_{\varphi\,s} \simeq v_{\parallel\,s} = \pm\sqrt{2\,\epsilon}\;|v_{\perp\,0\,s}|\left[\sin^2\left(\frac{\theta_b}{2}\right) - \sin^2\left(\frac{\theta}{2}\right)\right]^{1/2}, \tag{2.99}$$

where use has been made of equations (2.84), (2.86), and (2.88). Thus, employing equations (2.96) and (2.98),

$$\begin{aligned}\pm m_s\, R\sqrt{2\,\epsilon}\;|v_{\perp\,0\,s}|\left[\sin^2\left(\frac{\theta_b}{2}\right) - \sin^2\left(\frac{\theta}{2}\right)\right]^{1/2} + e_s\, R\, A_\varphi(r_b)\\ - e_s\, R\, B_\theta(r_b)\,(r - r_b) = \text{constant},\end{aligned} \tag{2.100}$$

which implies that the gyroaveraged particle orbit satisfies

$$r = r_b \pm \text{sgn}(e_s)\,\rho_{b\,s}\left[\sin^2\left(\frac{\theta_b}{2}\right) - \sin^2\left(\frac{\theta}{2}\right)\right]^{1/2}, \tag{2.101}$$

where

$$\rho_{b\,s} = \frac{\sqrt{2}\,q}{\epsilon^{\,1/2}}\,\rho_s. \tag{2.102}$$

Here, we have again used the large-aspect-ratio approximation that, on average, $|v_{\perp\,0\,s}| \simeq v_{t\,s}$ for a trapped particle. Note that the $\pm$ signs in equation (2.101) correspond to motion in the $\pm\theta$ and $\pm\varphi$ directions, respectively. As illustrated in figure 2.2, a trapped particle that oscillates between its bounce points makes radial excursions from the guiding magnetic flux surface that are of amplitude $\rho_{b\,s}\,\sin(\theta_b/2)$. In fact, the gyroaveraged orbit's poloidal cross section looks a little like a banana. Hence, such orbits are known as *banana orbits*, and $\rho_{b\,s}$ is known as the *banana width* [16]. (Note that ρ_{bs} is the banana width of a barely trapped particle (i.e. $\theta_b \simeq \pi$). Deeply trapped particles (i.e. $\theta_b \ll 1$) have much narrower banana widths.) Finally, as is clear from tables 2.1 and 2.4, although the electron and ion banana widths in a tokamak fusion reactor are much greater than the corresponding gyroradii, they are much smaller than the minor radius of the plasma.

2.8 The neoclassical closure scheme

As we have seen, plasmas that are confined in tokamak reactors are sufficiently collisionless that charged particles can travel around magnetic flux surfaces very many times before the accumulated effect of small-angle scattering events becomes significant. The classical collisional closure scheme described in section 2.6 fails under such circumstances. The aim of sections 2.8–2.21 is to describe an alternative scheme, known as a *neoclassical* closure scheme, which is appropriate to low-collisionality plasmas [16].

The fluid equations for species s take the form:

$$\frac{\partial n_s}{\partial t} = -\nabla \cdot (n_s \mathbf{V}_s), \tag{2.103}$$

$$m_s\, n_s \frac{\partial \mathbf{V}_s}{\partial t} = -m_s\, n_s\, (\mathbf{V}_s \cdot \nabla)\mathbf{V}_s + e_s\, n_s\, (\mathbf{E} + \mathbf{V}_s \times \mathbf{B}) - \nabla p_s - \nabla \cdot \boldsymbol{\pi}_s + \mathbf{F}_s, \tag{2.104}$$

$$\frac{3}{2}\frac{\partial p_s}{\partial t} = -\frac{3}{2}\, \mathbf{V}_s \cdot \nabla p_s - \frac{5}{2}\, p_s\, \nabla \cdot \mathbf{V}_s - \boldsymbol{\pi}_s\!: \nabla \mathbf{V}_s - \nabla \cdot \mathbf{q}_s + W_s. \tag{2.105}$$

(See section 2.3). However, the neoclassical closure scheme also requires the following third-order velocity-space moment of the kinetic equation [19]:

$$\frac{\partial \mathbf{Q}_s}{\partial t} = \frac{e_s}{m_s}\left[\mathbf{E} \cdot \left(\frac{5}{2}\, p_s\, \mathbf{I} + \boldsymbol{\pi}_s + m_s\, n_s\, \mathbf{V}_s\, \mathbf{V}_s\right) + \mathbf{Q}_s \times \mathbf{B}\right] - \nabla \cdot \mathbf{R}_s + \mathbf{G}_s, \tag{2.106}$$

where

$$\mathbf{Q}_s = \mathbf{q}_s + \frac{5}{2}\, p_s\, \mathbf{V}_s + \frac{1}{2}\, m_s\, n_s\, V_s^{\,2}\, \mathbf{V}_s + \mathbf{V}_s \cdot \boldsymbol{\pi}_s, \tag{2.107}$$

$$\mathbf{R}_s = \int \frac{1}{2}\, m_s\, v^2\, \mathbf{v}\, \mathbf{v} f_s\, d^3\mathbf{v}, \tag{2.108}$$

$$\mathbf{G}_s = \int \frac{1}{2}\, m_s\, v^2\, \mathbf{v}\, C_s(f_e, f_s)\, d^3\mathbf{v}. \tag{2.109}$$

2.9 Drift and transport orderings

It is helpful to define the species-s *toroidal magnetization parameter*,

$$\delta_s = \frac{\rho_s}{a}, \tag{2.110}$$

the species-s *poloidal magnetization parameter*,

$$\delta_{\theta\,s} = \frac{q}{\epsilon}\,\frac{\rho_s}{a}, \tag{2.111}$$

and the (modified) species-s *collisionality parameter*,

$$\Delta_s = \frac{q\,R_0}{v_{t\,s}\,\tau_s}. \tag{2.112}$$

As is apparent from table 2.4, these dimensionless parameters are all much less than unity for both electrons and ions in a tokamak fusion reactor.

In the following, we shall adopt the following fundamental ordering assumptions [20]:

$$\frac{E_\perp}{B\,v_{t\,s}} \sim \mathcal{O}(\delta_s), \tag{2.113}$$

$$\frac{a}{v_{t\,s}}\,\frac{\partial}{\partial t} \sim \mathcal{O}(\delta_{\theta\,s}^{\,2}\,\Delta_s). \tag{2.114}$$

Equation (2.113) is known as the *drift ordering* [1, 21], and serves to rule out the rapid plasma motions associated with ideal magnetohydrodynamic instabilities [22]. Equation (2.114) is known as the *transport ordering*. According to this ordering, any time dependence of plasma quantities is due to the action of perpendicular transport, viscous heating, and frictional heating. The drift and transport orderings, combined with the analysis contained in the following two sections, imply that $V_s/v_{t\,s} \sim q_s/(p_s\,v_{t\,s}) \sim \mathcal{O}(\delta_{\theta\,s})$, $\boldsymbol{\pi}_{\parallel\,s}/p_s \sim \mathcal{O}(\delta_{\theta\,s}\,\Delta_s)$, $E_\parallel/(B\,v_{t\,s}) \sim \mathcal{O}(\delta_s\,\delta_{\theta\,s}\,\Delta_s)$, $F_s/(p_s/a) \sim G_s/(p_s\,v_{t\,s}^{\,2}/a) \sim \mathcal{O}(\delta_{\theta\,s}\,\Delta_s)$, and $W_s/(p_s\,v_{t\,s}/a) \sim \mathcal{O}(\delta_{\theta\,s}^{\,2}\,\Delta_s)$. Equations (2.103)–(2.109) yield

$$[\delta_{\theta\,s}\,\Delta_s]\,\frac{\partial n_s}{\partial t} \simeq \nabla\cdot(n_s\,\mathbf{V}_s), \tag{2.115}$$

$$0 \simeq e_s\,n_s\,(\mathbf{E}_\perp + [\delta_{\theta\,s}\,\Delta_s]\,\mathbf{E}_\parallel + \mathbf{V}_s\times\mathbf{B}) - \nabla p_s - [\delta_{\theta\,s}\,\Delta_s]\,\nabla\cdot\boldsymbol{\pi}_{\parallel\,s} + [\delta_{\theta\,s}\,\Delta_s]\,\mathbf{F}_s, \tag{2.116}$$

$$\begin{aligned}[\delta_{\theta\,s}\,\Delta_s]\,\frac{3}{2}\,\frac{\partial p_s}{\partial t} \simeq{}& -\frac{3}{2}\,\mathbf{V}_s\cdot\nabla p_s - \left(\frac{5}{2}\,p_s\,\mathbf{I} + [\delta_{\theta\,s}\,\Delta_s]\,\boldsymbol{\pi}_{\parallel\,s}\right)\\ &:\nabla\mathbf{V}_s - \nabla\cdot\mathbf{q}_s + [\delta_{\theta\,s}\,\Delta_s]\,W_s,\end{aligned} \tag{2.117}$$

$$\begin{aligned}0 \simeq{}& \frac{e_s}{m_s}\left[(\mathbf{E}_\perp + [\delta_{\theta\,s}\,\Delta_s]\,\mathbf{E}_\parallel + \mathbf{V}_s\times\mathbf{B})\cdot\left(\frac{5}{2}\,p_s\,\mathbf{I} + [\delta_{\theta\,s}\,\Delta_s]\,\boldsymbol{\pi}_{\parallel\,s}\right) + \mathbf{q}_s\times\mathbf{B}\right]\\ &-\nabla\left(\frac{5}{2}\,\frac{T_s\,p_s}{m_s}\right) - [\delta_{\theta\,s}\,\Delta_s]\,\nabla\cdot\mathbf{r}_{\parallel\,s} + [\delta_{\theta\,s}\,\Delta_s]\,\mathbf{G}_s,\end{aligned} \tag{2.118}$$

where we have written [19]

$$\mathbf{R}_s \simeq \frac{5}{2}\frac{T_s\,p_s}{m_s}\,\mathbf{I} + [\delta_{\theta\,s}\,\Delta_s]\,\mathbf{r}_{\parallel\,s}. \tag{2.119}$$

In the previous five equations, we have neglected terms that are $\mathcal{O}(\delta_{\theta\,s}\,\delta_s)$, $\mathcal{O}(\delta_{\theta\,s}^{3}\,\Delta_s)$, or $\mathcal{O}\,(\delta_{\theta\,s}\,\Delta_s\,\rho_s/l_s)$ smaller than the leading-order terms, while retaining terms that are $\mathcal{O}(\delta_{\theta\,s}\,\Delta_s)$, which is only self-consistent if $\delta_s \ll \Delta_s$. As shown in table 2.4, the inequality $\delta_s \ll \Delta_s$ is well satisfied for electrons in a tokamak fusion reactor, but is only barely satisfied for ions. The $[\delta_{\theta\,s}\,\Delta_s]$ factors in the previous equations are there to remind us that the terms they precede are $\mathcal{O}(\delta_{\theta\,s}\,\Delta_s)$ smaller than the leading-order terms. Finally, it is helpful to define

$$\boldsymbol{\Theta}_{\parallel\,s} = \frac{5}{2}\,\boldsymbol{\pi}_{\parallel\,s} - \frac{m_s}{T_s}\,\mathbf{r}_{\parallel\,s}. \tag{2.120}$$

2.10 Toroidal plasma equilibrium

As before, let R, φ, Z be a set of right-handed cylindrical coordinates whose symmetry axis corresponds to that of the plasma equilibrium. On the other hand, let ψ, θ, φ be a set of right-handed flux coordinates such that $\psi(R, Z)$ labels the equilibrium magnetic flux surfaces, and θ increases by 2π for every poloidal circuit of a given flux surface. We can assume that $\theta = \theta(R, Z)$ without loss of generality. (Note that θ is a generalization of the poloidal angle introduced in section 2.7 that does not assume that the flux surfaces have circular cross sections.) As before, we shall set $\theta = 0$ on the outboard midplane. Note that $|\nabla\varphi| = 1/R$. The Jacobean of our flux-coordinate system is defined as

$$\mathcal{J}(R, Z) = (\nabla\psi \times \nabla\theta \cdot \nabla\varphi)^{-1}. \tag{2.121}$$

A general vector field, $\mathbf{A}$, can be written as

$$\mathbf{A} = A^{\psi}\,\mathcal{J}\,\nabla\theta \times \nabla\varphi + A^{\theta}\,\mathcal{J}\,\nabla\varphi \times \nabla\psi + A^{\varphi}\,\mathcal{J}\,\nabla\psi \times \nabla\theta. \tag{2.122}$$

Moreover [21],

$$\mathcal{J}\,\nabla \cdot \mathbf{A} = \frac{\partial(\mathcal{J}\,A^{\psi})}{\partial\psi} + \frac{\partial(\mathcal{J}\,A^{\theta})}{\partial\theta} + \frac{\partial(\mathcal{J}\,A^{\varphi})}{\partial\varphi}. \tag{2.123}$$

The axisymmetric equilibrium magnetic field of a tokamak can be expressed in the following manifestly divergence-free manner:

$$\mathbf{B} = \nabla\varphi \times \nabla\psi + \nabla\Psi \times \nabla\theta, \tag{2.124}$$

where $\Psi = \Psi(\psi)$. It follows that

$$\mathcal{J}\,B^{\psi} = 0, \tag{2.125}$$

$$\mathcal{J}\,B^{\theta} = 1, \tag{2.126}$$

$$\mathcal{J}\, B^{\varphi} = q, \tag{2.127}$$

where

$$q(\psi) = \frac{d\Psi}{d\psi} \tag{2.128}$$

is the safety-factor profile [21]. Note that the previous expression reduces to expression (1.76) in the large-aspect-ratio, circular magnetic flux-surface limit.

It is convenient to specialize the previous expressions to a coordinate system in which

$$\mathcal{J} = \frac{q(\psi)\, R^2}{I(\psi)}. \tag{2.129}$$

It follows that

$$\mathbf{B} = I\, \nabla\varphi + \nabla\varphi \times \nabla\psi. \tag{2.130}$$

The equilibrium electric field is written

$$\mathbf{E} = \dot{\psi}\, \nabla\varphi - \nabla\Phi. \tag{2.131}$$

Here, $\dot{\psi} \equiv \partial\psi/\partial t$, and $\Phi = \Phi(\psi)$. Note that the previous equation automatically satisfies $\nabla \times \mathbf{E} = -\partial\mathbf{B}/\partial t$.

Finally, we expect the plasma equilibrium to be characterized by number density, temperature, and pressure profiles that are flux-surface functions [1]. In other words, $n_s = n_s(\psi)$, $T_s = T_s(\psi)$, and $p_s = p_s(\psi)$ (see section 2.25).

2.11 Lowest-order flows

To the lowest order in the small parameter $\delta_{\theta\, s}\, \Delta_s$, equations (2.116) and (2.118) yield

$$0 \simeq e_s\, n_s\, (\mathbf{E}_\perp + \mathbf{V}_s \times \mathbf{B}) - \nabla p_s, \tag{2.132}$$

$$0 \simeq \frac{e_s}{m_s}\left[\frac{5}{2}\, p_s\, (\mathbf{E}_\perp + \mathbf{V}_s \times \mathbf{B}) + \mathbf{q}_s \times \mathbf{B}\right] - \nabla\left(\frac{5}{2}\, \frac{T_s\, p_s}{m_s}\right). \tag{2.133}$$

The previous two equations can be solved to give

$$\mathbf{V}_s = V_{\parallel\, s}\, \mathbf{b} + \mathbf{V}_{\perp\, s}, \tag{2.134}$$

$$\mathbf{q}_s = q_{\parallel\, s}\, \mathbf{b} + \mathbf{q}_{\perp\, s}, \tag{2.135}$$

where

$$\mathbf{V}_{\perp\, s} = \frac{\mathbf{E} \times \mathbf{B}}{B^2} + \frac{\mathbf{B} \times \nabla p_s}{e_s\, n_s\, B^2}, \tag{2.136}$$

$$\frac{\mathbf{q}_{\perp\,\mathrm{s}}}{p_\mathrm{s}} = \frac{5}{2}\,\frac{\mathbf{B}\times\nabla T_\mathrm{s}}{e_\mathrm{s}\,B^2}. \tag{2.137}$$

It can be seen that the lowest-order particle and heat flows are both confined to magnetic flux surfaces (i.e. $\mathbf{V}_\mathrm{s}\cdot\nabla\psi = \mathbf{q}_\mathrm{s}\cdot\nabla\psi = 0$). The first term on the right-hand side of equation (2.136) is the *E-cross-B velocity*,

$$\mathbf{V}_\mathrm{E} = \frac{\mathbf{E}\times\mathbf{B}}{B^2}, \tag{2.138}$$

which is common to all plasma species [1]. The second term is the *diamagnetic velocity*,

$$\mathbf{V}_{*\,\mathrm{s}} = \frac{\mathbf{B}\times\nabla p_\mathrm{s}}{e_\mathrm{s}\,n_\mathrm{s}\,B^2}, \tag{2.139}$$

which is different for electrons and ions and is a consequence of the rapid gyromotions of charged particles in the presence of equilibrium pressure gradients [1]. The drift ordering (2.113) ensures that the E-cross-B and diamagnetic velocities are similar in magnitude. It is clear from equation (2.137) that there is a diamagnetic flow of heat, as well as particles, around flux surfaces; this heat flow is the same as that associated with the cross thermal conductivities introduced in section 2.6.

To the lowest order in the small parameter $\delta_{\theta\,\mathrm{s}}\,\Delta_\mathrm{s}$, equations (2.115) and (2.117) yield

$$\nabla\cdot\mathbf{V}_\mathrm{s} = 0, \tag{2.140}$$

$$\nabla\cdot\mathbf{q}_\mathrm{s} = 0. \tag{2.141}$$

Here, use has been made of $\mathbf{V}_\mathrm{s}\cdot\nabla n_\mathrm{s} = \mathbf{V}_\mathrm{s}\cdot\nabla p_\mathrm{s} = 0$. Clearly, the lowest-order particle and heat flows are both divergence free. The fact that $\mathbf{V}_\mathrm{s}\cdot\nabla\psi = \mathbf{q}_\mathrm{s}\cdot\nabla\psi = 0$ implies that $V_\mathrm{s}^{\psi} = q_\mathrm{s}^{\psi} = 0$. Making use of these results, the previous two equations can be combined with equations (2.123) and (2.126), as well as the fact that $\partial/\partial\varphi \equiv 0$ in an axisymmetric equilibrium, to give [19]

$$\frac{\mathbf{V}_\mathrm{s}\cdot\nabla\theta}{\mathbf{B}\cdot\nabla\theta} = V_{\theta\,\mathrm{s}}(\psi), \tag{2.142}$$

$$\frac{\mathbf{q}_\mathrm{s}\cdot\nabla\theta}{\mathbf{B}\cdot\nabla\theta} = q_{\theta\,\mathrm{s}}(\psi). \tag{2.143}$$

Taking the scalar products of equations (2.134) and (2.135) with $\nabla\theta$, we obtain

$$V_{\theta\,\mathrm{s}}(\psi) = \frac{V_{\parallel\,\mathrm{s}}}{B} - \frac{V_{\psi\,\mathrm{s}}(\psi)}{B^2}, \tag{2.144}$$

$$q_{\theta\,\mathrm{s}}(\psi) = \frac{q_{\parallel\,\mathrm{s}}}{B} - \frac{q_{\psi\,\mathrm{s}}(\psi)}{B^2}, \tag{2.145}$$

where

$$V_{\psi\,s}(\psi) = -I\,\frac{d\Phi}{d\psi} - \frac{I}{e_s\,n_s}\,\frac{dp_s}{d\psi}, \tag{2.146}$$

$$\frac{q_{\psi\,s}(\psi)}{p_s(\psi)} = -\frac{5}{2}\,\frac{I}{e_s}\,\frac{dT_s}{d\psi}. \tag{2.147}$$

Here, use has been made of equations (2.136) and (2.137).

2.12 The flux-surface average operator

It is helpful to define the *flux-surface average operator*:

$$\langle\cdots\rangle = \oint R^2\,(\cdots)\,\frac{d\theta}{2\pi}\bigg/\oint R^2\,\frac{d\theta}{2\pi}. \tag{2.148}$$

According to equations (2.129) and (2.130),

$$\langle \mathbf{B}\cdot\nabla A\rangle = 0, \tag{2.149}$$

where $A(R, Z)$ is a general axisymmetric scalar field. Moreover, $\langle 1\rangle = 1$.

Making use of equations (2.121)–(2.123) and (2.129), it can easily be demonstrated that

$$\langle\nabla\cdot\mathbf{A}\rangle = \frac{d\langle\mathbf{A}\cdot\nabla\mathcal{V}\rangle}{d\mathcal{V}}, \tag{2.150}$$

where $\mathbf{A}$ is a general axisymmetric (i.e. $\partial/\partial\varphi = 0$) vector field, and

$$\mathcal{V}(\psi) = \int_0^{\psi}\oint\oint\mathcal{J}(\psi', \theta)\,d\psi'\,d\theta\,d\varphi = (2\pi)^2\int_0^{\psi}\frac{q}{I}\left(\oint R^2\,\frac{d\theta}{2\pi}\right)d\psi' \tag{2.151}$$

is the volume contained within the magnetic flux surface whose label is ψ.

2.13 Chew–Goldberger–Low forms

As is well known, the smallness of the toroidal magnetization parameter, δ_s, ensures that the dominant parallel viscosity tensors, $\boldsymbol{\pi}_{\parallel\,s}$ and $\boldsymbol{\Theta}_{\parallel\,s}$, take the so-called *Chew–Goldberger–Low* forms [21, 23]:

$$\boldsymbol{\pi}_{\parallel\,s} = \Delta\pi_{\parallel\,s}\left(\mathbf{b}\,\mathbf{b} - \frac{1}{3}\,\mathbf{I}\right), \tag{2.152}$$

$$\boldsymbol{\Theta}_{\parallel\,s} = \Delta\Theta_{\parallel\,s}\left(\mathbf{b}\,\mathbf{b} - \frac{1}{3}\,\mathbf{I}\right), \tag{2.153}$$

where $\langle\Delta\pi_{\parallel\,s}\rangle = \langle\Delta\Theta_{\parallel\,s}\rangle = 0$. Incidentally, it is clear from equation (2.60) that the parallel viscosity tensor in the classical closure scheme does indeed take this form.

If $\boldsymbol{\pi}_{\parallel\,s}$ takes the Chew–Goldberger–Low form, then

$$\nabla\psi\cdot\boldsymbol{\pi}_{\parallel\,s}\cdot\mathbf{b}=0, \tag{2.154}$$

$$\nabla\psi\cdot\boldsymbol{\pi}_{\parallel\,s}\cdot\nabla\varphi=0. \tag{2.155}$$

These results follow because $\mathbf{b}\cdot\nabla\psi=\nabla\varphi\cdot\nabla\psi=0$.

Furthermore,

$$R^2\,\nabla\varphi\cdot\nabla\cdot\boldsymbol{\pi}_{\parallel\,s}=\nabla\cdot\left(R^2\,\nabla\varphi\cdot\boldsymbol{\pi}_{\parallel\,s}\right)-\boldsymbol{\pi}_{\parallel\,s}\colon\tilde{\nabla}(R^2\,\nabla\varphi), \tag{2.156}$$

where, in Cartesian coordinates,

$$(\tilde{\mathbf{A}})_{jk}=\frac{1}{2}\left(\frac{\partial A_j}{\partial r_k}+\frac{\partial A_k}{\partial r_j}\right), \tag{2.157}$$

and use has been made of the fact that $\boldsymbol{\pi}_{\parallel\,s}$ is a symmetric tensor. In fact, it can be shown that [24]

$$\tilde{\nabla}(R^2\,\nabla\varphi)=0. \tag{2.158}$$

Hence, we deduce that

$$R^2\,\nabla\varphi\cdot\nabla\cdot\boldsymbol{\pi}_{\parallel\,s}=\nabla\cdot\left(R^2\,\nabla\varphi\cdot\boldsymbol{\pi}_{\parallel\,s}\right). \tag{2.159}$$

Making use of equation (2.150), we obtain

$$\left\langle R^2\,\nabla\varphi\cdot\nabla\cdot\boldsymbol{\pi}_{\parallel\,s}\right\rangle=\frac{d}{d\mathcal{V}}\left\langle R^2\,\nabla\varphi\cdot\boldsymbol{\pi}_{\parallel\,s}\cdot\nabla\mathcal{V}\right\rangle. \tag{2.160}$$

However, if $\boldsymbol{\pi}_{\parallel\,s}$ takes the Chew–Goldberger–Low form, then $\nabla\varphi\cdot\boldsymbol{\pi}_{\parallel\,s}\cdot\nabla\mathcal{V}=0$ because $\mathbf{b}\cdot\nabla\mathcal{V}=\nabla\varphi\cdot\nabla\mathcal{V}=0$. It follows that [19, 20]

$$\left\langle R^2\,\nabla\varphi\cdot\nabla\cdot\boldsymbol{\pi}_{\parallel\,s}\right\rangle=0. \tag{2.161}$$

Likewise, if $\boldsymbol{\Theta}_{\parallel\,s}$ takes the Chew–Goldberger–Low form, then

$$\left\langle R^2\,\nabla\varphi\cdot\nabla\cdot\boldsymbol{\Theta}_{\parallel\,s}\right\rangle=0. \tag{2.162}$$

2.14 Parallel flows

Equations (2.116), (2.118), and (2.120) yield

$$0\simeq e_s\,n_s\,(\mathbf{E}_\perp+[\delta_{\theta\,s}\,\Delta_s]\,\mathbf{E}_\parallel+\mathbf{u}_{s\,1}\times\mathbf{B})-\nabla p_s-[\delta_{\theta\,s}\,\Delta_s]\,\nabla\cdot\boldsymbol{\pi}_{\parallel\,s}+[\delta_{\theta\,s}\,\Delta_s]\,\mathbf{F}_{s\,1}, \tag{2.163}$$

$$0\simeq-[\delta_{\theta\,s}\,\Delta_s]\,\nabla p_s\cdot\boldsymbol{\pi}_s+\frac{5}{2}\,e_s\,n_s\,\mathbf{u}_{s\,2}\times\mathbf{B}-[\delta_{\theta\,s}\,\Delta_s]\,\nabla\cdot\boldsymbol{\Theta}_{\parallel\,s}+[\delta_{\theta\,s}\,\Delta_s]\,\mathbf{F}_{s\,2}, \tag{2.164}$$

where [19]

$$\mathbf{u}_{s\,1} = \mathbf{V}_s, \tag{2.165}$$

$$\mathbf{u}_{s\,2} = -\frac{2}{5}\,\frac{\mathbf{q}_s}{p_s}, \tag{2.166}$$

$$\mathbf{F}_{s\,1} = \mathbf{F}_s, \tag{2.167}$$

$$\mathbf{F}_{s\,2} = \frac{5}{2}\,\mathbf{F}_s - \frac{m_s}{T_s}\,\mathbf{G}_s. \tag{2.168}$$

Here, we have neglected terms that are second order in the small parameter $\delta_{\theta\,s}\,\Delta_s$. By analogy with the analysis given in section 2.11, we can write [19]

$$\mathbf{u}_{s\,j} = u_{\parallel\,s\,j}\,\mathbf{b} + \mathbf{u}_{\perp\,s\,j}, \tag{2.169}$$

$$u_{\theta\,s\,j}(\psi) \equiv \frac{\mathbf{u}_{s\,j}\cdot\nabla\theta}{\mathbf{B}\cdot\nabla\theta} = \frac{u_{\parallel\,s\,j}}{B} - \frac{V_{E\,j}(\psi) + V_{*\,s\,j}(\psi)}{B^2}, \tag{2.170}$$

for $j = 1, 2$, where

$$V_{E\,1}(\psi) = -I\,\frac{d\Phi}{d\psi}, \tag{2.171}$$

$$V_{E\,2}(\psi) = 0, \tag{2.172}$$

$$V_{*\,s\,1}(\psi) = -\frac{I}{e_s\,n_s}\,\frac{dp_s}{d\psi}, \tag{2.173}$$

$$V_{*\,s\,2}(\psi) = \frac{I}{e_s}\,\frac{dT_s}{d\psi}. \tag{2.174}$$

Taking the scalar product of equations (2.163) and (2.164) with $\mathbf{B}$ and flux-surface averaging, we annihilate the leading-order terms, leaving [19]

$$\langle \mathbf{B}\cdot\nabla\cdot\boldsymbol{\pi}_{\parallel\,s}\rangle = \langle \mathbf{B}\cdot\mathbf{F}_{s\,1}\rangle + e_s\,n_s\,\langle \mathbf{E}\cdot\mathbf{B}\rangle, \tag{2.175}$$

$$\langle \mathbf{B}\cdot\nabla\cdot\boldsymbol{\Theta}_{\parallel\,s}\rangle = \langle \mathbf{B}\cdot\mathbf{F}_{s\,2}\rangle. \tag{2.176}$$

Note that the annihilation of the term involving $\nabla p_s\cdot\boldsymbol{\pi}_{\parallel\,s}$ in equation (2.164) depends on the fact that $\boldsymbol{\pi}_{\parallel\,s}$ takes the Chew–Goldberger–Low form (see equation (2.154)).

According to equations (2.175) and (2.176), the parallel components of the friction force density and the electric force density drive both particle flows and heat flows along the magnetic field. These flows are opposed by the parallel viscous force density. We now need to solve the previous two equations to determine the parallel flows, which requires us to have expressions for the friction force density as well as the parallel viscous force density.

2.15 Useful results

In the following, (u) represents a 1×2 vector, while $[A]$ represents a 2×2 matrix. The elements of (u) and $[A]$ are denoted u_j and A_{kj}, respectively, where $k = 1, 2$ indexes the rows, and $j = 1, 2$ indexes the columns. Furthermore, (0) denotes a null vector, and [1] denotes an identity matrix. Equations (2.175) and (2.176) can be written in the compact form [25]

$$\langle \mathbf{B} \cdot \nabla \cdot (\pi_{\parallel\,\mathrm{s}}) \rangle = \langle \mathbf{B} \cdot (F_{\mathrm{s}}) \rangle + e_{\mathrm{s}}\, n_{\mathrm{s}}\, \langle B \rangle\, (E_{\parallel}), \tag{2.177}$$

where

$$(\pi_{\parallel\,\mathrm{s}}) = \begin{pmatrix} \boldsymbol{\pi}_{\parallel\,\mathrm{s}} \\ \boldsymbol{\Theta}_{\parallel\,\mathrm{s}} \end{pmatrix}, \tag{2.178}$$

$$(F_{\mathrm{s}}) = \begin{pmatrix} \mathbf{F}_{\mathrm{s}\,1} \\ \mathbf{F}_{\mathrm{s}\,2} \end{pmatrix}, \tag{2.179}$$

$$(E_{\parallel}) = \begin{pmatrix} \langle \mathbf{E} \cdot \mathbf{B} \rangle / \langle B \rangle \\ 0 \end{pmatrix}. \tag{2.180}$$

It is helpful to define

$$(u_{\parallel\,\mathrm{s}}) = \begin{pmatrix} \langle u_{\parallel\,\mathrm{s}\,1}\, B \rangle \\ \langle u_{\parallel\,\mathrm{s}\,2}\, B \rangle \end{pmatrix}, \tag{2.181}$$

$$(u_{\theta\,\mathrm{s}}) = \begin{pmatrix} u_{\theta\,\mathrm{s}\,1}\, \langle B^2 \rangle \\ u_{\theta\,\mathrm{s}\,2}\, \langle B^2 \rangle \end{pmatrix}. \tag{2.182}$$

Equations (2.170)–(2.174), (2.181), and (2.182) can be combined to give

$$(u_{\parallel\,\mathrm{s}}) = (u_{\theta\,\mathrm{s}}) + (V_{\mathrm{E}}) + (V_{*\,\mathrm{s}}), \tag{2.183}$$

where

$$(V_{\mathrm{E}}) = \begin{pmatrix} V_{\mathrm{E}\,1} \\ 0 \end{pmatrix}, \tag{2.184}$$

$$(V_{*\,\mathrm{s}}) = \begin{pmatrix} V_{*\,\mathrm{s}\,1} \\ V_{*\,\mathrm{s}\,2} \end{pmatrix}. \tag{2.185}$$

It is also helpful to define

$$(u_{\varphi\,\mathrm{s}}) = \frac{I}{\langle R^2 \rangle} \begin{pmatrix} \langle R^2\, \nabla\varphi \cdot \mathbf{u}_{\mathrm{s}\,1} \rangle \\ \langle R^2\, \nabla\varphi \cdot \mathbf{u}_{\mathrm{s}\,2} \rangle \end{pmatrix}. \tag{2.186}$$

It follows from equations (2.130), (2.134)–(2.137), (2.165), (2.166), and (2.171)–(2.174) that

$$(u_{\varphi\,s}) = \frac{I^2}{\langle R^2\rangle}\begin{pmatrix}\langle u_{\|\,s\,1}/B\rangle \\ \langle u_{\|\,s\,2}/B\rangle\end{pmatrix} + \frac{1}{\langle R^2\rangle}\left\langle\frac{|\nabla\psi|^2}{B^2}\right\rangle[(V_E) + (V_{*\,s})]. \tag{2.187}$$

However, equations (2.170) and (2.182) yield

$$\begin{pmatrix}\langle u_{\|s1}/B\rangle \\ \langle u_{\|s2}/B\rangle\end{pmatrix} = \frac{1}{\langle B^2\rangle}(u_\theta) + \left\langle\frac{1}{B^2}\right\rangle[(V_E) + (V_{*s})]. \tag{2.188}$$

Finally, the previous two equations can be combined to give

$$(u_{\varphi\,s}) = \left(\frac{I^2}{\langle R^2\rangle\,\langle B^2\rangle} - 1\right)(u_{\theta\,s}) + (u_{\|\,s}), \tag{2.189}$$

where use has been made of equations (2.130) and (2.183).

2.16 Friction force densities

It is helpful to define the modified collision time [19]

$$\tau_{ss} = \frac{6\sqrt{2}\,\pi^{3/2}\,\epsilon_0^{\,2}\,\sqrt{m_s}\,T_s^{3/2}}{e_s^{\,4}\,n_s\ln\Lambda}. \tag{2.190}$$

It can be seen, by comparison with equations (2.20) and (2.21), that $\tau_{ee} = \tau_e$ and $\tau_{ii} = \tau_i/\sqrt{2}$.

Our model expressions for the friction force densities, $\mathbf{F}_{s\,1}$ and $\mathbf{F}_{s\,2}$, are taken from the moment-based analysis of Hirschman & Sigmar [10, 19]. The ion friction force densities are written as

$$(F_i) = \frac{n_i\,m_i}{\tau_{ii}}\{-[F_{ii}]\,(u_i)\} + \frac{n_e\,m_e}{\tau_{ee}}\{-[E_{ii}]\,(u_i) + [E_{ie}]\,(u_e)\}, \tag{2.191}$$

where

$$(u_s) = \begin{pmatrix}\mathbf{u}_{s\,1} \\ \mathbf{u}_{s\,2}\end{pmatrix}, \tag{2.192}$$

$$[F_{ii}] = \begin{bmatrix}0, & 0 \\ 0, & \sqrt{2}\end{bmatrix}, \tag{2.193}$$

$$[E_{ii}] = \begin{bmatrix}1, & 0 \\ 0, & (15/2)\,T_e/T_i\end{bmatrix}, \tag{2.194}$$

$$[E_{ie}] = \begin{bmatrix}1, & 3/2 \\ 0, & 0\end{bmatrix}. \tag{2.195}$$

The electron friction force densities take the form

$$(F_e) = \frac{n_e m_e}{\tau_{ee}} \{-[F_{ee}]\,(u_e) + [F_{ei}]\,(u_i)\}, \tag{2.196}$$

where

$$[F_{ee}] = \begin{bmatrix} 1, & 3/2 \\ 3/2, & \sqrt{2} + 13/4 \end{bmatrix}, \tag{2.197}$$

$$[F_{ei}] = \begin{bmatrix} 1, & 0 \\ 3/2, & 0 \end{bmatrix}. \tag{2.198}$$

Note that $F_{ii\,1j} = 0$, $F_{ee\,1j} = E_{ie\,1j}$, and $F_{ei\,1j} = E_{ii\,1j}$, for $j = 1, 2$, ensuring that

$$\mathbf{F}_{i\,1} + \mathbf{F}_{e\,1} = 0, \tag{2.199}$$

which is a statement of collisional momentum conservation [1]. Note further that $F_{ii\,j1} = 0$, $E_{ii\,j1} = E_{ie\,j1}$, and $F_{ee\,j1} = F_{ei\,j1}$, for $j = 1, 2$, which ensures that $\mathbf{F}_{s\,1}$ and $\mathbf{F}_{s\,2}$ are invariant under a Galilean transformation of the coordinate system.

2.17 Parallel viscous force densities

It is helpful to define the *average fraction of passing particles* on a magnetic flux surface:

$$f_p = \frac{3}{4}\,\langle B^2\rangle \int_0^{1/B_{max}} \frac{\lambda\, d\lambda}{\langle\sqrt{1 - \lambda\, B}\rangle}, \tag{2.200}$$

where B_{max} is the maximum value of B on the surface [19]. For a flux surface with a circular poloidal cross section [25],

$$f_p = 1 - 1.46\,\epsilon^{1/2} + 0.46\,\epsilon^{3/2}. \tag{2.201}$$

Thus, the *average fraction of trapped particles* on the flux surface is

$$f_t \equiv 1 - f_p = 1.46\sqrt{\epsilon} - 0.46\,\epsilon^{3/2}. \tag{2.202}$$

Note that the previous expression differs slightly from the less exact expression (2.87).

In the so-called *banana collisionality regime*, $\nu_{*\,s} \ll 1$ (see equation (2.95)), the flux-surface averaged parallel viscous force densities take the form [19, 26, 27]

$$\left\langle \mathbf{B} \cdot \nabla \cdot (\pi_{\parallel\,s}) \right\rangle = \frac{n_s m_s}{\tau_{ss}} f_t\,[\mu_s]\,(u_{\theta\,s}). \tag{2.203}$$

The normalized viscosity coefficients for the ions are

$$\mu_{i\,11} = \int_0^\infty e^{-x}\left[\left(1 - \frac{1}{2\,x}\right)Y(\sqrt{x}) + Y'(\sqrt{x})\right]dx = 0.533, \tag{2.204}$$

$$\mu_{\mathrm{i}\,12} = \mu_{\mathrm{i}\,21} = \int_0^\infty \mathrm{e}^{-x}\left(\frac{5}{2} - x\right)\left[\left(1 - \frac{1}{2\,x}\right)Y(\sqrt{x}) + Y'(\sqrt{x})\right]dx = 0.625, \quad (2.205)$$

$$\mu_{\mathrm{i}\,22} = \int_0^\infty \mathrm{e}^{-x}\left(\frac{5}{2} - x\right)^2\left[\left(1 - \frac{1}{2\,x}\right)Y(\sqrt{x}) + Y'(\sqrt{x})\right]dx = 1.386, \quad (2.206)$$

where

$$Y(x) = \frac{2}{\sqrt{\pi}}\int_0^{x^2} \sqrt{t}\;\mathrm{e}^{-t}\,dt, \quad (2.207)$$

$$Y'(x) = \frac{2}{\sqrt{\pi}}\,x\,\mathrm{e}^{-x^2}. \quad (2.208)$$

The normalized viscosity coefficients for the electrons are

$$\mu_{\mathrm{e}\,11} = \mu_{\mathrm{i}\,11} + \int_0^\infty \mathrm{e}^{-x}\,dx = 1.533, \quad (2.209)$$

$$\mu_{\mathrm{e}\,12} = \mu_{\mathrm{e}\,21} = \mu_{\mathrm{i}\,12} + \int_0^\infty \left(\frac{5}{2} - x\right)\mathrm{e}^{-x}\,dx = 2.125, \quad (2.210)$$

$$\mu_{\mathrm{e}\,22} = \mu_{\mathrm{i}\,22} + \int_0^\infty \left(\frac{5}{2} - x\right)^2 \mathrm{e}^{-x}\,dx = 4.636. \quad (2.211)$$

In the banana collisionality regime, the parallel viscous force arises from collisional drag between passing and trapped particles [26, 27]. The origin of this drag is the fact that while passing particles can drift along magnetic field lines in one direction, trapped particles are forced to periodically reverse direction. Not surprisingly, the viscous force density is proportional to the fraction of trapped particles, f_{t} (see equation (2.203)).

2.18 The determination of ion flows

Making use of equations (2.169), (2.177), (2.191), and (2.203), we obtain

$$f_{\mathrm{t}}\,[\mu_{\mathrm{i}}]\,(u_{\theta\,\mathrm{i}}) = -[F_{\mathrm{ii}}]\,(u_{\parallel\,\mathrm{i}}). \quad (2.212)$$

Note that we have neglected terms that are $\mathcal{O}(\sqrt{m_{\mathrm{e}}/m_{\mathrm{i}}})$ smaller than the leading-order terms in the previous equation. This simplifying assumption allows the ion parallel flow to be decoupled from the electron parallel flow [25]. It follows from equations (2.183) and (2.212) that

$$([F_{\mathrm{ii}}] + f_{\mathrm{t}}\,[\mu_{\mathrm{i}}])\,(u_{\theta\,\mathrm{i}}) = -[F_{\mathrm{ii}}]\,(V_{\mathrm{E}}) - [F_{\mathrm{ii}}]\,(V_{*\,\mathrm{i}}). \quad (2.213)$$

However, $[F_{\mathrm{ii}}]\,(V_{\mathrm{E}}) = (0)$ because $F_{\mathrm{ii}\,j1} = 0$, for $j = 1, 2$, and $V_{\mathrm{E}\,2} = 0$. Thus, we get

$$(u_{\theta\,\mathrm{i}}) = -[L_{\mathrm{ii}}]\,(V_{*\,\mathrm{i}}), \quad (2.214)$$

where

$$[L_{\rm ii}] = ([F_{\rm ii} + f_{\rm t}\,[\mu_{\rm i}])^{-1}\,[F_{\rm ii}]. \tag{2.215}$$

Employing equation (2.193) and only retaining terms up to the first order in the small parameter $f_{\rm t}$, we obtain

$$[L_{\rm ii}] = \begin{bmatrix} 0, & -\,\alpha_1\,(1-\alpha_2\,f_{\rm t}) \\ 0, & 1-\alpha_2\,f_{\rm t} \end{bmatrix}, \tag{2.216}$$

where

$$\alpha_1 = \frac{\mu_{\rm i\,12}}{\mu_{\rm i\,11}} = 1.17, \tag{2.217}$$

$$\alpha_2 = \frac{\mu_{\rm i\,11}\,\mu_{\rm i\,22} - \mu_{\rm i\,12}\,\mu_{\rm i\,21}}{\sqrt{2}\,\mu_{\rm i\,11}} = 0.461. \tag{2.218}$$

Here, use has been made of equations (2.204)–(2.206). Incidentally, $f_{\rm t} \sim \epsilon^{1/2}$ is only a small parameter in the large-aspect-ratio limit $\epsilon \ll 1$.

It follows from equations (2.165), (2.166), (2.170), (2.181)–(2.185), (2.214), and (2.216) that

$$\frac{{\bf V}_{\rm i}\cdot\nabla\theta}{{\bf B}\cdot\nabla\theta} = \alpha_1\,(1-\alpha_2\,f_{\rm t})\,\frac{V_{*\,{\rm i}\,2}}{\langle B^2\rangle}, \tag{2.219}$$

$$\frac{{\bf q}_{\rm i}\cdot\nabla\theta}{p_{\rm i}\,{\bf B}\cdot\nabla\theta} = \frac{5}{2}\,(1-\alpha_2\,f_{\rm t})\,\frac{V_{*\,{\rm i}\,2}}{\langle B^2\rangle}, \tag{2.220}$$

and

$$\langle {\bf V}_{\rm i}\cdot{\bf B}\rangle = V_{\rm E\,1} + V_{*\,{\rm i}\,1} + \alpha_1\,(1-\alpha_2\,f_{\rm t})\,V_{*\,{\rm i}\,2}, \tag{2.221}$$

$$\frac{\langle {\bf q}_{\rm i}\cdot{\bf B}\rangle}{p_{\rm i}} = -\frac{5}{2}\,\alpha_2\,f_{\rm t}\,V_{*\,{\rm i}\,2}. \tag{2.222}$$

Consider the circular magnetic flux-surface limit. In this limit, $I \simeq R_0\,B$ and $d\psi/dr \simeq R_0\,B_\theta$, so

$$B_\theta\,V_{*\,{\rm s}\,1} \simeq -\frac{B}{e_{\rm s}\,n_{\rm s}}\,\frac{dp_{\rm s}}{dr}, \tag{2.223}$$

$$B_\theta\,V_{*\,{\rm s}\,2} \simeq \frac{B}{e_{\rm s}}\,\frac{dT_{\rm s}}{dr}, \tag{2.224}$$

$$B_\theta\,V_{\rm E\,1} \simeq -B\,\frac{d\Phi}{dr}, \tag{2.225}$$

where use has been made of equations (2.171)–(2.174). Thus, equations (2.219) and (2.220) reduce to [15, 16]

$$V_{\theta\,\mathrm{i}} \simeq \alpha_1\,(1-\alpha_2\,f_{\mathrm{t}})\,\frac{1}{e\,B}\,\frac{dT_{\mathrm{i}}}{dr} = 1.17\,(1-0.67\,\epsilon^{1/2})\,\frac{1}{e\,B}\,\frac{dT_{\mathrm{i}}}{dr}, \tag{2.226}$$

$$\frac{q_{\theta\,\mathrm{i}}}{p_{\mathrm{i}}} \simeq \frac{5}{2}\,(1-\alpha_2\,f_{\mathrm{t}})\,\frac{1}{e\,B}\,\frac{dT_{\mathrm{i}}}{dr} = \frac{5}{2}\,(1-0.67\,\epsilon^{1/2})\,\frac{1}{e\,B}\,\frac{dT_{\mathrm{i}}}{dr}, \tag{2.227}$$

while equations (2.221) and (2.222) yield

$$V_{\parallel\,\mathrm{i}} \simeq -\frac{1}{B_\theta}\,\frac{d\Phi}{dr} - \frac{1}{e\,n_{\mathrm{e}}\,B_\theta}\,\frac{dp_{\mathrm{i}}}{dr} + \alpha_1\,(1-\alpha_2\,f_{\mathrm{t}})\,\frac{1}{e\,B_\theta}\,\frac{dT_{\mathrm{i}}}{dr}, \tag{2.228}$$

$$\frac{q_{\parallel\,\mathrm{i}}}{p_{\mathrm{i}}} = -\frac{5}{2}\,\alpha_2\,f_{\mathrm{t}}\,\frac{1}{e\,B_\theta}\,\frac{dT_{\mathrm{i}}}{dr}, \tag{2.229}$$

where use has been made of equations (2.202), (2.217), and (2.218).

According to equation (2.227), the poloidal component of the ion heat flux is equivalent to the poloidal component of the diamagnetic heat flux specified in equation (2.137), except that it is reduced by a factor of $1 - 0.67\,\epsilon^{1/2}$. The explanation for this reduction is that trapped ions cannot carry a net poloidal heat flux around a magnetic flux surface, because they are forced to periodically reverse their poloidal drift direction. Hence, the diamagnetic ion heat flux is reduced by a factor that is roughly equal to the fraction of passing particles (see equation (2.201)). This reduction is accomplished by the parallel heat flux specified in equation (2.229).

Equation (2.226) shows that the poloidal component of the ion fluid velocity is directly proportional to the poloidal component of the ion heat flux. The most surprising feature of equation (2.226) is the absence of a contribution from the E-cross-B velocity, despite the fact that this velocity possesses a nonzero poloidal component (see equation (2.138)). The explanation for this absence is that parallel friction forces drive an ion flow parallel to the magnetic field that cancels out the poloidal component of the E-cross-B velocity as well as the poloidal component of the ion diamagnetic velocity. The requisite flow, which is specified in equation (2.228), is quite large, being a factor of $q/\epsilon \sim 9$ larger than a typical diamagnetic velocity. Equation (2.226) implies that a tokamak plasma is not free to rotate in the poloidal direction. In fact, any deviation of the ion poloidal flow velocity from the so-called *neoclassical velocity* specified by equation (2.226) is opposed by friction between trapped and passing ions and is consequently damped away on a timescale that is proportional to τ_{i}. This effect is known as *poloidal flow damping* [28] (see section 2.24).

Incidentally, in the large-aspect-ratio limit, equation (2.189) reduces to

$$(u_{\varphi\,\mathrm{s}}) \simeq (u_{\parallel\,\mathrm{s}}), \tag{2.230}$$

because the omitted term is $\mathcal{O}[(\epsilon/q)^2]$ smaller than the other terms. Thus, it follows from equations (2.130), (2.165), (2.166), (2.181), and (2.186) that

$$V_{\varphi\,s} \simeq V_{\|\,s}, \tag{2.231}$$

$$q_{\varphi\,s} \simeq q_{\|\,s}. \tag{2.232}$$

In other words, the toroidal flow velocities and heat fluxes are approximately equal to the parallel flow velocities and heat fluxes, respectively. This is true for both ions and electrons.

2.19 The determination of electron flows

Making use of equations (2.169), (2.177), (2.196), and (2.203), we obtain

$$f_t\,[\mu_e]\,(u_{\theta\,e}) = -[F_{ee}]\,(u_{\|\,e}) + [F_{ei}]\,(u_{\|\,i}) + \langle\Omega_e\rangle\,\tau_{ee}\,(E_\|). \tag{2.233}$$

It follows from equations (2.183) and (2.233) that

$$\begin{aligned}([F_{ee}] + f_t\,[\mu_e])\,(u_{\theta\,e}) = {} & (-[F_{ee}] + [F_{ei}])\,(V_E) - [F_{ee}]\,(V_{*\,e})\\ & + [F_{ei}]\,\{(u_{\theta\,i}) + (V_{*\,i})\}\\ & + \langle\Omega_e\rangle\,\tau_{ee}\,(E_\|).\end{aligned} \tag{2.234}$$

However, $(-[F_{ee}] + [F_{ei}])\,(V_E) = (0)$ because $F_{ee\,j1} = F_{ei\,j1}$, for $j = 1, 2$, and $V_{E\,2} = 0$. Hence, we get

$$([F_{ee}] + f_t\,[\mu_e])\,(u_{\theta\,e}) = -[F_{ee}]\,(V_{*\,e}) + [F_{ei}]\,([1] - [L_{ii}])\,(V_{*\,i}) + \langle\Omega_e\rangle\,\tau_{ee}\,(E_\|), \tag{2.235}$$

where use has been made of equation (2.214). It follows that

$$(u_{\theta\,e}) = -[L_{ee}]\,(V_{*\,e}) + [L_{ei}]\,(V_{*\,i}) + \langle\Omega_e\rangle\,\tau_{ee}\,[Q_{ee}]\,(E_\|), \tag{2.236}$$

where

$$[L_{ee}] = ([F_{ee}] + f_t\,[\mu_e])^{-1}\,[F_{ee}], \tag{2.237}$$

$$[L_{ei}] = ([F_{ee}] + f_t\,[\mu_e])^{-1}\,[F_{ei}]\,([1] - [L_{ii}]), \tag{2.238}$$

$$[Q_{ee}] = ([F_{ee}] + f_t\,[\mu_e])^{-1}. \tag{2.239}$$

Making use of equations (2.197), (2.198), and (2.216), and only retaining terms up to the first order in the small parameter f_t, we get

$$[L_{ee}] = \begin{bmatrix} 1 - \beta_{11} f_t, & -\beta_{12} f_t \\ -\beta_{21} f_t, & 1 - \beta_{22} f_t \end{bmatrix}, \tag{2.240}$$

$$[L_{\rm ei}] = \begin{bmatrix} 1 - \beta_{11} f_{\rm t}, & \alpha_1 (1 - [\alpha_2 + \beta_{11}] f_{\rm t}) \\ -\beta_{21} f_{\rm t}, & -\alpha_1 \beta_{21} f_{\rm t} \end{bmatrix}, \tag{2.241}$$

$$[Q_{\rm ee}] = \begin{bmatrix} \gamma_{11} - \delta_{11} f_{\rm t}, & \gamma_{12} - \delta_{12} f_{\rm t} \\ \gamma_{21} - \delta_{21} f_{\rm t}, & \gamma_{22} - \delta_{22} f_{\rm t} \end{bmatrix}, \tag{2.242}$$

where

$$\beta_{11} = \frac{(\sqrt{2} + 13/4)\, \mu_{\rm e\,11} - (3/2)\, \mu_{\rm e\,12}}{\sqrt{2} + 1} = 1.64, \tag{2.243}$$

$$\beta_{12} = \frac{(\sqrt{2} + 13/4)\, \mu_{\rm e\,12} - (3/2)\, \mu_{\rm e\,22}}{\sqrt{2} + 1} = 1.23, \tag{2.244}$$

$$\beta_{21} = \frac{-(3/2)\, \mu_{\rm e\,11} + \mu_{\rm e\,12}}{\sqrt{2} + 1} = -0.0722, \tag{2.245}$$

$$\beta_{22} = \frac{-(3/2)\, \mu_{\rm e\,12} + \mu_{\rm e\,22}}{\sqrt{2} + 1} = 0.600, \tag{2.246}$$

$$\gamma_{11} = \frac{\sqrt{2} + 13/4}{\sqrt{2} + 1} = 1.93, \tag{2.247}$$

$$\gamma_{12} = -\frac{3/2}{\sqrt{2} + 1} = -0.621, \tag{2.248}$$

$$\gamma_{21} = -\frac{3/2}{\sqrt{2} + 1} = -0.621, \tag{2.249}$$

$$\gamma_{22} = \frac{1}{\sqrt{2} + 1} = 0.414, \tag{2.250}$$

$$\delta_{11} = \beta_{11}\, \gamma_{11} + \beta_{12}\, \gamma_{21} = 2.41, \tag{2.251}$$

$$\delta_{12} = \beta_{11}\, \gamma_{12} + \beta_{12}\, \gamma_{22} = -0.512, \tag{2.252}$$

$$\delta_{21} = \beta_{21}\, \gamma_{11} + \beta_{22}\, \gamma_{21} = -0.512, \tag{2.253}$$

$$\delta_{22} = \beta_{21}\, \gamma_{12} + \beta_{22}\, \gamma_{22} = 0.293. \tag{2.254}$$

Here, use has been made of equations (2.209)–(2.211), (2.217), and (2.218).

It follows from equations (2.165), (2.166), (2.170), (2.181)–(2.185), (2.236), and (2.240)–(2.242) that

$$\frac{\mathbf{V}_e \cdot \nabla\theta}{\mathbf{B} \cdot \nabla\theta} = -(1 - \beta_{11} f_t)\,\frac{V_{*\,e\,1}}{\langle B^2\rangle} + \beta_{12} f_t\, V_{*\,e\,2} + (1 - \beta_{11} f_t)\,\frac{V_{*\,i\,1}}{\langle B^2\rangle}$$
$$+ \alpha_1\,[1 - (\alpha_2 + \beta_{11}) f_t]\,\frac{V_{*\,i\,2}}{\langle B^2\rangle} - \frac{e\,\tau_e}{m_e}\,(\gamma_{11} - \delta_{11} f_t)\,\frac{\langle \mathbf{E}\cdot\mathbf{B}\rangle}{\langle B^2\rangle}, \tag{2.255}$$

$$\frac{\mathbf{q}_e \cdot \nabla\theta}{p_e\,\mathbf{B} \cdot \nabla\theta} = -\frac{5}{2}\,\beta_{21} f_t\, V_{*\,e\,1} + \frac{5}{2}\,(1 - \beta_{22} f_t)\,\frac{V_{*\,e\,2}}{\langle B^2\rangle} + \frac{5}{2}\,\beta_{21} f_t\,\frac{V_{*\,i\,1}}{\langle B^2\rangle}$$
$$+ \frac{5}{2}\,\alpha_1\,\beta_{21} f_t\,\frac{V_{*\,i\,2}}{\langle B^2\rangle} + \frac{5}{2}\,\frac{e\,\tau_e}{m_e}\,(\gamma_{21} - \delta_{21} f_t)\,\frac{\langle \mathbf{E}\cdot\mathbf{B}\rangle}{\langle B^2\rangle}, \tag{2.256}$$

and

$$\langle \mathbf{V}_e \cdot \mathbf{B}\rangle = V_{E\,1} + \beta_{11} f_t\, V_{*\,e\,1} + \beta_{12} f_t\, V_{*\,e\,2} + (1 - \beta_{11} f_t)\, V_{*\,i\,1}$$
$$+ \alpha_1\,[1 - (\alpha_2 + \beta_{11}) f_t]\, V_{*\,i\,2} - \frac{e\,\tau_e}{m_e}\,(\gamma_{11} - \delta_{11} f_t)\,\langle \mathbf{E}\cdot\mathbf{B}\rangle, \tag{2.257}$$

$$\frac{\langle \mathbf{q}_e \cdot \mathbf{B}\rangle}{p_e} = -\frac{5}{2}\,\beta_{21} f_t\, V_{*\,e\,1} - \frac{5}{2}\,\beta_{22} f_t\, V_{*\,e\,2} + \frac{5}{2}\,\beta_{21} f_t\, V_{*\,i\,1} + \frac{5}{2}\,\alpha_1\,\beta_{21} f_t\, V_{*\,i\,2}$$
$$+ \frac{5}{2}\,\frac{e\,\tau_e}{m_e}\,(\gamma_{21} - \delta_{21} f_t)\,\langle \mathbf{E}\cdot\mathbf{B}\rangle. \tag{2.258}$$

Here, we have made use of the fact that $\tau_{ee} = \tau_e$.

In the circular magnetic flux-surface limit, equations (2.255) and (2.256) reduce to

$$V_{\theta\,e} \simeq -(1 - \beta_{11} f_t)\,\frac{1}{e\,n_e\,B}\,\frac{dp_e}{dr} - \beta_{12} f_t\,\frac{1}{e\,B}\,\frac{dT_e}{dr} - (1 - \beta_{11} f_t)\,\frac{1}{e\,n_e\,B}\,\frac{dp_i}{dr}$$
$$+ \alpha_1\,[1 - (\alpha_2 + \beta_{11}) f_t]\,\frac{1}{e\,B}\,\frac{dT_i}{dr} - \frac{e\,\tau_e}{m_e}\,\frac{B_\theta}{B}\,(\gamma_{11} - \delta_{11} f_t)\,E_\parallel, \tag{2.259}$$

$$\frac{q_{\theta\,e}}{p_e} \simeq -\frac{5}{2}\,\beta_{21} f_t\,\frac{1}{e\,n_e\,B}\,\frac{dp_e}{dr} - \frac{5}{2}\,(1 - \beta_{22} f_t)\,\frac{1}{e\,B}\,\frac{dT_e}{dr} - \frac{5}{2}\,\beta_{21} f_t\,\frac{1}{e\,n_e\,B}\,\frac{dp_i}{dr}$$
$$+ \frac{5}{2}\,\alpha_1\,\beta_{21} f_t\,\frac{1}{e\,B}\,\frac{dT_i}{dr} + \frac{5}{2}\,\frac{e\,\tau_e}{m_e}\,\frac{B_\theta}{B}\,(\gamma_{21} - \delta_{21} f_t)\,E_\parallel, \tag{2.260}$$

while equations (2.257)–(2.258) yield

$$V_{\parallel\,e} \simeq -\frac{1}{B_\theta}\,\frac{d\Phi}{dr} + \beta_{11} f_t\,\frac{1}{e\,n_e\,B_\theta}\,\frac{dp_e}{dr} - \beta_{12} f_t\,\frac{1}{e\,B_\theta}\,\frac{dT_e}{dr}$$
$$- (1 - \beta_{11} f_t)\,\frac{1}{e\,n_e\,B_\theta}\,\frac{dp_i}{dr}$$
$$+ \alpha_1\,[1 - (\alpha_2 + \beta_{11}) f_t]\,\frac{1}{e\,B_\theta}\,\frac{dT_i}{dr} - \frac{e\,\tau_e}{m_e}\,(\gamma_{11} - \delta_{11} f_t)\,E_\parallel, \tag{2.261}$$

$$\frac{q_{\parallel\,e}}{p_e} \simeq -\frac{5}{2}\,\beta_{21} f_t\,\frac{1}{e\,n_e\,B_\theta}\,\frac{dp_e}{dr} + \frac{5}{2}\,\beta_{22} f_t\,\frac{1}{e\,B_\theta}\,\frac{dT_e}{dr} - \frac{5}{2}\,\beta_{21} f_t\,\frac{1}{e\,n_e\,B_\theta}\,\frac{dp_i}{dr}$$
$$+ \frac{5}{2}\,\alpha_1\,\beta_{21} f_t\,\frac{1}{e\,B_\theta}\,\frac{dT_i}{dr} + \frac{5}{2}\,\frac{e\,\tau_e}{m_e}\,(\gamma_{21} - \delta_{21} f_t)\,E_\parallel. \tag{2.262}$$

Expressions (2.259)–(2.262) have many features in common with the corresponding expressions for the ions (i.e. (2.226)–(2.229)), despite being much more complicated in nature. In particular, the E-cross-B velocity makes no contribution to the poloidal component of the electron fluid velocity. Moreover, the poloidal component of the electron fluid velocity is similar in magnitude to a diamagnetic velocity, while the parallel component is larger by a factor q/ϵ.

2.20 Parallel current density

The parallel current density takes the form

$$\langle \mathbf{j}\cdot\mathbf{B}\rangle = n_e\,e(\langle \mathbf{V}_i\cdot\mathbf{B}\rangle - \langle \mathbf{V}_e\cdot\mathbf{B}\rangle). \tag{2.263}$$

It follows from equations (2.221) and (2.257) that

$$\langle \mathbf{j}\cdot\mathbf{B}\rangle = (\gamma_{11} - \delta_{11} f_t)\,\sigma_\perp\,\langle \mathbf{E}\cdot\mathbf{B}\rangle$$
$$+ n_e\,e\,f_t\big(-\beta_{11}\,V_{*\,e\,1} - \beta_{12}\,V_{*\,e\,2} + \beta_{11}\,V_{*\,i\,1} + \alpha_1\,\beta_{11}\,V_{*\,i\,2}\big), \tag{2.264}$$

where $\sigma_\perp$ is specified in equation (2.41). In the circular magnetic flux-surface limit, the previous expression reduces to [15, 16]

$$j_\parallel \simeq \left(1 - \frac{\delta_{11}}{\gamma_{11}} f_t\right)\sigma_\parallel\,E_\parallel - \frac{f_t}{B_\theta}\left(\beta_{11}\,\frac{dp_e}{dr} - \beta_{12}\,n_e\,\frac{dT_e}{dr} + \beta_{11}\,\frac{dp_i}{dr} - \alpha_1\,\beta_{11}\,n_e\frac{dT_i}{dr}\right)$$
$$= (1 - 1.82\,\epsilon^{1/2})\,\sigma_\parallel\,E_\parallel - \frac{\epsilon^{1/2}}{B_\theta}\left[2.40\,(T_e + T_i)\,\frac{dn_e}{dr} + 0.61\,n_e\,\frac{dT_e}{dr} - 0.41\,n_e\,\frac{dT_i}{dr}\right], \tag{2.265}$$

where

$$\sigma_\parallel = \gamma_{11}\,\sigma_\perp = 1.93\,\sigma_\perp. \tag{2.266}$$

Here, use has been made of equations (2.202), (2.217), (2.243), (2.244), (2.247), and (2.251). Note that the previous formula agrees fairly well with the more accurate formula (2.40). (The reason for the slight discrepancy is that the moment-based expressions for the friction force densities given in section 2.16 are only approximate.)

The first term on the right-hand side of equation (2.265) represents the conventional ohmic parallel current driven by the inductive parallel electric field. Note, however, that the ohmic current is smaller than that predicted by the classical fluid closure scheme (see section 2.6) by a factor $1 - 1.82\,\epsilon^{1/2}$ [29]. The explanation for this reduction is that trapped electrons cannot carry a net parallel current, because they are forced to periodically reverse their drift direction parallel to the magnetic field. Hence, the parallel electrical conductivity of the plasma is reduced by a factor that is roughly equal to the fraction of passing particles [20] (see equation (2.201)).

This effect is known as the *neoclassical enhancement of (parallel) electrical resistivity* and has been observed in experiments [30–32].

The second term on the right-hand side of equation (2.265) represents a non-inductive parallel current driven by equilibrium density and temperature gradients that is known as the *bootstrap current* [33]. The bootstrap current has also been observed in experiments [34, 35]. The ultimate origin of the bootstrap current is collisional friction between trapped and passing particles. It is, therefore, not surprising that the current is proportional to the fraction of trapped particles, f_t.

2.21 Neoclassical transport

Taking the scalar product of equations (2.163) and (2.164) with $R^2\,\nabla\varphi$ and flux-surface averaging, we obtain

$$\Gamma_{s\,1} = -n_s\,\dot{\psi} - \frac{1}{e_s}\,\langle R^2\,\nabla\varphi\cdot\mathbf{F}_{s\,1}\rangle, \tag{2.267}$$

$$\Gamma_{s\,2} = -\frac{1}{e_s}\,\langle R^2\,\nabla\varphi\cdot\mathbf{F}_{s\,2}\rangle, \tag{2.268}$$

where

$$\Gamma_{s\,1} = n_s\,\langle\mathbf{u}_{s\,1}\cdot\nabla\psi\rangle = n_s\,\langle\mathbf{V}_s\cdot\nabla\psi\rangle, \tag{2.269}$$

$$\Gamma_{s\,2} = \frac{5}{2}\,n_s\,\langle\mathbf{u}_{s\,2}\cdot\nabla\psi\rangle = -\frac{1}{T_s}\,\langle\mathbf{q}_s\cdot\nabla\psi\rangle, \tag{2.270}$$

and use has been made of equations (2.130), (2.131), (2.155), (2.161), (2.162), (2.165), and (2.166). Note that $\Gamma_{s\,1}$ and $-T_s\,\Gamma_{s\,2}$ are proportional to the species-s particle and heat fluxes, respectively, across (i.e. perpendicular to) magnetic flux surfaces. Note further that the cross-flux-surface components of the particle and heat flow velocities are $\mathcal{O}(\delta_{\theta\,s}\,\Delta_s)$ smaller than the in-flux-surface flow velocities discussed in section 2.11.

We can now write

$$\Gamma_{i\,1} - \Gamma_{e\,1} = -\frac{1}{e}\,\langle R^2\,\nabla\varphi\cdot(\mathbf{F}_{i\,1}+\mathbf{F}_{e\,1})\rangle = 0, \tag{2.271}$$

where use has been made of equations (2.16) and (2.199). It follows that the cross-flux-surface particle fluxes in tokamaks are automatically *ambipolar* as a consequence of axisymmetry, quasi-neutrality, and collisional momentum conservation [36]. This result is of great significance because it implies that there is no preferred radial electric field, $E_r = -d\Phi/dr$, in a tokamak plasma, which means that the plasma can rotate freely in the toroidal direction (see equations (2.228) and (2.231)). The situation is very different in a non-axisymmetric magnetic confinement system, such as a stellarator, in which ambipolarity is only attained at a certain value of the radial electric field. It follows that a stellarator plasma cannot rotate freely in the toroidal direction [37].

Equations (2.267) and (2.268) can be combined with equations (2.186), (2.191), and (2.196) to give

$$(\Gamma_{\rm i}) = -n_{\rm e}\,(\dot{\psi}) + \frac{\langle R^2\rangle}{I}\,\frac{n_{\rm e}\,m_{\rm i}}{e\,\tau_{\rm ii}}\,[F_{\rm ii}]\,(u_{\varphi\,{\rm i}}) + \frac{\langle R^2\rangle}{I}\,\frac{n_{\rm e}\,m_{\rm e}}{e\,\tau_{\rm ee}}\Big\{[E_{\rm ii}]\,(u_{\varphi\,{\rm i}}) - [E_{\rm ie}]\,(u_{\varphi\,{\rm e}})\Big\}, \tag{2.272}$$

$$(\Gamma_{\rm e}) = -n_{\rm e}\,(\dot{\psi}) - \frac{\langle R^2\rangle}{I}\,\frac{n_{\rm e}\,m_{\rm e}}{e\,\tau_{\rm ee}}\Big\{[F_{\rm ee}]\,(u_{\varphi\,{\rm e}}) - [F_{\rm ei}]\,(u_{\varphi\,{\rm i}})\Big\}, \tag{2.273}$$

where

$$(\Gamma_{\rm s}) = \begin{pmatrix}\Gamma_{{\rm s}\,1}\\ \Gamma_{{\rm s}\,2}\end{pmatrix}, \tag{2.274}$$

$$(\dot{\psi}) = \begin{pmatrix}\dot{\psi}\\ 0\end{pmatrix}. \tag{2.275}$$

Making use of equations (2.193)–(2.195), (2.197), and (2.198), as well as the large-aspect-ratio approximation (2.230), we obtain

$$(\Gamma_{\rm i}) = -n_{\rm e}\,(\dot{\psi}) + \frac{\langle R^2\rangle}{I}\,\frac{n_{\rm e}\,m_{\rm i}}{e\,\tau_{\rm i}}\begin{pmatrix}0\\ 2\,u_{\|\,{\rm i}\,2}\end{pmatrix} + \frac{\langle R^2\rangle}{I}\,\frac{n_{\rm e}\,m_{\rm e}}{e\,\tau_{\rm e}}\begin{pmatrix}u_{\|\,{\rm i}\,1} - u_{\|\,{\rm e}\,1} - (3/2)\,u_{\|\,{\rm e}\,2}\\ (15/2)\,(T_{\rm e}/T_{\rm i})\,u_{\|\,{\rm i}\,2}\end{pmatrix}, \tag{2.276}$$

$$(\Gamma_{\rm e}) = -n_{\rm e}\,(\dot{\psi}) + \frac{\langle R^2\rangle}{I}\,\frac{n_{\rm e}\,m_{\rm e}}{e\,\tau_{\rm e}}\begin{pmatrix}u_{\|\,{\rm i}\,1} - u_{\|\,{\rm e}\,1} - (3/2)\,u_{\|\,{\rm e}\,2}\\ (3/2)\,u_{\|\,{\rm i}\,1} - (3/2)\,u_{\|\,{\rm e}\,1} - (\sqrt{2} + 13/4)\,u_{\|\,{\rm e}\,2}\end{pmatrix}. \tag{2.277}$$

Here, use has been made of the results $\tau_{\rm ee} = \tau_{\rm e}$ and $\tau_{\rm ii} = \tau_{\rm i}/\sqrt{2}$. According to the previous two equations, the large parallel particle and heat flows present in a low-collisionality plasma (see sections 2.18 and 2.19) give rise to a transport of particles and heat across magnetic flux surfaces. This transport, which is far larger than the cross-flux-surface transport predicted by the classical closure scheme (see section 2.6), is known as *neoclassical transport* [26, 27].

According to equations (2.130), (2.131), (2.165), (2.166), (2.181), (2.221), (2.222), (2.243)–(2.250), (2.257), and (2.258),

$$u_{\|\,{\rm i}\,2} = \alpha_2\,f_{\rm t}\;V_{*\,{\rm i}\,2}, \tag{2.278}$$

$$u_{\parallel\,i\,1} - u_{\parallel\,e\,1} - \frac{3}{2}\,u_{\parallel\,e\,2} = -\,\mu_{e\,11} f_t\, V_{*\,e\,1} - \mu_{e\,12} f_t\, V_{*\,e\,2} + \mu_{e\,11} f_t\, V_{*\,i\,1} + \alpha_1\,\mu_{e\,11} f_t\, V_{*\,i\,2} + I\left\langle\frac{1}{R^2}\right\rangle\frac{e\,\tau_e}{m_e}\,(1 - \beta_{11} f_t)\,\dot{\psi}, \tag{2.279}$$

$$\frac{3}{3}\,u_{\parallel\,i\,1} - \frac{3}{2}\,u_{\parallel\,e\,1} - (\sqrt{2} + 13/4)\,u_{\parallel\,e\,2} = -\,\mu_{e\,12} f_t\, V_{*\,e\,1} - \mu_{e\,22} f_t\, V_{*\,e\,2} + \mu_{e\,12} f_t\, V_{*\,i\,1} + \alpha_1\,\mu_{e\,12} f_t\, V_{*\,i\,2} - I\left\langle\frac{1}{R^2}\right\rangle\frac{e\,\tau_e}{m_e}\,\beta_{12} f_t\,\dot{\psi}. \tag{2.280}$$

It follows that

$$\langle \mathbf{V}_i \cdot \nabla\psi\rangle = \langle \mathbf{V}_e \cdot \nabla\psi\rangle = -\beta_{11} f_t\,\dot{\psi} + \frac{\langle R^2\rangle}{I}\,\frac{m_e}{e\,\tau_e}\,f_t \left(-\mu_{e\,11}\, V_{*\,e\,1} - \mu_{e\,12}\, V_{*\,e\,2} + \mu_{e\,11}\, V_{*\,i\,1} + \alpha_1\,\mu_{e\,11}\, V_{*\,i\,2}\right), \tag{2.281}$$

$$\frac{\langle \mathbf{q}_i \cdot \nabla\psi\rangle}{p_i} = -\frac{\langle R^2\rangle}{I}\,\frac{m_i}{e\,\tau_i}\,2\,\alpha_2\, f_t\, V_{*\,i\,2}, \tag{2.282}$$

$$\frac{\langle \mathbf{q}_e \cdot \nabla\psi\rangle}{p_e} = \beta_{12} f_t\,\dot{\psi} + \frac{\langle R^2\rangle}{I}\,\frac{m_e}{e\,\tau_e}\,f_t\left(\mu_{e\,12}\, V_{*\,e\,1} + \mu_{e\,22}\, V_{*\,e\,2} - \mu_{e\,12}\, V_{*\,i\,1} - \alpha_1\,\mu_{e\,12}\, V_{*\,i\,2}\right), \tag{2.283}$$

where use has been made of equations (2.269) and (2.270). Note that we have also made use of the large-aspect-ratio approximation that $|\langle R^2\rangle\,\langle 1/R^2\rangle - 1| \ll f_t$. Furthermore, we have neglected a term that is $\mathcal{O}[(m_e/m_i)^{1/2}]$ smaller than the leading-order term in equation (2.282). In the circular magnetic flux-surface limit, the previous three expressions reduce to [16]

$$\begin{aligned} V_{r\,i} = V_{r\,e} &\simeq -\beta_{11} f_t\,\frac{E_\parallel}{B_\theta} + \left(\frac{q}{\epsilon}\right)^2 \frac{\rho_e^{\,2}}{2\,\tau_e}\, f_t\left(-\mu_{e\,11}\,\frac{1}{p_e}\,\frac{dp_e}{dr} + \mu_{e\,12}\,\frac{1}{T_e}\,\frac{dT_e}{dr} - \mu_{e\,11}\,\frac{1}{p_e}\,\frac{dp_i}{dr} + \alpha_1\,\mu_{e\,11}\,\frac{1}{T_e}\,\frac{dT_i}{dr}\right) \\ &= -\,2.40\,\epsilon^{1/2}\,\frac{E_\parallel}{B_\theta} + \frac{q^2\,\rho_e^{\,2}}{\epsilon^{3/2}\,\tau_e}\left[-1.12\left(1 + \frac{T_i}{T_e}\right)\frac{1}{n_e}\,\frac{dn_e}{dr} + 0.43\,\frac{1}{T_e}\,\frac{dT_e}{dr} + 0.19\,\frac{1}{T_e}\,\frac{dT_i}{dr}\right], \end{aligned} \tag{2.284}$$

and

$$\frac{q_{r\,i}}{p_i} \simeq -\left(\frac{q}{\epsilon}\right)^2 \frac{\rho_i^{\,2}}{2\,\tau_i}\,\alpha_2\, f_t\,\frac{1}{T_i}\,\frac{dT_i}{dr} = -0.67\,\frac{q^2\,\rho_i^{\,2}}{\epsilon^{3/2}\,\tau_i}\,\frac{1}{T_i}\,\frac{dT_i}{dr}, \tag{2.285}$$

and

$$
\begin{aligned}
\frac{q_{r\,\mathrm{e}}}{p_\mathrm{e}} &\simeq \beta_{12}\,f_\mathrm{t}\,\frac{E_\parallel}{B_\theta} + \left(\frac{q}{\epsilon}\right)^2 \frac{\rho_\mathrm{e}^{\,2}}{2\,\tau_\mathrm{e}}\,f_\mathrm{t}\left(\mu_{\mathrm{e}\,12}\,\frac{1}{p_\mathrm{e}}\,\frac{dp_\mathrm{e}}{dr} - \mu_{\mathrm{e}\,22}\,\frac{1}{T_\mathrm{e}}\,\frac{dT_\mathrm{e}}{dr}\right.\\
&\quad\left. + \mu_{\mathrm{e}\,12}\,\frac{1}{p_\mathrm{e}}\,\frac{dp_\mathrm{i}}{dr} - \alpha_1\,\mu_{\mathrm{e}\,12}\,\frac{1}{T_\mathrm{e}}\,\frac{dT_\mathrm{i}}{dr}\right)\\
&= 1.79\,\epsilon^{1/2}\,\frac{E_\parallel}{B_\theta} + \frac{q^2\,\rho_\mathrm{e}^{\,2}}{\epsilon^{3/2}\,\tau_\mathrm{e}}\left[1.55\left(1+\frac{T_\mathrm{i}}{T_\mathrm{e}}\right)\frac{1}{n_\mathrm{e}}\,\frac{dn_\mathrm{e}}{dr} - 1.83\,\frac{1}{T_\mathrm{e}}\,\frac{dT_\mathrm{e}}{dr}\right.\\
&\quad\left. - 0.27\,\frac{1}{T_\mathrm{e}}\,\frac{dT_\mathrm{i}}{dr}\right],
\end{aligned}
\tag{2.286}
$$

where use has been made of equations (2.202), (2.209)–(2.211), (2.243), and (2.244).

The first term on the right-hand side of equation (2.284) represents an inward flow of trapped particles driven by the inductive parallel component of the electric field. This effect is known as the *Ware pinch* [38]. Consider the toroidal component of the equation of motion of a particle of species s:

$$
m_\mathrm{s}\,\frac{dv_{\varphi\,\mathrm{s}}}{dt} = e_\mathrm{s}\,(E_\varphi + v_{r\,\mathrm{s}}\,B_\theta). \tag{2.287}
$$

(Note that $B_r = 0$.) For a passing particle, $e_\mathrm{s}\,E_\varphi$ is balanced by $m_\mathrm{s}\,dv_{\varphi\,\mathrm{s}}/dt$. In other words, the toroidal electric field causes the toroidal velocity of the particle to continuously increase or decrease. This increase or decrease is ultimately limited by collisions and parallel viscosity (see equation (2.175)). The net result is that passing particles of both species carry a net toroidal current that is driven by the inductive electric field. For a trapped particle, on the other hand, the integral between bounces of the left-hand side of the previous equation is zero. Thus, we obtain

$$
\langle v_{r\,\mathrm{s}}\,B_\theta\rangle = -E_\varphi. \tag{2.288}
$$

It follows that the time-averaged radial velocity of a trapped particle is

$$
\langle v_r\rangle = -\frac{E_\varphi}{B_\theta}, \tag{2.289}
$$

independent of the particle's mass or charge. Hence, given that the trapped-particle fraction is $\mathcal{O}(\epsilon^{1/2})$, we expect both plasma species to exhibit a mean radial flow of the form

$$
V_{r\,\mathrm{i}} = V_{r\,\mathrm{e}} \sim -\epsilon^{1/2}\,\frac{E_\varphi}{B_\theta}. \tag{2.290}
$$

The Ware pinch does indeed have this form (when we take into account the fact that $E_\parallel \simeq E_\varphi$ in a large-aspect-ratio tokamak). It is clear from equation (2.286) that the inward flow of trapped electrons is associated with an outward electron heat flux.

The remaining terms in equations (2.284)–(2.286) are all diffusive in nature. According to Fick's law [39], we expect the diffusive component of the species-s radial velocity to take the form

$$V_{r\,s} = -D_{\perp\,s}\,\frac{1}{n_s}\,\frac{dn_s}{dr}, \tag{2.291}$$

where $D_{\perp\,s}$ is the perpendicular (i.e. cross-flux-surface) particle diffusivity. If we ignore the Ware pinch and temperature gradient terms in equation (2.284) and make the simplifying assumption that $T_e = T_i$, then we obtain

$$V_{r\,s} \simeq -2.24\,\frac{q^2\,\rho_e^{\,2}}{\epsilon^{3/2}\,\tau_e}\,\frac{1}{n_s}\,\frac{dn_s}{dr}. \tag{2.292}$$

This leads to the following estimate for the perpendicular particle diffusivity (which is the same for ions and electrons):

$$D_\perp \simeq 2.24\,\frac{q^2\,\rho_e^{\,2}}{\epsilon^{3/2}\,\tau_e}. \tag{2.293}$$

The previous equation can also be written

$$D_\perp \sim f_t\,\frac{\nu_e}{\epsilon}\,\rho_{b\,e}^{\,2}, \tag{2.294}$$

where use has been made of equations (2.102) and (2.202). The previous formula can be interpreted as follows: the collisions that scatter electrons out of their trapped orbits displace the electrons across magnetic flux surfaces by a random distance that is of the order of the banana width, $\rho_{b\,e}$. Such collisions take place at the frequency ν_e/ϵ (see equation (2.94)). Hence, the trapped electrons have a diffusivity of $(\nu_e/\epsilon)\,\rho_{b\,e}^{\,2}$ [2]. However, trapped electrons only make up a fraction f_t of the total number of electrons, so the overall electron diffusivity is $f_t\,(\nu_e/\epsilon)\,\rho_{b\,e}^{\,2}$ [16, 20, 26]. Note that the ion diffusivity is limited to be the same as the electron diffusivity by the constraint that the cross-flux-surface particle fluxes be ambipolar.

According to Fick's law, we expect the diffusive component of the ion heat flux to take the form

$$q_{r\,i} = -\chi_{\perp\,i}\,n_i\,\frac{dT_i}{dr}. \tag{2.295}$$

We can see that equation (2.285) does indeed take this form. Our estimate for the perpendicular ion energy diffusivity is thus

$$\chi_{\perp\,i} = 0.67\,\frac{q^2\,\rho_i^{\,2}}{\epsilon^{3/2}\,\tau_i}. \tag{2.296}$$

The previous expression can also be written as

$$\chi_{\perp\,i} \sim f_t\,\frac{\nu_i}{\epsilon}\,\rho_{b\,i}^{\,2}. \tag{2.297}$$

We interpret the previous formula as saying that the neoclassical perpendicular ion energy diffusivity corresponds to the random-walk motions of trapped ions with step frequency ν_i/ϵ and step length $\rho_{b\,i}$ [16, 20, 26]. Note that the neoclassical diffusivity specified in the previous equation is much larger, by a factor $q^2/\epsilon^{3/2}$, that that predicted by the classical closure scheme (see section 2.6).

According to Fick's law [39], we expect the diffusive component of the electron heat flux to take the form

$$q_{r\,e} = -\chi_{\perp\,e}\, n_e\, \frac{dT_e}{dr}. \tag{2.298}$$

If we neglect the Ware pinch, density gradient, and ion temperature gradient terms in equation (2.286), then we get

$$q_{r\,e} = -1.83\, \frac{q^2\, \rho_e^{\ 2}}{\epsilon^{3/2}\, \tau_e}\, n_e\, \frac{dT_e}{dr}. \tag{2.299}$$

Hence, our estimate for the electron energy diffusivity becomes

$$\chi_{\perp\,e} = 1.83\, \frac{q^2\, \rho_e^{\ 2}}{\epsilon^{3/2}\, \tau_e}. \tag{2.300}$$

The previous expression can also be written

$$\chi_{\perp\,e} \sim f_t\, \frac{\nu_e}{\epsilon}\, \rho_{b\,e}^{\ 2}. \tag{2.301}$$

We interpret the previous formula as saying that the neoclassical perpendicular electron energy diffusivity corresponds to the random-walk motions of trapped electrons with step frequency ν_e/ϵ and step length $\rho_{b\,e}$ [16, 20, 26]. As before, the neoclassical diffusivity specified in the previous equation is much larger, by a factor $q^2/\epsilon^{3/2}$, that that predicted by the classical closure scheme (see section 2.6).

It turns out that the neoclassical cross-flux-surface momentum diffusivities only exceed the classical momentum diffusivities by a factor of q^2. For example [16],

$$\Xi_{\perp\,i} = 0.1\, q^2\, \nu_i\, \rho_i^{\ 2}. \tag{2.302}$$

Table 2.5 shows estimates for the neoclassical cross-flux-surface particle, heat, and momentum diffusivities in a tokamak fusion reactor. Note that although the diffusivities are much larger than the classical cross-flux-surface diffusivities shown in table 2.3, they are still all much smaller than the experimentally observed cross-flux-surface diffusivities, which are $D_\perp \sim 0.2\ \text{m}^2\,\text{s}^{-1}$ and $\chi_s \sim \Xi_s \sim 1\ \text{m}^2\,\text{s}^{-1}$ [16].

2.22 The perpendicular closure scheme

As we have seen, the neoclassical cross-flux-surface particle, heat, and momentum diffusivities are all much smaller than the experimentally observed diffusivities. The additional, or *anomalous*, cross-flux-surface transport that is found in tokamaks is due to the action of small-scale plasma *turbulence* [40]. Turbulent eddies in a tokamak plasma are the nonlinearly saturated states of micro-instabilities driven by temperature gradients [41]. As mentioned previously, turbulent eddies are localized

Table 2.5. Neoclassical diffusivities in a low-field tokamak reactor and a high-field tokamak reactor. Here, B is the toroidal magnetic field strength, $D_\perp$ the perpendicular particle diffusivity, $\chi_{\perp\,\mathrm{e}}$ the perpendicular electron energy diffusivity, $\chi_{\perp\,\mathrm{i}}$ the perpendicular ion energy diffusivity, and $\Xi_{\perp\,\mathrm{i}}$ the perpendicular ion momentum diffusivity.

	Low field	High field
B (T)	5.0	12.0
$D_\perp$ ($\mathrm{m^2\,s^{-1}}$)	2.3×10^{-3}	2.3×10^{-3}
$\chi_{\perp\,\mathrm{e}}$ ($\mathrm{m^2\,s^{-1}}$)	1.9×10^{-3}	1.9×10^{-3}
$\chi_{\perp\,\mathrm{i}}$ ($\mathrm{m^2\,s^{-1}}$)	3.4×10^{-2}	3.4×10^{-2}
$\Xi_{\perp\,\mathrm{i}}$ ($\mathrm{m^2\,s^{-1}}$)	1.0×10^{-3}	1.0×10^{-3}

on toroidal magnetic flux surfaces in flux tubes that run parallel to the magnetic field and whose radial extents are a few ion gyroradii. It follows that turbulent eddies are characterized by the so-called *flute ordering*, $|k_\parallel|/|k_\perp| \ll 1$, where $k_\parallel$ and $k_\perp$ are the wave numbers of the underlying micro-instabilities parallel and perpendicular to the equilibrium magnetic field, respectively. An immediate consequence of the flute ordering is that turbulent eddies generate comparatively little transport parallel to magnetic field lines. Consequently, we shall assume that turbulence does not upset the parallel force and heat flux balance described in equations (2.175) and (2.176). This implies that the expressions for the in-flux-surface neoclassical ion and electron flows given in sections 2.18 and 2.19, well as the expression for the parallel current density given in section 2.20, remain valid in the presence of turbulence. We also expect the expressions for the cross heat fluxes and the gyroviscous tensors given in section 2.6 to remain valid, because these effects are merely a consequence of the rapid gyration of charged particles perpendicular to magnetic field lines, which is not significantly affected by turbulence. On the other hand, we shall replace our previous expressions for the perpendicular viscous force densities and the perpendicular heat fluxes by phenomenological terms of the forms

$$\nabla \cdot \boldsymbol{\pi}_{\perp\,\mathrm{s}} = -\nabla \cdot (n_\mathrm{e}\, m_\mathrm{s}\, \Xi_{\perp\,\mathrm{s}}\, \mathbf{W}_\mathrm{s}), \tag{2.303}$$

$$\mathbf{q}_{\perp\,\mathrm{s}} = -n_\mathrm{e}\, \chi_{\perp\,\mathrm{s}}\, \nabla_\perp T_\mathrm{s}, \tag{2.304}$$

respectively, where $\Xi_{\perp\,\mathrm{s}}$ and $\chi_{\perp\,\mathrm{s}}$ are taken from experimental measurements.

Finally, if axisymmetric tokamak plasmas are to retain their freedom to rotate in the toroidal direction, then anomalous transport needs to be intrinsically ambipolar. Fortunately, good theoretical arguments can be made that this is the case [21, 42, 43].

2.23 The parallel closure scheme

There are a number of other terms that appear in the so-called *Braginskii equations*, described in section 2.6, that are subject to modification in a long mean-free-path plasma.

Consider the thermal force density (see equation (2.36)):

$$\mathbf{F}_{\rm T} = \mathbf{F}_{\rm T\,\|} + \mathbf{F}_{\rm T\,\perp}, \tag{2.305}$$

where

$$\mathbf{F}_{\rm T\,\|} = -0.71\, n_{\rm e}\, \nabla_{\|} T_{\rm e}, \tag{2.306}$$

$$\mathbf{F}_{\rm T\,\perp} = \frac{3\, n_{\rm e}}{2\,\Omega_{\rm e}\,\tau_{\rm e}}\, \mathbf{b}\times\nabla_{\perp} T_{\rm e}. \tag{2.307}$$

Both the parallel and the perpendicular components of this force density are a consequence of the fact that the friction force density is inversely proportional to the temperatures of electrons that are scattered by ions [1, 13].

Let us first examine the parallel component of the thermal force density. Suppose that x denotes distance along a magnetic field line. We can write [1, 13]

$$\mathbf{F}_{\rm T\,\|}(x) = -\frac{\sqrt{2}\, m_{\rm e}}{\tau_{\rm e}}\int_{-\infty}^{\infty}\frac{v_{\|\,\rm e}\, T_{\rm e}(x)}{T_{\rm e}(x - v_{\|\,\rm e}\,\tau_{\rm e})}\, f(v_{\|\,\rm e})\, dv_{\|\,\rm e}\,\mathbf{b}, \tag{2.308}$$

where

$$f(v_{\|\,\rm e}) = \frac{n_{\rm e}}{\sqrt{\pi}\, v_{\rm t\,e}}\exp\left(-\frac{v_{\|\,\rm e}^{2}}{v_{\rm t\,e}^{2}}\right). \tag{2.309}$$

Here, we are making use of the fact that $\mathbf{F}_{\rm T\,\|}$ is the sum of $\sqrt{2}\, m_{\rm e}\, v_{\|\,\rm e}/\tau_{\rm e}$ (the $\sqrt{2}$ factor is to get agreement with the collisional result) taken over all electrons moving parallel to the field line and weighted by the inverse of the electron temperature at the last scattering position. Thus, electrons with parallel velocity $v_{\|\,\rm e}$ that are scattered by ions at position x were last scattered at position $x - v_{\|\,\rm e}\,\tau_{\rm e}$ and are, therefore, characterized by the electron temperature at the latter position. Suppose that a tearing mode gives rise to a perturbation of the electron temperature along the field line of the form

$$T_{\rm e}(x) = T_{\rm e} + \delta T_{\rm e}\cos(k_{\|}\, x), \tag{2.310}$$

where $\delta T_{\rm e}/T_{\rm e} \ll 1$. It follows that

$$\mathbf{F}_{\rm T\,\|}(x) \simeq \frac{\sqrt{2}\, n_{\rm e}\, m_{\rm e}\, v_{\rm t\,e}}{\sqrt{\pi}\,\tau_{\rm e}\, T_{\rm e}}\,\delta T_{\rm e}\sin(k_{\|}\, x)\int_{-\infty}^{\infty} y\sin(k_{\|}\, l_{\rm e}\, y)\exp(-y^{2})dy\,\mathbf{b}, \tag{2.311}$$

which reduces to

$$\mathbf{F}_{\rm T\,\|}(x) = -\frac{1}{\sqrt{2}}\, n_{\rm e}\exp\left[-\frac{(k_{\|}\, l_{\rm e})^{2}}{4}\right]\nabla_{\|} T_{\rm e}. \tag{2.312}$$

Here, use has been made of the result

$$\int_{-\infty}^{\infty} y\sin(\alpha\, y)\exp(-y^{2})dy = \frac{\sqrt{\pi}}{4}\,\alpha\exp\left(-\frac{\alpha^{2}}{4}\right). \tag{2.313}$$

It can be seen that, in the short mean-free-path limit, $k_\parallel\, l_e \ll 1$, equation (2.312) is identical to equation (2.306). However, in the long mean-free-path limit, $k_\parallel\, l_e \gg 1$, the parallel thermal force density is attenuated by a factor of $\exp[-(k_\parallel\, l_e)^2/4]$. As is clear from table 2.1, unless the parallel wavelength of the tearing mode exceeds a few kilometers (i.e. the electron mean free path), this attenuation is very strong in a tokamak fusion reactor. Tearing modes are global instabilities of tokamak plasmas, so they have wavelengths that do not differ greatly from the dimensions of the plasma, which is a few meters. Thus, we conclude that there is no parallel component of the thermal force density in a tokamak fusion reactor. To be slightly more exact, $k_\parallel \sim \Delta r/(R_0\, r_s)$ for a tearing mode in a tokamak plasma of minor radius r_s (where Δr measures radial distance from the rational surface), at which it resonates with the equilibrium magnetic field (see section 1.11). Thus, the attenuation of the parallel thermal force is strong unless $\Delta r/r_s \lesssim R_0/l_e \sim 10^{-3}$.

Let us now consider the perpendicular component of the thermal force density. The appropriate calculation is analogous to the calculation just performed, except that electrons that move perpendicular to magnetic field lines and are scattered by ions at a given point in the plasma can only have originated from points that are, at most, a few electron gyroradii away from the given point. Hence, the appropriate attenuation factor is $\exp[-(k_\perp\, \rho_e)^2/4]$. As is clear from table 2.1, unless the perpendicular wavelength of the tearing mode falls below about 0.1 mm (i.e. the electron gyroradius), there is no attenuation of the perpendicular thermal force density in a tokamak fusion reactor. Thus, we conclude that equation (2.307) remains valid in such a reactor.

The electron heat flux also has a thermal component, which is specified in equation (2.47). Similar arguments to those that we just made reveal that the parallel component of this heat flux is strongly attenuated in a tokamak fusion reactor, due to long mean-free-path effects, while the perpendicular component is unaffected.

Consider the diffusive parallel heat flux associated with the electron temperature perturbation (2.310). In the short mean-free-path limit, we have

$$\mathbf{q}_{\parallel\,e} = -n_e\, \chi_{\parallel\,e}^{\rm smfp}\, \nabla_\parallel T_e, \tag{2.314}$$

where

$$\chi_{\parallel\,e}^{\rm smfp} = 1.58\, v_{t\,e}\, l_e. \tag{2.315}$$

(See equations (2.44) and (2.48).) On the other hand, in the long mean-free-path (i.e. collisionless) limit [44, 45]

$$\mathbf{q}_{\parallel\,e}(x) = \frac{n_e\, v_{t\,e}}{\pi^{3/2}} \int_0^\infty \frac{T_e(x - x') - T_e(x + x')}{x'}\, dx'\, \mathbf{b}. \tag{2.316}$$

It follows from equation (2.310) that

$$\mathbf{q}_{\parallel\,e}(x) = \frac{2\, n_e\, v_{t\,e}}{\pi^{3/2}}\, \delta T_0\, \sin(k_\parallel\, x) \int_0^\infty \frac{\sin y}{y}\, dy\, \mathbf{b}, \tag{2.317}$$

which yields

$$\mathbf{q}_{\parallel\,\mathrm{e}} = -n_{\mathrm{e}}\,\chi_{\parallel\,\mathrm{e}}^{\mathrm{lmfp}}\,\nabla_{\parallel}T_{\mathrm{e}}, \tag{2.318}$$

where the long mean-free-path thermal conductivity takes the form [46]

$$\chi_{\parallel\,\mathrm{e}}^{\mathrm{lmfp}} = \frac{v_{\mathrm{t\,e}}}{\pi^{1/2}\,k_{\parallel}}. \tag{2.319}$$

Here, use has been made of the result $\int_0^{\infty}(\sin y/y)\,dy = \pi/2$. It can be seen from a comparison between equations (2.315) and (2.319) that the short mean-free-path parallel electron thermal diffusivity exceeds the collisionless diffusivity when $l_{\mathrm{e}}\,k_{\parallel}$ exceeds about unity. However, this is impossible, because the collisionless diffusivity corresponds to the free flow of electrons along magnetic field lines, completely unimpeded by collisions, and therefore sets an upper limit on the diffusivity. (Of course, in this limit, the parallel heat flow is not diffusive at all; instead, it is convective.) Hence, we conclude that in a tokamak fusion reactor, which is characterized by $l_{\mathrm{e}}\,k_{\parallel} \gg 1$, the appropriate expression for the parallel electron thermal 'diffusivity' is given by equation (2.319). Analogous arguments reveal that the ion parallel thermal 'diffusivity' in a tokamak fusion reactor takes the form [46]

$$\chi_{\parallel\,\mathrm{i}}^{\mathrm{lmfp}} = \frac{v_{\mathrm{t\,i}}}{\pi^{1/2}\,k_{\parallel}}. \tag{2.320}$$

2.24 The derivation of the neoclassical fluid equations

It is helpful to define the so-called *magnetohydrodynamic (MHD) velocity* [21],

$$\mathbf{V} = \mathbf{V}_{\mathrm{E}} + V_{\parallel\,\mathrm{i}}\,\mathbf{b}. \tag{2.321}$$

which is the fluid velocity associated with the ion guiding centers. Here, a *guiding center* is the imaginary point about which a charged particle gyrates in the plane perpendicular to the local magnetic field. According to section 2.11, the lowest-order ion and electron fluid velocities can be written as

$$\mathbf{V}_{\mathrm{i}} = \mathbf{V} + \mathbf{V}_{*\,\mathrm{i}}, \tag{2.322}$$

$$\mathbf{V}_{\mathrm{e}} = \mathbf{V} + \mathbf{V}_{*\,\mathrm{i}} - \frac{\mathbf{j}}{n_{\mathrm{e}}\,e}, \tag{2.323}$$

respectively, where

$$\mathbf{j} = \frac{\mathbf{b}\times\nabla p}{B} + j_{\parallel}\,\mathbf{b} \tag{2.324}$$

is the lowest-order electrical current density. Quasi-neutrality demands that

$$\nabla\cdot\mathbf{j} \simeq 0. \tag{2.325}$$

The electron and ion continuity equations, (2.26) and (2.29), can be combined with equations (2.322), (2.323), and (2.325) to give an *electron number density continuity equation* of the form [21]:

$$\frac{\partial n_e}{\partial t} + \nabla \cdot [n_e (\mathbf{V} + \mathbf{V}_{*\,i})] = 0. \tag{2.326}$$

If we sum the electron and the ion equations of motion, (2.27) and (2.30), then we obtain

$$m_i\, n_e \frac{\partial \mathbf{V}_i}{\partial t} + m_i\, n_e\, (\mathbf{V}_i \cdot \nabla)\mathbf{V}_i - \mathbf{j} \times \mathbf{B} + \nabla p + \nabla \cdot \boldsymbol{\pi}_{\parallel\, i} + \nabla \cdot \boldsymbol{\pi}_{\times i} + \nabla \cdot \boldsymbol{\pi}_{\perp\, i} = 0. \tag{2.327}$$

Here, $p = p_e + p_i$ is the total plasma pressure, and use has been made of equations (2.16), (2.39), and (2.59). In the previous equation, we neglected electron inertia with respect to ion inertia because the former is $\mathcal{O}(m_e/m_i)$ times smaller than the latter. We also neglected electron viscosity with respect to ion viscosity because the former is at least $\mathcal{O}[(m_e/m_i)^{1/2}]$ times smaller than the latter. It can be demonstrated that [21, 47]:

$$\begin{aligned} m_i\, n_e \frac{\partial \mathbf{V}_i}{\partial t} + m_i\, n_e\, (\mathbf{V}_i \cdot \nabla)\mathbf{V}_i + \nabla \cdot \boldsymbol{\pi}_{\times i} = m_i \\ n_e\left[\frac{\partial \mathbf{V}}{\partial t} + (\mathbf{V} \cdot \nabla)\mathbf{V} + (\mathbf{V}_{*\,i} \cdot \nabla)\mathbf{V}_E\right]. \end{aligned} \tag{2.328}$$

This important result is known as the *gyroviscous cancellation*. Furthermore, it is clear from equations (2.152) and (2.161) that $\nabla \cdot \boldsymbol{\pi}_{\parallel\, i}$ acts predominately in the $\mathbf{e}_\theta$ direction (i.e. parallel to $\nabla\theta$). A model form for $\nabla \cdot \boldsymbol{\pi}_{\parallel\, i}$ that is consistent with the analysis of section 2.18 is

$$\nabla \cdot \boldsymbol{\pi}_{\parallel\, i} = \sqrt{2}\, \mu_{i\,11} f_t \left(\frac{q}{\epsilon}\right)^2 \frac{m_i\, n_e}{\tau_i} (V_{\theta\, i} - V_{\theta\, i}^{nc})\, \mathbf{e}_\theta, \tag{2.329}$$

where

$$V_{\theta\, i}^{nc} = \alpha_1 (1 - \alpha_2 f_t) \frac{1}{e\, B} \frac{\partial T_i}{\partial r}. \tag{2.330}$$

Thus, the neoclassical ion parallel viscosity tensor acts to relax the ion poloidal velocity toward the neoclassical velocity specified in equation (2.330) on a timescale that is roughly $(\epsilon^{3/2}/q^2)\, \tau_i$. As has already been mentioned, this effect is known as poloidal flow damping. Equations (2.328) and (2.329) can be combined with equation (2.303) to give an *MHD equation of motion* of the form:

$$\begin{aligned} m_i\, n_e\left[\frac{\partial \mathbf{V}}{\partial t} + (\mathbf{V} \cdot \nabla)\mathbf{V} + (\mathbf{V}_{*\,i} \cdot \nabla)\mathbf{V}_E\right] - \mathbf{j} \times \mathbf{B} + \nabla p + \frac{m_i\, n_e}{\tau_{\theta\, i}}(V_{\theta\, i} - V_{\theta\, i}^{nc})\mathbf{e}_\theta \\ - \nabla \cdot (m_i\, n_e\, \Xi_{\perp\, i}\, \mathbf{W}_i) = 0, \end{aligned} \tag{2.331}$$

where

$$\tau_{\theta\,\mathrm{i}} = \frac{1}{\sqrt{2}\,\mu_{\mathrm{i}\,11} f_{\mathrm{t}}} \left(\frac{\epsilon}{q}\right)^2 \tau_{\mathrm{i}} \tag{2.332}$$

is the *poloidal flow-damping time*. Of course, the value of the anomalous perpendicular momentum diffusivity, $\Xi_{\perp\,\mathrm{i}}$, must be taken from experiments.

The electron equation of motion, (2.27), can be combined with equations (2.59), (2.139), and (2.323) to give

$$\mathbf{E} + \mathbf{V} \times \mathbf{B} + \frac{1}{e\,n_{\mathrm{e}}}\left(\nabla p - \nabla_{\parallel} p_{\mathrm{i}} - \mathbf{j} \times \mathbf{B}\right) + \frac{\nabla \cdot \boldsymbol{\pi}_{\parallel\,\mathrm{e}} - \mathbf{F}_{\mathrm{e}}}{e\,n_{\mathrm{e}}} = 0. \tag{2.333}$$

Here, we have neglected electron inertia and gyroviscosity because these terms are $\mathcal{O}(\delta_{\mathrm{e}}^{\,2})$ times smaller than the leading-order terms We have also neglected anomalous electron perpendicular viscosity because there are no available experimental measurements of this effect in tokamak plasmas. Let us adopt the following model form for $\nabla \cdot \boldsymbol{\pi}_{\parallel\,\mathrm{e}}$ which is consistent with the analysis of sections 2.16, 2.19, and 2.20:

$$\frac{\nabla \cdot \boldsymbol{\pi}_{\parallel\,\mathrm{e}} - \mathbf{F}_{\mathrm{e}}}{e\,n_{\mathrm{e}}} = -\eta_{\perp}^{\,\mathrm{nc}}\left(\mathbf{j}_{\perp} - \frac{\mathbf{b} \times \nabla p}{B}\right) - \eta_{\parallel}^{\,\mathrm{nc}}\,(j_{\parallel} - j_{\parallel}^{\,\mathrm{nc}})\,\mathbf{b}, \tag{2.334}$$

where

$$\eta_{\perp}^{\,\mathrm{nc}} = \mu_{\mathrm{e}\,11} f_{\mathrm{t}} \left(\frac{q}{\epsilon}\right)^2 \frac{m_{\mathrm{e}}}{n_{\mathrm{e}}\,e^2\,\tau_{\mathrm{e}}}, \tag{2.335}$$

$$\eta_{\parallel}^{\,\mathrm{nc}} = \frac{1}{\gamma_{11} - \delta_{11} f_{\mathrm{t}}} \frac{m_{\mathrm{e}}}{n_{\mathrm{e}}\,e^2\,\tau_{\mathrm{e}}}, \tag{2.336}$$

$$j_{\parallel}^{\,\mathrm{nc}} = -f_{\mathrm{t}}\,\frac{q}{\epsilon}\left(\beta_{11}\,\frac{1}{B}\frac{\partial p}{\partial r} - \beta_{12}\,\frac{n_{\mathrm{e}}}{B}\frac{\partial T_{\mathrm{e}}}{\partial r} - \alpha_1\,\beta_{11}\,\frac{n_{\mathrm{e}}}{B}\frac{\partial T_{\mathrm{i}}}{\partial r}\right). \tag{2.337}$$

Thus, the neoclassical electron parallel viscosity tensor acts to relax the perpendicular current to its lowest-order value (see equation (2.324)), enhances the parallel electrical resistivity of the plasma, and also generates the non-inductive parallel current specified in equation (2.337). As has already been mentioned, this non-inductive current is known as the bootstrap current. Note that in equation (2.334) we have neglected the parallel thermal force density, (2.306), in accordance with the discussion in section 2.23. We have also neglected the perpendicular component of the thermal force density, (2.307) because it is $\mathcal{O}(\epsilon^{3/2}/q^2)$ smaller than the similar term in the previous equation that involves $\eta_{\perp}^{\,\mathrm{nc}}$. Thus, equations (2.333) and (2.334) yield the following *generalized Ohm's law* for the plasma:

$$\begin{aligned} &\mathbf{E} + \mathbf{V} \times \mathbf{B} + \frac{1}{e\,n_{\mathrm{e}}}\left(\nabla p - \nabla_{\parallel} p_{\mathrm{i}} - \mathbf{j} \times \mathbf{B}\right) \\ &= \eta_{\perp}^{\,\mathrm{nc}}\left(\mathbf{j}_{\perp} - \frac{\mathbf{b} \times \nabla p}{B}\right) + \eta_{\parallel}^{\,\mathrm{nc}}\,(j_{\parallel} - j_{\parallel}^{\,\mathrm{nc}})\,\mathbf{b}. \end{aligned} \tag{2.338}$$

The electron and ion energy conservation equations, (2.28) and (2.31), yield the following *electron and ion energy conservation equations*:

$$\begin{aligned}\frac{3}{2}\,\frac{\partial p_e}{\partial t} + \frac{3}{2}\,\mathbf{V}\cdot\nabla p_e + \frac{5}{2}\,p_e\,\nabla\cdot\left(\mathbf{V} + \mathbf{V}_{*\,e} + \mathbf{V}_{*\,e}^{T} + V_{\parallel\,e}^{\rm nc}\,\mathbf{b}\right)&\\ + \nabla\cdot(q_{\parallel\,e}^{\rm nc}\,\mathbf{b}) - \nabla\cdot(n_e\,\chi_{\parallel\,e}\,\nabla_\parallel\,T_e)&\\ - \nabla\cdot(n_e\,\chi_{\perp\,e}\,\nabla_\perp\,T_e) = 0,&\end{aligned} \tag{2.339}$$

$$\begin{aligned}\frac{3}{2}\,\frac{\partial p_i}{\partial t} + \frac{3}{2}\,\mathbf{V}\cdot\nabla p_i + \frac{5}{2}\,p_i\,\nabla\cdot(\mathbf{V} + \mathbf{V}_{*\,i} + \mathbf{V}_{*\,i}^{T}) + \nabla\cdot(q_{\parallel\,i}^{\rm nc}\,\mathbf{b})&\\ - \nabla\cdot(n_e\,\chi_{\parallel\,i}\,\nabla_\parallel\,T_i) - \nabla\cdot(n_e\,\chi_{\perp\,i}\,\nabla_\perp\,T_i) = 0,&\end{aligned} \tag{2.340}$$

where

$$\mathbf{V}_{*\,s}^{T} = \frac{\mathbf{b}\times\nabla T_s}{e_s\,B}, \tag{2.341}$$

$$\begin{aligned}V_{\parallel\,e}^{\rm nc} = \beta_{11}\,f_t\,\frac{1}{e\,n_e\,B_\theta}\,\frac{\partial p}{\partial r} - \beta_{12}\,f_t\,\frac{1}{e\,B_\theta}\,\frac{\partial T_e}{\partial r} - \alpha_1\,\beta_{11}\,f_t\,\frac{1}{e\,B_\theta}\,\frac{\partial T_i}{\partial r}&\\ - (\gamma_{11} - \delta_{11}\,f_t)\,\frac{\sigma_\perp}{e\,n_e}\,E_\parallel,&\end{aligned} \tag{2.342}$$

$$\begin{aligned}\frac{q_{\parallel\,e}^{\rm nc}}{p_e} = \epsilon_1\,f_t\,\frac{1}{e\,n_e\,B_\theta}\,\frac{\partial p}{\partial r} - \epsilon_2\,f_t\,\frac{1}{e\,B_\theta}\,\frac{\partial T_e}{\partial r} - \alpha_1\,\epsilon_1\,f_t\,\frac{1}{e\,B_\theta}\,\frac{\partial T_i}{\partial r}&\\ - (\epsilon_3 - \epsilon_4\,f_t)\,\frac{\sigma_\perp}{e\,n_e}\,E_\parallel,&\end{aligned} \tag{2.343}$$

$$\frac{q_{\parallel\,i}^{\rm nc}}{p_i} = -\frac{5}{2}\,\alpha_2\,f_t\,\frac{1}{e\,B_\theta}\,\frac{\partial T_i}{\partial r}, \tag{2.344}$$

and

$$\epsilon_1 = \frac{3}{2}\,\beta_{11} - \frac{5}{2}\,\beta_{21} = 2.64, \tag{2.345}$$

$$\epsilon_2 = \frac{3}{2}\,\beta_{12} - \frac{5}{2}\,\beta_{22} = 0.338, \tag{2.346}$$

$$\epsilon_3 = \frac{3}{2}\,\gamma_{11} - \frac{5}{2}\,\gamma_{21} = 4.45, \tag{2.347}$$

$$\epsilon_4 = \frac{3}{2}\,\delta_{11} - \frac{5}{2}\,\delta_{21} = 4.89. \tag{2.348}$$

In writing equations (2.339) and (2.258), use has been made of equations (2.139), (2.226)–(2.229), (2.259)–(2.262), (2.322), and (2.323). We have neglected electron–ion energy exchange (i.e. W_i) because this is a small effect, due to the smallness of the mass ratio m_e/m_i (see equation (2.37)). We have also neglected ohmic heating (i.e. $\mathbf{j}\cdot\mathbf{F}_e$), because this effect is $\mathcal{O}[(\rho_e/l_e)/\beta^2] \sim 10^{-4}$ (see tables 1.2 and 2.1) smaller than the leading-order terms. In addition, we have neglected viscous heating because this effect is at least ρ_s/l_s times smaller than the leading-order terms. In accordance with the discussion in section 2.23, we have neglected the parallel component of the thermal heat flux, (2.47). We have also neglected the perpendicular component of the thermal heat flux because this term is the same size as the classical electron perpendicular heat flux (and therefore much smaller than the anomalous electron perpendicular heat flux). The parallel heat 'diffusivities,' $\chi_{\parallel\,e}$ and $\chi_{\parallel\,i}$, are given the collisionless values specified in equations (2.319) and (2.320), respectively. Of course, the values of the anomalous perpendicular heat diffusivities, $\chi_{\perp\,e}$ and $\chi_{\perp\,i}$, must be taken from experiments.

Note, finally, that when combined with the following subset of Maxwell's equations,

$$\nabla\cdot\mathbf{B} = 0, \tag{2.349}$$

$$\nabla\times\mathbf{B} = \mu_0\,\mathbf{j}, \tag{2.350}$$

$$\nabla\times\mathbf{E} = -\frac{\partial\mathbf{B}}{\partial t}, \tag{2.351}$$

our final set of *neoclassical fluid equations*, (2.326), (2.331), and (2.338)–(2.340), form a complete set.

2.25 The normalization of the neoclassical fluid equations

In this section, we shall estimate the relative sizes of the various terms appearing in our neoclassical fluid equations. Let us assume that

$$\nabla_\perp \sim \frac{1}{a}, \tag{2.352}$$

$$\nabla_\parallel \sim \frac{1}{R_0\,q}. \tag{2.353}$$

As before, we shall adopt the drift ordering (2.113). It is helpful to define the following timescales:

$$\tau_s = \frac{a}{v_{t\,i}}, \tag{2.354}$$

$$\tau_{E\,e} = \frac{a^2}{\chi_{\perp\,e}}, \tag{2.355}$$

$$\tau_{\mathrm{E\,i}} = \frac{a^2}{\chi_{\perp\,\mathrm{i}}}, \tag{2.356}$$

$$\tau_{\mathrm{M}} = \frac{a^2}{\Xi_{\perp\,\mathrm{i}}}, \tag{2.357}$$

$$\tau_{\mathrm{R}\,\perp} = \frac{\mu_0\, a^2}{\eta_\perp^{\mathrm{nc}}}, \tag{2.358}$$

$$\tau_{\mathrm{R}\,\parallel} = \frac{\mu_0\, a^2}{\eta_\parallel^{\mathrm{nc}}}, \tag{2.359}$$

$$\tau_{\parallel\,\mathrm{e}} = \frac{(R_0\, q)^2}{\chi_{\parallel\,\mathrm{e}}}, \tag{2.360}$$

$$\tau_{\parallel\,\mathrm{i}} = \frac{(R_0\, q)^2}{\chi_{\parallel\,\mathrm{i}}}. \tag{2.361}$$

Here, τ_{s} is the typical time required for a sound wave to traverse the plasma, $\tau_{\mathrm{E\,e}}$ is the typical time required for electron energy to diffuse out of the plasma, $\tau_{\mathrm{E\,i}}$ is the typical time required for ion energy to diffuse out of the plasma, τ_{M} is the typical time required for momentum to diffuse out of the plasma, $\tau_{\mathrm{R}\,\perp}$ is the typical time required for the perpendicular current to diffuse out of the plasma, $\tau_{\mathrm{R}\,\parallel}$ is the typical time required for the parallel current to diffuse out of the plasma, $\tau_{\parallel\,\mathrm{e}}$ is the typical time required for the electron temperature to attain equilibrium on magnetic flux surfaces, and $\tau_{\parallel\,\mathrm{i}}$ is the typical time required for the ion temperature to attain equilibrium on magnetic flux surfaces. We shall assume that $\tau_{\mathrm{E\,e}} \sim \tau_{\mathrm{E\,i}} \sim \tau_{\mathrm{M}} \sim \tau_\perp$. We shall also assume that

$$\frac{a}{v_{\mathrm{t\,i}}}\,\frac{\partial}{\partial t} \sim \delta_{\mathrm{i}}. \tag{2.362}$$

The previous equation is an extension of the drift ordering, (2.113), which takes into account the fact that tearing modes usually propagate with respect to the MHD fluid at diamagnetic velocities [48].

It is helpful to define the following dimensionless parameters:

$$\Delta_{\mathrm{i}} = \frac{1}{\delta_{\mathrm{i}}^{\,2}}, \tag{2.363}$$

$$\Delta_\theta = \frac{\tau_{\mathrm{s}}}{\tau_{\theta\,\mathrm{i}}\,\delta_{\mathrm{i}}}, \tag{2.364}$$

$$\Delta_\perp = \frac{\tau_s}{\tau_\perp\,\delta_i}, \tag{2.365}$$

$$\Delta_{R\,\perp} = \frac{\beta\,\tau_s}{\tau_{R\,\perp}\,\delta_i}, \tag{2.366}$$

$$\Delta_{R\,\parallel} = f_t\left(\frac{q}{\epsilon}\right)\frac{\beta\,\tau_s}{\tau_{R\,\parallel}\,\delta_i}, \tag{2.367}$$

$$\Delta_{\parallel\,e} = \frac{\tau_s}{\tau_{\parallel\,e}\,\delta_e}, \tag{2.368}$$

$$\Delta_{\parallel\,i} = \frac{\tau_s}{\tau_{\parallel\,i}\,\delta_i}. \tag{2.369}$$

Here, δ_i is the ion magnetization parameter defined in equation (2.110), while β is as defined in equation (1.23). Table 2.6 gives estimates for the dimensionless parameters defined in equations (2.363)–(2.369) for a low-field fusion reactor and a high-field fusion reactor. As before, these estimates are made assuming that $\beta = 0.02$, $\chi_{\perp\,e} = \chi_{\perp\,i} = \Xi_{\perp\,i} = 1\ \mathrm{m^2\,s^{-1}}$, and $q = 3$.

Our neoclassical fluid equations, (2.326), (2.331), and (2.338)–(2.340), can be written

$$\frac{\partial n_e}{\partial t} + \nabla\cdot[n_e\,(\mathbf{V} + \mathbf{V}_{*\,i})] = 0, \tag{2.370}$$

$$\begin{aligned} &m_i\,n_e\,\frac{\partial \mathbf{V}}{\partial t} + m_i\,n_e[(\mathbf{V}\cdot\nabla)\mathbf{V} + (\mathbf{V}_{*\,i}\cdot\nabla)\mathbf{V}_E] + [\Delta_i](\nabla p - \mathbf{j}\times\mathbf{B})\\ &+ [\Delta_\theta]\,\frac{m_i\,n_e}{\tau_{\theta\,i}}(V_{\theta\,i} - V_{\theta\,i}^{nc})\mathbf{e}_\theta - [\Delta_\perp]\,\nabla\cdot(m_i\,n_e\,\Xi_{\perp\,i}\,\mathbf{W}_i) = 0, \end{aligned} \tag{2.371}$$

Table 2.6. Dimensionless parameters used in the normalization of the neoclassical fluid equations in a low-field tokamak reactor and a high-field tokamak reactor. See equations (2.363)–(2.374).

	Low field	High field
B (T)	5.0	12.0
Δ_i	4.35×10^{5}	4.35×10^{5}
Δ_θ	1.17×10^{-3}	2.81×10^{-3}
$\Delta_\perp$	3.57×10^{-4}	8.57×10^{-4}
$\Delta_{R\,\perp}$	7.50×10^{-7}	1.80×10^{-6}
$\Delta_{R\,\parallel}$	4.85×10^{-7}	1.12×10^{-6}
$\Delta_{\parallel\,e}$	2.80×10^{3}	2.80×10^{3}
$\Delta_{\parallel\,i}$	4.14×10^{1}	4.14×10^{1}

$$\mathbf{E} + \mathbf{V} \times \mathbf{B} + \frac{1}{e\, n_e}(\nabla p - \nabla_\parallel p_i - \mathbf{j} \times \mathbf{B}) = [\Delta_{R\,\perp}]\, \eta_\perp^{nc}\left(\mathbf{j}_\perp - \frac{\mathbf{b} \times \nabla p}{B}\right) \tag{2.372}$$
$$+ [\Delta_{R\,\parallel}]\, \eta_\parallel^{nc}\, (j_\parallel - j_\parallel^{nc})\, \mathbf{b},$$

$$\frac{3}{2}\frac{\partial p_e}{\partial t} + \frac{3}{2}\mathbf{V} \cdot \nabla p_e + \frac{5}{2} p_e\, \nabla \cdot \left(\mathbf{V} + \mathbf{V}_{*\,e} + \mathbf{V}_{*\,e}^T + V_{\parallel\,e}^{nc}\, \mathbf{b}\right) + \nabla \cdot (q_{\parallel\,e}^{nc}\, \mathbf{b}) \tag{2.373}$$
$$- [\Delta_{\parallel\,e}]\, \nabla \cdot (n_e\, \chi_{\parallel\,e}\, \nabla_\parallel T_e) - [\Delta_\perp]\, \nabla \cdot (n_e\, \chi_{\perp\,e}\, \nabla_\perp T_e) = 0,$$

$$\frac{3}{2}\frac{\partial p_i}{\partial t} + \frac{3}{2}\mathbf{V} \cdot \nabla p_i + \frac{5}{2} p_i\, \nabla \cdot (\mathbf{V} + \mathbf{V}_{*\,i} + \mathbf{V}_{*\,i}^T) + \nabla \cdot (q_{\parallel\,i}^{nc}\, \mathbf{b}) \tag{2.374}$$
$$- [\Delta_{\parallel\,i}]\nabla \cdot (n_e\, \chi_{\parallel\,i}\, \nabla_\parallel T_i) - [\Delta_\perp]\nabla \cdot (n_e\, \chi_{\perp\,i}\, \nabla_\perp T_i) = 0.$$

Here, the factors $[\Delta_i]$, $[\Delta_\theta]$, $[\Delta_\perp]$, etc. indicate that the terms they precede are larger or smaller than terms preceded by no factor by the dimensionless parameter contained within the square brackets.

According to table 2.6, the dominant parallel diffusivity term in the electron energy conservation equation, (2.373), yields

$$\nabla_\parallel T_e \simeq 0. \tag{2.375}$$

In other words, the parallel electron energy diffusivity in a tokamak fusion reactor is sufficiently large to ensure that the electron temperature is uniform on magnetic flux surfaces. Likewise, according to table 2.6, the dominant parallel diffusivity term in the ion energy conservation equation, (2.374), gives

$$\nabla_\parallel T_i \simeq 0. \tag{2.376}$$

In other words, the parallel ion energy diffusivity in a tokamak fusion reactor is sufficiently large to ensure that the ion temperature is uniform on magnetic flux surfaces. According to table 2.6, the dominant terms in the plasma equation of motion, (2.371), yield

$$\mathbf{j} \times \mathbf{B} - \nabla p \simeq 0. \tag{2.377}$$

In other words, the plasma in a tokamak fusion reactor exists in a state of approximate force balance. The previous equation suggests that $\mathbf{B} \cdot \nabla p = 0$. When combined with equations (2.375) and (2.376), this relation gives

$$\nabla_\parallel n_e = \nabla_\parallel p_e = \nabla_\parallel p_i \simeq 0. \tag{2.378}$$

We conclude that the electron number density, the electron pressure, and the ion pressure are all uniform on magnetic flux surfaces in a tokamak fusion reactor. According to table 2.6, the dominant terms in the plasma Ohm's law, (2.372), yield

$$\mathbf{E} + \mathbf{V} \times \mathbf{B} \simeq 0, \tag{2.379}$$

where use has been made of equations (2.377) and (2.378). Thus, we conclude that the plasma in a tokamak fusion reactor satisfies the so-called *perfect conductivity* or *flux-freezing* constraint. As is well known, this constraint forbids any change in the topology of magnetic field lines [1]. Finally, according to table 2.6, the dominant term in the electron number density continuity equation, (2.370), gives

$$\frac{\partial n_e}{\partial t} + \nabla \cdot [n_e (\mathbf{V} + \mathbf{V}_{*\,i})] \simeq 0. \tag{2.380}$$

Equations (2.375)–(2.380) are known collectively as the equations of *marginally stable ideal MHD* [22].

2.26 Discussion

Our derivation of the neoclassical fluid equations, (2.370)–(2.374), is premised on the following assumptions:

- The plasma is axisymmetric.
- The plasma is in the banana collisionality regime.
- The drift ordering (2.113) holds good.
- The transport ordering (2.114) holds good.
- Plasma turbulence does not affect force and heat balance parallel to magnetic field lines.

These assumptions are all fairly reasonable when discussing the evolution of an axisymmetric tokamak plasma equilibrium.

In this book, however, we intend to use the neoclassical fluid equations, (2.370)–(2.374), to describe the evolution of a tokamak plasma equilibrium that is perturbed by a non-axisymmetric tearing mode. This approach requires some justification. The first justification is that tearing modes in tokamak plasmas usually saturate at a fairly low amplitudes (typically $\delta B/B \sim 10^{-4}$) [16]. Hence, the departure from axisymmetry associated with the tearing modes is comparatively small. The second justification is that tearing modes do not affect the plasma collisionality, and, in particular, do not significantly modify particle trapping. The third justification is that tearing modes are sufficiently slow growing that they do not give rise to perturbed E-cross-**B** flows that are large enough to violate the drift ordering [21, 47]. It is harder to argue that tearing modes are sufficiently slow growing that they do not violate the transport ordering. It is certainly the case that large-amplitude tearing modes grow on timescales that are comparable to a transport timescale (because such modes are effectively slightly helical plasma equilibria) [49]. However, low-amplitude tearing modes can grow on somewhat shorter timescales [50]. Furthermore, diamagnetic effects cause tearing modes to propagate with respect to the MHD fluid [48], which is the origin of the modified ordering (2.362). We are essentially hoping that this slightly faster time evolution does not invalidate our analysis. The final justification is that, to a first approximation, tearing modes do not modify, and are not directly affected by, plasma turbulence. Of course, this is not strictly true, because a sufficiently large-amplitude tearing mode can flatten the density and pressure profiles and thereby stabilize micro-instabilities—which has the effect of reducing turbulent transport—in the immediate vicinity of its rational surface [51–54]. Under certain circumstance, turbulence can also directly influence tearing-mode evolution [53, 55, 56]. Unfortunately, the mutual interaction between plasma turbulence and a tearing mode is almost impossible to model accurately with fluid equations, and will, therefore, be neglected in this book.

As we saw in the previous section, under normal circumstances the neoclassical fluid equations, (2.370)–(2.374), reduce to the much simpler equations of marginally stable ideal MHD, (2.375)–(2.380). However, as we shall demonstrate in the next chapter, when applied to tearing-mode dynamics, the equations of marginally stable ideal MHD become singular at the associated rational magnetic flux surface (see section 3.7). The singularity is resolved in a thin current sheet characterized by $\nabla_\perp \gg 1/a$ and $\nabla_\parallel \ll 1/(R_0\, q)$ (see equations (2.352) and (2.353)). When we reorder the neoclassical fluid equations using appropriate estimates for $\nabla_\perp$ and $\nabla_\parallel$ within the layer, we shall discover that most of the terms in the equations need to be retained.

References

[1] Fitzpatrick R 2014 *Plasma Physics: An Introduction* (Boca Raton, FL: CRC Press)

[2] Reif F 1965 *Fundamentals of Statistical and Thermal Physics* (New York: McGraw-Hill)

[3] Brizard A J and Hahm T S 2007 Foundations of nonlinear gyrokinetic theory *Rev. Mod. Phys.* **2** 421

[4] Catto P J 1978 Linearized gyro-kinetics *Plasma Phys.* **20** 719

[5] Hahm T S 1988 Nonlinear gyrokinetic equations for tokamak microturbulence *Phys. Fluids* **31** 2670

[6] Garbet X, Idomura Y, Villard L and Watanabe T H 2010 Gyrokinetic simulations of turbulent transport *Nucl. Fusion* **50** 043002

[7] Kotschenreuther M, Dorland W, Beer M A and Hammett G W 1995 Quantitative predictions of tokamak energy confinement from first-principles simulations with kinetic effects *Phys. Plasmas* **2** 2381

[8] Waltz R E, Candy J and Rosenbluth M N 2002 Gyrokinetic turbulence simulation of profile shear stabilization and broken gyrobohm scaling *Phys. Plasmas* **9** 1938

[9] Richardson A S 2019 *NRL Plasma Formulary* (Washington, DC: Naval Research Laboratory) https://www.nrl.navy.mil/Portals/38/PDF%20Files/NRL_Formulary_2019.pdf

[10] Grad H 1958 Principles of the kinetic theory of gases *Handbuch der Physik* vol 3 (Berlin: Springer) pp 205–94

[11] Hunana P, Passot T, Khomenko E, Martínez-Gómez D, Collados M, Tenerani A, Zank G P, Maneva Y, Goldstein M L and Webb G M 2022 Generalized fluid models of the Braginskii type *Astrophys. J. Suppl. Ser.* **260** 26

[12] Chapman S and Cowling T G 1953 *The Mathematical Theory of Non-Uniform Gases* 2nd edn (Cambridge: Cambridge University Press)

[13] Braginskii S I 1965 *Transport processes in a plasma Reviews of Plasma Physics* vol 1 (New York: Consultants Bureau) p 205

[14] Spitzer L Jr and Härm R 1953 Transport phenomena in a completely ionized gas *Phys. Rev. Lett.* **89** 977

[15] Helander P and Sigmar D J 2002 *Collisional Transport in Magnetized Plasmas* (Cambridge: Cambridge University Press) https://www.cambridge.org/gb/academic/subjects/physics/plasma-physics-and-fusion-physics/collisional-transport-magnetized-plasmas?format=HB
[16] Wesson J A 2011 *Tokamaks* 4th edn (Oxford: Oxford University Press) https://global.oup.com/academic/product/tokamaks-9780199592234
[17] Kruskal M D 1962 Asymptotic theory of Hamiltonian and other systems with all solutions nearly periodic *J. Math. Phys.* **3** 806
[18] Abramowitz M and Stegun I A (ed) 1964 *Handbook of Mathematical Functions with Formulas, Graphs, and Mathematical tables* (New York: Dover) https://store.doverpublications.com/0486612724.html
[19] Hirschman S P and Sigmar D J 1981 Neoclassical transport of impurities in tokamak plasmas *Nucl. Fusion* **21** 1079
[20] Hinton F L and Hazeltine R D 1976 Theory of plasma transport in toroidal confinement systems *Rev. Mod. Phys.* **48** 239
[21] Hazeltine R D and Meiss J D 2003 *Plasma Confinement* (New York: Dover)
[22] Freidberg J P 1987 *Ideal Magnetohydrodynamics* (New York: Plenum)
[23] Chew G F, Goldberger M L and Low F E 1956 The Boltzmann equation and the one-fluid hydromagnetic equations in the absence of particle collisions *Proc. R. Soc.* A **236** 112
[24] Fitzpatrick R 2017 *Theoretical Fluid Mechanics* (Bristol: IOP Publishing)
[25] Kim Y B, Diamond P H and Groebner R J 1991 Neoclassical poloidal and toroidal rotation in tokamaks *Phys. Fluids* B **3** 2050
[26] Galeev A A and Sagdeev R Z 1968 Transport phenomena in a collisionless plasma in a toroidal magnetic system *Sov. Phys. - JETP* **26** 233 http://jetp.ras.ru/cgi-bin/e/index/e/26/1/p233?a=list
[27] Rosenbluth M N, Hazeltine R D and Hinton F L 1972 Plasma transport in toroidal confinement systems *Phys. Fluids* **15** 116
[28] Stix T H 1973 Decay of poloidal rotation in a tokamak plasma *Phys. Fluids* **16** 1260
[29] Hinton F L and Oberman C 1969 Electrical conductivity of plasma in a spatially inhomogeneous magnetic field *Nucl. Fusion* **9** 319
[30] Etzweiler J F and Brouchous D A 1980 Observations of neoclassical and anomalous resistivity in toroidal discharges *Phys. Fluids* **23** 2547
[31] Kikuchi M, Azumi M, Tsuji D, Tani K and Kubo H 1990 Bootstrap current during perpendicular neutral injection in JT-60 *Nucl. Fusion* **30** 343
[32] Zarnstorff M C, McGuire K, Bell M G, Grek B, Johnson D, McCune H, Park A and Taylor G 1990 Parallel electric resistivity in the TFTR tokamak *Phys. Fluids* B **2** 1852
[33] Bickerton R J, Connor J W and Taylor J B 1971 Diffusion driven plasma currents and bootstrap tokamak *Nat. Phys. Sci.* **229** 110
[34] Zarnstorff M C and Prager S C 1984 Experimental observation of neoclassical currents in a plasma *Phys. Rev. Lett.* **53** 454
[35] Zarnstorff M C *et al* 1988 Bootstrap current in TFTR *Phys. Rev. Lett.* **60** 1306
[36] Rutherford P H 1970 Collisional diffusion in an axisymmetric torus *Phys. Plasmas* **13** 482
[37] Helander P and Simakov A N 2008 Intrinsic ambipolarity and rotation in stellarators *Phys. Rev. Lett.* **101** 145003
[38] Ware A A 1970 Pinch effect for trapped particles in a tokamak *Phys. Rev. Lett.* **25** 15
[39] Fick A 1855 Über diffusion *Ann. Phys., Lpz.* **94** 59

[40] Wooton A J, Carreras B A, Matsumoto H, McGuire K, Peebles W A, Ritz C P, Terry P W and Zweben S J 1990 Fluctuations and anomalous transport in tokamaks *Phys. Fluids* B **2** 2879

[41] Connor J W and Wilson H R 1994 Survey of theories of anomalous transport *Plasma Phys. Control. Fusion* **36** 719

[42] Calvo I and Parra F I 2012 Long-wavelength limit of gyrokinetics in a turbulent tokamak and its intrinsic ambipolarity *Plasma Phys. Control. Fusion* **54** 115007

[43] Waltz R E 1982 Magnetic fluctuations, ambipolarity, charge filamentation, and plasma rotation in tokamaks *Phys. Plasmas* **25** 1269

[44] Hammett B W and Perkins F W 1990 Fluid moment models for Landau damping with application to the ion-temperature-gradient instability *Phys. Rev. Lett.* **64** 3019

[45] Hazeltine R D 1998 Transport in the collisionless limit *Phys. Plasmas* **5** 3282

[46] Fitzpatrick R 1995 Helical temperature perturbations associated with tearing modes in tokamak plasmas *Phys. Plasmas* **2** 825

[47] Hazeltine R D and Meiss J D 1985 Shear-Alfvén Dynamics of toroidally confined plasmas *Phys. Rep.* **121** 1

[48] Ara G, Basu B, Coppi B, Laval G, Rosenbluth M N and Waddell B V 1978 Magnetic reconnection and $m = 1$ oscillations in current carrying plasmas *Ann. Phys., NY* **112** 443

[49] Rutherford P H 1973 Nonlinear growth of the tearing mode *Phys. Fluids* **16** 1906

[50] Furth H P, Killeen J and Rosenbluth M N 1963 Finite-resistivity instabilities of a sheet pinch *Phys. Fluids* **6** 459

[51] Bardóczi L, Rhodes T L, Carter T A, Crocker N A, Peebles W A and Grierson B A 2016 Non-perturbative measurement of cross-field thermal diffusivity reduction at the O-point of 2/1 neoclassical tearing mode islands in the DIII-D tokamak *Phys. Plasmas* **23** 052507

[52] Bardóczi L, Rhodes T L, Bañón Navarro A, Sung C, Carter T A, La Haye R J, McKee G R, Petty C C, Chrystal C and Jenko F 2017 Multi-field/-scale interactions of turbulence with neoclassical tearing mode magnetic islands in the DIII-D tokamak *Phys. Plasmas* **24** 056106

[53] Choi M J 2021 Interaction between a magnetic island and turbulence *Rev. Mod. Plasma Phys.* **5** 9

[54] Hornsby W A, Peeters A G, Snodin A P, Casson F J, Camenen Y, Szepesi G, Siccinio M and Poli E 2010 The nonlinear coupling between gyroradius scale turbulence and mesoscale magnetic islands in fusion plasmas *Phys. Plasmas* **17** 092301

[55] Choi M J *et al* 2021 Effects of plasma turbulence on the nonlinear evolution of magnetic island in tokamak *Nat. Commun.* **12** 375

[56] Sen A, Singh R, Chandra D, Kaw P and Raju D 2009 ETG turbulence effects on the evolution of an NTM *Nucl. Fusion* **49** 115012

IOP Publishing

Tearing Mode Dynamics in Tokamak Plasmas

Richard Fitzpatrick

Chapter 3

Cylindrical tearing-mode theory

3.1 Introduction

The aim of this chapter is to describe the simplest theory of tearing-mode dynamics in tokamak plasmas, according to which, the plasma equilibrium is approximated by a periodic cylinder.

3.2 Cylindrical tokamak equilibrium

Consider a low-β, large-aspect-ratio, tokamak plasma equilibrium whose magnetic flux surfaces map out (almost) concentric circles in the poloidal plane. Such an equilibrium can be approximated by a periodic cylinder [1, 2]. Let us employ a conventional set of right-handed cylindrical coordinates, r, θ, z. The equilibrium magnetic flux surfaces lie on surfaces of constant r. The system is assumed to be periodic in the z ('toroidal') direction and has a periodicity length of $2\pi\, R_0$, where R_0 is the simulated major radius of the plasma. Let a be the minor radius of the plasma. The equilibrium magnetic field is written

$$\mathbf{B} = B_\theta(r)\,\mathbf{e}_\theta + B_z(r)\,\mathbf{e}_z, \tag{3.1}$$

where $B_\theta(r)$ is the poloidal magnetic field strength, and $B_z(r)$ the toroidal magnetic field strength. Here, $\mathbf{e}_\theta \equiv \nabla\theta/|\nabla\theta|$ and $\mathbf{e}_z \equiv \nabla z/|\nabla z|$. The safety-factor profile takes the form

$$q(r) = \frac{r\,B_z(r)}{R_0\,B_\theta(r)}. \tag{3.2}$$

(See equation (1.76).) It is assumed that $q \sim \mathcal{O}(1)$. The equilibrium current density is written

$$\mathbf{j} = j_\theta(r)\,\mathbf{e}_\theta + j_z(r)\,\mathbf{e}_z, \tag{3.3}$$

where the poloidal and toroidal current densities take the respective forms

$$\mu_0\, j_\theta = -B_z', \tag{3.4}$$

$$\mu_0\, j_z = \frac{(r\, B_\theta)'}{r}, \tag{3.5}$$

and $'$ denotes d/dr. The plasma equilibrium satisfies the force balance criterion (see equation (2.377)),

$$\mathbf{j} \times \mathbf{B} - \nabla p = 0, \tag{3.6}$$

where $p(r)$ is the total plasma pressure. It follows from equations (3.1) and (3.3)–(3.5) that

$$\frac{d}{dr}\left(p + \frac{B_\theta^{\,2} + B_z^{\,2}}{2\,\mu_0}\right) + \frac{B_\theta^{\,2}}{\mu_0\, r} = 0. \tag{3.7}$$

3.3 Magnetic field and current density perturbations

Consider a tearing-mode perturbation that has m periods in the poloidal direction and n periods in the toroidal direction, where $m > 0$, $n > 0$, and $m \sim n \sim \mathcal{O}(1)$. We shall assume that all perturbed scalar and vector quantities vary as

$$\delta A(r,\, \theta,\, z,\, t) = \delta A(r,\, t)\, \exp[\mathrm{i}\,(m\,\theta - n\,\varphi)], \tag{3.8}$$

$$\delta \mathbf{A}(r,\, \theta,\, z,\, t) = \delta \mathbf{A}(r,\, t)\, \exp[\mathrm{i}\,(m\,\theta - n\,\varphi)], \tag{3.9}$$

respectively, where $\varphi = z/R_0$ is a simulated toroidal angle.

Given that tearing modes in tokamak plasmas are relatively low amplitude (i.e. $\delta B/B \sim 10^{-4}$) [3], global (i.e. $\nabla_\perp \sim 1/a$) (see equation (2.352)), relatively slow growing (i.e. $\partial/\partial t \sim \delta_\mathrm{i}\, v_{\mathrm{t\,i}}/a$) (see equation (2.362)) instabilities, it follows from the analysis of section 2.25 that they are governed by the linearized forms of the equations of marginally stable ideal magnetohydrodynamics (MHD), (2.375)–(2.380). In particular, the linearized form of the curl of the force balance criterion, (2.377), combined with the linearized forms of Maxwell's equations, (2.349)–(2.351), give

$$(\mathbf{B}\cdot\nabla)\,\delta\mathbf{j} + (\delta\mathbf{B}\cdot\nabla)\,\mathbf{j} - (\mathbf{j}\cdot\nabla)\,\delta\mathbf{B} - (\delta\mathbf{j}\cdot\nabla)\,\mathbf{B} = 0, \tag{3.10}$$

$$\nabla\cdot\delta\mathbf{B} = 0, \tag{3.11}$$

$$\mu_0\,\delta\mathbf{j} = \nabla\times\delta\mathbf{B}, \tag{3.12}$$

where $\delta\mathbf{B}$ and $\delta\mathbf{j}$ are the perturbed magnetic field and current density, respectively. Equations (3.1), (3.3), and (3.9)–(3.12) yield

$$\mathrm{i}\,F\,r\,\delta j_r + \mathrm{i}\,(n\,\epsilon\,j_z - m\,j_\theta)\,\delta B_r = 0, \tag{3.13}$$

$$\mathrm{i}\,F\,r\,\delta j_z - B_z'\,r\,\delta j_r + r\,j_z'\,\delta B_r + \mathrm{i}\,(n\,\epsilon\,j_z - m\,j_\theta)\,\delta B_z = 0, \tag{3.14}$$

and

$$(r\,\delta B_r)' + \mathrm{i}\,m\,\delta B_\theta - \mathrm{i}\,n\,\epsilon\,\delta B_z = 0, \tag{3.15}$$

with

$$\mu_0\,r\,\delta j_r = \mathrm{i}\,m\,\delta B_z + \mathrm{i}\,n\,\epsilon\,\delta B_\theta, \tag{3.16}$$

$$\mu_0\,r\,\delta j_z = (r\,\delta B_\theta)' - \mathrm{i}\,m\,\delta B_r, \tag{3.17}$$

where

$$\epsilon(r) = \frac{r}{R_0}, \tag{3.18}$$

$$F(r) = \frac{m}{r}\,B_\theta - \frac{n}{R_0}\,B_z. \tag{3.19}$$

If we write

$$\delta B_r = \mathrm{i}\,\frac{m}{r}\,\delta\psi(r,\,t), \tag{3.20}$$

then, after some algebra, equations (3.13)–(3.17) reduce to

$$\delta B_\theta = -\frac{m^2}{m^2 + (n\,\epsilon)^2}\,\delta\psi' - \frac{m}{r}\,\frac{n\,\epsilon}{m^2 + (n\,\epsilon)^2}\,\frac{\mu_0\,(n\,\epsilon\,j_z - m\,j_\theta)}{F}\,\delta\psi, \tag{3.21}$$

$$\delta B_z = \frac{m\,(n\,\epsilon)}{m^2 + (n\,\epsilon)^2}\,\delta\psi' - \frac{m}{r}\,\frac{m}{m^2 + (n\,\epsilon)^2}\,\frac{\mu_0\,(n\,\epsilon\,j_z - m\,j_\theta)}{F}\,\delta\psi, \tag{3.22}$$

and [4, 5]

$$\frac{1}{r}\,\frac{\partial}{\partial r}\!\left(f\,r\,\frac{\partial\delta\psi}{\partial r}\right) - g\,\delta\psi = 0, \tag{3.23}$$

where

$$f(r) = \frac{m^2}{m^2 + (n\,\epsilon)^2}, \tag{3.24}$$

$$\begin{aligned} g(r) = \frac{m}{r}\Bigg\{&\frac{m}{r} + \frac{\mu_0\,j_z'}{F} - \left[\frac{n\,\epsilon}{m^2 + (n\,\epsilon)^2}\,\frac{\mu_0\,(n\,\epsilon\,j_z - m\,j_\theta)}{F}\right]' \\ &- \frac{m}{r}\,\frac{n\,\epsilon}{m^2 + (n\,\epsilon)^2}\,\frac{\mu_0\,(n\,\epsilon\,j_z - m\,j_\theta)}{F}\,\frac{\mu_0\,(m\,j_z + n\,\epsilon\,j_\theta)}{F}\Bigg\}. \end{aligned} \tag{3.25}$$

A global tearing instability in a low-β, large-aspect-ratio, tokamak plasma is characterized by [4]

$$\frac{\epsilon}{q} = \frac{B_\theta}{B_z} \ll 1, \tag{3.26}$$

$$\beta \equiv \frac{2\,\mu_0\,p}{B_z^{\,2}} \sim \mathcal{O}\!\left(\frac{\epsilon}{q}\right)^2, \tag{3.27}$$

$$\frac{n\,\epsilon}{m} \sim \mathcal{O}\!\left(\frac{\epsilon}{q}\right). \tag{3.28}$$

It follows from equations (3.4), (3.5), (3.7), and (3.19) that

$$\frac{r\,B_z'}{B_z} \sim \mathcal{O}\!\left(\frac{\epsilon}{q}\right)^2, \tag{3.29}$$

$$\frac{j_\theta}{j_z} \sim \mathcal{O}\!\left(\frac{\epsilon}{q}\right), \tag{3.30}$$

$$\frac{\mu_0\,j_z}{F} \sim \mathcal{O}(1). \tag{3.31}$$

Thus, in the low-β, large-aspect-ratio limit, equations (3.20)–(3.25) simplify considerably to give

$$\delta B_r = \mathrm{i}\,\frac{m}{r}\,\delta\psi, \tag{3.32}$$

$$\delta B_\theta \simeq -\delta\psi', \tag{3.33}$$

$$\delta B_z \simeq \frac{n\,\epsilon}{m}\,\delta\psi' - \frac{\mu_0\,(n\,\epsilon\,j_z - m\,j_\theta)}{r\,F}\,\delta\psi, \tag{3.34}$$

and

$$\frac{1}{r}\frac{\partial}{\partial r}\!\left(r\,\frac{\partial\delta\psi}{\partial r}\right) - \frac{m^2}{r^2}\,\delta\psi - \frac{m\,\mu_0\,j_z'}{r\,F}\,\delta\psi \simeq 0. \tag{3.35}$$

It can also easily be demonstrated that

$$\mu_0\,\delta j_r \simeq \mathrm{i}\,\frac{m}{r}\frac{\mu_0}{F}\!\left(\frac{m}{r}\,j_\theta - \frac{n}{R_0}\,j_z\right)\!\delta\psi, \tag{3.36}$$

$$\mu_0\,\delta j_\theta \simeq \frac{m}{r}\frac{n}{R_0}\,\delta\psi - \frac{n}{R_0\,m}\,(r\,\delta\psi')' - \left[\frac{\mu_0}{F}\!\left(\frac{m}{r}\,j_\theta - \frac{n}{R_0}\,j_z\right)\!\delta\psi\right]', \tag{3.37}$$

$$\mu_0\,\delta j_z \simeq \frac{m^2}{r^2}\,\delta\psi - \frac{1}{r}\,(r\,\delta\psi')'. \tag{3.38}$$

Hence, we conclude that the magnetic field and current density perturbations associated with a tearing mode in a low-β, large-aspect-ratio, tokamak plasma are specified by equations (3.32)–(3.38). From now on, we shall treat B_z as approximately independent of r, in accordance with equation (3.29).

3.4 Density and temperature perturbations

According to the equations of marginally stable ideal MHD, (2.375)–(2.380), the electron number density, the electron temperature, and the ion temperature all satisfy equations of the form

$$\nabla_{\|}A \propto \mathbf{B}\cdot\nabla A = 0. \tag{3.39}$$

Linearization of the previous equation yields

$$\delta\mathbf{B}\cdot\nabla A + \mathbf{B}\cdot\nabla\delta A = 0, \tag{3.40}$$

where $A(r)$ denotes an equilibrium quantity. It follows from equations (3.8), (3.19), and (3.32) that

$$\delta A = -\frac{m}{r}\,\frac{A'}{F}\,\delta\psi. \tag{3.41}$$

More explicitly, we conclude that the perturbations in the electron number density, the electron temperature, and the ion temperature that are associated with a tearing mode in a low-β, large-aspect-ratio, tokamak plasma take the respective forms

$$\delta n_e = -\frac{m}{r}\,\frac{n_e{}'}{F}\,\delta\psi, \tag{3.42}$$

$$\delta T_e = -\frac{m}{r}\,\frac{T_e'}{F}\,\delta\psi, \tag{3.43}$$

$$\delta T_i = -\frac{m}{r}\,\frac{T_i'}{F}\,\delta\psi. \tag{3.44}$$

3.5 Fluid continuity

According to the equations of marginally stable ideal MHD, the electron number density continuity equation takes the form

$$\frac{\partial n_e}{\partial t} + \nabla\cdot\left(n_e\,\mathbf{V} + \frac{\mathbf{b}\times\nabla p_i}{e\,B_z}\right) = 0, \tag{3.45}$$

where use has been made of equations (2.139), (2.380), and (3.26). Here, $\mathbf{b} \equiv \mathbf{B}/B \simeq (\epsilon/q)\, \mathbf{e}_\theta + \mathbf{e}_z$. It follows that $\nabla\cdot\mathbf{b} = 0$ and $r\,\nabla\times\mathbf{b} \sim \mathcal{O}(\epsilon/q)$. Thus,

$$\nabla\cdot\left(\frac{\mathbf{b}\times\nabla p_{\rm i}}{e\,B_z}\right) \simeq 0, \tag{3.46}$$

where use has been made of equation (3.29).

The equilibrium plasma flow is written

$$\mathbf{V} = V_\theta(r)\,\mathbf{e}_\theta + V_z(r)\,\mathbf{e}_z. \tag{3.47}$$

It follows that

$$\nabla\cdot\mathbf{V} = 0. \tag{3.48}$$

The linearized form of equation (3.45) is

$$\frac{\partial \delta n_{\rm e}}{\partial t} + \mathbf{V}\cdot\nabla\delta n_{\rm e} + \delta\mathbf{V}\cdot\nabla n_{\rm e} + n_{\rm e}\,\nabla\cdot\delta\mathbf{V} = 0, \tag{3.49}$$

where $\delta\mathbf{V}$ is the perturbed plasma velocity, and use has been made of equations (3.46) and (3.48). Let us write

$$\delta\mathbf{V} = \nabla\phi\times\mathbf{e}_z + \delta v\,\mathbf{B}. \tag{3.50}$$

It follows that

$$\nabla\cdot\delta\mathbf{V} = \mathbf{B}\cdot\nabla\delta v = {\rm i}\,F\,\delta v, \tag{3.51}$$

where use has been made of equations (3.8) and (3.19). Hence, writing

$$\frac{\partial}{\partial t} = -{\rm i}\,\omega, \tag{3.52}$$

$$\Omega(r) = \frac{m}{r}\,V_\theta - \frac{n}{R_0}\,V_\phi, \tag{3.53}$$

equation (3.49) reduces to

$$\delta v = -\frac{m}{r\,F}\left[\phi + \left(\frac{\omega-\Omega}{F}\right)\delta\psi\right]\frac{n_{\rm e}'}{n_{\rm e}}, \tag{3.54}$$

where use has been made of equation (3.42).

3.6 Velocity perturbation

The flux-freezing constraint of marginally stable ideal MHD takes the form

$$\mathbf{E} + \mathbf{V}\times\mathbf{B} = 0. \tag{3.55}$$

(See equation (2.379).) Taking the curl of the previous equation, combining with Maxwell's equations, and linearizing, we obtain

$$\frac{\partial \delta \mathbf{B}}{\partial t} = (\delta \mathbf{B} \cdot \nabla)\, \mathbf{V} - (\mathbf{V} \cdot \nabla)\, \delta \mathbf{B} + (\mathbf{B} \cdot \nabla)\, \delta \mathbf{V}_{\perp} - (\delta \mathbf{V}_{\perp} \cdot \nabla)\, \mathbf{B}, \tag{3.56}$$

where

$$\delta \mathbf{V}_{\perp} = \nabla \phi \times \mathbf{e}_z, \tag{3.57}$$

and use has been made of equation (3.50) as well as the facts that $\nabla \cdot \mathbf{B} = \nabla \cdot \delta \mathbf{B} = \nabla \cdot \mathbf{V} = \nabla \cdot \delta \mathbf{V}_{\perp} = 0$. The radial component of the previous equation yields

$$\phi = -\left(\frac{\omega - \Omega}{F}\right) \delta \psi, \tag{3.58}$$

where use has been made of equations (3.9), (3.19), (3.32), (3.52), and (3.53). Equations (3.54) and (3.58) imply that

$$\delta v = 0. \tag{3.59}$$

Hence, we deduce that the perturbed plasma flow associated with a tearing mode in a low-β, large-aspect-ratio, tokamak plasma is divergence free (see equations (3.51) and (3.59)) and is specified by equations (3.57) and (3.58).

3.7 The cylindrical tearing-mode equation

The magnetic structure of a tearing perturbation is determined by the so-called *cylindrical tearing-mode equation*, (3.35), which can be written in the form [1, 2]

$$\frac{\partial^2 \delta \psi}{\partial r^2} + \frac{1}{r} \frac{\partial \delta \psi}{\partial r} - \frac{m^2}{r^2} \delta \psi - \frac{J_z' \, \delta \psi}{r\,(1/q - n/m)} = 0, \tag{3.60}$$

where

$$J_z(r) = \frac{R_0 \, \mu_0 \, j_z(r)}{B_z} \tag{3.61}$$

is a dimensionless measure of the toroidal current density profile. Note that equation (3.60) is singular at the so-called *rational* magnetic flux surface, radius $r = r_s$, at which

$$q(r_s) = \frac{m}{n}. \tag{3.62}$$

At the rational surface, $\mathbf{k} \cdot \mathbf{B} = 0$, where $\mathbf{B}$ is the equilibrium magnetic field and $\mathbf{k} = (k_r, m/r, -n/R_0)$ is the wavevector of the tearing perturbation.

3.8 The solution in the presence of a perfectly conducting wall

Suppose that the plasma occupies the region $0 \leqslant r \leqslant a$, where a is the plasma's minor radius. It follows that $p(r) = j_\theta(r) = j_z(r) = J_z(r) = 0$ for $r > a$. Let the plasma be surrounded by a concentric, rigid, radially thin, perfectly conducting wall of radius $r_w > a$. (In most circumstances, the wall represents the metallic vacuum vessel that surrounds the plasma.) An appropriate physical solution of the cylindrical tearing-mode equation, (3.60), takes the separable form

$$\delta\psi(r, t) = \Psi_s(t)\, \hat{\psi}_s(r), \tag{3.63}$$

where the real function $\hat{\psi}_s(r)$ is a solution of

$$\frac{d^2\hat{\psi}_s}{dr^2} + \frac{1}{r}\frac{d\hat{\psi}_s}{dr} - \frac{m^2}{r^2}\hat{\psi}_s - \frac{J_z'\, \hat{\psi}_s}{r\,(1/q - n/m)} = 0 \tag{3.64}$$

that satisfies

$$\hat{\psi}_s(0) = 0, \tag{3.65}$$

$$\hat{\psi}_s(r_s) = 1, \tag{3.66}$$

$$\hat{\psi}_s(r \geqslant r_w) = 0. \tag{3.67}$$

Equation (3.65) ensures that the perturbed magnetic field associated with the tearing mode remains finite at the magnetic axis ($r = 0$), while equation (3.67) represents the physical constraint that the perturbed magnetic field cannot penetrate a perfectly conducting wall.

Let $\rho = (r - r_s)/r_s$. The solution of equation (3.64) in the vicinity of the rational surface is

$$\hat{\psi}_s(\rho) = 1 + \Delta_{s+}\, \rho + \alpha_s\, \rho\, \ln|\rho| + \mathcal{O}(\rho^2) \tag{3.68}$$

for $\rho > 0$, and

$$\hat{\psi}_s(\rho) = 1 + \Delta_{s-}\, \rho + \alpha_s\, \rho\, \ln|\rho| + \mathcal{O}(\rho^2) \tag{3.69}$$

for $\rho < 0$. Here,

$$\alpha_s = -\left(\frac{r\, q\, J_z'}{s}\right)_{r=r_s}, \tag{3.70}$$

$$s(r) = \frac{r\, q'}{q}. \tag{3.71}$$

Moreover, the real parameters Δ_{s+} and Δ_{s-} are fully determined by equation (3.64) and the boundary conditions (3.65)–(3.67). Note that, in general, $\delta\psi \propto \delta B_r$ is continuous across the rational surface (in accordance with Maxwell's equations), whereas $\partial\delta\psi/\partial r$ is discontinuous. The discontinuity in $\partial\delta\psi/\partial r$ implies the presence of a radially thin, helical current sheet at the rational surface. This current sheet is resolved in a thin resistive layer that, in principle, can only be described by employing the full set of neoclassical fluid equations, (2.370)–(2.374), rather than the reduced set of marginally stable ideal MHD equations, (2.375)–(2.380).

The value of $\delta\psi(r,\, t)$ at the rational surface,

$$\Psi_s(t) = \delta\psi(r_s,\, t), \tag{3.72}$$

is known as the *reconnected magnetic flux* [1]. Note that Ψ_s is, in general, a complex quantity. The complex quantity

$$\Delta\Psi_s(t) = \left[r\, \frac{\partial \delta\psi}{\partial r}\right]_{r_{s-}}^{r_{s+}} \tag{3.73}$$

parameterizes the amplitude and phase of the current sheet flowing (parallel to the equilibrium magnetic field) at the rational surface. By asymptotically matching the solutions in the so-called *inner region* (i.e. the region of the plasma in the immediate vicinity of the rational surface) and the so-called *outer region* (i.e. everywhere in the plasma other than the inner region) with the help of equations (3.63), (3.68), (3.69), and (3.73), we obtain

$$\Delta\Psi_s = E_{ss}\, \Psi_s, \tag{3.74}$$

where

$$E_{ss} = \left[r\, \frac{d\hat{\psi}_s}{dr}\right]_{r_{s-}}^{r_{s+}} = \Delta_{s+} - \Delta_{s-} \tag{3.75}$$

is a real dimensionless quantity that is known as the *tearing stability index* [6].

3.9 The solution in the presence of a resistive wall

Suppose, now, that the wall at $r = r_w$ possesses nonzero electrical resistivity but is surrounded by a perfectly conducting wall located at radius $r_c > r_w$. The most general solution to the cylindrical tearing-mode equation, (3.60), in the outer region can be written

$$\delta\psi(r,\, t) = \Psi_s(t)\, \hat{\psi}_s(r) + \Psi_w(t)\, \hat{\psi}_w(r), \tag{3.76}$$

where the real function $\hat{\psi}_s(r)$ is specified in section 3.8, and the real function $\hat{\psi}_w(r)$ is a solution of

$$\frac{d^2\hat{\psi}_w}{dr^2} + \frac{1}{r}\frac{d\hat{\psi}_w}{dr} - \frac{m^2}{r^2}\,\hat{\psi}_w - \frac{J_z'\, \hat{\psi}_w}{r\,(1/q - n/m)} = 0 \tag{3.77}$$

that satisfies

$$\hat{\psi}_w(r \leqslant r_s) = 0, \tag{3.78}$$

$$\hat{\psi}_w(r_w) = 1, \tag{3.79}$$

$$\hat{\psi}_w(r \geqslant r_c) = 0. \tag{3.80}$$

Again, equation (3.80) ensures that the perturbed magnetic field associated with the tearing mode cannot penetrate the perfectly conducting wall. It can easily be seen that

$$\hat{\psi}_{\rm w}(r_{\rm w} < r < r_{\rm c}) = \frac{(r/r_{\rm c})^{-m} - (r/r_{\rm c})^{m}}{(r_{\rm w}/r_{\rm c})^{-m} - (r_{\rm w}/r_{\rm c})^{m}}. \tag{3.81}$$

In general, $\delta\psi$ is continuous across the resistive wall (in accordance with Maxwell's equations), whereas $\partial\delta\psi/\partial r$ is discontinuous. The discontinuity in $\partial\delta\psi/\partial r$ is caused by a helical current sheet induced in the wall. The complex quantity

$$\Psi_{\rm w}(t) = \delta\psi(r_{\rm w}, t), \tag{3.82}$$

determines the amplitude and phase of the perturbed magnetic flux that penetrates the resistive wall. The complex quantity

$$\Delta\Psi_{\rm w}(t) = \left[r\,\frac{\partial\delta\psi}{\partial r}\right]_{r_{\rm w-}}^{r_{\rm w+}} \tag{3.83}$$

parameterizes the amplitude and phase of the helical current sheet flowing in the wall. Simultaneously matching the outer solution (3.76) across the rational surface and the resistive wall yields [1]

$$\Delta\Psi_{\rm s} = E_{\rm ss}\,\Psi_{\rm s} + E_{\rm sw}\,\Psi_{\rm w}, \tag{3.84}$$

$$\Delta\Psi_{\rm w} = E_{\rm ws}\,\Psi_{\rm s} + E_{\rm ww}\,\Psi_{\rm w}. \tag{3.85}$$

Here,

$$E_{\rm ww} = \left[r\,\frac{d\hat{\psi}_{\rm w}}{dr}\right]_{r_{\rm w-}}^{r_{\rm w+}}, \tag{3.86}$$

$$E_{\rm sw} = \left[r\,\frac{d\hat{\psi}_{\rm w}}{dr}\right]_{r=r_{\rm s+}}, \tag{3.87}$$

$$E_{\rm ws} = -\left[r\,\frac{d\hat{\psi}_{\rm s}}{dr}\right]_{r=r_{\rm w-}} \tag{3.88}$$

are real quantities determined by the solutions of equations (3.64) and (3.77) in the outer region.

Equations (3.64) and (3.77) can be combined to give

$$\frac{d}{dr}\left(\hat{\psi}_{\rm s}\,r\,\frac{d\hat{\psi}_{\rm w}}{dr} - \hat{\psi}_{\rm w}\,r\,\frac{d\hat{\psi}_{\rm s}}{dr}\right) = 0. \tag{3.89}$$

If we integrate the previous equation from $r = r_{\rm s+}$ to $r = r_{\rm w-}$, making use of equations (3.66), (3.67), (3.78), (3.79), (3.87), and (3.88), then we obtain [1]

$$E_{\rm sw} = E_{\rm ws}. \tag{3.90}$$

3.10 Resistive-wall physics

Outside the plasma, the perturbed electric field induced by the tearing mode satisfies

$$\delta\mathbf{E} = \nabla \times \delta\mathbf{B}, \tag{3.91}$$

which yields

$$\mathrm{i}\,\frac{m}{r}\,\delta E_z + \mathrm{i}\,\frac{n}{R_0}\,\delta E_\theta = -\frac{\partial\delta B_r}{\partial t}, \tag{3.92}$$

$$-\mathrm{i}\,\frac{n}{R_0}\,\delta E_r - \frac{\partial\delta E_z}{\partial r} = -\frac{\partial\delta B_\theta}{\partial t}, \tag{3.93}$$

$$\frac{1}{r}\,\frac{\partial}{\partial r}(r\,\delta E_\theta) - \mathrm{i}\,\frac{m}{r}\,\delta E_r = -\frac{\partial\delta B_z}{\partial t}. \tag{3.94}$$

Making use of equations (3.32)–(3.34), as well as the ordering $n\,\epsilon \ll m$, the previous three equations imply that

$$\delta E_r \simeq 0, \tag{3.95}$$

$$\delta E_\theta \simeq -\frac{n\,\epsilon_{\mathrm{w}}}{m}\,\frac{\partial\delta\psi}{\partial t}, \tag{3.96}$$

$$\delta E_z \simeq -\frac{\partial\delta\psi}{\partial t}, \tag{3.97}$$

where $\epsilon_{\mathrm{w}} = r_{\mathrm{w}}/R_0$. Here, we have assumed that $\delta_{\mathrm{w}} \ll r_{\mathrm{w}}$, where δ_{w} is the radial thickness of the wall. We have also made use of the fact that $j_\theta = j_z = 0$ outside the plasma.

Inside the wall,

$$\delta j_r = 0, \tag{3.98}$$

$$\delta j_\theta = \frac{\delta E_\theta}{\eta_{\mathrm{w}}}, \tag{3.99}$$

$$\delta j_z = \frac{\delta E_z}{\eta_{\mathrm{w}}}, \tag{3.100}$$

where η_{w} is the wall's electrical resistivity. If we make use of equations (3.37), (3.38), (3.96), and (3.97), as well as the fact that $\delta_{\mathrm{w}} \ll r_{\mathrm{w}}$, then both of the previous two equations reduce to

$$\frac{\partial^2\delta\psi}{\partial r^2} \simeq \frac{\mu_0}{\eta_{\mathrm{w}}}\,\frac{\partial\delta\psi}{\partial t}. \tag{3.101}$$

Let us adopt the so-called *thin-wall limit*, according to which $\delta\psi$ is assumed to vary only weakly in r across the wall. In this limit, integration of the previous equation across the wall yields

$$\Delta\Psi_w = \tau_w \frac{d\Psi_w}{dt}, \tag{3.102}$$

where

$$\tau_w = \frac{\mu_0\, r_w\, \delta_w}{\eta_w} \tag{3.103}$$

is the so-called wall *time constant* [1, 7]. Here, use has been made of equations (3.82) and (3.83). The thin-wall limit is valid as long as the wall thickness is less than the resistive skin depth in the wall material. In other words, provided that

$$\frac{r_w}{\delta_w} \gg \tau_w \left|\frac{d\ln\Psi_w}{dt}\right|. \tag{3.104}$$

In the thin-wall limit, equations (3.82) and (3.95)–(3.100) yield

$$\delta j_r = 0, \tag{3.105}$$

$$\delta j_\theta = -\frac{n\,\epsilon_w}{m\,\eta_w}\frac{d\Psi_w}{dt}, \tag{3.106}$$

$$\delta j_z = -\frac{1}{\eta_w}\frac{d\Psi_w}{dt} \tag{3.107}$$

inside the wall. Note that $\nabla\cdot\delta\mathbf{j} = 0$, as required by charge conservation.

3.11 Resistive-layer physics

By analogy with equation (3.102), we can write

$$\Delta\Psi_s = \tau_s\left(\frac{d\Psi_s}{dt} + \mathrm{i}\,\omega\,\Psi_s\right). \tag{3.108}$$

Here, τ_s is the *reconnection time* (i.e. the typical timescale on which magnetic reconnection takes place in the resistive layer surrounding the rational surface), while ω is the *angular rotation frequency* of the tearing mode in the laboratory frame. The rotation frequency is nonzero because the reconnected flux in the resistive layer is convected by the local plasma flow [1].

3.12 The solution in the presence of an external magnetic field coil

Suppose that the perfectly conducting wall at $r = r_c$ is replaced by a radially thin, magnetic field coil that carries a helical current possessing m periods in the poloidal direction and n periods in the toroidal direction. Let the current density in the field coil take the form

$$\delta j_r = 0, \tag{3.109}$$

$$\delta j_\theta = \frac{n\,\epsilon_c}{m}\,\frac{I_c(t)}{r_c}\,\delta(r - r_c), \tag{3.110}$$

$$\delta j_z = \frac{I_c(t)}{r_c}\,\delta(r - r_c), \tag{3.111}$$

where $\epsilon_c = r_c/R_0$. Here, the complex quantity $I_c(t)$ specifies the amplitude and phase of the helical current flowing in the field coil. Note that $\nabla \cdot \delta\mathbf{j} = 0$, as required by charge conservation.

The most general solution to the cylindrical tearing-mode equation, (3.60), in the outer region can now be written

$$\delta\psi(r, t) = \Psi_s(t)\,\hat{\psi}_s(r) + \Psi_w(t)\,\hat{\psi}_w(r) + \Psi_c(t)\,\hat{\psi}_c(r), \tag{3.112}$$

where the real functions $\hat{\psi}_s(r)$ and $\hat{\psi}_w(r)$ are specified in sections 3.8 and 3.9, respectively. Moreover, the real function $\hat{\psi}_c(r)$ is a solution of

$$\frac{d^2\hat{\psi}_c}{dr^2} + \frac{1}{r}\frac{d\hat{\psi}_c}{dr} - \frac{m^2}{r^2}\,\hat{\psi}_c = 0 \tag{3.113}$$

that satisfies

$$\hat{\psi}_c(r \leqslant r_w) = 0, \tag{3.114}$$

$$\hat{\psi}_c(r_c) = 1, \tag{3.115}$$

$$\hat{\psi}_c(\infty) = 0. \tag{3.116}$$

It can easily seen that

$$\hat{\psi}_c(r \leqslant r_w) = 0, \tag{3.117}$$

$$\hat{\psi}_c(r_w < r \leqslant r_c) = \frac{(r/r_w)^m - (r/r_w)^{-m}}{(r_c/r_w)^m - (r_c/r_w)^{-m}}, \tag{3.118}$$

$$\hat{\psi}_c(r > r_c) = \left(\frac{r}{r_c}\right)^{-m}. \tag{3.119}$$

In general, $\delta\psi$ is continuous across the field coil (in accordance with Maxwell's equations), whereas $\partial\delta\psi/\partial r$ is discontinuous. The discontinuity in $\partial\delta\psi/\partial r$ is caused by the helical current flowing in the field coil. The complex quantity

$$\Psi_c(t) = \delta\psi(r_c, t) \tag{3.120}$$

determines the amplitude and phase of the perturbed magnetic flux at the field coil. The complex quantity

$$\Delta\Psi_c(t) = \left[r\,\frac{\partial\delta\psi}{\partial r}\right]_{r_{c-}}^{r_{c+}} \tag{3.121}$$

parameterizes the amplitude and phase of the helical current sheet flowing in the field coil. It follows from equations (3.37), (3.38), (3.110), and (3.111) that

$$\Delta\Psi_c = -\mu_0\, I_c. \tag{3.122}$$

By simultaneously matching the outer solution, (3.112), across the rational surface, the resistive wall, and the field coil, we obtain

$$\Delta\Psi_s = E_{ss}\,\Psi_s + E_{sw}\,\Psi_w, \tag{3.123}$$

$$\Delta\Psi_w = E_{ws}\,\Psi_s + E_{ww}\,\Psi_w + E_{wc}\,\Psi_c, \tag{3.124}$$

$$\Delta\Psi_c = E_{cw}\,\Psi_w + E_{cc}\,\Psi_c. \tag{3.125}$$

Here,

$$E_{cc} = \left[r\,\frac{d\hat{\psi}_c}{dr}\right]_{r_{c-}}^{r_{c+}} = -\frac{2m}{1-(r_w/r_c)^{2m}}, \tag{3.126}$$

$$E_{wc} = \left[r\,\frac{d\hat{\psi}_c}{dr}\right]_{r_{w+}} = \frac{2m\,(r_w/r_c)^m}{1-(r_w/r_c)^{2m}}, \tag{3.127}$$

$$E_{cw} = -\left[r\,\frac{d\hat{\psi}_w}{dr}\right]_{r_{c-}} = \frac{2m\,(r_w/r_c)^m}{1-(r_w/r_c)^{2m}}, \tag{3.128}$$

where use has been made of equations (3.81), (3.118), and (3.119).

3.13 Electromagnetic torques

The flux-surface integrated poloidal and toroidal electromagnetic torque densities acting on the plasma can be written

$$T_\theta(r,t) = \{r\,\mathbf{j}\times\mathbf{B}\cdot\mathbf{e}_\theta\} \tag{3.129}$$

$$T_z(r,t) = \{R_0\,\mathbf{j}\times\mathbf{B}\cdot\mathbf{e}_z\}, \tag{3.130}$$

respectively, where

$$\{\cdots\} \equiv \oint\oint r\,R_0\,\cdots\,d\theta\,d\varphi \tag{3.131}$$

is a flux-surface integration operator. However, according to the equations of marginally stable ideal MHD, (2.375)–(2.380), both the plasma equilibrium and the tearing perturbation satisfy the force balance criterion

$$\mathbf{j} \times \mathbf{B} \simeq \nabla p. \tag{3.132}$$

Given that the scalar pressure is a single-valued function of θ and φ, it immediately follows that $T_\theta = T_z = 0$ throughout the plasma [1]. The only exception to this rule occurs in the immediate vicinity of the rational surface, where equation (3.132) breaks down. It follows that we can write

$$T_\theta(r,\, t) = T_{\theta\,\mathrm{s}}(t)\, \delta(r - r_\mathrm{s}), \tag{3.133}$$

$$T_z(r,\, t) = T_{z\,\mathrm{s}}(t)\, \delta(r - r_\mathrm{s}), \tag{3.134}$$

where

$$T_{\theta\,\mathrm{s}} = \frac{1}{4}\int_{r_{\mathrm{s}-}}^{r_{\mathrm{s}+}} \oint\oint R_0\, r^{\,2}\, (\delta j_z\, \delta B_r^* + \delta j_z^*\, \delta B_r - \delta j_r \delta B_z^* - \delta j_r^* \delta B_z)\, dr\, d\theta\, d\varphi, \tag{3.135}$$

$$T_{z\,\mathrm{s}} = \frac{1}{4}\int_{r_{\mathrm{s}-}}^{r_{\mathrm{s}+}} \oint\oint R_0^2\, r\, (\delta j_r\, \delta B_\theta^* + \delta j_r^*\, \delta B_\theta - \delta j_\theta\, \delta B_r^* - \delta j_\theta^*\, \delta B_r)\, dr\, d\theta\, d\varphi \tag{3.136}$$

are the net poloidal and toroidal torques, respectively, acting at the rational surface. Note that the zeroth-order (in perturbed quantities) torques are zero because $B_r = j_r = 0$. Furthermore, the linear (in perturbed quantities) torques average to zero over the flux surface. Hence, the largest nonzero torques are *quadratic* in perturbed quantities.

It follows from equations (3.32)–(3.38) that

$$\delta j_z\, \delta B_r^* + \delta j_z^*\, \delta B_r - \delta j_r\, \delta B_z^* - \delta j_r^*\, \delta B_z \simeq \frac{\mathrm{i}\, m}{\mu_0\, r^2} \frac{\partial}{\partial r}\left(r\, \frac{\partial \delta\psi}{\partial r}\, \delta\psi^* - r\, \frac{\partial \delta\psi^*}{\partial r}\, \delta\psi\right) + \mathcal{O}(n\,\epsilon)^2, \tag{3.137}$$

$$\delta j_r\, \delta B_\theta^* + \delta j_r^*\, \delta B_\theta - \delta j_\theta\, \delta B_r^* - \delta j_\theta^*\, \delta B_r \simeq -\frac{\mathrm{i}\, n\, \epsilon}{\mu_0\, r^2} \frac{\partial}{\partial r}\left(r\, \frac{\partial \delta\psi}{\partial r}\, \delta\psi^* - r\, \frac{\partial \delta\psi^*}{\partial r}\, \delta\psi\right). \tag{3.138}$$

The previous four equations yield [1]

$$T_{\theta\,\mathrm{s}} = -\frac{2\pi^2\, R_0\, m}{\mu_0}\, \mathrm{Im}(\Delta\Psi_\mathrm{s}\, \Psi_\mathrm{s}^*), \tag{3.139}$$

$$T_{z\,\mathrm{s}} = \frac{2\pi^2\, R_0\, n}{\mu_0}\, \mathrm{Im}(\Delta\Psi_\mathrm{s}\, \Psi_\mathrm{s}^*). \tag{3.140}$$

where use has been made of equations (3.72) and (3.73).

The net poloidal and toroidal electromagnetic torques acting on the resistive wall can be written

$$T_{\theta\,\mathrm{w}} = \frac{1}{4}\int_{r_{\mathrm{w}-}}^{r_{\mathrm{w}+}} \oint\oint R_0\, r^2\, (\delta j_z\, \delta B_r^* + \delta j_z^*\, \delta B_r - \delta j_r\, \delta B_z^* - \delta j_r^*\, \delta B_z)\, dr\, d\theta\, d\varphi, \tag{3.141}$$

$$T_{z\,\mathrm{w}} = \frac{1}{4}\int_{r_{\mathrm{w}-}}^{r_{\mathrm{w}+}} \oint\oint R_0^{\,2}\, r\, (\delta j_r\, \delta B_\theta^* + \delta j_r^*\, \delta B_\theta - \delta j_\theta\, \delta B_r^* - \delta j_\theta^*\, \delta B_r)\, dr\, d\theta\, d\varphi. \tag{3.142}$$

Making use of equations (3.32), (3.82), and (3.105)–(3.107), we obtain [1]

$$T_{\theta\,\mathrm{w}} = -\frac{2\pi^2\, R_0\, m}{\mu_0}\, \mathrm{Im}(\Delta\Psi_{\mathrm{w}}\, \Psi_{\mathrm{w}}^*), \tag{3.143}$$

$$T_{z\,\mathrm{w}} = \frac{2\pi^2\, R_0\, n}{\mu_0}\, \mathrm{Im}(\Delta\Psi_{\mathrm{w}}\, \Psi_{\mathrm{w}}^*). \tag{3.144}$$

The net poloidal and toroidal electromagnetic torques acting on the magnetic field coil can be written as

$$T_{\theta\,\mathrm{c}} = \frac{1}{4}\int_{r_{\mathrm{c}-}}^{r_{\mathrm{c}+}} \oint\oint R_0\, r^2\, (\delta j_z\, \delta B_r^* + \delta j_z^*\, \delta B_r - \delta j_r\, \delta B_z^* - \delta j_r^*\, \delta B_z)\, dr\, d\theta\, d\varphi, \tag{3.145}$$

$$T_{z\,\mathrm{c}} = \frac{1}{4}\int_{r_{\mathrm{c}-}}^{r_{\mathrm{c}+}} \oint\oint R_0^{\,2}\, r\, (\delta j_r\, \delta B_\theta^* + \delta j_r^*\, \delta B_\theta - \delta j_\theta\, \delta B_r^* - \delta j_\theta^*\, \delta B_r)\, dr\, d\theta\, d\varphi. \tag{3.146}$$

Making use of equations (3.32), (3.109)–(3.111), (3.120), and (3.122), we obtain [1]

$$T_{\theta\,\mathrm{c}} = -\frac{2\pi^2\, R_0\, m}{\mu_0}\, \mathrm{Im}(\Delta\Psi_{\mathrm{c}}\, \Psi_{\mathrm{c}}^*), \tag{3.147}$$

$$T_{z\,\mathrm{c}} = \frac{2\pi^2\, R_0\, n}{\mu_0}\, \mathrm{Im}(\Delta\Psi_{\mathrm{c}}\, \Psi_{\mathrm{c}}^*). \tag{3.148}$$

It follows from equations (3.123)–(3.125) that

$$\mathrm{Im}(\Delta\Psi_{\mathrm{s}}\, \Psi_{\mathrm{s}}^*) = E_{\mathrm{sw}}\, \mathrm{Im}(\Psi_{\mathrm{w}}\, \Psi_{\mathrm{s}}^*), \tag{3.149}$$

$$\mathrm{Im}(\Delta\Psi_{\mathrm{w}}\, \Psi_{\mathrm{w}}^*) = E_{\mathrm{ws}}\, \mathrm{Im}(\Psi_{\mathrm{s}}\, \Psi_{\mathrm{w}}^*) + E_{\mathrm{wc}}\, \mathrm{Im}(\Psi_{\mathrm{c}}\, \Psi_{\mathrm{w}}^*), \tag{3.150}$$

$$\mathrm{Im}(\Delta\Psi_{\mathrm{c}}\, \Psi_{\mathrm{c}}^*) = E_{\mathrm{cw}}\, \mathrm{Im}(\Psi_{\mathrm{w}}\, \Psi_{\mathrm{c}}^*). \tag{3.151}$$

Thus,

$$\begin{aligned}\mathrm{Im}(\Delta\Psi_{\mathrm{s}}\, \Psi_{\mathrm{s}}^*) + \mathrm{Im}(\Delta\Psi_{\mathrm{w}}\, \Psi_{\mathrm{w}}^*) + \mathrm{Im}(\Delta\Psi_{\mathrm{c}}\, \Psi_{\mathrm{c}}^*) &= (E_{\mathrm{sw}} - E_{\mathrm{ws}})\, \mathrm{Im}(\Psi_{\mathrm{w}}\, \Psi_{\mathrm{s}}^*)\\ &\quad + (E_{\mathrm{wc}} - E_{\mathrm{cw}})\, \mathrm{Im}(\Psi_{\mathrm{c}}\, \Psi_{\mathrm{w}}^*).\end{aligned} \tag{3.152}$$

However, according to equations (3.90), (3.127), and (3.128), $E_{sw} = E_{ws}$ and $E_{wc} = E_{cw}$. We deduce that

$$\mathrm{Im}(\Delta\Psi_s\,\Psi_s^*) + \mathrm{Im}(\Delta\Psi_w\,\Psi_w^*) + \mathrm{Im}(\Delta\Psi_c\,\Psi_c^*) = 0. \tag{3.153}$$

Hence, equations (3.139), (3.140), (3.143), (3.144), (3.147), and (3.148) yield [1]

$$T_{\theta\,s} + T_{\theta\,w} + T_{\theta\,c} = 0, \tag{3.154}$$

$$T_{z\,s} + T_{z\,w} + T_{z\,c} = 0. \tag{3.155}$$

In other words, the plasma/resistive-wall/field-coil system exerts zero net poloidal electromagnetic torque and zero net toroidal electromagnetic torque on itself.

3.14 The plasma angular equations of motion

The full plasma equation of motion, (2.331), can be written as

$$\begin{aligned}&\rho\left[\frac{\partial \mathbf{V}_i}{\partial t} + (\mathbf{V}\cdot\nabla)\mathbf{V} + (\mathbf{V}_{*\,i}\cdot\nabla)\mathbf{V}_E\right] - \mathbf{j}\times\mathbf{B} + \nabla p + \frac{\rho}{\tau_{\theta\,i}}(V_{\theta\,i} - V_{\theta\,i}^{nc})\mathbf{e}_\theta\\ &-\nabla\cdot\left[\rho\,\Xi_{\perp\,i}\left(\nabla\mathbf{V}_i + \nabla\mathbf{V}_i^\dagger - \frac{2}{3}\nabla\cdot\mathbf{V}_i\,\mathbf{I}\right)\right] = 0,\end{aligned} \tag{3.156}$$

where $\rho = n_e\,m_i$ is the plasma mass density, $\mathbf{V}_E$ the E-cross-B velocity (see equation (2.138)), $\mathbf{V}_{*\,i}$ the ion diamagnetic velocity (see equation (2.139)), $\mathbf{V}$ the MHD velocity (see equation (2.321)), $\mathbf{V}_i = \mathbf{V} + \mathbf{V}_{*\,i}$ the ion fluid velocity, $V_{\theta\,i}^{nc}$ the neoclassical poloidal velocity (see equation (2.330)), $\tau_{\theta\,i}$ the poloidal flow-damping time (see equation (2.332)), and $\Xi_{\perp\,i}$ the perpendicular ion momentum diffusivity. In writing the previous equation, we have assumed that $\partial\mathbf{V}/\partial t = \partial\mathbf{V}_i/\partial t$ because changes in the ion fluid rotation induced by tearing modes are mostly due to changes in the MHD velocity rather than changes in the ion diamagnetic velocity (which is fixed by the ion pressure profile).

According to the analysis given in section 2.25, the dominant terms in the previous equation are $\mathbf{j}\times\mathbf{B}$ and ∇p. However, if we either take the poloidal flux-surface integral, $\{r\ (3.156)\ \cdot\mathbf{e}_\theta\}$, or the toroidal flux-surface integral, $\{R_0\ (3.156)\ \cdot\mathbf{e}_z\}$, of this equation, then the ∇p term is completely annihilated (because p is a single-valued function of θ and φ), and the $\mathbf{j}\times\mathbf{B}$ term is largely annihilated. In fact, as demonstrated in the previous section, the residual $\mathbf{j}\times\mathbf{B}$ term is quadratic in perturbed quantities and localized in the vicinity of the rational surface. In these circumstances, it makes sense to include contributions from the other smaller terms in equation (3.156). We shall calculate these contributions using the lowest-order (i.e. neglecting the contribution of the tearing perturbation) ion flow,

$$\mathbf{V}_i = r\,\Omega_\theta(r,\,t)\,\mathbf{e}_\theta + R_0\,\Omega_z(r,\,t)\,\mathbf{e}_z. \tag{3.157}$$

Here, $\Omega_\theta(r,\,t)$ and $\Omega_z(r,\,t)$ are the ion poloidal and toroidal angular velocity profiles, respectively.

Taking $\{r\ (3.156)\ \cdot\mathbf{e}_\theta\}$, where the flux-surface integration operator is defined in equation (3.131), we find that [8]

$$4\pi^2 R_0\left[\rho\, r^3 \frac{\partial\Omega_\theta}{\partial t} + \frac{\rho\, r^3}{\tau_{\theta\,\mathrm{i}}}(\Omega_\theta - \Omega_\theta^{\mathrm{nc}}) - \frac{\partial}{\partial r}\left(\rho\,\Xi_{\perp\,\mathrm{i}}\, r^3 \frac{\partial\Omega_\theta}{\partial r}\right)\right] \tag{3.158}$$
$$= T_{\theta\,\mathrm{s}}\,\delta(r - r_\mathrm{s}) + S_\theta,$$

where use has been made of equations (3.133) and (3.157). Note that the advective inertia terms in equation (3.156) make no contribution to the previous equation, because $\mathbf{V}\cdot\mathbf{e}_r = \mathbf{V}_\mathrm{E}\cdot\mathbf{e}_r = \mathbf{V}_{*\,\mathrm{i}}\cdot\mathbf{e}_r = 0$. Here, $\mathbf{e}_r = \nabla r/|\nabla r|$. Moreover, $\Omega_\theta^{\mathrm{nc}} = V_{\theta\,\mathrm{i}}^{\mathrm{nc}}/r$. Here, we have added a source term, $S_\theta(r)$, to the previous equation in order to account for the equilibrium flow. Let $\Omega_{\theta\,0}(r)$ be the unperturbed (by the tearing mode) ion poloidal angular velocity profile. It follows that

$$S_\theta = 4\pi^2 R_0\left[\frac{\rho}{\tau_{\theta\,\mathrm{i}}}\, r^3\,(\Omega_{\theta\,0} - \Omega_\theta^{\mathrm{nc}}) - \frac{d}{dr}\left(\rho\,\Xi_{\perp\,\mathrm{i}}\, r^3 \frac{d\Omega_{\theta\,0}}{dr}\right)\right]. \tag{3.159}$$

Hence, writing

$$\Omega_\theta(r, t) = \Omega_{\theta\,0}(r) + \Delta\Omega_\theta(r, t), \tag{3.160}$$

where $\Delta\Omega_\theta(r, t)$ is the modification to the ion poloidal angular velocity profile induced by the tearing mode, equation (3.158) yields [1]

$$4\pi^2 R_0\left[\rho\, r^3 \frac{\partial\Delta\Omega_\theta}{\partial t} + \frac{\rho}{\tau_{\theta\,\mathrm{i}}}\, r^3\,\Delta\Omega_\theta - \frac{\partial}{\partial r}\left(\rho\,\Xi_{\perp\,\mathrm{i}}\, r^3 \frac{\partial\Delta\Omega_\theta}{\partial r}\right)\right] = T_{\theta\,\mathrm{s}}\,\delta(r - r_\mathrm{s}). \tag{3.161}$$

Taking $\{R_0\ (3.156)\cdot\mathbf{e}_z\}$, we find that [8]

$$4\pi^2 R_0^3\left[\rho\, r \frac{\partial\Omega_z}{\partial t} - \frac{\partial}{\partial r}\left(\rho\,\Xi_{\perp\,\mathrm{i}}\, r \frac{\partial\Omega_z}{\partial r}\right)\right] = T_{z\,\mathrm{s}}\,\delta(r - r_\mathrm{s}) + S_z, \tag{3.162}$$

where use has been made of equations (3.134) and (3.157). We have again added a source term, $S_z(r)$, to the previous equation in order to account for the equilibrium flow. Let $\Omega_{z\,0}(r)$ be the unperturbed (by the tearing mode) ion toroidal angular velocity profile. It follows that

$$S_z = -4\pi^2 R_0^3 \frac{d}{dr}\left(\rho\,\Xi_{\perp\,\mathrm{i}}\, r \frac{d\Omega_{\theta\,0}}{dr}\right). \tag{3.163}$$

Hence, writing

$$\Omega_z(r, t) = \Omega_{z\,0}(r) + \Delta\Omega_z(r, t), \tag{3.164}$$

where $\Delta\Omega_z(r, t)$ is the modification to the ion toroidal angular velocity profile induced by the tearing mode, equation (3.162) yields [1]

$$4\pi^2 R_0^3\left[\rho\, r \frac{\partial\Delta\Omega_z}{\partial t} - \frac{\partial}{\partial r}\left(\rho\,\Xi_{\perp\,\mathrm{i}}\, r \frac{\partial\Delta\Omega_z}{\partial r}\right)\right] = T_{z\,\mathrm{s}}\,\delta(r - r_\mathrm{s}). \tag{3.165}$$

Equations (3.161) and (3.165) are subject to the boundary conditions

$$\frac{\partial \Delta\Omega_\theta(0, t)}{\partial r} = \frac{\partial \Delta\Omega_z(0, t)}{\partial r} = 0, \tag{3.166}$$

$$\Delta\Omega_\theta(a, t) = \Delta\Omega_z(a, t) = 0. \tag{3.167}$$

The boundary conditions (3.166) merely ensure that the ion angular velocities remain finite at the magnetic axis. On the other hand, the boundary conditions (3.167) are a consequence of the action of *charge exchange* with electrically neutral particles isotropically emitted from the wall in the edge regions of the plasma [1, 9, 10]. Charge exchange with neutrals gives rise to dominant damping torques acting at the edge of the plasma that relax the edge ion angular velocities toward particular values. Moreover, the electromagnetic torques that develop at the rational surface are not large enough, compared with the charge-exchange torques, to significantly modify the edge ion angular velocities [1].

3.15 The solution of the plasma angular equations of motion

In many situations of interest, the perturbed angular velocity profiles, $\Delta\Omega_\theta(r, t)$ and $\Delta\Omega_z(r, t)$, are localized in the vicinity of the rational surface [1]. Hence, it is reasonable to express the perturbed angular equations of motion, (3.161) and (3.165), in the simplified forms

$$4\pi^2 R_0\left[\rho_s\, r^3 \frac{\partial \Delta\Omega_\theta}{\partial t} + \frac{\rho_s}{\tau_\theta}\, r^3\, \Delta\Omega_\theta - \rho_s\, \Xi_{\perp\, s}\frac{\partial}{\partial r}\left(r^3\, \frac{\partial \Delta\Omega_\theta}{\partial r}\right)\right] = T_{\theta\, s}\, \delta(r - r_s), \tag{3.168}$$

$$4\pi^2 R_0^{\,3}\left[\rho_s\, r\, \frac{\partial \Delta\Omega_z}{\partial t} - \rho_s\, \Xi_{\perp\, s}\frac{\partial}{\partial r}\left(r\, \frac{\partial \Delta\Omega_z}{\partial r}\right)\right] = T_{z\, s}\, \delta(r - r_s), \tag{3.169}$$

where $\rho_s = \rho(r_s)$, $\tau_\theta = \tau_{\theta\, i}(r_s)$, and $\Xi_{\perp\, s} = \Xi_{\perp\, i}(r_s)$.

Let us write [11, 12]

$$\Delta\Omega_\theta(r, t) = -\frac{1}{m}\sum_{p=1,\infty} \alpha_p(t)\, \frac{y_p(r)}{y_p(r_s)}, \tag{3.170}$$

$$\Delta\Omega_z(r, t) = \frac{1}{n}\sum_{p=1,\infty} \beta_p(t)\, \frac{z_p(r)}{z_p(r_s)}, \tag{3.171}$$

where

$$y_p(r) = \frac{J_1(j_{1p}\, r/a)}{r/a}, \tag{3.172}$$

$$z_p(r) = J_0(j_{0p}\, r/a). \tag{3.173}$$

Here, $J_m(z)$ is a Bessel function, and j_{mp} denotes its pth zero [13]. Note that equations (3.170)–(3.173) automatically satisfy the boundary conditions (3.166) and (3.167).

It can easily be demonstrated that [14]

$$\frac{d}{dr}\left(r^3\,\frac{dy_{\rm p}}{dr}\right) = -\frac{j_{1\rm p}^{\,2}\,r^3\,y_{\rm p}}{a^2}, \tag{3.174}$$

$$\frac{d}{dr}\left(r\,\frac{dz_{\rm p}}{dr}\right) = -\frac{j_{0\rm p}^{\,2}\,r\,z_{\rm p}}{a^2}, \tag{3.175}$$

and

$$\int_0^a r^3\,y_{\rm p}(r)\,y_{\rm q}(r)\,dr = \frac{a^4}{2}\,[J_2(j_{1\rm p})]^2\,\delta_{\rm pq}, \tag{3.176}$$

$$\int_0^a r\,z_{\rm p}(r)\,z_{\rm q}(r)\,dr = \frac{a^2}{2}\,[J_1(j_{0\rm p})]^2\,\delta_{\rm pq}. \tag{3.177}$$

Equations (3.139), (3.140), and (3.168)–(3.177) yield

$$\frac{d\alpha_{\rm p}}{dt} + \left(\frac{1}{\tau_\theta} + \frac{j_{1\rm p}^{\,2}}{\tau_M}\right)\alpha_{\rm p} = \frac{m^2\,[J_1(j_{1\rm p}\,r_{\rm s}/a)]^2}{\tau_{\rm A}^{\,2}\,\epsilon_{\rm s}^{\,2}\,[J_2(j_{1\rm p})]^2}\,{\rm Im}(\Delta\hat{\Psi}_{\rm s}\,\hat{\Psi}_{\rm s}^*), \tag{3.178}$$

$$\frac{d\beta_{\rm p}}{dt} + \frac{j_{0\rm p}^{\,2}}{\tau_M}\,\beta_{\rm p} = \frac{n^2\,[J_0(j_{0\rm p}\,r_{\rm s}/a)]^2}{\tau_{\rm A}^{\,2}\,[J_1(j_{0\rm p})]^2}\,{\rm Im}(\Delta\hat{\Psi}_{\rm s}\,\hat{\Psi}_{\rm s}^*). \tag{3.179}$$

Here,

$$\tau_{\rm M} = \frac{a^{\,2}}{\Xi_{\perp\,\rm s}}, \tag{3.180}$$

is the *momentum confinement time*,

$$\tau_{\rm A} = \left(\frac{\mu_0\,\rho_{\rm s}\,a^{\,2}}{B_z^{\,2}}\right)^{1/2} \tag{3.181}$$

is the *Alfvén time*,

$$\epsilon_{\rm s} = \frac{r_{\rm s}}{R_0} \ll 1 \tag{3.182}$$

is the *inverse aspect ratio* of the rational surface, and

$$\Delta\hat{\Psi}_{\rm s} = \frac{\Delta\Psi_{\rm s}}{R_0\,B_z}, \tag{3.183}$$

$$\hat{\Psi}_{\rm s} = \frac{\Psi_{\rm s}}{R_0\,B_z}. \tag{3.184}$$

3.16 Modification of the rotational frequency

The reconnected magnetic flux at the rational surface is assumed to convected by the local plasma flow (see section 3.11). Hence, changes in the plasma flow at the rational surface induced by the electromagnetic torques that develop in the vicinity of this surface modify the convection velocity. Consequently, the tearing mode's rotational frequency can be written as

$$\omega(t) = \omega_0 + (\mathbf{k}\cdot\Delta\mathbf{V}_{\rm i})_{r=r_{\rm s}} = \omega_0 + m\,\Delta\Omega_\theta(r_{\rm s},\,t) - n\,\Delta\Omega_z(r_{\rm s},\,t), \tag{3.185}$$

where ω_0 is the *natural frequency* of the tearing mode. The natural frequency is defined as the rotational frequency of a tearing mode that develops spontaneously and does not interact with any external structure (such as a resistive wall or an error field). The previous expression is known as the *no-slip constraint* [1] and has been verified experimentally [15]. It follows from equations (3.170) and (3.171) that

$$\omega(t) = \omega_0 - \sum_{p=1,\infty}\left[\alpha_{\rm p}(t) + \beta_{\rm p}(t)\right]. \tag{3.186}$$

3.17 The tearing-mode evolution equations

Equations (3.90), (3.102), (3.108), (3.122)–(3.128), (3.178), (3.179), and (3.186) can be combined to give the following complete set of equations that determine the time evolution of the reconnected magnetic flux at the rational surface in the presence of a resistive vacuum vessel and an external magnetic field coil:

$$\Delta\hat{\Psi}_{\rm s} \equiv \tau_{\rm s}\left(\frac{d\hat{\Psi}_{\rm s}}{dt} + {\rm i}\,\omega\,\hat{\Psi}_{\rm s}\right) = E_{\rm ss}\,\hat{\Psi}_{\rm s} + E_{\rm sw}\,\hat{\Psi}_{\rm w}, \tag{3.187}$$

$$\Delta\hat{\Psi}_{\rm w} \equiv \tau_{\rm w}\,\frac{d\hat{\Psi}_{\rm w}}{dt} = E_{\rm ws}\,\hat{\Psi}_{\rm s} + \tilde{E}_{\rm ww}\,\hat{\Psi}_{\rm w} + \hat{I}_{\rm c}, \tag{3.188}$$

$$\omega(t) = \omega_0 - \sum_{p=1,\infty}\left[\alpha_{\rm p}(t) + \beta_{\rm p}(t)\right], \tag{3.189}$$

$$\frac{d\alpha_{\rm p}}{dt} + \left(\frac{1}{\tau_\theta} + \frac{j_{1\rm p}^{\,2}}{\tau_{\rm M}}\right)\alpha_{\rm p} = \frac{m^2\,[J_1(j_{1\rm p}\,r_{\rm s}/a)]^2}{\tau_{\rm A}^{\,2}\,\epsilon_{\rm s}^{\,2}\,[J_2(j_{1\rm p})]^2}\,{\rm Im}(\Delta\hat{\Psi}_{\rm s}\,\hat{\Psi}_{\rm s}^*), \tag{3.190}$$

$$\frac{d\beta_{\rm p}}{dt} + \frac{j_{0\rm p}^{\,2}}{\tau_{\rm M}}\,\beta_{\rm p} = \frac{n^2\,[J_0(j_{0\rm p}\,r_{\rm s}/a)]^2}{\tau_{\rm A}^{\,2}\,[J_1(j_{0\rm p})]^2}\,{\rm Im}(\Delta\hat{\Psi}_{\rm s}\,\hat{\Psi}_{\rm s}^*). \tag{3.191}$$

Here,

$$\hat{\Psi}_{\rm w} = \frac{\Psi_{\rm w}}{R_0\,B_z}, \tag{3.192}$$

$$\Delta\hat{\Psi}_{\rm w} = \frac{\Delta\Psi_{\rm w}}{R_0\,B_z}, \tag{3.193}$$

$$\hat{I}_{\rm c}(t) = \left(\frac{r_{\rm w}}{r_{\rm c}}\right)^m \frac{\mu_0\,I_{\rm c}(t)}{R_0\,B_z}, \tag{3.194}$$

$$\tilde{E}_{\rm ww} = E_{\rm ww} - \frac{E_{\rm wc}\,E_{\rm cw}}{E_{\rm cc}} = E_{\rm ww} + \frac{2m\,(r_{\rm w}/r_{\rm c})^{2m}}{1-(r_{\rm w}/r_{\rm c})^{2m}}. \tag{3.195}$$

References

[1] Fitzpatrick R 1993 Interaction of tearing modes with external structures in cylindrical geometry *Nucl. Fusion* **33** 1049

[2] Wesson J A 1978 Hydrodynamic stability of tokamaks *Nucl. Fusion* **18** 87

[3] Wesson J A 2011 *Tokamaks* 4th edn (Oxford: Oxford University Press) https://global.oup.com/academic/product/tokamaks-9780199592234

[4] Freidberg J P 1987 *Ideal Magnetohydrodynamics* (New York: Plenum)

[5] Newcomb W A 1960 Hydromagnetic stability of a diffuse linear pinch *Ann. Phys., NY* **10** 232

[6] Furth H P, Killeen J and Rosenbluth M N 1963 Finite-resistivity instabilities of a sheet pinch *Phys. Fluids* **6** 459

[7] Nave M F F and Wesson J A 1990 Mode locking in tokamaks *Nucl. Fusion* **30** 2575

[8] Richardson A S 2019 *2019 NRL Plasma formulary* (Washington, DC: Naval Research Laboratory) https://www.nrl.navy.mil/Portals/38/PDF%20Files/NRL_Formulary_2019.pdf

[9] Brau K, Bitter M, Goldston R J, Manos D, McGuire K and Suckewer S 1983 Plasma rotation in the PDX tokamak *Nucl. Fusion* **23** 1643

[10] Monier-Garbet P, Burrel K H, Hinton F L, Kim J, Garbet X and Groebner R J 1997 Effects of neutrals on plasma rotation in DIII-D *Nucl. Fusion* **37** 403

[11] Chapman B E, Fitzpatrick R, Craig D, Martin P and Spizzo G 2004 Observation of tearing mode deceleration and locking due to eddy currents induced in a conducting shell *Phys. Plasmas* **11** 2156

[12] Fitzpatrick R, Rossi E and Yu E P 2001 Improved evolution equations for magnetic island chains in toroidal pinch plasmas subject to externally applied resonant magnetic perturbations *Phys. Plasmas* **8** 4489

[13] Abramowitz M and Stegun I A (ed) 1964 *Handbook of Mathematical Functions with Formulas, Graphs, and Mathematical tables* (New York: Dover) https://store.doverpublications.com/0486612724.html

[14] Watson G N 1922 *A Treatise on the Theory of Bessel Functions* 1st edn. (Cambridge: Cambridge University Press)

[15] Vahala G, Vahala L, Harris J H, Bateman G, Waddell B V, Dunlap J L, Paré V K and Burris R D 1980 Perturbed magnetic-field phase slip for tokamaks *Nucl. Fusion* **20** 17

IOP Publishing

Richard Fitzpatrick

Chapter 4

Reduced resonant response model

4.1 Introduction

We saw in the previous chapter that, when we describe the response of a tokamak plasma to a tearing-mode perturbation, the plasma can be divided into two regions [5]. In the so-called *outer region*, which comprises most of the plasma, the tearing perturbation is governed by the relatively simple equations of marginally stable ideal magnetohydrodynamics (MHD), (2.375)–(2.380). However, these equations become singular at the so-called *rational* magnetic flux surface, where the tearing mode resonates with the equilibrium magnetic field. The resonance condition is $\mathbf{k} \cdot \mathbf{B} = 0$, where $\mathbf{k}$ is the wave number of the tearing mode and $\mathbf{B}$ the equilibrium magnetic field. A radially thin resistive layer forms around the rational magnetic flux surface and constitutes the so-called *inner region*. The tearing-mode solution in the inner region needs to be asymptotically matched to that in the outer region [5]. In principle, the relatively complicated neoclassical fluid equations, (2.370)–(2.374), must be used to describe the resonant plasma response in the inner region. Note, however, that the set of tearing-mode evolution equations, (3.187)–(3.191), derived in the previous chapter, only contains two quantities that actually depend on the plasma response in the inner region: namely, the reconnection time, τ_s, and the natural frequency, ω_0. This fact suggests that the true set of equations that describe the plasma response in the inner region are almost certainly considerably simpler than the neoclassical fluid equations. Let us search for some simplifying approximations.

One simplifying approximation that we can make is to ignore the specifically neoclassical terms (e.g. the terms associated with ion poloidal flow damping and the bootstrap current) in the neoclassical fluid equations. As we saw in chapter 2, all of these terms have their origin in friction between trapped and passing particles. However, trapped ions make radial excursions from magnetic flux surfaces that are of the order of the ion banana width, $\rho_{b\,i}$ (see section 2.7). As shown in table 2.4, this width is a few centimeters in a tokamak fusion reactor. In situations in which the

doi:10.1088/978-0-7503-5367-0ch4

width of the inner region is much less than the ion banana width, we would not expect ion neoclassical effects to contribute to the plasma response in the inner region (because the trapped ions would average over the spatial structure of the inner region). According to table 2.4, the electron banana width is of the order of a few tenths of a centimeter in a tokamak fusion reactor. Hence, we would not expect electron neoclassical effects to contribute to the plasma response in very thin resistive layers. When our neoclassical fluid equations are stripped of specifically neoclassical terms, we shall refer to them as *drift-MHD fluid equations*. The 'drift' appellation merely indicates that we are still taking diamagnetic flows into account, which necessarily entails treating the ion and electron flows separately.

A second simplifying approximation that we can make is to assume that the perturbed electron temperature and ion temperature profiles are functions of the perturbed electron number density profile, which reduces the number of independent equations that we have to deal with (see section 4.2).

A final simplification is associated with the fact that a tearing mode is a modified shear Alfvén wave [8] and has very little connection with a compressible Alfvén wave [8]. For instance, the resonance condition, $\mathbf{k}\cdot\mathbf{B} = 0$, which determines the position of the rational surface, is identical to the resonance condition for a shear Alfvén wave but differs substantially from that of a compressible Alfvén wave [2]. Hence, one very effective way of simplifying the resonant layer equations is to remove the irrelevant physics of compressible Alfvén waves from them altogether, in the process converting a drift-MHD response model into a so-called *reduced* drift-MHD response model [7, 9, 10]. The aim of this chapter is to describe the reduction process and to derive the reduced drift-MHD fluid equations that will subsequently be used to model the resonant response of the plasma.

4.2 The drift-MHD fluid equations

Let r, θ, z be conventional right-handed cylindrical coordinates, and let the rational magnetic flux surface lie at $r = r_s$. It is helpful to define the following parameters:

$$n_0 = n_e(r_s), \tag{4.1}$$

$$p_0 = p(r_s), \tag{4.2}$$

$$\eta_e = \left.\frac{d\ln T_e}{d\ln n_e}\right|_{r=r_s}, \tag{4.3}$$

$$\eta_i = \left.\frac{d\ln T_i}{d\ln n_e}\right|_{r=r_s}, \tag{4.4}$$

$$\tau = \left(\frac{T_e}{T_i}\right)_{r=r_s}\left(\frac{1+\eta_e}{1+\eta_i}\right), \tag{4.5}$$

$$\tau' = \left(\frac{T_e}{T_i}\right)_{r=r_s}, \tag{4.6}$$

where $n_e(r)$, $p(r) \equiv n_e\,(T_e + T_i)$, $T_e(r)$, and $T_i(r)$ refer to electron number density, total pressure, electron temperature, and ion temperature profiles, respectively, that are unperturbed by the tearing mode.

As has already been suggested, it is convenient to make the simplifying approximation that the perturbed electron and ion temperature profiles in the vicinity of the resonant layer are functions of the perturbed electron number density profile. In other words, $T_e = T_e(n_e)$ and $T_i = T_i(n_e)$. It follows that the perturbed total pressure profile is also a function of the perturbed electron number density profile: that is, $p = p(n_e)$.

When stripped of specifically neoclassical terms (i.e. the ion poloidal flow-damping terms and any other terms involving the superscript *nc*), our neoclassical fluid equations, (2.370)–(2.374), reduce to the following set of *drift-MHD fluid equations*:

$$\begin{aligned} &m_i\, n_0\left[\frac{\partial \mathbf{V}}{\partial t} + (\mathbf{V}\cdot\nabla)\mathbf{V} + \frac{1}{1+\tau}\,(\mathbf{V}_*\cdot\nabla)\mathbf{V}_E\right] + \nabla p - \mathbf{j}\times\mathbf{B} \\ &- m_i\, n_0\, \Xi_\perp\, \nabla\cdot\left(\nabla\mathbf{V}_i + \nabla\mathbf{V}_i^\dagger - \frac{2}{3}\,\nabla\cdot\mathbf{V}_i\,\mathbf{I}\right) = 0, \end{aligned} \tag{4.7}$$

$$\begin{aligned} \mathbf{E} + \mathbf{V}\times\mathbf{B} + \frac{1}{e\, n_0}\left[\nabla p - \frac{1}{1+\tau}\,(\mathbf{b}\cdot\nabla p) - \mathbf{j}\times\mathbf{B}\right] \\ = \eta_\perp\left(\mathbf{j}_\perp - \frac{\mathbf{b}\times\nabla p}{B_z}\right) + \eta_\parallel\,\mathbf{j}_\parallel, \end{aligned} \tag{4.8}$$

$$\begin{aligned} \frac{3}{2}\frac{\partial p}{\partial t} + \frac{3}{2}\,\mathbf{V}\cdot\nabla p + \frac{5}{2}\,p_0\,\nabla\cdot\mathbf{V} + \frac{5}{2}\,\lambda\, p_0\,\nabla\cdot\mathbf{V}_* - \chi_\parallel\,\mathbf{b}\cdot\nabla(\mathbf{b}\cdot\nabla p) - \chi_\perp\,\nabla^2 p \\ = 0, \end{aligned} \tag{4.9}$$

where

$$\mathbf{V}_E = \frac{\mathbf{E}\times\mathbf{b}}{B_z}, \tag{4.10}$$

$$\mathbf{V}_* = \frac{\mathbf{b}\times\nabla p}{e\, n_0\, B_z}, \tag{4.11}$$

$$\mathbf{V} = \mathbf{V}_E + V_\parallel\,\mathbf{b}, \tag{4.12}$$

$$\mathbf{V}_i = \mathbf{V} + \frac{1}{1+\tau}\,\mathbf{V}_*, \tag{4.13}$$

and

$$\eta_\perp = \frac{1}{\sigma_\perp(r_s)}, \tag{4.14}$$

$$\eta_\parallel = \frac{1}{\sigma_\parallel(r_s)}, \tag{4.15}$$

$$\lambda = \left(\frac{1}{1+\tau'}\right)\left(\frac{1}{1+\tau}\right)\left(1+\frac{\eta_i}{1+\eta_i}\right) - \left(\frac{\tau'}{1+\tau'}\right)\left(\frac{\tau}{1+\tau}\right)\left(1+\frac{\eta_e}{1+\eta_e}\right), \tag{4.16}$$

$$\Xi_\perp = \Xi_{\perp\,i}(r_s), \tag{4.17}$$

$$\chi_\parallel = \left(\frac{\tau}{1+\tau}\right)\left(\frac{\eta_e}{1+\eta_e}\right)\chi_{\parallel\,e}(r_s) + \left(\frac{1}{1+\tau}\right)\left(\frac{\eta_i}{1+\eta_i}\right)\chi_{\parallel\,i}(r_s), \tag{4.18}$$

$$\chi_\perp = \left(\frac{\tau}{1+\tau}\right)\left(\frac{\eta_e}{1+\eta_e}\right)\chi_{\perp\,e}(r_s) + \left(\frac{1}{1+\tau}\right)\left(\frac{\eta_i}{1+\eta_i}\right)\chi_{\perp\,i}(r_s). \tag{4.19}$$

Here, $\mathbf{B}$ is the magnetic field strength, $\mathbf{E}$ the electric field strength, $\mathbf{j}$ the current density, $\mathbf{V}_i$ the ion fluid velocity, m_i the ion mass, $\sigma_\perp$ the classical (see section 2.6) perpendicular (to the equilibrium magnetic field) electrical conductivity, $\sigma_\parallel$ the classical parallel electrical conductivity, $\Xi_{\perp\,i}$ the ion perpendicular momentum diffusivity, $\chi_{\parallel\,e}$ the electron parallel energy diffusivity, $\chi_{\parallel\,i}$ the ion parallel energy diffusivity, $\chi_{\perp\,e}$ the electron perpendicular energy diffusivity, $\chi_{\perp\,i}$ the ion perpendicular energy diffusivity, and $\mathbf{b} = \mathbf{B}/B_z$. Note that the simplifying assumption that $p = p(n_e)$ negates the need for a separate electron number density conservation equation. (In other words, equation (2.370) is redundant.) Moreover, equation (4.9) is the sum of the electron and ion energy conservation equations, (2.373) and (2.374). In writing equations (4.7)–(4.9), we have made a number of additional simplifying assumptions. For instance, we have replaced B by B_z, which is approximated as a spatial constant (see section 3.3), n_e by n_0 (when a spatial or temporal derivative of the electron number density is not being taken), and p by p_0 (under the same circumstances). Finally, we have neglected any spatial variation in transport coefficients (e.g. $\Xi_{\perp\,i}$, $\eta_\parallel$, $\chi_{\parallel\,e}$) across the inner region.

Our drift-MHD fluid equations are closed by the following subset of Maxwell's equations:

$$\nabla \cdot \mathbf{B} = 0, \tag{4.20}$$

$$\nabla \times \mathbf{E} = -\frac{\partial \mathbf{B}}{\partial t}, \tag{4.21}$$

$$\mu_0\, \mathbf{j} = \nabla \times \mathbf{B}. \tag{4.22}$$

4.3 The normalization scheme

The *Alfvén speed*, which is the typical phase velocity of a compressible Alfvén wave, is defined as

$$V_{\rm A} = \frac{a}{\tau_{\rm A}} = \frac{B_z}{\sqrt{\mu_0\, n_0\, m_{\rm i}}}. \tag{4.23}$$

(See equation (3.181).) It is helpful to define the *collisionless ion skin depth*,

$$d_{\rm i} = \left(\frac{m_{\rm i}}{n_0\, e^2\, \mu_0}\right)^{1/2}. \tag{4.24}$$

Note that $d_{\rm i} = c/\omega_{\rm p\,i}$, where c is the velocity of light in vacuum and $\omega_{\rm p\,i}$ the ion plasma frequency [2].

Let l be a typical variation length scale in the resonant layer. It is convenient to adopt the following normalization scheme that renders all quantities in the drift-MHD fluid equations dimensionless: $\hat{\nabla} = l\,\nabla$, $\hat{t} = t/(l/V_{\rm A})$, $\hat{d}_{\rm i} = d_{\rm i}/l$, $\hat{\mathbf{E}} = \mathbf{E}/(B_z\, V_{\rm A})$, $\hat{\mathbf{j}} = \mathbf{j}/(B_z/\mu_0\, l)$, $\hat{p} = p/(B_z^{\,2}/\mu_0)$, $\hat{p}_0 = p_0/(B_z^{\,2}/\mu_0)$, $\hat{\mathbf{V}} = \mathbf{V}/V_{\rm A}$, $\hat{\mathbf{V}}_{{\rm E},*,{\rm i}} = \mathbf{V}_{{\rm E},*,{\rm i}}/V_{\rm A}$, $\hat{V}_\parallel = V_{\parallel\,{\rm i}}/V_{\rm A}$, $\hat{\Xi}_\perp = \Xi_\perp/(l\, V_{\rm A})$, $\hat{\eta}_{\parallel,\perp} = \eta_{\parallel,\perp}/(\mu_0\, l\, V_{\rm A})$, and $\hat{\chi}_{\parallel,\perp} = \chi_{\parallel,\perp}/(l\, V_{\rm A})$. Equations (4.7)–(4.13) and (4.20)–(4.22) yield the following set of normalized drift-MHD fluid equations:

$$\begin{aligned} &\frac{\partial\hat{\mathbf{V}}}{\partial\hat{t}} + \hat{\nabla}\left(\frac{\hat{V}^2}{2}\right) - \hat{\mathbf{V}}\times(\hat{\nabla}\times\hat{\mathbf{V}}) \\ &\quad + \frac{1}{1+\tau}\,\frac{1}{2}[\hat{\nabla}(\hat{\mathbf{V}}_{\rm E}\cdot\hat{\mathbf{V}}_*) - \hat{\mathbf{V}}_{\rm E}\times(\hat{\nabla}\times\hat{\mathbf{V}}_*) - \hat{\mathbf{V}}_*\times(\hat{\nabla}\times\hat{\mathbf{V}}_{\rm E}) \\ &\quad + (\hat{\nabla}\cdot\hat{\mathbf{V}}_{\rm E})\,\hat{\mathbf{V}}_* - (\hat{\nabla}\cdot\hat{\mathbf{V}}_*)\,\hat{\mathbf{V}}_{\rm E} + \hat{\nabla}\times(\hat{\mathbf{V}}_{\rm E}\times\hat{\mathbf{V}}_*)] + \hat{\nabla}\hat{p} - \hat{\mathbf{j}}\times\mathbf{b} \\ &\quad - \hat{\Xi}_\perp\hat{\nabla}\cdot\left(\hat{\nabla}\hat{\mathbf{V}}_{\rm i} + \hat{\nabla}\hat{\mathbf{V}}_{\rm i}^{\dagger} - \frac{2}{3}\,\hat{\nabla}\cdot\hat{\mathbf{V}}_{\rm i}\,\mathbf{I}\right) = 0, \end{aligned} \tag{4.25}$$

$$\begin{aligned} &\hat{\mathbf{E}} + \hat{\mathbf{V}}\times\mathbf{b} + \hat{d}_{\rm i}\left[\hat{\nabla}\hat{p} - \frac{1}{1+\tau}\,(\mathbf{b}\cdot\hat{\nabla}\hat{p})\,\mathbf{b} - \hat{\mathbf{j}}\times\mathbf{b}\right] \\ &\qquad = \hat{\eta}_\perp\,(\hat{\mathbf{j}}_\perp - \mathbf{b}\times\hat{\nabla}\hat{p}) + \hat{\eta}_\parallel\,\hat{\mathbf{j}}_\parallel, \end{aligned} \tag{4.26}$$

$$\begin{aligned} &\frac{3}{2}\,\frac{\partial\hat{p}}{\partial\hat{t}} + \frac{3}{2}\,\hat{\mathbf{V}}\cdot\hat{\nabla}\hat{p} + \frac{5}{2}\,\hat{p}_0\,\hat{\nabla}\cdot\hat{\mathbf{V}} + \frac{5}{2}\,\lambda\,\hat{p}_0\,\hat{\nabla}\cdot\hat{\mathbf{V}}_* - \hat{\chi}_\parallel\,\mathbf{b}\cdot\hat{\nabla}(\mathbf{b}\cdot\hat{\nabla}\hat{p}) \\ &\qquad - \hat{\chi}_\perp\,\hat{\nabla}^2\hat{p} = 0, \end{aligned} \tag{4.27}$$

where

$$\hat{\mathbf{V}}_{\rm E} = \hat{\mathbf{E}}\times\mathbf{b}, \tag{4.28}$$

$$\hat{\mathbf{V}}_* = \hat{d}_{\rm i}\,\mathbf{b}\times\hat{\nabla}\hat{p}, \tag{4.29}$$

$$\hat{\mathbf{V}} = \hat{\mathbf{V}}_{\rm E} + \hat{V}_\parallel\, \mathbf{b}, \tag{4.30}$$

$$\hat{\mathbf{V}}_{\rm i} = \hat{\mathbf{V}} + \frac{1}{1+\tau}\,\hat{\mathbf{V}}_*, \tag{4.31}$$

and

$$\hat{\nabla}\cdot\mathbf{b} = 0, \tag{4.32}$$

$$\hat{\nabla}\times\hat{\mathbf{E}} = -\frac{\partial\mathbf{b}}{\partial\hat{t}}, \tag{4.33}$$

$$\hat{\mathbf{j}} = \hat{\nabla}\times\mathbf{b}. \tag{4.34}$$

Here, use has been made of some standard vector identities.

4.4 The reduction process

All variables in the resonant layer are assumed to be functions of

$$\hat{x} = \frac{r - r_{\rm s}}{l}, \tag{4.35}$$

$$\zeta = m\,\theta - n\,\varphi, \tag{4.36}$$

and $\hat{t}$, only. Here, m and n are the poloidal and toroidal mode numbers, respectively, of the tearing mode, $\varphi = z/R_0$ is a simulated toroidal angle, and R_0 is the simulated major radius of the plasma (see chapter 3).

Let

$$\mathbf{n} = \frac{\epsilon^{(1)}}{q_{\rm s}}\,\mathbf{e}_\theta + \mathbf{e}_z, \tag{4.37}$$

where $\epsilon(r) = r/R_0$ and $q_{\rm s} \equiv q(r_{\rm s}) = m/n$. Here, $q(r)$ is the safety-factor profile (see equation (3.2)). Moreover, the superscript (1) indicates a quantity that is first order in our ordering scheme. (Zeroth-order terms are left without superscripts, whereas second-order terms are given the superscript (2).)

It follows that

$$\mathbf{n}\cdot\hat{\nabla}A = 0 \tag{4.38}$$

for any $A(\hat{x}, \zeta, \hat{t})$. Furthermore,

$$\hat{\nabla}\cdot\mathbf{n} = 0, \tag{4.39}$$

$$\hat{\nabla}\times\mathbf{n} \simeq \frac{2\,\hat{\epsilon}^{(1)}}{q_{\rm s}}\,\mathbf{n}, \tag{4.40}$$

where $\hat{\epsilon}^{(1)} = l/R_0$.

We can automatically satisfy equation (4.32) by writing

$$\mathbf{b} = (1 + \delta b^{(1)})\mathbf{n} + \hat{\nabla} \times (\psi^{(1)}\,\mathbf{n}) \simeq \left(1 + \delta b^{(1)} + \frac{2\,\epsilon^{(1)}}{q_s}\,\psi^{(1)}\right)\mathbf{n} + \hat{\nabla}\psi^{(1)} \times \mathbf{n}. \tag{4.41}$$

It follows that

$$\psi^{(1)}(\hat{x}, \zeta, \hat{t}) = \frac{\psi_0(\hat{x}) + \delta\psi(\hat{x}, \hat{t})\,\mathrm{e}^{\,\mathrm{i}\,\zeta}}{l\,B_z}, \tag{4.42}$$

where (see equations (3.1) and (3.2))

$$\psi_0(\hat{x}) = \frac{B_z}{R_0}\int_{r_s}^{r} r'\left[\frac{1}{q_s} - \frac{1}{q(r')}\right]dr', \tag{4.43}$$

and $\delta\psi$ is defined in equation (3.20). Note that

$$\mathbf{b}\cdot\hat{\nabla}A^{(1)} = [A^{(1)}, \psi^{(1)}], \tag{4.44}$$

where

$$[A, B] \equiv \hat{\nabla}A \times \hat{\nabla}B \cdot \mathbf{n}. \tag{4.45}$$

Equations (4.33) and (4.41) give

$$\hat{\mathbf{E}} = \hat{\nabla}\phi^{(1)} + \left(\hat{E}_{\parallel}^{(2)} - \frac{\partial^{(1)}\psi^{(1)}}{\partial\hat{t}}\right)\mathbf{n} + \hat{\nabla} \times \left(\frac{\partial^{(1)}\chi^{(1)}}{\partial\hat{t}}\,\mathbf{n}\right), \tag{4.46}$$

where

$$\hat{\nabla}^2\chi^{(1)} = \delta b^{(1)}. \tag{4.47}$$

Here, we have explicitly separated out the time-dependent, spatially uniform part of $\psi^{(1)}$ which generates the (normalized) inductive electric field, $\hat{E}_{\parallel}^{(2)}$, that is responsible for maintaining the equilibrium parallel current density in the inner region against ohmic decay. Equations (4.34) and (4.41) yield

$$\hat{\mathbf{j}} = \left(-J^{(1)} + \frac{2\,\epsilon^{(1)}}{q_s}\,\delta b^{(1)}\right)\mathbf{n} + \hat{\nabla}\delta b^{(1)} \times \mathbf{n} - \frac{2\,\hat{\epsilon}^{(1)}\,\epsilon^{(1)}}{q_s^{\,2}}\,\mathbf{e}_\theta, \tag{4.48}$$

where

$$J^{(1)} = -\frac{2\,\epsilon^{(1)}}{q_s} + \hat{\nabla}^2\psi^{(1)}. \tag{4.49}$$

Let us write

$$\hat{p} = \hat{p}_0 + \delta p^{(1)}. \tag{4.50}$$

Note that the ordering of the plasma pressure adopted here is somewhat different to that adopted in section 3.3. In fact, in section 3.3, $\hat{p}$ was assumed to be second order. Here, we are assuming that $\hat{p}$ has a spatially constant component that is zeroth order and a spatially varying component that is first order. This high-pressure ordering is merely an artifice to aid the extraction of the compressible Alfvén (i.e. fast magnetosonic) wave from the system of equations [3, 4]. In fact, the ordering ensures that the compressible Alfvén wave has a substantially different phase velocity than the shear Alfvén and slow magnetosonic waves [2].

Equations (4.28)–(4.31), (4.41), (4.46), and (4.50) yield

$$\hat{\mathbf{V}}_{\rm E} = \hat{\nabla}\phi^{(1)} \times \mathbf{b} - \hat{\nabla}\left(\frac{\partial^{(1)}\chi^{(1)}}{\partial \hat{t}}\right), \tag{4.51}$$

$$\hat{\mathbf{V}}_{*} = \hat{d}_{\rm i}\, \mathbf{b} \times \hat{\nabla}\delta p^{(1)}, \tag{4.52}$$

$$\hat{\mathbf{V}} = \hat{\mathbf{V}}_{\rm E} + \hat{V}_{\parallel}^{(1)}\, \mathbf{b} + \hat{\nabla}\Upsilon^{(2)}, \tag{4.53}$$

$$\hat{\mathbf{V}}_{\rm i} = \hat{\mathbf{V}} + \frac{1}{1+\tau}\, \hat{\mathbf{V}}_{*}. \tag{4.54}$$

Here, the additional factor involving $\Upsilon^{(2)}$ in equation (4.53) is needed because the plasma flow is slightly compressible. In fact, the velocity fields $\hat{\mathbf{V}}_{\rm E}$, $\hat{\mathbf{V}}_{*}$, $\hat{\mathbf{V}}$, and $\hat{\mathbf{V}}_{\rm i}$ all have normalized divergences that are second order.

Evaluating the normalized drift-MHD fluid equations, (4.25)–(4.27), up to the second order, we obtain

$$\begin{aligned}
&\hat{\nabla}\left(\delta p^{(1)} + \delta b^{(1)}\right) + \left(\frac{\partial^{(1)}\hat{V}_{\parallel}^{(1)}}{\partial \hat{t}} + \left[\hat{V}_{\parallel}^{(1)}, \phi^{(1)}\right] + \left[\psi^{(1)}, \delta b^{(1)}\right]\right)\mathbf{n} \\
&+ \hat{\nabla}\left(\frac{\partial^{(1)}\phi^{(1)}}{\partial \hat{t}} - \frac{\hat{d}_{\rm i}\left[\phi^{(1)}, \delta p^{(1)}\right]}{2\,(1+\tau)}\right) \times \mathbf{n} + \frac{1}{2}\,\hat{\nabla}\left(\hat{\mathbf{V}}_{\perp}^{(1)}\cdot\hat{\mathbf{V}}_{\perp}^{(1)} + \frac{\hat{\mathbf{V}}_{\rm E}^{(1)}\cdot\hat{\mathbf{V}}_{*}^{(1)}}{1+\tau} + \delta b^{(1)}\,\delta b^{(1)} + \frac{2\,\epsilon^{(1)}\,\epsilon^{(1)}}{q_{\rm s}^{2}}\right) \\
&- \hat{\nabla}^{2}\phi^{(1)}\,\hat{\nabla}\phi^{(1)} + \frac{\hat{d}_{\rm i}}{2\,(1+\tau)}\left(\hat{\nabla}^{2}\phi^{(1)}\,\hat{\nabla}\delta p^{(1)} + \hat{\nabla}^{2}\delta p^{(1)}\,\hat{\nabla}\phi^{(1)}\right) + J^{(1)}\,\hat{\nabla}\psi^{(1)} - \Xi_{\perp}^{(1)}\,\hat{\nabla}^{2}\mathbf{V}_{\rm i}^{(1)} = 0,
\end{aligned} \tag{4.55}$$

$$\begin{aligned}
&\hat{d}_{\rm i}\,\hat{\nabla}\left(\delta p^{(1)} + \delta b^{(1)}\right) \\
&+ \left(\hat{E}_{\parallel}^{(2)} - \frac{\partial^{(1)}\psi^{(1)}}{\partial \hat{t}} - \left[\psi^{(1)}, \phi^{(1)}\right] + \frac{\hat{d}_{\rm i}}{1+\tau}\left[\psi^{(1)}, \delta p^{(1)}\right] + \hat{d}_{\rm i}\left[\psi^{(1)}, \delta b^{(1)}\right] + \hat{\eta}_{\parallel}^{(1)}\,J^{(1)}\right)\mathbf{n} \\
&+ \hat{\nabla}\left[\Upsilon^{(2)} - \hat{\eta}_{\perp}^{(1)}\left(\delta b^{(1)} - \delta p^{(1)}\right)\right] \times \mathbf{n} + \frac{\hat{d}_{\rm i}}{2}\,\hat{\nabla}\left(\delta b^{(1)}\,\delta b^{(1)} + \frac{2\,\epsilon^{(1)}\,\epsilon^{(1)}}{q_{\rm s}^{2}}\right) \\
&+ \hat{d}_{\rm i}\,J^{(1)}\,\hat{\nabla}\psi^{(1)} - 2\,\delta b^{(1)}\,\hat{\nabla}\phi^{(1)} = 0,
\end{aligned} \tag{4.56}$$

$$\frac{3}{2}\frac{\partial^{(1)}\delta p^{(1)}}{\partial\hat{t}} + \frac{3}{2}\left[\delta p^{(1)},\, \phi^{(1)}\right] + \frac{5}{2}$$
$$\hat{p}_0\left(-\frac{\partial^{(1)}\delta b^{(1)}}{\partial\hat{t}} + \left[\delta b^{(1)},\, \delta\phi^{(1)}\right] + \left[\hat{V}_\parallel^{(1)},\, \psi^{(1)}\right] + \hat{\nabla}^2\Upsilon^{(2)}\right) \tag{4.57}$$
$$+ \frac{5}{2}\lambda\,\hat{p}_0\,\hat{d}_{\rm i}\,[\delta p^{(1)},\, \delta b^{(1)}] - \hat{\chi}_\parallel^{(-1)}\left[\left[\delta p^{(1)},\, \psi^{(1)}\right],\, \psi^{(1)}\right] - \hat{\chi}_\perp^{(1)}\,\hat{\nabla}^2\delta p^{(1)} = 0.$$

To the first order, equations (4.55) and (4.56) both yield

$$\delta b^{(1)} = -\delta p^{(1)}, \tag{4.58}$$

which is simply an expression of lowest-order equilibrium force balance [3, 7].

Taking the scalar product of equation (4.55) with **n** annihilates the first-order terms, leaving

$$\frac{\partial^{(1)}\hat{V}_\parallel^{(1)}}{\partial\hat{t}} = \left[\phi^{(1)},\, \hat{V}_\parallel^{(1)}\right] - \left[\delta p^{(1)},\, \psi^{(1)}\right] + \hat{\Xi}_\perp^{(1)}\,\hat{\nabla}^2\,\hat{V}_\parallel^{(1)}. \tag{4.59}$$

Moreover, taking the scalar product of equation (4.56) with **n** annihilates the first-order terms, leaving

$$\frac{\partial^{(1)}\psi^{(1)}}{\partial\hat{t}} = \left[\phi^{(1)},\, \psi^{(1)}\right] + \hat{d}_{\rm i}\left(\frac{\tau}{1+\tau}\right)\left[\delta p^{(1)},\, \psi^{(1)}\right] + \hat{\eta}_\parallel^{(1)}\,J^{(1)} + \hat{E}_\parallel^{(2)}. \tag{4.60}$$

Furthermore, taking the scalar product of the curl of equation (4.55) with **n** annihilates the first-order terms, leaving

$$\frac{\partial^{(1)}\,U^{(1)}}{\partial\hat{t}} = \left[\phi^{(1)},\, U^{(1)}\right] + \frac{\hat{d}_{\rm i}}{2\,(1+\tau)}\left(\hat{\nabla}^2\left[\phi^{(1)},\, \delta p^{(1)}\right] + \left[\hat{\nabla}^2\phi^{(1)},\, \delta p^{(1)}\right] + \left[\hat{\nabla}^2\delta p^{(1)},\, \phi^{(1)}\right]\right)$$
$$+ \left[J^{(1)},\, \psi^{(1)}\right] + \hat{\Xi}_\perp^{(1)}\,\hat{\nabla}^4\left(\phi^{(1)} - \frac{\hat{d}_{\rm i}}{1+\tau}\,\delta p^{(1)}\right), \tag{4.61}$$

where

$$U^{(1)} = \hat{\nabla}^2\phi^{(1)}. \tag{4.62}$$

Finally, taking the scalar product of the curl of equation (4.56) with **n** annihilates the first-order terms, leaving

$$\hat{\nabla}^2\Upsilon^{(2)} = 2\left[\delta p^{(1)},\, \phi^{(1)}\right] + \hat{d}_{\rm i}\left[J^{(1)},\, \psi^{(1)}\right]. \tag{4.63}$$

The previous equation can be combined with equations (4.57) and (4.58) to give

$$\frac{\partial^{(1)}\delta p^{(1)}}{\partial\hat{t}} = \left[\phi^{(1)},\, \delta p^{(1)}\right] - c_\beta^2\left[\hat{V}_\parallel^{(1)},\, \psi^{(1)}\right] - c_\beta^2\,\hat{d}_{\rm i}\left[J^{(1)},\, \psi^{(1)}\right]$$
$$+ \frac{2}{3}(1 - c_\beta^2)\,\hat{\chi}_\parallel^{(-1)}\left[\left[\delta p^{(1)},\, \psi^{(1)}\right],\, \psi^{(1)}\right] + \frac{2}{3}(1 - c_\beta^2)\,\hat{\chi}_\perp^{(1)}\,\hat{\nabla}^2\delta p^{(1)}, \tag{4.64}$$

where

$$c_\beta = \sqrt{\frac{\beta_*}{1+\beta_*}}, \tag{4.65}$$

$$\beta_* = \frac{5}{3}\,\frac{\mu_0\,p_0}{B_z^{\,2}}. \tag{4.66}$$

4.5 The reduced drift-MHD model

Let us define

$$V = \hat{d}_{\rm i}\,\hat{V}_\parallel, \tag{4.67}$$

$$N = -\hat{d}_{\rm i}\,\delta p. \tag{4.68}$$

Equations (4.49), (4.59)–(4.62), and (4.64) yield the following closed set of equations [4]:

$$\frac{\partial\psi}{\partial\hat{t}} = [\phi,\,\psi] - \left(\frac{\tau}{1+\tau}\right)[N,\,\psi] + \hat{\eta}_\parallel\,J + \hat{E}_\parallel, \tag{4.69}$$

$$\begin{aligned}\frac{\partial N}{\partial\hat{t}} &= [\phi,\,N] + c_\beta^{\,2}\,[V,\,\psi] + \hat{d}_\beta^{\,2}\,[J,\,\psi] + \frac{2}{3}\,(1-c_\beta^2)\,\hat{\chi}_\parallel\,[[N,\,\psi],\,\psi]\\ &\quad + \frac{2}{3}\,(1-c_\beta^2)\,\hat{\chi}_\perp\,\hat{\nabla}^{\,2}N,\end{aligned} \tag{4.70}$$

$$\begin{aligned}\frac{\partial U}{\partial\hat{t}} &= [\phi,\,U] - \frac{1}{2\,(1+\tau)}\left(\hat{\nabla}^2[\phi,\,N] + [U,\,N] + [\hat{\nabla}^2 N,\,\phi]\right) + [J,\,\psi]\\ &\quad + \hat{\Xi}_\perp\,\hat{\nabla}^4\left(\phi + \frac{N}{1+\tau}\right),\end{aligned} \tag{4.71}$$

$$\frac{\partial V}{\partial\hat{t}} = [\phi,\,V] + [N,\,\psi] + \hat{\Xi}_\perp\,\hat{\nabla}^2\,V, \tag{4.72}$$

$$J = -\frac{2\,\epsilon}{q_{\rm s}} + \hat{\nabla}^2\psi, \tag{4.73}$$

$$U = \hat{\nabla}^2\phi. \tag{4.74}$$

Here, we have suppressed the ordering superscripts. Moreover, the quantity

$$d_\beta = c_\beta\,d_{\rm i} \tag{4.75}$$

is usually referred to as the *ion sound radius*, and $\hat{d}_\beta = d_\beta/l$. Note that equations (4.69)–(4.74) contain the physics of the shear Alfvén and parallel ion sound waves but do not contain the physics of the compressible Alfvén wave.

Equations (4.69)–(4.74) constitute our reduced drift-MHD resonant plasma response model. Our model is very similar to the so-called *four-field model* derived by Hazeltine *et al* [7]. Equation (4.69) is the *generalized Ohm's law* that governs the time evolution of the (normalized) helical magnetic flux, ψ. (Note that this ψ should not be confused with the equilibrium poloidal flux defined in equation (2.124).) Equation (4.70) is the *energy equation* that governs the time evolution of the (normalized) perturbed total plasma pressure, N. Equation (4.71) is the *ion parallel vorticity equation* that governs the time evolution of the (normalized) ion parallel vorticity, U. Finally, equation (4.72) is the *ion parallel equation of motion* that governs the time evolution of the (normalized) ion parallel velocity, V.

Strictly speaking, the reduced resonant response model derived in this chapter is only valid when the ion gyroradius is smaller than the typical radial variation length scale in the resonant layer. If this is not the case, then it is necessary to adopt a so-called *gyrofluid* approach in which the ion response is averaged over the ion gyro-orbits [1, 6, 11].

Finally, the crucial element of the reduction process, by which our original drift-MHD response model is converted into a reduced drift-MHD model, is the set of ordering assumptions that render $\hat{\nabla} \cdot \hat{\mathbf{V}}$ a second-order quantity while leaving $\hat{\mathbf{V}}$ a first-order quantity. This ordering implies that the MHD fluid is *incompressible* to the lowest order (see section 3.6). It should be noted, however, that the small but finite compressibility of the MHD fluid has a significant influence on the form of the energy equation (4.70).

References

[1] Cowley S C, Kulsrud R M and Hahm T S 1986 Linear stability of tearing modes *Phys. Fluids* **29** 3230

[2] Fitzpatrick R 2014 *Plasma Physics: An Introduction* (Boca Raton, FL: CRC Press)

[3] Fitzpatrick R and Porcelli F 2004 Collisionless magnetic reconnection with arbitrary guide field *Phys. Plasmas* **11** 4713

[4] Fitzpatrick R and Waelbroeck F L 2005 Two-fluid magnetic island dynamics in slab geometry. I. Isolated islands *Phys. Plasmas* **12** 022307

[5] Furth H P, Killeen J and Rosenbluth M N 1963 Finite-resistivity instabilities of a sheet pinch *Phys. Fluids* **6** 459

[6] Hammett G W, Dorland W and Perkins F W 1992 Fluid models of phase mixing, Landau damping, and nonlinear gyrokinetic dynamics *Phys. Fluids* B **4** 2052

[7] Hazeltine R D, Kotschenreuther M and Morrison P G 1985 A four-field model for tokamak plasma dynamics *Phys. Fluids* **28** 2466

[8] Hazeltine R D and Meiss J D 1985 Shear-Alfvén Dynamics of toroidally confined plasmas *Phys. Rep.* **121** 1

[9] Kadomtsev B B and Pogutse O P 1974 Nonlinear helical perturbations of a plasma in the tokamak *Sov. Phys. - JETP* **38** 283 http://jetp.ras.ru/cgi-bin/e/index/e/38/2/p283?a=list

[10] Strauss H R 1976 Nonlinear, three-dimensional magnetohydrodynamics of noncircular tokamaks *Phys. Fluids* **19** 134
[11] Waelbroeck F L, Hazeltine R D and Morrison P J 2009 A Hamiltonian electromagnetic gyrofluid model *Phys. Plasmas* **16** 032109

IOP Publishing

Tearing Mode Dynamics in Tokamak Plasmas

Richard Fitzpatrick

Chapter 5

Linear resonant response model

5.1 Introduction

As was explained in the previous two chapters, when we describe the response of a tokamak plasma to a tearing-mode perturbation, the plasma can be divided into two regions [1]. In the so-called *outer region*, which comprises most of the plasma, the tearing perturbation is governed by the equations of marginally stable ideal magnetohydrodynamics (MHD). However, these equations become singular at the so-called *rational* magnetic flux surface where the safety factor (see section 3.2) takes the rational value m/n. Here, m and n are the poloidal and toroidal mode numbers, respectively, of the tearing mode (see chapter 3). A radially thin resistive layer forms around the rational magnetic flux surface and constitutes the so-called *inner region*. The aim of this chapter is to employ the reduced drift-MHD model derived in the previous chapter to determine the *linear* response of the plasma in the inner region to a rotating tearing perturbation in the outer region.

5.2 The reduced drift-MHD model

It is convenient to set the normalization scale length, l, (see section 4.3) in our reduced drift-MHD model equal to the minor radius of the rational surface, r_s. The model evolves four scalar fields. These fields are the (normalized) helical magnetic flux, ψ, the (normalized) perturbed total plasma pressure, N, the (normalized) MHD fluid stream function, ϕ, and the (normalized) ion parallel velocity, V. Our four fields have the following definitions:

$$\hat{\nabla}\psi(\hat{r}, \zeta, \hat{t}) = \frac{\mathbf{n} \times \mathbf{B}}{B_z}, \tag{5.1}$$

$$N(\hat{r}, \zeta, \hat{t}) = -\hat{d}_i\left(\frac{p - p_0}{B_z^{\,2}/\mu_0}\right), \tag{5.2}$$

$$\hat{\nabla}\phi(\hat{r}, \zeta, \hat{t}) = \frac{\mathbf{n}\times\mathbf{V}}{V_A}, \tag{5.3}$$

$$V(\hat{r}, \zeta, \hat{t}) = \hat{d}_i\left(\frac{\mathbf{n}\cdot\mathbf{V}}{V_A}\right), \tag{5.4}$$

where

$$\mathbf{n} = \frac{\epsilon(r)}{q_s}\,\mathbf{e}_\theta + \mathbf{e}_z, \tag{5.5}$$

$$\epsilon(r) = \frac{r}{R_0}, \tag{5.6}$$

$$q_s \equiv q(r_s) = \frac{m}{n}. \tag{5.7}$$

(See equations (4.37), (4.38), (4.41), (4.50), (4.51), (4.53), (4.67), and (4.68).) Here, r, θ, z are conventional cylindrical coordinates, $\hat{r} = r/r_s$, $\hat{\nabla} = r_s\,\nabla$, $\hat{t} = t/(r_s/V_A)$, $\zeta = m\,\theta - n\,\varphi$ a helical angle, $\varphi = z/R_0$ a simulated toroidal angle (see chapter 3), $\mathbf{B}$ the magnetic field strength, B_z the (spatially constant) equilibrium magnetic field strength, V_A the Alfvén speed (see equation (4.23)), $\mathbf{V}$ the MHD fluid velocity (see equation (2.321)), d_i the collisionless ion skin depth (see equation (4.24)), $\hat{d}_i = d_i/r_s$, p the total plasma pressure, p_0 the equilibrium total plasma pressure at the rational surface, $q(r)$ the safety-factor profile, and R_0 the simulated major radius of the plasma. The reduced drift-MHD response model also employs the auxiliary fields

$$J(\hat{r}, \zeta, \hat{t}) = -2\,\frac{\epsilon_s}{q_s} + \hat{\nabla}^2\psi, \tag{5.8}$$

$$U(\hat{r}, \zeta, \hat{t}) = \hat{\nabla}^2\phi, \tag{5.9}$$

where $\epsilon_s = \epsilon(r_s)$.

The reduced drift-MHD model takes the form (see section 4.5)

$$\frac{\partial\psi}{\partial\hat{t}} = [\phi, \psi] - \left(\frac{\tau}{1+\tau}\right)[N, \psi] + \hat{\eta}_\parallel\, J + \hat{E}_\parallel, \tag{5.10}$$

$$\frac{\partial N}{\partial\hat{t}} = [\phi, N] + c_\beta^{\,2}\,[V, \psi] + \hat{d}_\beta^{\,2}\,[J, \psi] + \hat{\chi}_\parallel\,[[N, \psi], \psi] + \hat{\chi}_\perp\hat{\nabla}^2 N, \tag{5.11}$$

$$\begin{aligned}\frac{\partial U}{\partial\hat{t}} = {} & [\phi, U] - \frac{1}{2\,(1+\tau)}\left(\hat{\nabla}^2[\phi, N] + [\hat{\nabla}^2\phi, N] + [\hat{\nabla}^2 N, \phi]\right) + [J, \psi] \\ & + \hat{\Xi}_\perp\,\hat{\nabla}^4\left(\phi + \frac{N}{1+\tau}\right),\end{aligned} \tag{5.12}$$

$$\frac{\partial V}{\partial \hat{t}} = [\phi, V] + [N, \psi] + \hat{\Xi}_\perp \hat{\nabla}^2 V. \tag{5.13}$$

Here, $[A, B] \equiv \hat{\nabla} A \times \hat{\nabla} B \cdot \mathbf{n}$, $\hat{E}_\parallel$ is the (normalized) parallel inductive electric field that maintains the parallel plasma current at the rational surface against ohmic decay, τ the ratio of the electron to the ion equilibrium pressure gradient at the rational surface (see equation (4.5)), c_β a dimensionless measure of the plasma pressure at the rational surface (see equation (4.65)), $d_\beta = c_\beta\, d_i$ the ion sound radius at the rational surface, and $\hat{d}_\beta = d_\beta / r_s$. Finally,

$$\hat{\eta}_\parallel = \frac{1}{\mu_0\, \sigma_\parallel(r_s)\, r_s\, V_A}, \tag{5.14}$$

$$\hat{\Xi}_\perp = \frac{\Xi_{\perp\, i}(r_s)}{r_s\, V_A}, \tag{5.15}$$

$$\hat{\chi}_\parallel = \frac{2}{3}\left(\frac{1 - c_\beta^2}{r_s\, V_A}\right)\left[\left(\frac{\tau}{1+\tau}\right)\left(\frac{\eta_e}{1+\eta_e}\right)\chi_{\parallel\, e}(r_s) + \left(\frac{1}{1+\tau}\right)\left(\frac{\eta_i}{1+\eta_i}\right)\chi_{\parallel\, i}(r_s)\right], \tag{5.16}$$

$$\hat{\chi}_\perp = \frac{2}{3}\left(\frac{1 - c_\beta^2}{r_s\, V_A}\right)\left[\left(\frac{\tau}{1+\tau}\right)\left(\frac{\eta_e}{1+\eta_e}\right)\chi_{\perp\, e}(r_s) + \left(\frac{1}{1+\tau}\right)\left(\frac{\eta_i}{1+\eta_i}\right)\chi_{\perp\, i}(r_s)\right], \tag{5.17}$$

are the (normalized) perpendicular (to the equilibrium magnetic field) magnetic flux diffusivity, ion perpendicular momentum diffusivity, effective parallel energy diffusivity, and effective perpendicular energy diffusivity, respectively (see equations (4.15), (4.17), (4.18), and (4.19)). Here, the dimensionless parameters η_e and η_i are defined in equations (4.3) and (4.4), respectively. Moreover, $\sigma_\parallel$, $\Xi_{\perp\, i}$, $\chi_{\parallel\, e}$, $\chi_{\parallel\, i}$, $\chi_{\perp\, e}$, and $\chi_{\perp\, i}$ are the (classical) parallel electrical conductivity, the ion perpendicular momentum diffusivity, the electron parallel energy diffusivity, the ion parallel energy diffusivity, the electron perpendicular energy diffusivity, and the ion perpendicular energy diffusivity, respectively.

5.3 Plasma equilibrium

The unperturbed (by the tearing mode) plasma equilibrium is such that

$$\mathbf{B} = \frac{r\, B_z}{R_0\, q(r)}\, \mathbf{e}_\theta + B_z\, \mathbf{e}_z, \tag{5.18}$$

$$\mathbf{V} \simeq V_E(r)\, \mathbf{e}_\theta + V_{\parallel\, i}(r)\, \mathbf{e}_z, \tag{5.19}$$

$$p = p(r). \tag{5.20}$$

(see sections 2.11, 2.24, and 3.2). Here,

$$V_{\rm E}(r) \simeq -\frac{E_{\rm r}}{B_z} \tag{5.21}$$

is the (dominant θ-component of the) E-cross-B velocity (see equation (2.138)) and $V_{\parallel\,{\rm i}}$ the ion parallel fluid velocity.

The resonant layer is assumed to have a radial thickness that is much smaller than $r_{\rm s}$ (see section 5.5). Hence, we only need to evaluate plasma equilibrium quantities in the immediate vicinity of the rational surface. Equations (4.1), (4.23), (4.24), and (5.1)–(5.4) imply that

$$\psi(\hat{x}) = \frac{\hat{x}^{\,2}}{2\,\hat{L}_{\rm s}}, \tag{5.22}$$

$$N(\hat{x}) = -\hat{V}_*\,\hat{x}, \tag{5.23}$$

$$\phi(\hat{x}) = -\hat{V}_{\rm E}\,\hat{x}, \tag{5.24}$$

$$V(\hat{x}) = \hat{V}_{\parallel}, \tag{5.25}$$

where

$$\hat{x} = \frac{r - r_{\rm s}}{r_{\rm s}}, \tag{5.26}$$

$\hat{L}_{\rm s} = L_{\rm s}/r_{\rm s}$,

$$L_{\rm s} = \frac{R_0\,q_{\rm s}}{s_{\rm s}} \tag{5.27}$$

is the *magnetic shear length*, $s_{\rm s} = s(r_{\rm s})$,

$$s(r) = \frac{d\ln q}{d\ln r} \tag{5.28}$$

the *magnetic shear*, $\hat{V}_{\rm E} = V_{\rm E}(r_{\rm s})/V_{\rm A}$, $\hat{V}_* = V_*(r_{\rm s})/V_{\rm A}$,

$$V_*(r) = \frac{1}{e\,n_0\,B_z}\,\frac{dp}{dr} \tag{5.29}$$

the (dominant θ-component of the) diamagnetic velocity (see equation (4.11)), and $\hat{V}_{\parallel} = \hat{d}_{\rm i}\;V_{\parallel\,{\rm i}}(r_{\rm s})/V_{\rm A}$. We also have

$$J(\hat{x}) = -\left(\frac{2}{s_{\rm s}} - 1\right)\frac{1}{\hat{L}_{\rm s}}, \tag{5.30}$$

$$U(\hat{x}) = 0, \tag{5.31}$$

$$\hat{E}_{\parallel}(\hat{x}) = \left(\frac{2}{s_{\rm s}} - 1\right)\frac{\hat{\eta}_{\parallel}}{\hat{L}_{\rm s}}. \tag{5.32}$$

Note that we are neglecting any radial gradients in the equilibrium MHD fluid velocity because, in conventional tokamak plasmas, such gradients do not significantly affect the linear response of a resonant layer, due to the fact that the gradients are comparatively weak (i.e. $|dV/dr| \ll V_A/r_s$) combined with the fact that a linear resonant layer is very narrow [2] (see section 5.5).

5.4 Linearized reduced drift-MHD equations

Let us now take the tearing-mode perturbation into account. In accordance with equations (5.22)–(5.25) and (5.30)–(5.31), we can write

$$\psi(\hat{x}, \zeta, \hat{t}) = \frac{\hat{x}^{2}}{2\,\hat{L}_{\rm s}} + \tilde{\psi}(\hat{x})\,{\rm e}^{\,{\rm i}\,(\zeta-\hat{\omega}\,\hat{t})}, \tag{5.33}$$

$$N(\hat{x}, \zeta, \hat{t}) = -\hat{V}_{*}\,\hat{x} + \left(\frac{1+\tau}{\tau}\right)\tilde{N}(\hat{x})\,{\rm e}^{\,{\rm i}\,(\zeta-\hat{\omega}\,\hat{t})}, \tag{5.34}$$

$$\phi(\hat{x}, \zeta, \hat{t}) = -\hat{V}_{\rm E}\,\hat{x} + \tilde{\phi}(\hat{x})\,{\rm e}^{\,{\rm i}\,(\zeta-\hat{\omega}\,\hat{t})}, \tag{5.35}$$

$$V(\hat{x}, \zeta, \hat{t}) = \hat{V}_{\parallel} + \left(\frac{1+\tau}{\tau}\right)\tilde{V}(\hat{x})\,{\rm e}^{\,{\rm i}\,(\zeta-\hat{\omega}\,\hat{t})}, \tag{5.36}$$

$$J(\hat{x}, \zeta, \hat{t}) = -\left(\frac{2}{s_{\rm s}} - 1\right)\frac{1}{\hat{L}_{\rm s}} + \hat{\nabla}^{2}\tilde{\psi}(\hat{x})\,{\rm e}^{\,{\rm i}\,(\zeta-\hat{\omega}\,\hat{t})}, \tag{5.37}$$

$$U(\hat{x}, \zeta, \hat{t}) = \hat{\nabla}^{2}\tilde{\phi}(\hat{x})\,{\rm e}^{\,{\rm i}\,(\zeta-\hat{\omega}\,\hat{t})}, \tag{5.38}$$

where $\hat{\omega} = r_{\rm s}\,\omega/V_{\rm A}$, and ω is the rotational frequency of the tearing perturbation in the laboratory frame. Here, $\tilde{A}$ denotes a perturbed quantity. By substituting equations (5.33)–(5.38) into the reduced drift-MHD model, (5.8)–(5.13), and only retaining first-order terms in perturbed quantities, we obtain the following set of linear equations:

$$-{\rm i}\,(\omega - \omega_{\rm E} - \omega_{*\,{\rm e}})\,\tau_{\rm H}\,\tilde{\psi} = -{\rm i}\,\hat{x}\,(\tilde{\phi}-\tilde{N}) + S^{-1}\,\hat{\nabla}^{2}\tilde{\psi}, \tag{5.39}$$

$$\begin{aligned} -{\rm i}\,(\omega - \omega_{\rm E})\,\tau_{\rm H}\,\tilde{N} = &-{\rm i}\,\omega_{*\,{\rm e}}\,\tau_{\rm H}\,\tilde{\phi} - {\rm i}\,c_{\beta}^{2}\,\hat{x}\,\tilde{V} - {\rm i}\left(\frac{\tau}{1+\tau}\right)\hat{d}_{\beta}^{\,2}\,\hat{x}\,\hat{\nabla}^{2}\tilde{\psi} \\ &+ S^{-1}\,P_{\parallel}(\omega_{*\,{\rm e}}\,\tau_{\rm H}\,\hat{x}\,\tilde{\psi} - \hat{x}^{2}\,\tilde{N}) + S^{-1}\,P_{\perp}\,\hat{\nabla}_{\perp}^{2}\tilde{N}, \end{aligned} \tag{5.40}$$

$$-{\rm i}\,(\omega - \omega_{\rm E} - \omega_{*\,{\rm i}})\,\tau_{\rm H}\,\hat{\nabla}^{2}\tilde{\phi} = -{\rm i}\,\hat{x}\,\hat{\nabla}^{2}\tilde{\psi} + S^{-1}\,P_{\varphi}\,\hat{\nabla}^{4}\left(\tilde{\phi}+\frac{\tilde{N}}{\tau}\right), \tag{5.41}$$

$$-\mathrm{i}\,(\omega - \omega_\mathrm{E})\,\tau_\mathrm{H}\,\tilde{V} = \mathrm{i}\,\omega_{*\,\mathrm{e}}\,\tau_\mathrm{H}\,\tilde{\psi} - \mathrm{i}\,\hat{x}\,\tilde{N} + S^{-1}\,P_\varphi\,\hat{\nabla}^2\tilde{V}. \tag{5.42}$$

Here,

$$\tau_\mathrm{H} = \frac{L_\mathrm{s}}{m\,V_\mathrm{A}} \tag{5.43}$$

is the *hydromagnetic time*,

$$\omega_\mathrm{E} = \frac{m}{r_\mathrm{s}}\,V_\mathrm{E}(r_\mathrm{s}) \tag{5.44}$$

the *E-cross-B frequency*,

$$\omega_{*\,\mathrm{e}} = \left(\frac{\tau}{1+\tau}\right)\omega_*, \tag{5.45}$$

the *electron diamagnetic frequency*,

$$\omega_{*\,\mathrm{i}} = -\left(\frac{1}{1+\tau}\right)\omega_*, \tag{5.46}$$

the *ion diamagnetic frequency*,

$$\omega_* = -\frac{m}{r_\mathrm{s}}\,V_*(r_\mathrm{s}) = -\frac{m}{r_\mathrm{s}}\frac{1}{e\,n_0\,B_z}\left.\frac{dp}{dr}\right|_{r_\mathrm{s}}, \tag{5.47}$$

the *(total) diamagnetic frequency*,

$$S = \frac{\tau_\mathrm{R}}{\tau_\mathrm{H}} \tag{5.48}$$

the *Lundquist number* (note that this is a slightly different definition to that given in equation (1.84)),

$$\tau_\mathrm{R} = \mu_0\,r_\mathrm{s}^{\,2}\,\sigma_\parallel(r_\mathrm{s}) \tag{5.49}$$

the *resistive diffusion time* (note that this is a slightly different definition to that given in equation (1.83)),

$$\tau_\varphi = \frac{r_\mathrm{s}^{\,2}}{\Xi_{\perp\,\mathrm{i}}(r_\mathrm{s})} \tag{5.50}$$

the *toroidal momentum confinement time*,

$$\tau_\parallel = r_\mathrm{s}^{\,2}\Bigg/\left\{\frac{2}{3}\,(1-c_\beta^2)\left[\left(\frac{\tau}{1+\tau}\right)\left(\frac{\eta_\mathrm{e}}{1+\eta_\mathrm{e}}\right)\chi_{\parallel\,\mathrm{e}}(r_\mathrm{s}) + \left(\frac{1}{1+\tau}\right)\left(\frac{\eta_\mathrm{i}}{1+\eta_\mathrm{i}}\right)\chi_{\parallel\,\mathrm{i}}(r_\mathrm{s})\right]\right\} \tag{5.51}$$

the effective *parallel energy equilibration time*, and

$$\tau_\perp = r_s^{\,2} \Bigg/ \left\{\frac{2}{3}\,(1 - c_\beta^2)\left[\left(\frac{\tau}{1+\tau}\right)\left(\frac{\eta_e}{1+\eta_e}\right)\chi_{\perp\,e}(r_s) + \left(\frac{1}{1+\tau}\right)\left(\frac{\eta_i}{1+\eta_i}\right)\chi_{\perp\,i}(r_s)\right]\right\} \tag{5.52}$$

the effective *energy confinement time*. Furthermore,

$$P_\varphi = \frac{\tau_R}{\tau_\varphi}, \tag{5.53}$$

$$P_\perp = \frac{\tau_R}{\tau_\perp}, \tag{5.54}$$

$$P_\parallel = (n\,\epsilon_s\,s_s)^2\,\frac{\tau_R}{\tau_\parallel}. \tag{5.55}$$

5.5 Resonant-layer equations

In a conventional tokamak plasma, the Lundquist number, S, which is the nominal ratio of the plasma inertia term to the resistive diffusion term in the plasma Ohm's law [3], is very much greater than unity. In fact, according to table 1.5, S typically exceeds 10^8 in a tokamak fusion reactor. However, a resonant layer is characterized by a balance between plasma inertia and resistive diffusion [1]. Such a balance is only possible if the layer is very narrow in the radial direction (because a narrow layer enhances radial derivatives and thereby enhances resistive diffusion). Let us define the stretched radial variable [4]

$$X = S^{1/3}\,\hat{x}. \tag{5.56}$$

Assuming that $X \sim \mathcal{O}(1)$ in the layer (i.e. assuming that the layer thickness is roughly of the order of $S^{-1/3}\,r_s$) and making use of the fact that $S \gg 1$, we deduce that $\hat{\nabla}^2 \simeq \hat{\nabla}_\perp^2 \simeq S^{2/3}\,d^2/dX^2$. Hence, the linear equations (5.39)–(5.42) reduce to the following set of resonant-layer equations [5, 6]:

$$-\mathrm{i}\,(Q - Q_E - Q_e)\,\tilde{\psi} = -\mathrm{i}\,X\,(\tilde{\phi} - \tilde{N}) + \frac{d^2\tilde{\psi}}{dX^2}, \tag{5.57}$$

$$-\mathrm{i}\,(Q - Q_E)\,\tilde{N} = -\mathrm{i}\,Q_e\,\tilde{\phi} - \mathrm{i}\,c_\beta^2\,X\,\tilde{V} - \mathrm{i}\,D^2\,X\,\frac{d^2\tilde{\psi}}{dX^2} + \hat{P}_\parallel\,X\,(Q_e\,\tilde{\psi} - X\,\tilde{N}) + P_\perp\,\frac{d^2\tilde{N}}{dX^2}, \tag{5.58}$$

$$-\mathrm{i}\,(Q - Q_{\mathrm{E}} - Q_{\mathrm{i}})\,\frac{d^2\tilde{\phi}}{dX^2} = -\mathrm{i}\,X\,\frac{d^2\tilde{\psi}}{dX^2} + P_\varphi\,\frac{d^4}{dX^4}\left(\tilde{\phi}+\frac{\tilde{N}}{\tau}\right), \tag{5.59}$$

$$-\mathrm{i}\,(Q - Q_{\mathrm{E}})\,\tilde{V} = \mathrm{i}\,Q_{\mathrm{e}}\,\tilde{\psi} - \mathrm{i}\,X\,\tilde{N} + P_\varphi\,\frac{d^2\tilde{V}}{dX^2}. \tag{5.60}$$

Here,

$$Q = S^{1/3}\,\omega\,\tau_{\mathrm{H}}, \tag{5.61}$$

$$Q_{\mathrm{E}} = S^{1/3}\,\omega_{\mathrm{E}}\,\tau_{\mathrm{H}}, \tag{5.62}$$

$$Q_{\mathrm{e}} = S^{1/3}\,\omega_{*\,\mathrm{e}}\,\tau_{\mathrm{H}} = \left(\frac{\tau}{1+\tau}\right) Q_*, \tag{5.63}$$

$$Q_{\mathrm{i}} = S^{1/3}\,\omega_{*\,\mathrm{i}}\,\tau_{\mathrm{H}} = -\left(\frac{1}{1+\tau}\right) Q_*, \tag{5.64}$$

$$Q_* = S^{1/3}\,\omega_*\,\tau_{\mathrm{H}}, \tag{5.65}$$

$$D = S^{1/3}\left(\frac{\tau}{1+\tau}\right)^{1/2}\hat{d}_\beta, \tag{5.66}$$

$$\hat{P}_\parallel = \frac{P_\parallel}{S^{4/3}}. \tag{5.67}$$

Table 5.1 gives estimates for the values of the dimensionless parameters that characterize the resonant-layer equations, (5.57)–(5.60), in a low-field fusion reactor and a high-field tokamak fusion reactor (see chapter 1). These estimates are made using the following assumptions: $B = 5$ T (low field) or $B = 12$ T (high field), $\beta = 0.02$, $T_{\mathrm{e}} = T_{\mathrm{i}} = 7$ keV, $m_{\mathrm{i}} = (m_{\mathrm{D}} + m_{\mathrm{T}})/2$ (where m_{D} and m_{T} are the deuteron and triton masses, respectively), $\Xi_{\perp\,\mathrm{i}} = \chi_{\perp\,\mathrm{e}} = \chi_{\perp\,\mathrm{i}} = 1\ \mathrm{m^2\,s^{-1}}$, $m = 2$, $n = 1$, $r_{\mathrm{s}} = a/2$ (where a is the minor radius of the plasma), $s_{\mathrm{s}} = 1$, $\tau = 1$, $\eta_{\mathrm{e}} = \eta_{\mathrm{i}} = 1$, and $dp/dr = -p/a$. The parallel energy diffusivities, $\chi_{\parallel\,\mathrm{e}}$ and $\chi_{\parallel\,\mathrm{e}}$, are estimated from equations (2.319) and (2.320), respectively, using $k_\parallel \simeq (m/L_{\mathrm{s}})\,S^{-1/3}$, which is the typical parallel wave number of the tearing perturbation at the edge of a resistive layer whose characteristic thickness is $S^{-1/3}\,r_{\mathrm{s}}$. Note that $\hat{P}_\parallel \ll P_\perp$, which allows us to neglect the parallel transport terms (i.e. the terms involving $\hat{P}_\parallel$) in equation (5.58). The neglect of these terms is justified because the linear layer width is much less than the critical width,

$$W_d = \sqrt{8}\left(\frac{\tau_\parallel}{\tau_\perp}\right)^{1/4}\left(\frac{1}{\epsilon_{\mathrm{s}}\,s_{\mathrm{s}}\,n}\right)^{1/2} r_{\mathrm{s}}, \tag{5.68}$$

Table 5.1. Dimensionless parameters that appear in resonant-layer equations for a low-field tokamak reactor and a high-field tokamak reactor. See equations (4.5), (5.65), (5.66), (5.53), (5.54), and (4.65).

	Low field	High field
B (T)	5.0	12.0
τ	1.0	1.0
Q_*	2.00	1.50
D	3.02	2.26
P_φ	874	874
$P_\perp$	287	287
$\hat{P}_\parallel$	0.635	0.635
c_β	0.128	0.128

below which parallel transport is unable to constrain the perturbed electron temperature to be constant along magnetic field lines [7].

5.6 Asymptotic matching

The resonant-layer equations, (5.57)–(5.60), possess independent *tearing parity* solutions, characterized by the symmetry $\tilde{\psi}(-X) = \tilde{\psi}(X)$, $\tilde{\phi}(-X) = -\tilde{\phi}(X)$, $\tilde{N}(-X) = -\tilde{N}(X)$, $\tilde{V}(-X) = \tilde{V}(X)$ and *twisting parity* solutions, characterized by the symmetry $\tilde{\psi}(-X) = -\tilde{\psi}(X)$, $\tilde{\phi}(-X) = \tilde{\phi}(X)$, $\tilde{N}(-X) = \tilde{N}(X)$, $\tilde{V}(-X) = -\tilde{V}(X)$ [8]. However, only the tearing parity solutions can be asymptotically matched to tearing-mode solutions in the outer region. If we assume that the asymptotic behavior of the tearing parity layer solutions is such that

$$\tilde{\psi}(X) \to \psi_0\left[1 + \frac{\hat{\Delta}}{2}\,|X| + \mathcal{O}(X^2)\right] \tag{5.69}$$

as $|X| \to \infty$, where ψ_0 is an arbitrary constant, then asymptotic matching to the outer solution yields

$$\Delta\Psi_s = S^{1/3}\,\hat{\Delta}\,\Psi_s. \tag{5.70}$$

(See equations (3.72), (3.73), and (5.56).) Furthermore, it follows from equations (3.19) and (3.58) that the asymptotic behavior of $\tilde{\phi}(X)$ is

$$\tilde{\phi}(X) \to \phi_0\left[\frac{1}{X} + \frac{\hat{\Delta}}{2}\,\mathrm{sgn}(X) + \mathcal{O}(X)\right] \tag{5.71}$$

as $|X| \to \infty$, where ϕ_0 is an arbitrary constant.

5.7 Fourier transformation

Equations (5.57)–(5.60) are most conveniently solved in Fourier transform space [5, 6, 9]. Let

$$\bar{\phi}(p) = \int_{-\infty}^{\infty} \tilde{\phi}(X)\, \mathrm{e}^{-\mathrm{i}\, p\, X}\, dX, \tag{5.72}$$

etc. The Fourier transformed layer equations become

$$-\mathrm{i}\,(Q - Q_{\mathrm{E}} - Q_{\mathrm{e}})\,\bar{\psi} = \frac{d}{dp}(\bar{\phi} - \bar{N}) - p^2\,\bar{\psi}, \tag{5.73}$$

$$-\mathrm{i}\,(Q - Q_{\mathrm{E}})\,\bar{N} = -\mathrm{i}\,Q_{\mathrm{e}}\,\bar{\phi} + c_\beta^{\,2}\,\frac{d\bar{V}}{dp} - D^2\,\frac{d(p^2\,\bar{\psi})}{dp} - P_\perp\,p^2\,\bar{N}, \tag{5.74}$$

$$-\mathrm{i}\,(Q - Q_{\mathrm{E}} - Q_{\mathrm{i}})\,p^2\,\bar{\phi} = \frac{d(p^2\,\bar{\psi})}{dp} - P_\varphi\,p^4\left(\bar{\phi} + \frac{\bar{N}}{\tau}\right), \tag{5.75}$$

$$-\mathrm{i}\,(Q - Q_{\mathrm{E}})\,\bar{V} = \mathrm{i}\,Q_{\mathrm{e}}\,\bar{\psi} + \frac{d\bar{N}}{dp} - P_\varphi\,p^2\,\bar{V}, \tag{5.76}$$

where, for a tearing parity solution [10],

$$\bar{\phi}(p) \to \bar{\phi}_0\left[\frac{\hat{\Delta}}{\pi\,p} + 1 + \mathcal{O}(p)\right] \tag{5.77}$$

as $p \to 0$. Here, $\bar{\phi}_0$ is an arbitrary constant, and use has been made of equation (5.71).

In accordance with our previous discussion, we have neglected the parallel transport terms in equation (5.74). Let us also ignore the term $c_\beta^{\,2}\,d\bar{V}/dp$. This approximation can be justified *a posteriori*. It is equivalent to the neglect of the contribution of ion parallel dynamics to the linear plasma response in the resonant layer and effectively decouples equation (5.76) from equations (5.73)–(5.75) [11]. Equations (5.73)–(5.75) reduce to [5, 6]

$$\frac{d}{dp}\left[A(p)\,\frac{dY_{\mathrm{e}}}{dp}\right] - \frac{B(p)}{C(p)}\,p^2\;Y_{\mathrm{e}} = 0, \tag{5.78}$$

where

$$A(p) = \frac{p^2}{-\mathrm{i}\,(Q - Q_{\mathrm{E}} - Q_{\mathrm{e}}) + p^2}, \tag{5.79}$$

$$B(p) = -(Q - Q_{\mathrm{E}})\,(Q - Q_{\mathrm{E}} - Q_{\mathrm{i}}) - \mathrm{i}\,(Q - Q_{\mathrm{E}} - Q_{\mathrm{i}})\,(P_\varphi + P_\perp)\,p^2 + P_\varphi\,P_\perp\,p^4, \tag{5.80}$$

$$\begin{aligned} C(p) = {} & -\mathrm{i}\,(Q - Q_{\mathrm{E}} - Q_{\mathrm{e}}) + \left[P_\perp - \mathrm{i}\,(Q - Q_{\mathrm{E}} - Q_{\mathrm{i}})\,D^2\right]p^2 \\ & + (1 + 1/\tau)\,P_\varphi\,D^2\,p^4, \end{aligned} \tag{5.81}$$

and $Y_{\mathrm{e}} = \bar{\phi} - \bar{N}$ is the normalized, Fourier transformed, electron fluid stream function. It can easily be demonstrated that

$$Y_e(p) \to \left(\frac{Q - Q_E - Q_e}{Q - Q_E}\right)\bar{\phi}(p) \tag{5.82}$$

as $p \to 0$. Hence, the boundary conditions on equation (3.19) are that $Y_e(p)$ is bounded as $p \to \infty$, and

$$Y_e(p) \to Y_0\left[\frac{\hat{\Delta}}{\pi\, p} + 1 + \mathcal{O}(p)\right] \tag{5.83}$$

as $p \to 0$, where Y_0 is an arbitrary constant. Here, use has been made of equation (5.77). In the following, we shall assume that $|Q| \sim |Q_E| \sim Q_*$, $P_\varphi \sim P_\perp \sim P$, and $\tau \sim \mathcal{O}(1)$, for the sake of simplicity.

5.8 The constant-ψ limit

Let us suppose that there are two layers in p space. In the small-p layer, suppose that equation (3.19) reduces to

$$\frac{d}{dp}\left[\frac{p^2}{-\mathrm{i}\,(Q - Q_E - Q_e) + p^2}\,\frac{dY_e}{dp}\right] \simeq 0 \tag{5.84}$$

when $p \sim Q^{1/2}$. By direct integration, we find that

$$Y_e(p) \simeq Y_0\left\{\frac{\hat{\Delta}}{\pi}\left[\frac{1}{p} + \frac{p}{\mathrm{i}\,(Q - Q_E - Q_e)}\right] + 1 + \mathcal{O}(p^2)\right\} \tag{5.85}$$

for $p \lesssim \mathcal{O}(Q^{1/2})$, where use has been made of equation (5.83). The two-layer approximation is equivalent to the well-known *constant-ψ approximation* [1].

In the large-p layer, for $p \gg \mathcal{O}(Q^{1/2})$, we obtain

$$\frac{d^2 Y_e}{dp^2} - \frac{B(p)}{C(p)}\,p^2\, Y_e \simeq 0, \tag{5.86}$$

with $Y_e(p)$ bounded as $p \to \infty$. Asymptotic matching to the small-p layer solution (5.85) yields the boundary condition

$$Y_e(p) \simeq Y_0\left[1 + \frac{\hat{\Delta}}{\pi}\,\frac{p}{\mathrm{i}\,(Q - Q_E - Q_e)} + \mathcal{O}(p^2)\right] \tag{5.87}$$

as $p \to 0$.

In the various constant-ψ linear response regimes considered in section 5.9, equation (5.86) reduces to an equation of the form

$$\frac{d^2 Y_e}{dp^2} - G\,p^k\, Y_e \simeq 0, \tag{5.88}$$

where k is real and nonnegative, and G is a complex constant. Let $Y_e = \sqrt{p}\, Z$ and $q = \sqrt{G}\, p^j/j$, where $j = (k+2)/2$. The previous equation transforms into a modified Bessel equation of general order,

$$q^2\,\frac{d^2Z}{dq^2} + q\,\frac{dZ}{dq} - (q^2+\nu^2)\,Z = 0, \tag{5.89}$$

where $\nu = 1/(k+2)$. The solution that is bounded as $q\to\infty$ has the small-q expansion [12]

$$K_\nu(q) = \frac{1}{\Gamma(1-\nu)}\left(\frac{2}{q}\right)^{\nu} - \frac{1}{\Gamma(1+\nu)}\left(\frac{q}{2}\right)^{\nu} + \mathcal{O}(q^{2-\nu}), \tag{5.90}$$

where $\Gamma(z)$ is a gamma function. A comparison of this expression with equation (5.87) reveals that

$$\hat{\Delta} = \frac{\nu^{2\nu-1}\,\pi\,\Gamma(1-\nu)}{\Gamma(\nu)}\left[-\mathrm{i}\left(Q - Q_E - Q_e\right)\right]G^{\nu}. \tag{5.91}$$

Note, finally, that $p_* \sim |G|^{-\nu}$, where p_* denotes the width of the large-p layer in p space. This width must be larger than $Q^{1/2}$ (i.e. the width of the small-p layer) in order for the constant-ψ approximation to hold. Finally, it can easily be demonstrated that the neglect of the term involving c_β in equation (5.74) is justified provided that $c_\beta \ll (Q + P\,p_*^2)\,p_*$.

5.9 Constant-ψ linear resonant response regimes

Suppose that $Q \gg P\,p^2$ and $D^2\,p^2 \ll 1$. It follows that $k=2$, $j=2$, $\nu = 1/4$, and

$$G = \frac{[-\mathrm{i}\,(Q-Q_E)]\,[-\mathrm{i}\,(Q-Q_E-Q_i)]}{-\mathrm{i}\,(Q-Q_E-Q_e)}. \tag{5.92}$$

Hence, we deduce that

$$\hat{\Delta} = \frac{2\pi\,\Gamma(3/4)}{\Gamma(1/4)}\,[-\mathrm{i}\,(Q-Q_E-Q_e)]^{3/4}\,[-\mathrm{i}\,(Q-Q_E)]^{1/4}\,[-\mathrm{i}\,(Q-Q_E-Q_i)]^{1/4}. \tag{5.93}$$

This response regime is known as the *resistive–inertial regime* because the layer response is dominated by plasma resistivity and ion inertia [4, 13]. The characteristic layer width is $p_* \sim Q^{-1/4}$, which implies that the regime is valid when $P \ll Q^{3/2}$, $Q \gg D^4$, $Q \ll 1$, and $c_\beta \ll Q^{3/4}$.

Suppose that $Q \ll P\,p^2$ and $D^2\,p^2 \ll 1$. It follows that $k=4$, $j=3$, $\nu = 1/6$, and $G = P_\varphi$. Hence, we deduce that

$$\hat{\Delta} = \frac{6^{2/3}\,\pi\,\Gamma(5/6)}{\Gamma(1/6)}\,[-\mathrm{i}\left(Q-Q_E-Q_e\right)]\,P_\varphi^{1/6}. \tag{5.94}$$

This response regime is known as the *viscous–resistive regime* because the layer response is dominated by ion perpendicular viscosity and plasma resistivity [14, 15]. The characteristic layer width is $p_* \sim P^{-1/6}$, which implies that the regime is valid when $P \gg Q^{3/2}$, $P \gg D^6$, $P \ll Q^{-3}$, and $c_\beta \ll P^{1/2}$.

Suppose that $Q \gg P\,p^2$ and $D^2\,p^2 \gg 1$. It follows that $k = 0$, $j = 1$, $\nu = 1/2$, and

$$G = \frac{-\mathrm{i}\,(Q - Q_\mathrm{E})}{D^2}. \tag{5.95}$$

Hence, we deduce that

$$\hat{\Delta} = \frac{\pi\,[-\mathrm{i}\,(Q - Q_\mathrm{E} - Q_\mathrm{e})]\,[-\mathrm{i}\,(Q - Q_\mathrm{E})]^{1/2}}{D}. \tag{5.96}$$

This response regime is known as the *semi-collisional regime* [16, 17]. The characteristic layer width is $p_* \sim Q^{-1/2}\,D$, which implies that the regime is valid when $Q \ll D^4$, $P \ll Q^2/D^2$, $Q \ll D$, and $c_\beta \ll Q^{1/2}\,D$.

Suppose, finally, that $Q \ll P\,p^2$ and $D^2\,p^2 \gg 1$. It follows that $k = 2$, $j = 2$, $\nu = 1/4$, and

$$G = \frac{P_\perp}{(1 + 1/\tau)\,D^2}. \tag{5.97}$$

Hence, we deduce that

$$\hat{\Delta} = \frac{2\pi\,\Gamma(3/4)}{\Gamma(1/4)}\,\frac{[-\mathrm{i}\,(Q - Q_\mathrm{E} - Q_\mathrm{e})]\,P_\perp^{1/4}}{(1 + 1/\tau)^{1/4}\,D^{1/2}}. \tag{5.98}$$

This response regime is known as the *diffusive–resistive regime* because the layer response is dominated by perpendicular energy diffusivity and plasma resistivity [6]. The characteristic layer width is $p_* \sim P^{-1/4}\,D^{1/2}$, which implies that the regime is valid when $P \gg Q^2/D^2$, $P \ll D^6$, $P \ll D^2/Q^2$, and $c_\beta \ll P^{1/4}\,D^{3/2}$.

5.10 The nonconstant-ψ limit

Suppose that $p \ll Q^{1/2}$. In this limit, equation (3.19) reduces to

$$\frac{d}{dp}\left(p^2\,\frac{dY_\mathrm{e}}{dp}\right) - [-\mathrm{i}\,(Q - Q_\mathrm{E} - Q_\mathrm{e})]\,\frac{B(p)}{C(p)}\,p^2\,Y_\mathrm{e} = 0. \tag{5.99}$$

In the various nonconstant-ψ regimes considered in section 5.11, the previous equation takes the form

$$\frac{d}{dp}\left(p^2\,\frac{dY_\mathrm{e}}{dp}\right) - G\,p^{k+2}\,Y_\mathrm{e} = 0, \tag{5.100}$$

where k is real and nonnegative, and G is a complex constant. Let $U = p\,Y_\mathrm{e}$. The previous equation yields

$$\frac{d^2U}{dp^2} - G\,p^k\,U = 0. \tag{5.101}$$

This equation is identical in form to equation (5.88), which we have already solved. Indeed, the solution that is bounded as $p \to \infty$ has the small-p expansion (5.90), where $q = \sqrt{G}\,p^j/j$, $j = (k+2)/2$, and $\nu = 1/(k+2)$. Matching to equation (5.83) yields

$$\hat{\Delta} = -\frac{\pi\,\nu^{1-2\nu}\Gamma(\nu)}{\Gamma(1-\nu)}\,G^{-\nu}. \tag{5.102}$$

The layer width in p space again scales as $p_* \sim |G|^{-\nu}$. This width must be less than $Q^{1/2}$. As before, the neglect of the term involving c_β in equation (5.74) is justified provided that $c_\beta \ll (Q + P\,p_*^2)\,p_*$.

5.11 Nonconstant-ψ linear resonant response regimes

Suppose that $Q \gg P\,p^2$ and $D^2\,p^2 \ll 1$. It follows that $k = 0$, $j = 1$, $\nu = 1/2$, and

$$G = [-\mathrm{i}\,(Q - Q_E)]\,[-\mathrm{i}\,(Q - Q_E - Q_i)]. \tag{5.103}$$

Hence, we deduce that

$$\hat{\Delta} = -\frac{\pi}{[-\mathrm{i}\,(Q - Q_E)]^{1/2}\,[-\mathrm{i}\,(Q - Q_E - Q_i)]^{1/2}}. \tag{5.104}$$

This response regime is known as the *inertial regime*, because the layer response is dominated by ion inertia [4, 13]. Note that the plasma response in the inertial regime is equivalent to that of two closely spaced shear-Alfvén resonances that straddle the rational surface [18]. In fact, it can easily be demonstrated that in real space,

$$\hat{\Delta} = \int_{-\infty}^{\infty} \frac{dX}{(Q - Q_E)\,(Q - Q_E - Q_i) - X^2}, \tag{5.105}$$

which suggests that the resonances lie at $X = \pm\sqrt{(Q - Q_E)\,(Q - Q_E - Q_i)}$. The characteristic layer width is $p_* \sim Q^{-1}$, which implies that the regime is valid when $P \ll Q^3$, $Q \gg D$, $Q \gg 1$, and $c_\beta \ll 1$.

Suppose that $Q \ll P\,p^2$ and $D^2\,p^2 \ll 1$. It follows that $k = 2$, $j = 2$, $\nu = 1/4$, and

$$G = [-\mathrm{i}\,(Q - Q_E - Q_e)]\,P_\varphi. \tag{5.106}$$

Hence, we deduce that

$$\hat{\Delta} = -\frac{\pi}{2}\,\frac{\Gamma(1/4)}{\Gamma(3/4)}\,[-\mathrm{i}\,(Q - Q_E - Q_e)]^{-1/4}\,P_\varphi^{-1/4}. \tag{5.107}$$

This response regime is known as the *viscous–inertial regime* because the layer response is dominated by ion perpendicular viscosity and ion inertia [13]. The characteristic layer width is $p_* \sim Q^{-1/4}\,P^{-1/4}$, which implies that the regime is valid when $P \gg Q^3$, $P \gg D^4/Q$, $P \gg Q^{-3}$, and $c_\beta \ll Q^{-3/4}\,P^{1/4}$.

Suppose, finally, that $Q \ll P\,p^2$ and $D^2\,p^2 \gg 1$. It follows that $k = 0$, $j = 1$, $\nu = 1/2$, and

$$G = \frac{[-\mathrm{i}\,(Q - Q_E - Q_e)]\,P_\perp}{(1 + 1/\tau)\,D^2}. \tag{5.108}$$

Hence, we deduce that

$$\hat{\Delta} = -\frac{\pi\,(1 + 1/\tau)^{1/2}\,D}{[-\mathrm{i}\,(Q - Q_E - Q_e)]^{1/2}\,P_\perp^{1/2}}. \tag{5.109}$$

This response regime is known as the *diffusive–inertial regime*, because the layer response is dominated by perpendicular energy diffusivity and ion inertia [6]. The characteristic layer width is $p_* \sim Q^{-1/2}\,P^{-1/2}\,D$, which implies that the regime is valid when $Q \ll D$, $P \ll D^4/Q$, $P \gg D^2/Q^2$, and $c_\beta \ll Q^{-3/2}\,P^{-1/2}\,D^3$.

5.12 Linear resonant response regimes

Table 5.2 summarizes the properties of the various linear resonant response regimes found in sections 5.9 and 5.11. Here, we have made use of the abbreviations

$$R_e = -\mathrm{i}\,(Q - Q_E - Q_e), \tag{5.110}$$

$$R_E = -\mathrm{i}\,(Q - Q_E), \tag{5.111}$$

$$R_i = -\mathrm{i}\,(Q - Q_E - Q_i), \tag{5.112}$$

$$P_* = \frac{P_\perp}{1 + 1/\tau}. \tag{5.113}$$

In addition, figures 5.1 and 5.2 show the extents of the various different response regimes in Q–P space for the cases $D < 1$ and $D > 1$, respectively.

Table 5.2. The various linear resonant response regimes. See equations (4.65), (5.61), (5.69), (5.66), and (5.110)–(5.113). The response regimes are the resistive–inertial (RI), the viscous–resistive (VR), the semi-collisional (SC), the diffusive–resistive (DR), the inertial (I), the viscous–inertial (VI), and the diffusive–inertial (DI).

Regime	$\hat{\Delta}$	Extent in Q–P space			c_β limit
RI	$+2.124\,R_e^{3/4}\,R_E^{1/4}\,R_i^{1/4}$	$Q \ll 1$	$P \ll Q^{3/2}$	$Q \gg D^4$	$c_\beta \ll Q^{3/4}$
VR	$+0.670\,R_e\,P_\varphi^{1/6}$	$P \ll Q^{-3}$	$P \gg Q^{3/2}$	$P \gg D^6$	$c_\beta \ll P^{1/2}$
SC	$+3.142\,R_e\,R_E^{1/2}\,D^{-1}$	$Q \ll D$	$P \ll Q^2\,D^{-2}$	$Q \ll D^4$	$c_\beta \ll Q^{1/2}\,D$
DR	$+2.124\,R_e\,P_*^{1/4}\,D^{-1/2}$	$P \ll D^2\,Q^{-2}$	$P \gg Q^2\,D^{-2}$	$P \ll D^6$	$c_\beta \ll P^{1/4}\,D^{3/2}$
I	$-3.142\,R_E^{-1/2}\,R_i^{-1/2}$	$Q \gg 1$	$P \ll Q^3$	$Q \gg D$	$c_\beta \ll 1$
VI	$-4.647\,R_e^{-1/4}\,P_\varphi^{-1/4}$	$P \gg Q^{-3}$	$P \gg Q^3$	$P \gg D^4\,Q^{-1}$	$c_\beta \ll Q^{-3/4}\,P^{1/4}$
DI	$-3.142\,R_e^{-1/2}\,P_*^{-1/2}\,D$	$P \gg D^2\,Q^{-2}$	$Q \ll D$	$P \ll D^4\,Q^{-1}$	$c_\beta \ll Q^{-3/2}\,P^{-1/2}\,D^3$

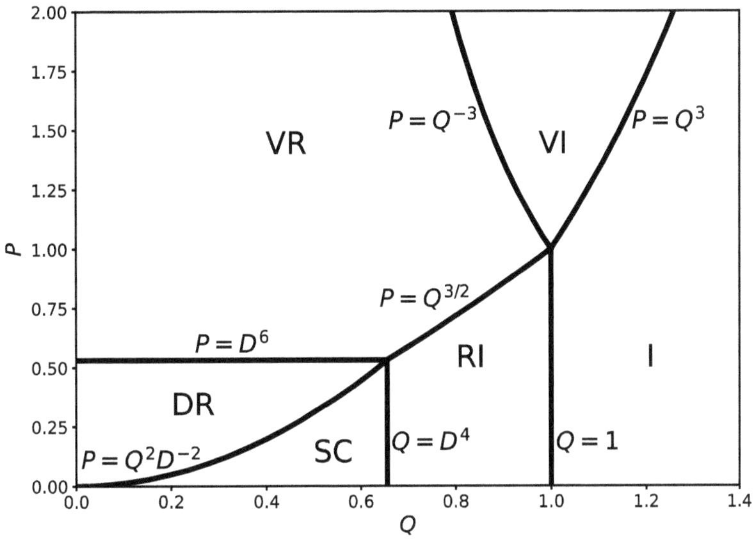

Figure 5.1. Linear resonant plasma response regimes in Q–P space for the case $D = 0.9$. The various regimes are the diffusive–resistive (DR), the semi-collisional (SC), the resistive–inertial (RI), the viscous–resistive (VR), the viscous–inertial (VI), and the inertial (I).

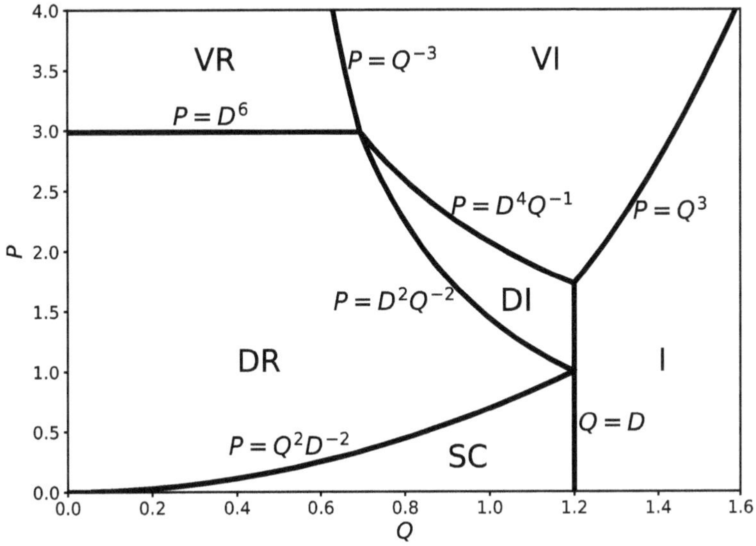

Figure 5.2. Linear resonant plasma response regimes in Q–P space for the case $D = 1.2$. The various regimes are the diffusive–resistive (DR), the semi-collisional (SC), the diffusive–inertial (DI), the viscous–resistive (VR), the viscous–inertial (VI), and the inertial (I).

Let $\hat{\delta}_* \sim p_*^{-1}$ be the normalized radial thickness of the resonant layer. Of course, the true thickness is $\delta_s = S^{-1/3}\,\hat{\delta}_*\,r_s$. It follows from equation (5.69) that the relative change in the perturbed helical magnetic flux, $\tilde{\psi}$, across the layer is

$$\left.\frac{d\ln\tilde{\psi}}{dX}\right|_{-\hat{\delta}/2}^{+\hat{\delta}/2} = \hat{\Delta}\,\hat{\delta}_* \sim \hat{\Delta}\,p_*^{-1}. \tag{5.114}$$

According to the analysis given in section 5.9, $\hat{\Delta}\,p_*^{-1}$ takes the respective values $Q^{1/2}$, $Q\,P^{1/3}$, Q^2/D^2, and $Q\,P^{1/2}/D$ in the resistive–inertial, viscous–resistive, semi-collisional, and diffusive–resistive response regimes. Moreover, it is clear from figures 5.1 and 5.2 that these values are all much less than unity. In other words, it is indeed the case that $\tilde{\psi}$ does not vary substantially across a 'constant-ψ' resonant layer. On the other hand, according to the analysis given in section 5.11, $\hat{\Delta}\,p_*^{-1} \sim 1$ in the inertial, viscous–inertial, and diffusive–inertial response regimes, which implies that $\tilde{\psi}$ does vary substantially across a 'nonconstant-ψ' layer.

5.13 Response regimes in tokamak fusion reactors

Figure 5.3 shows the linear response regimes in a low-field tokamak fusion reactor. The boundaries between the regimes are calculated with $D = 3.021$ (see table 5.1). If we assume that $P \simeq \sqrt{P_\varphi\,P_\perp} \simeq 500$ and $Q \lesssim Q_*$, which implies that $Q \lesssim 2$ (see table 5.1), then it is clear from the figure that the relevant response regimes are the viscous–inertial and diffusive–resistive regimes. Note that the respective inequalities, $c_\beta \ll Q^{-3/4}\,P^{1/4}$ and $c_\beta \ll P^{1/4}\,D^{3/2}$, that must be satisfied in order to justify our neglect of the ion parallel dynamics in the layer equation are very easily satisfied in the viscous–resistive and resistive–diffusive regimes (see section 5.7 and tables 5.1

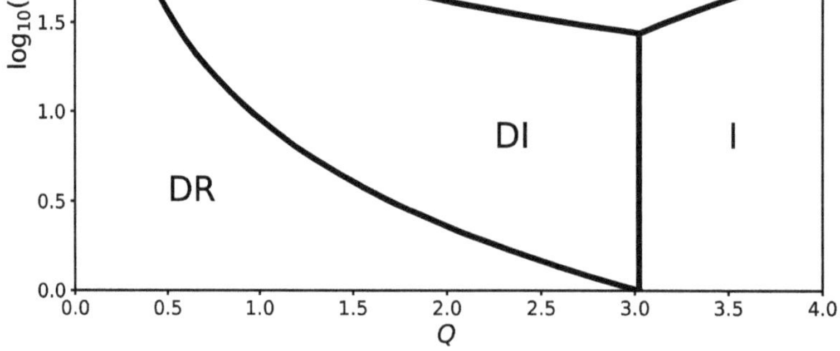

Figure 5.3. Linear resonant plasma response regimes in Q–P space for a low-field tokamak fusion reactor. The various regimes are the diffusive–resistive (DR), the diffusive–inertial (DI), the viscous–resistive (VR), the viscous–inertial (VI), and the inertial (I). The range of probable Q and P values in the reactor is indicated by the × markers.

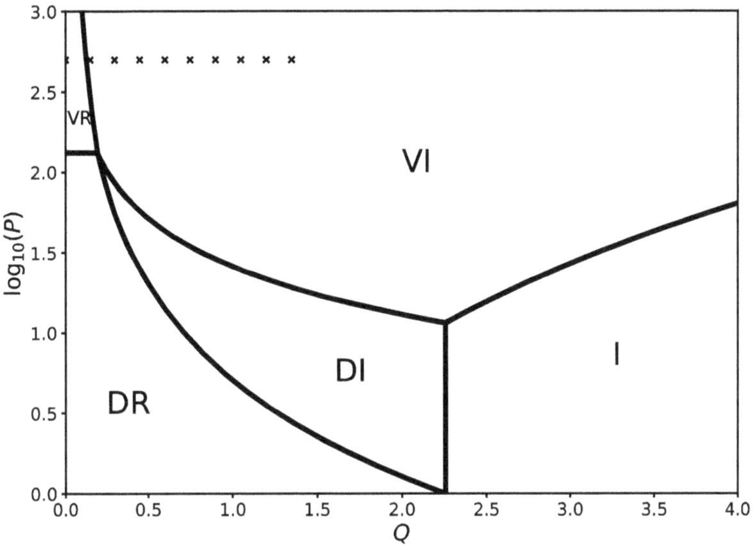

Figure 5.4. Linear resonant plasma response regimes in Q–P space for a high-field tokamak fusion reactor. The various regimes are the diffusive–resistive (DR), the diffusive–inertial (DI), the viscous–resistive (VR), the viscous–inertial (VI), and the inertial (I). The range of probable Q and P values in the reactor is indicated by the × markers.

and 5.2). Figure 5.4 shows the linear response regimes in a high-field tokamak fusion reactor. The boundaries between the regimes are calculated with $D = 2.257$ (see table 5.1). If we assume that $P \simeq \sqrt{P_\varphi\, P_\perp} \simeq 500$ and $Q \lesssim Q_*$, which implies that $Q \lesssim 1.5$ (see table 5.1), then it is clear from the figure that the relevant response regimes are the viscous–inertial and viscous–resistive regimes. Note that the respective inequalities, $c_\beta \ll Q^{-3/4}\, P^{1/4}$ and $c_\beta \ll P^{1/2}$, that must be satisfied in order to justify our neglect of the ion parallel dynamics in the layer equation are very easily satisfied in the viscous–inertial and viscous–resistive regimes.

5.14 The numerical solution of the resonant-layer equations

It is possible to solve the resonant-layer equation (3.19) numerically. We already know that in the small-p limit, the solution to this equation takes the form

$$Y_e(p) \to Y_0\left[\frac{\hat{\Delta}}{\pi\, p} + 1 + \mathcal{O}(p)\right]. \tag{5.115}$$

(See equation (5.83).) In the large-p limit, equations (3.19)–(5.81) reduce to

$$\frac{d^2 Y_e}{dp^2} - \frac{P_\perp}{(1 + 1/\tau)\, D^2}\, p^2\, Y_e \simeq 0. \tag{5.116}$$

This is a parabolic cylinder equation [12] whose most general large-p solution is

$$Y_e(p) = \frac{A\,e^{-\alpha\,p^2/2} + B\,e^{+\alpha\,p^2/2}}{p^{1/2}}\left[1 + \mathcal{O}\left(\frac{1}{p^2}\right)\right], \tag{5.117}$$

where A and B are arbitrary constants, and

$$\alpha = \left[\frac{P_\perp}{(1 + 1/\tau)\,D^2}\right]^{1/2}. \tag{5.118}$$

Obviously, the physical solution of equation (5.116) does not blow up at large p. Hence, we must select $B = 0$ in equation (5.117), which implies that

$$Y_e(p) = \frac{A\,e^{-\alpha\,p^2/2}}{p^{1/2}}\left[1 + \mathcal{O}\left(\frac{1}{p^2}\right)\right], \tag{5.119}$$

at large p.

Let us make use of the so-called *Riccati transformation* [9, 19],

$$W(p) = \frac{p}{Y_e}\frac{dY_e}{dp}. \tag{5.120}$$

Equation (5.78) yields

$$\begin{aligned}\frac{dW}{dp} = &\left[\frac{2\,p}{-\mathrm{i}\,(Q - Q_E - Q_e) + p^2} - \frac{1}{p}\right]\\ &W - \frac{W^2}{p} + p\left[-\mathrm{i}\,(Q - Q_E - Q_e) + p^2\right]\frac{B(p)}{C(p)}.\end{aligned} \tag{5.121}$$

According to equation (5.115), the small-p behavior of the solution to the previous equation is

$$W(p) = -1 + \frac{\pi\,p}{\hat{\Delta}} + \mathcal{O}(p^2). \tag{5.122}$$

Likewise, according to equation (5.119), the large-p behavior of the solution is

$$W(p) = -\alpha\,p^2 - \frac{1}{2} + \mathcal{O}\left(\frac{1}{p^2}\right). \tag{5.123}$$

Equation (5.121) can be conveniently solved numerically by launching a solution of the form (5.123) at large p and then integrating backward to small p [9]. Equation (5.122) yields

$$\hat{\Delta} = \lim_{p\to 0}\left(\frac{\pi}{dW/dp}\right). \tag{5.124}$$

Figure 5.5 shows a numerical solution of the resonant-layer equation for a low-field tokamak fusion reactor. This calculation is made with $Q_* = 2.002$, $D = 3.021$,

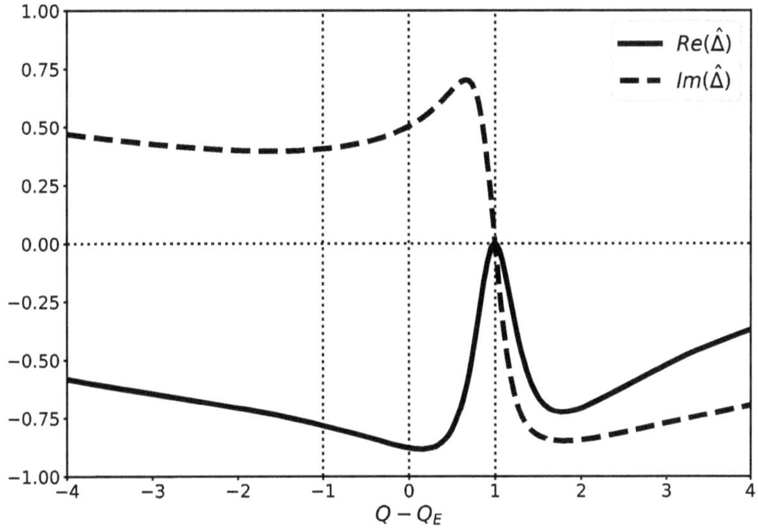

Figure 5.5. The numerical solution of the resonant-layer equations for a low-field tokamak fusion reactor. The vertical dashed lines correspond to $Q - Q_{\rm E} = Q_{\rm i}$, 0, and $Q_{\rm e}$, respectively, in order from left to right.

$P_\varphi = 874$, $P_\perp = 287$, and $\tau = 1$, assuming that Q is real (see table 5.1). Note that $\hat{\Delta}$ parameterizes the amplitude and phase of a shielding current that is driven inductively at the rational surface, in response to a rotating tearing perturbation in the outer region, and acts to suppress magnetic reconnection at the surface [15]. It can be seen that the shielding current is zero when $Q = Q_{\rm E} + Q_{\rm e}$, which is equivalent to $\omega = \omega_{\rm E} + \omega_{*\,{\rm e}}$. In other words, the shielding current is zero when the tearing perturbation in the outer region rotates at the frequency of a naturally unstable tearing mode at the rational surface [4, 15] (see chapter 6). The shielding current clearly increases linearly with $Q - Q_{\rm E} - Q_{\rm e}$ when $|Q - Q_{\rm E} - Q_{\rm e}| \ll 1$, but saturates in magnitude as $|Q - Q_{\rm E} - Q_{\rm e}| \to \mathcal{O}(1)$.

Figure 5.6 shows a numerical solution of the resonant-layer equation for a high-field tokamak fusion reactor. This calculation is made with $Q_* = 1.496$, $D = 2.257$, $P_\varphi = 874$, $P_\perp = 287$, and $\tau = 1$, assuming that Q is real (see table 5.1). Note that the figure is very similar to figure 5.5, indicating that the resonant-layer responses in low-field and high-field tokamak fusion reactors do not differ substantially from one another.

5.15 Plasma rotation

In this chapter, we have made the assumption that the E-cross-B velocity, $V_{\rm E}(r)$, is similar in magnitude to the diamagnetic velocity, $V_*(r)$, in a tokamak fusion reactor (see equations (5.21) and (5.29)). Given that most existing tokamak plasmas are characterized by levels of plasma rotation such that $|V_{\rm E}| \gg |V_*|$, this assumption needs justification.

As explained in section 1.8, a tokamak fusion reactor requires auxiliary heating in order to attain plasma temperatures sufficient for nuclear fusion. Most existing

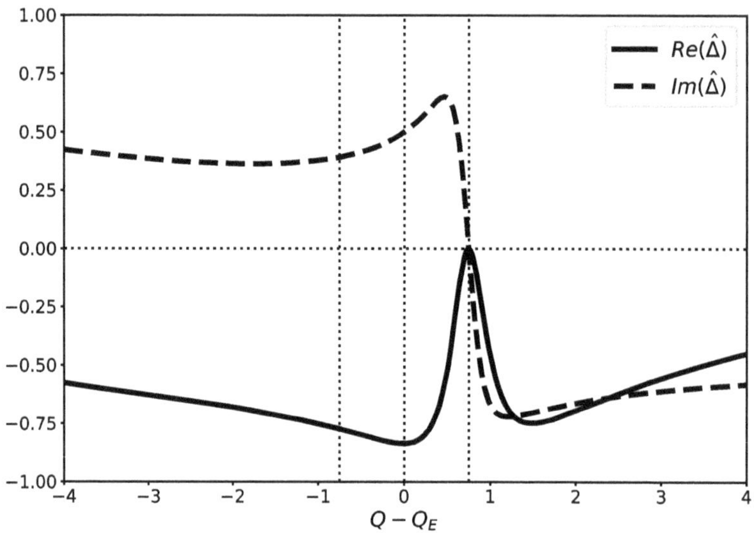

Figure 5.6. The numerical solution of the resonant-layer equations for a high-field tokamak fusion reactor. The vertical dashed lines correspond to $Q - Q_{\rm E} = Q_{\rm i}$, 0, and $Q_{\rm e}$, respectively, in order from left to right.

tokamaks employ *neutral beam injection* as their primary auxiliary heating method [20]. In this heating scheme, high-energy (i.e. 100 keV) neutral particles are injected into the plasma. The neutral particles are ionized within the plasma to form energetic ions and low-energy electrons. The energetic ions are confined by the tokamak's magnetic field and subsequently slow down and heat the plasma. It is vitally important that the neutral particles are ionized within the plasma, otherwise they pass completely through it and damage the plasma-facing components. In order to avoid this problem, which is known as *shine-through*, most existing neutral beam injection systems angle the beam so that it passes through the plasma tangentially rather than radially, in order to maximize the path length of the beam through the plasma. Neutral particle ions have relatively large gyroradii and banana radii before they slow down. In order to minimize fast ion losses, the neutral particle beam is usually angled through the plasma in the direction such that, immediately after ionization, the interaction between the fast ions and the tokamak's magnetic field causes the ions to move radially inwards, rather than outwards. It follows from the analysis given in section 2.7 that, in a tokamak in which the toroidal magnetic field and plasma current are both directed in the $+\varphi$ direction, the beam needs to be angled such that fast ions are injected in the $+\varphi$ direction rather than the $-\varphi$ direction. This scheme is known as *co-injection* because the beam injection direction is parallel to the toroidal plasma current.

A co-injected neutral beam imparts $+\varphi$ toroidal momentum, as well as energy, to a tokamak plasma. Roughly speaking, the ratio of the momentum to the energy injected by the beam is $2/v \propto 1/\mathcal{E}^{1/2}$, where v and $\mathcal{E}$ are the speed and energy, respectively, of the injected particles. The injected toroidal momentum causes the plasma to spin rapidly in the *ion diamagnetic direction* (i.e. in the same direction as those of ion diamagnetic flows). Plasmas in most existing tokamak experiments are characterized by $|V_E| \gg |V_*|$ as a consequence of co-injected neutral beams.

The low-field tokamak fusion reactor (i.e. ITER–see section 1.5 and [21]) considered in this book will be sufficiently large and will have sufficiently high density that only exceptionally high-energy (i.e. 1000 keV) neutral particles will be able to penetrate to its core. Given that the ratio of the momentum to the energy injected by a neutral beam system is inversely proportional to the square root of the neutral particle injection energy, it follows that only a relatively small amount of momentum will be injected into ITER plasmas by its neutral beam heating system, compared to the amounts injected into existing tokamaks.

The high-field tokamak fusion reactor (i.e. SPARC [22]) considered in this book will not possess a neutral particle heating system at all. Instead, it will rely on solely on auxiliary heating provided by radio-frequency electromagnetic waves injected into the plasma. Such heating schemes do not inject any appreciable momentum into the plasma.

In the absence of a significant toroidal momentum source, there is no reason to suppose that a tokamak plasma will rotate in such a manner that $|V_E|$ greatly exceeds $|V_*|$. Given that the two planned tokamak fusion reactors lack significant toroidal momentum sources, the assumption that $|V_E| \sim |V_*|$ in tokamak fusion reactors is reasonable, and will adopted throughout the remainder of this book.

5.16 Magnetic reconnection

Suppose that the resonant layer lies in a constant-ψ response regime. According to equation (5.33), the perturbed helical magnetic flux in the layer takes the form:

$$\psi(\hat{x}, \zeta, \hat{t}) = \frac{\hat{x}^2}{2\hat{L}_s} + \tilde{\psi}\, e^{\,i(\zeta - \hat{\omega}\,\hat{t})}, \tag{5.125}$$

where $\tilde{\psi}$ is a complex constant. Of course, the physical flux is the real part of the previous expression; that is,

$$\psi(\hat{x}, \xi) = \frac{\hat{x}^2}{2\hat{L}_s} + |\tilde{\psi}| \cos \xi, \tag{5.126}$$

where

$$\xi = \zeta + \arg(\tilde{\psi}) - \hat{\omega}\,\hat{t} = m\,\theta - n\,\varphi + \arg(\tilde{\psi}) - \omega\,t. \tag{5.127}$$

It is clear from equation (5.1) that $\mathbf{B} \cdot \hat{\nabla}\psi = 0$. In other words, the contours of $\psi(\hat{x}, \xi)$ map out the perturbed magnetic flux surfaces in the vicinity of the resonant layer. Let $\hat{W} = 4\,(\hat{L}_s\,|\tilde{\psi}|)^{1/2}$. Equation (5.126) can be written

$$\frac{\psi(\hat{x}, \xi)}{|\tilde{\psi}|} = 8\left(\frac{\hat{x}}{\hat{W}}\right)^2 + \cos \xi. \tag{5.128}$$

Figure 5.7 shows the contours of $\psi(\hat{x}, \xi)$ specified by the previous equation. It can be seen that the tearing mode has changed the topology of the magnetic field in the

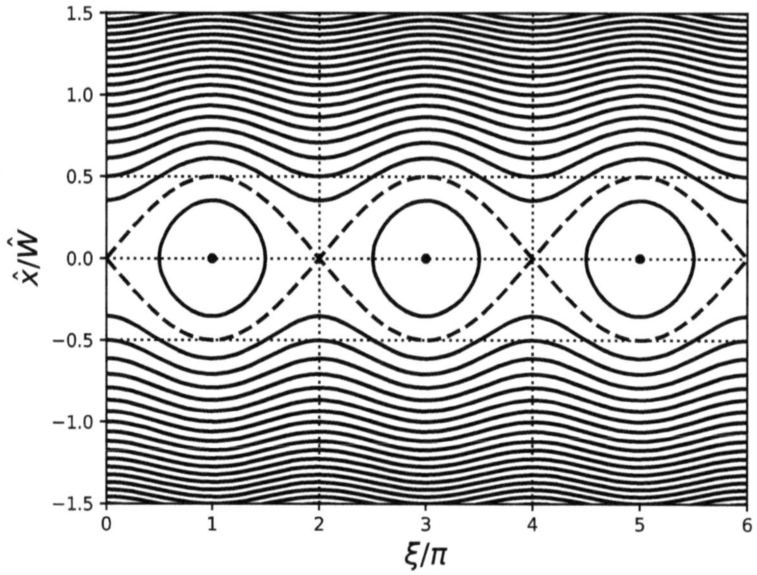

Figure 5.7. Equally spaced contours of $\psi(\hat{x}, \xi)$. The magnetic separatrix is shown as a dashed line.

immediate vicinity of the rational surface, $\hat{x} = 0$. In fact, as the tearing mode grows in amplitude (i.e. as $|\tilde{\psi}|$ increases), magnetic field lines pass through the magnetic 'X-points' (which are located at $\hat{x} = 0$, $\xi = j\,2\pi$, where j is an integer), at which time they break (or 'tear') and then reconnect to form new field lines that do not extend over all values of ξ. The magnetic field line that forms the boundary between the unreconnected and reconnected regions is known as the *magnetic separatrix;* it corresponds to the contour $\hat{\psi}(\hat{x}, \xi) = |\tilde{\psi}|$. The reconnected regions within the magnetic separatrix are termed *magnetic islands.* The tearing mode clearly generates a chain of magnetic islands centered on the rational surface, with m periods in the poloidal direction and n periods in the toroidal direction, which propagates in the laboratory frame at the helical phase velocity ω. The maximum full (as opposed to half) radial width of the magnetic separatrix is

$$W = \hat{W}\, r_s = 4\,(\hat{L}_s\,|\tilde{\psi}|)^{1/2}\, r_s = 4\left(L_s\, R_0\, |\hat{\Psi}_s|\right)^{1/2}, \tag{5.129}$$

where $\Psi_s = r_s\, B_z\, \tilde{\psi}$ is the reconnected magnetic flux defined in equation (3.72), and $\hat{\Psi}_s = \Psi_s/(R_0\, B_z)$.

Finally, our reduced drift-MHD model, (5.10)–(5.13), contains many terms of the general form $[\psi, \tilde{A}]$, where $\tilde{A}(\hat{x}, \zeta, \hat{t})$ is a perturbed quantity. According to equation (5.126), such terms can be written

$$[\psi, \tilde{A}] = \mathrm{i}\,\frac{m}{r_s}\left(\frac{\hat{x}}{\hat{L}_s}\,\tilde{A} - \tilde{\psi}\,\frac{\partial \tilde{A}}{\partial \hat{x}}\right) \sim \mathrm{i}\,\frac{m}{r_s}\,\frac{\hat{\delta}_s}{\hat{L}_s}\,\tilde{A} - \mathrm{i}\,\frac{m}{r_s}\,\frac{\tilde{\psi}}{\hat{\delta}_s}\,\tilde{A}, \tag{5.130}$$

where $\delta_s = \hat{\delta}_s\, r_s$ is the radial width of the linear layer. Linear layer theory is only valid when the second term on the extreme right-hand side of the previous equation is negligible compared to the first (because the second term is quadratic in perturbed quantities, whereas the first is linear). In other words, we require $\hat{\delta}_s^{\,2} \gg |\tilde{\psi}|\,\hat{L}_s$, which reduces to

$$\delta_s \gg W. \tag{5.131}$$

(See equation (5.129).) We conclude that the linear resonant response theory presented in this chapter is only valid when the radial width of the magnetic island chain that develops in the vicinity of the rational surface is *much less* than the radial width of the linear layer. Given that linear layers are very narrow in conventional tokamak plasmas (see chapter 6), this is a stringent constraint.

References

[1] Furth H P, Killeen J and Rosenbluth M N 1963 Finite-resistivity instabilities of a sheet pinch *Phys. Fluids* **6** 459

[2] Chen X L and Morrison P J 1990 Resistive tearing instability with equilibrium shear flow *Phys. Plasmas* **2** 495

[3] Fitzpatrick R 2014 *Plasma Physics: An Introduction* (Boca Raton, FL: CRC Press)

[4] Ara G, Basu B, Coppi B, Laval G, Rosenbluth M N and Waddell B V 1978 Magnetic reconnection and $m = 1$ oscillations in current carrying plasmas *Ann. Phys., NY* **112** 443

[5] Cole A and Fitzpatrick R 2006 Drift-magnetohydrodynamical model of error-field penetration in tokamak plasmas *Phys. Plasmas* **13** 032503

[6] Fitzpatrick R 2022 Influence of anomalous perpendicular transport on linear tearing mode dynamics in tokamak plasmas *Phys. Plasmas* **29** 032507

[7] Fitzpatrick R 1995 Helical temperature perturbations associated with tearing modes in tokamak plasmas *Phys. Plasmas* **2** 825

[8] Fitzpatrick R and Hender T C 1994 Effect of a static external magnetic perturbation on resistive mode stability in tokamaks *Phys. Plasmas* **1** 3337

[9] Park J-K 2022 Parametric dependencies of resonant layer responses across linear, two-fluid, drift-MHD regimes *Phys. Plasmas* **29** 072506

[10] Erdélyi A, Magnus W, Oberhettinger F and Tricomi F G 1954 *Tables of Intergal Transforms* (New York: McGraw-Hill)

[11] Waelbroeck F L, Joseph I, Nardon E, Bécoulet M and Fitzpatrick R 2012 Role of singular layers in the plasma response to resonant magnetic perturbations *Nucl. Fusion* **52** 074004

[12] Abramowitz M and Stegun I A (ed) 1964 *Handbook of Mathematical Functions with Formulas, Graphs, and Mathematical tables* (New York: Dover) https://store.doverpublications.com/0486612724.html

[13] Fitzpatrick R 1998 Bifurcated states of a rotating tokamak plasma in the presence of a static error-field *Phys. Plasmas* **5** 3325

[14] Bondeson A and Persson M 1986 Resistive tearing modes in the presence of equilibrium flows *Phys. Fluids* **29** 2997

[15] Fitzpatrick R 1993 Interaction of tearing modes with external structures in cylindrical geometry *Nucl. Fusion* **33** 1049

[16] Drake J F and Lee Y C 1977 Kinetic theory of tearing instabilities *Phys. Fluids* **20** 1341

[17] Waelbroeck F L 2003 Shielding of resonant magnetic perturbations in the long mean-free path regime *Phys. Plasmas* **10** 4040

[18] Boozer A H 1996 Shielding of resonant magnetic perturbations by rotation *Phys. Plasmas* **3** 4620
[19] Brennan D P, Cole A J, Akcay C and Finn J M 2019 Two fluid stability of rotating tokamak plasmas using machine learning *Bull. Am. Phys. Soc.* **64** 249 https://meetings.aps.org/Meeting/DPP19/Session/PP10.93
[20] Wesson J A 2011 *Tokamaks* 4th edn (Oxford: Oxford University Press) https://global.oup.com/academic/product/tokamaks-9780199592234
[21] Shimada M *et al* 2007 Chapter 1: Overview and summary *Nucl. Fusion* **41** S1
[22] Creely A J *et al* 2020 Overview of the SPARC tokamak *J. Plasma Phys.* **86** 865860502

Chapter 6

Linear tearing-mode stability

6.1 Introduction

The aim of this chapter is to combine the analyses given in chapters 4 and 5 to determine the linear stability of a typical tearing mode in a tokamak fusion reactor. In this study, we shall assume that the rigid wall that surrounds the plasma is perfectly conductive.

6.2 The linear dispersion relation

Let us write (see equation (5.33))

$$\omega = \mathrm{i}\,\gamma + \omega_\mathrm{r}, \tag{6.1}$$

where γ is the growth rate of the tearing mode, while ω_r is the real frequency of the mode in the laboratory frame. We shall assume that $|\gamma| \ll |\omega_\mathrm{r}|$. This assumption can easily be verified *a posteriori* (see table 6.2). Asymptotic matching (see section 3.8) between the tearing-mode solution in the outer region and that in the thin resistive layer surrounding the rational surface (see section 3.7), which constitutes the inner region, yields a linear tearing-mode dispersion relation of the form

$$S^{1/3}\,\hat{\Delta}(\gamma,\,\omega_\mathrm{r}) = E_\mathrm{ss}, \tag{6.2}$$

where use has been made of equations (3.74) and (5.70). Here, S is the Lundquist number at the rational surface (see equation (5.48)), E_ss is the real tearing stability index (see section 3.8), while the complex layer-matching parameter, $\hat{\Delta}$, is defined in equation (5.69).

In a conventional tokamak fusion reactor (assuming that $m \sim \mathcal{O}(1)$, where m is the poloidal mode number of the tearing perturbation), $E_\mathrm{ss} \sim \mathcal{O}(1)$[1], and $S^{1/3} \gg 1$

[1] Note that we are neglecting $m = 1$ modes, for which $|E_\mathrm{ss}| \gg 1$, because such modes are not really tearing modes [6].

doi:10.1088/978-0-7503-5367-0ch6

(see table 5.1). Hence, the previous dispersion relation can only be satisfied if ω_r takes a value that renders $|\hat{\Delta}| \ll 1$. It is clear from equations (5.61)–(5.65), (5.93), (5.94), (5.96), and (5.98) that this goal can be achieved in all of the constant-ψ linear response regimes if

$$\omega_r = \omega_{\perp\,e} \equiv \omega_E + \omega_{*\,e}. \tag{6.3}$$

Here, ω_E and $\omega_{*\,e}$ are the E-cross-B and electron diamagnetic frequencies, respectively, at the rational surface. (See equations (5.21), (5.29), (5.44), and (5.45).) The previous equation implies that the tearing mode corotates with the electron fluid at the rational surface [1] (see section 2.24).

6.3 The determination of linear growth rates

Reusing the analysis of sections 5.7–5.9, let us again suppose that there are two layers in p space (i.e. Fourier space). The small-p layer turns out to be of width $|\hat{\gamma}|^{1/2}$, where

$$\hat{\gamma} = S^{1/3}\,\gamma\,\tau_H. \tag{6.4}$$

Here, τ_H is the hydromagnetic timescale defined in equation (5.43). Given that we are effectively assuming that $|\hat{\gamma}| \ll 1$, the condition for the separation of the layer solution into two layers (i.e. that the width of the small-p layer is less than that of the large-p layer) is always satisfied. The large-p layer is governed by the equation

$$\frac{d^2 Y_e}{dp^2} - \frac{E(p)}{F(p)}\,Y_e = 0, \tag{6.5}$$

where

$$E(p) \simeq -(1 + 1/\tau)^{-1}\,Q_*^{\,2} - i\,Q_*\,(P_\varphi + P_\perp)\,p^2 + P_\varphi\,P_\perp\,p^4, \tag{6.6}$$

$$F(p) \simeq P_\perp - i\,Q_*\,D^2 + (1 + 1/\tau)\,P_\varphi\,D^2\,p^2. \tag{6.7}$$

Here, Q_* is the normalized diamagnetic frequency (see equation (5.65)), P_φ and $P_\perp$ are magnetic Prandtl numbers (see equations (5.53) and (5.54)), and τ is the ratio of the electron to the ion pressure gradient at the rational surface (see equation (4.5)). The boundary conditions on equation (6.5) are that Y_e is bounded as $p \to \infty$, and

$$Y_e(p) = Y_0\left[1 - \frac{\hat{\Delta}\,p}{\pi\,\hat{\gamma}} + \mathcal{O}(p^2)\right] \tag{6.8}$$

as $p \to 0$. Here, Y_0 is an arbitrary constant.

In the various constant-ψ linear growth-rate regimes considered in the next section, equation (6.5) reduces to an equation of the form

$$\frac{d^2 Y_e}{dp^2} - G\,p^k\,Y_e = 0, \tag{6.9}$$

where k is real and nonnegative, and G is a complex constant. As described in section 5.8, the solution of this equation that is bounded as $p \to \infty$ can be matched to the small-p asymptotic form (6.8) to give

$$\hat{\Delta} = \frac{\nu^{2\nu-1}\,\pi\,\Gamma(1-\nu)}{\Gamma(\nu)}\,G^{\nu}\,\hat{\gamma}, \tag{6.10}$$

where $\nu = 1/(k+2)$. The width of the large-p layer in p space is $|G|^{-\nu}$.

6.4 Linear growth-rate regimes

As before (see section 5.7), we shall assume that $P_\varphi \sim P_\perp \sim P$ and $\tau \sim \mathcal{O}(1)$ for the sake of simplicity.

Suppose that $Q_* \gg P\,p^2$ and $P \gg Q_*\,D^2$. It follows that $k = 0$, $\nu = 1/2$, and

$$G = -\frac{Q_*^{\,2}}{(1+1/\tau)\,P_\perp}. \tag{6.11}$$

Hence,

$$\hat{\Delta} = \mathrm{e}^{-\mathrm{i}\,\pi/2}\,\pi\,\frac{Q_*}{(1+1/\tau)^{1/2}\,P_\perp^{1/2}}\,\hat{\gamma}, \tag{6.12}$$

and $p_* \sim P^{1/2}/Q_*$. This so-called *resistive–inertial growth-rate regime* is valid when $P \ll Q_*^{\,3/2}$ and $P \gg Q_*\,D^2$. Making use of equations (5.48), (5.54), (5.65), (6.2), and (6.4), the corresponding tearing-mode growth rate is [3]

$$\gamma = \mathrm{e}^{\,\mathrm{i}\,\pi/2}\,\frac{E_{\rm ss}}{\pi}\,\frac{(1+1/\tau)^{1/2}}{\omega_*\,\tau_{\rm H}\,\tau_{\rm R}^{1/2}\,\tau_\perp^{1/2}}. \tag{6.13}$$

Here, ω_* is the total diamagnetic frequency (see equation (5.47)), $\tau_{\rm R}$ is the resistive diffusion timescale (see equation (5.49)), and $\tau_\perp$ is the energy confinement timescale (see equation (5.52)). Note that τ and ω_* are both assumed to be positive quantities (which is always the case if the electron and ion equilibrium pressure profiles are monotonically decreasing functions of the minor radius). According to the previous equation, the tearing mode is purely oscillatory in the resistive–inertial growth-rate regime.

Suppose that $Q_* \ll P\,p^2$ and $D^2\,p^2 \ll 1$. It follows that $k = 3$, $\nu = 1/6$, and

$$G = P_\varphi. \tag{6.14}$$

Hence,

$$\hat{\Delta} = \frac{6^{2/3}\,\pi\,\Gamma(5/6)}{\Gamma(1/6)}\,P_\varphi^{1/6}\,\hat{\gamma}, \tag{6.15}$$

and $p_* \sim P^{-1/6}$. This so-called *viscous–resistive growth-rate regime* is valid when $P \gg Q_*^{\,3/2}$ and $P \gg D^6$. Making use of equation (5.53), the corresponding tearing-mode growth rate is [3]

$$\gamma = \frac{E_{ss}}{[6^{2/3}\,\pi\,\Gamma(5/6)/\Gamma(1/6)]}\,\frac{\tau_\varphi^{1/6}}{\tau_H^{1/3}\,\tau_R^{5/6}}, \tag{6.16}$$

where τ_φ is the momentum confinement timescale (see equation (5.50)). We conclude that, in the viscous–resistive growth-rate regime, the tearing mode is a purely growing mode (in a frame of reference that is co-moving with the electron fluid at the rational surface) when the tearing stability index, E_{ss}, is positive; however, it is a purely decaying mode otherwise [4].

Suppose that $Q_* \gg P\,p^2$ and $P \ll Q_*\,D^2$. It follows that $k = 0$, $\nu = 1/2$, and

$$G = -{\rm i}\,\frac{Q_*}{(1 + 1/\tau)\,D^2}. \tag{6.17}$$

Hence,

$$\hat{\Delta} = {\rm e}^{-{\rm i}\,\pi/4}\,\pi\,\frac{Q_*^{1/2}}{(1 + 1/\tau)^{1/2}\,D}\,\hat{\gamma}, \tag{6.18}$$

and $p_* \sim D/Q_*^{1/2}$. This so-called *semi-collisional growth-rate regime* is valid when $P \ll Q_*^{\,2}/D^2$ and $P \ll Q_*\,D^2$. Making use of equation (5.53), the corresponding tearing-mode growth rate is [3]

$$\gamma = {\rm e}^{\,{\rm i}\,\pi/4}\,\frac{E_{ss}}{\pi}\,\frac{d_\beta/r_s}{\omega_*^{1/2}\,\tau_H\,\tau_R^{1/2}}, \tag{6.19}$$

where d_β is the ion sound radius (see equation (4.75)), and r_s is the minor radius of the rational surface. We conclude that, in the semi-collisional growth-rate regime, the tearing mode is a growing oscillatory mode (in the electron fluid reference frame) when the tearing stability index is positive, but it is a decaying oscillatory mode otherwise.

Suppose, finally, that $Q_* \ll P\,p^2$ and $D^2\,p^2 \gg 1$. It follows that $k = 2$, $\nu = 1/4$, and

$$G = \frac{P_\perp}{(1 + 1/\tau)\,D^2}. \tag{6.20}$$

Hence,

$$\hat{\Delta} = \frac{2\pi\,\Gamma(3/4)}{\Gamma(1/4)}\,\frac{P_\perp^{1/4}}{(1 + 1/\tau)^{1/4}\,D^{1/2}}\,\hat{\gamma}, \tag{6.21}$$

and $p_* \sim D^{1/2}/P^{1/4}$. This so-called *diffusive–resistive growth-rate regime* is valid when $P \gg Q_*^{2}/D^2$ and $P \ll D^6$. The corresponding tearing-mode growth rate is [3]

$$\gamma = \frac{E_{ss}}{[2\pi\,\Gamma(3/4)/\Gamma(1/4)]}\,\frac{(d_\beta/r_s)^{1/2}\,\tau_\perp^{1/4}}{\tau_H^{1/2}\,\tau_R^{3/4}}. \tag{6.22}$$

We conclude that, in the diffusive–resistive growth-rate regime, the tearing mode is a purely growing mode (in the electron fluid reference frame) when the tearing stability index is positive, but a purely decaying mode otherwise.

Table 5.1 gives estimates for all of the normalized quantities that appear in the layer equation, (6.5), for a low-field fusion reactor and a high-field fusion reactor (see chapter 1). Likewise, table 6.1 gives estimates for all of the unnormalized quantities that appear in the growth-rate formulae (6.13), (6.16), (6.19), and (6.22) (except for E_{ss}, which is $\mathcal{O}(1)$) for a low-field fusion reactor and a high-field fusion reactor (see chapter 1).

According to equations (6.1), (6.3), (6.13), (6.16), (6.19), and (6.22), a linear tearing mode is unstable (except in the resistive–inertial growth-rate regime, in which it is marginally stable) when the tearing stability index, E_{ss}, is positive, but it is stable otherwise [4]. Moreover, the perturbed magnetic field associated with the mode corotates with the electron fluid at the resonant surface [1]. Finally, the mode grows on a hybrid timescale that is much greater than the hydromagnetic time, τ_H, but much less than the resistive evolution time, τ_R. Note that, in all cases, the growth rate goes to zero as $\tau_R \to \infty$. This is not surprising because, as is clear from equation (5.39), the perturbed helical magnetic flux at the resonant surface, $\tilde{\psi}(\hat{x} = 0)$, is constrained to take the value zero in the limit that $S \to \infty$. In other words, magnetic reconnection at the resonant surface (which corresponds to a finite $\tilde{\psi}$ at $\hat{x} = 0$) is impossible in the absence of plasma resistivity [4].

There are three main factors, other than plasma inertia and resistivity, that affect the growth rate of a tearing mode in a conventional tokamak plasma. First, the strength of diamagnetic flows in the plasma, which is parameterized by the diamagnetic frequency, ω_* (and by the normalized diamagnetic frequency, Q_*). Second, the anomalous perpendicular diffusion of momentum and particles, which is

Table 6.1. Quantities controlling tearing-mode growth rates in a low-field tokamak fusion reaction and a high-field tokamak fusion reaction. Here, τ is the ratio of the electron to the ion pressure gradients at the rational surface, τ_H the hydromagnetic timescale, τ_R the resistive diffusion timescale, τ_φ the momentum confinement timescale, $\tau_\perp$ the energy confinement timescale, ω_* the diamagnetic frequency, r_s the minor radius of the rational surface, and d_β the ion sound radius.

	Low field	High field
B (T)	5.0	12.0
τ	1.0	1.0
τ_H (s)	1.04×10^{-6}	4.31×10^{-7}
τ_R (s)	1.40×10^{3}	2.43×10^{2}
τ_φ (s)	1.60	0.278
$\tau_\perp$(s)	4.89	0.848
ω_* (krad s^{-1})	1.75	4.19
r_s (m)	1.23	0.527
d_β (m)	4.89×10^{-3}	2.04×10^{-3}

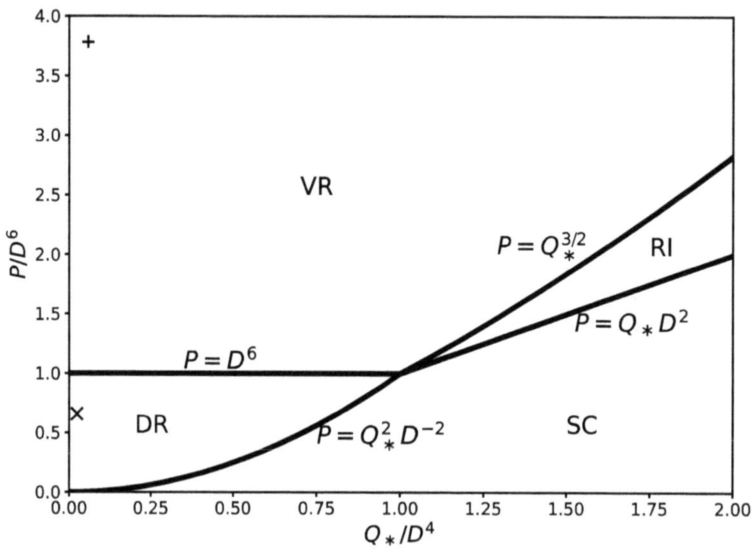

Figure 6.1. Linear tearing-mode growth-rate regimes in Q_*–P space. The various regimes are the diffusive–resistive (DR), the viscous–resistive (VR), the resistive–inertial (RI), and the semi-collisional (SC) regimes. The × and + markers indicate the location of a low-field and a high-field tokamak fusion reactor, respectively, in Q_*–P space.

parameterized by the momentum and particle confinement timescales, τ_φ and $\tau_\perp$ (and by the magnetic Prandtl numbers, P_φ and $P_\perp$). Third, finite ion sound radius effects, which are parameterized by the ion sound radius, d_β (and by the normalized ion sound radius D).

There are four tearing-mode growth-rate regimes—the resistive–inertial, the viscous–resistive, the semi-collisional, and the diffusive–resistive—and their extents in Q_*–P space are illustrated in figure 6.1 [3]. Note that figure 6.1 differs somewhat from figures 5.1 and 5.2 because in the latter two figures it is assumed that $|Q - Q_E - Q_e| \sim |Q - Q_E| \sim |Q - Q_E - Q_i|$, whereas in the former figure it is assumed that $|Q - Q_E - Q_e| \ll |Q - Q_E| \sim |Q - Q_E - Q_i|$. This refined ordering eliminates the nonconstant-ψ response regimes and significantly modifies the extent of the resistive–inertial growth-rate regime. It is clear from figure 6.1 that a low-field tokamak fusion reactor lies in the diffusive–resistive growth-rate regime, whereas a high-field tokamak fusion reactor lies in the viscous–resistive growth-rate regime (see section 5.13).

The absence of nonconstant-ψ response regimes in figure 6.1 should come as no surprise. As we saw in section 5.12, nonconstant-ψ resonant layers are characterized by $\hat{\Delta} \sim \hat{\delta}_*^{\,-1} = S^{-1/3}\,(r_s/\delta_s)$, where δ_s is the radial layer thickness. Hence, according to equation (6.2), asymptotic matching of such a layer to the outer solution is only possible if $|E_{ss}| \sim r_s/\delta_s \gg 1$ (given that resonant layers in tokamak plasmas are invariably very thin compared to the minor radius of the plasma). However, low-m tearing modes in conventional tokamak plasmas are characterized by $|E_{ss}| \sim \mathcal{O}(1)$ rather than $|E_{ss}| \gg 1$ [6]. (As before, we are neglecting $m = 1$ modes, which are characterized by $|E_{ss}| \gg 1$, because they are not really tearing modes.)

6.5 Resonant-layer thickness

According to equations (3.102) and (3.103), the dispersion relation of a magnetic perturbation interacting with a thin, rigid, resistive wall can be written as

$$\Delta\Psi_{\rm w} = \gamma\,\tau_{\rm r}\,\frac{\delta_{\rm w}}{r_{\rm w}}\,\Psi_{\rm w}, \tag{6.23}$$

where $\Delta\Psi_{\rm w}$ is a measure of the current flowing in the wall, $\Psi_{\rm w}$ the perturbed magnetic flux that penetrates the wall, $r_{\rm w}$ the wall minor radius, $\delta_{\rm w}$ the wall radial thickness, and $\tau_{\rm r} = \mu_0\,r_{\rm s}^{\,2}\,\sigma_{\rm w}$ the time required for magnetic flux to diffuse a distance $r_{\rm w}$ through the wall material. Here, $\sigma_{\rm w}$ is the electrical conductivity of the wall material. By analogy with the previous result, we would expect the dispersion relation of a magnetic perturbation interacting with the thin resistive layer that surrounds the rational surface to take the form

$$\Delta\Psi_{\rm s} = \gamma\,\tau_{\rm R}\,\frac{\delta_{\rm s}}{r_{\rm s}}\,\Psi_{\rm s}, \tag{6.24}$$

where $\Delta\Psi_{\rm s}$ is a measure of the current flowing in the layer, $\Psi_{\rm s}$ the perturbed magnetic flux that penetrates the layer, $\delta_{\rm s}$ the radial thickness of the layer, and $\tau_{\rm R}$ the time required for magnetic flux to diffuse a distance $r_{\rm s}$ through the plasma (see equation (5.49)). Note that, in general, $\delta_{\rm s}$ is a complex quantity. In fact, the true layer thickness is $|\delta_{\rm s}|$. It follows from equation (3.74) that

$$\gamma\,\tau_{\rm R}\,\frac{\delta_{\rm s}}{r_{\rm s}} = E_{\rm ss}, \tag{6.25}$$

where $E_{\rm ss}$ is the real tearing stability index. Finally, equations (5.48), (5.53), (6.2), and (6.4) can be combined with the previous equation to give

$$\frac{\hat{\Delta}}{\hat{\gamma}\,D} = (1 + 1/\tau)^{1/2}\,\frac{\delta_{\rm s}}{d_\beta}, \tag{6.26}$$

where d_β is the ion sound radius at the rational surface (see equation (4.75)).

6.6 The numerical solution of the resonant-layer equations

Let

$$q = p\,D. \tag{6.27}$$

Equations (6.5)–(6.7) yield

$$\frac{d^2 Y_{\rm e}}{dq^2} - \frac{\hat{E}(q)}{\hat{F}(q)}\,Y_{\rm e} = 0, \tag{6.28}$$

where

$$\hat{E}(q) = -(1 + 1/\tau)^{-1}\,\hat{Q}_*^{\,2} - {\rm i}\,\hat{Q}_*\,(\hat{P}_\varphi + \hat{P}_\perp)\,q^2 + \hat{P}_\varphi\,\hat{P}_\perp\,q^4, \tag{6.29}$$

$$\hat{F}(q) \simeq \hat{P}_\perp - \mathrm{i}\,\hat{Q}_* + (1 + 1/\tau)\,\hat{P}_\varphi\,q^2, \tag{6.30}$$

and

$$\hat{Q}_* = \frac{Q_*}{D^4}, \tag{6.31}$$

$$\hat{P}_\varphi = \frac{P_\varphi}{D^6}, \tag{6.32}$$

$$\hat{P}_\perp = \frac{P_\perp}{D^6}. \tag{6.33}$$

Equation (6.28) must be solved subject to the constraint that $Y_e(q)$ is bounded as $q \to \infty$, and

$$Y_e(q) = Y_0\left[1 - \frac{(1 + 1/\tau)^{1/2}\,\delta_s\,q}{\pi\,d_\beta} + \mathcal{O}(q^2)\right] \tag{6.34}$$

as $q \to 0$. Here, use has been made of equations (6.8), (6.26), and (6.27). As can easily be demonstrated (see section 5.14), the solution of equation (6.28) that is bounded as $q \to \infty$ is

$$Y_e(q) = \frac{A\,\mathrm{e}^{-\alpha\,q^2/2}}{q^{1/2}}\left[1 + \mathcal{O}\left(\frac{1}{q^2}\right)\right], \tag{6.35}$$

where

$$\alpha = \left(\frac{\hat{P}_\perp}{1 + 1/\tau}\right)^{1/2}, \tag{6.36}$$

and A is an arbitrary constant.

Let us again (see section 5.14) use the Riccati transformation [2, 5]

$$W(q) = \frac{q}{Y_e}\,\frac{dY_e}{dq}. \tag{6.37}$$

Equation (6.28) yields

$$\frac{dW}{dq} = \frac{W}{q} - \frac{W^2}{q} + \frac{q\,\hat{E}(q)}{\hat{F}(q)}. \tag{6.38}$$

According to equation (6.34), the small-q behavior of the solution to the previous equation is

$$W(q) = -\frac{(1 + 1/\tau)^{1/2}\,\delta_s\,q}{\pi\,d_\beta} + \mathcal{O}(q^2). \tag{6.39}$$

Likewise, according to equation (6.35), the large-q behavior of the solution is

$$W(q) = -\alpha\, q^2 - \frac{1}{2} + \mathcal{O}\left(\frac{1}{q^2}\right). \tag{6.40}$$

Equation (6.38) can be conveniently solved numerically by launching a solution of the form (6.40) at large q and then integrating backward to small q [5]. It follows from equation (6.39) that

$$\frac{\delta_s}{d_\beta} = -\lim_{q\to 0}\left[\frac{\pi}{(1 + 1/\tau)^{1/2}}\,\frac{dW}{dq}\right]. \tag{6.41}$$

Table 6.2 gives estimates for the normalized resonant-layer parameters, $\hat{Q}_*$, $\hat{P}_\varphi$, and $\hat{P}_\perp$, that appear in equations (6.28)–(6.30), in a low-field tokamak fusion reactor and a high-field tokamak fusion reactor. These estimates are made using the data shown in table 6.1. Table 6.2 also gives estimates for the linear layer thicknesses and tearing-mode growth rates in such reactors. These estimates are obtained via numerical solution of the resonant-layer equation. It can be seen that the typical radial thickness of a linear tearing layer in a tokamak fusion reactor is only a few millimeters. Furthermore, linear tearing modes in tokamak fusion reactors grow on timescales that typically lie between a tenth of a second and a second (assuming that $E_{ss} \sim \mathcal{O}(1)$). Finally, the real frequencies of such modes, in a frame of reference that corotates with the electron fluid at the resonant surface (i.e. the imaginary components of γ), are very much smaller (by a factor of the order of 10^6) than a typical diamagnetic frequency (see table 6.1). In other words, linear tearing modes do indeed corotate with the electron fluid at the resonant surface to a very high degree of fidelity.

Table 6.2. Normalized dimensionless layer parameters, layer thicknesses, and tearing-mode growth rates in a low-field tokamak fusion reactor and a high-field tokamak fusion reactor. See equations (6.31)–(6.33). Here, δ_s is the complex layer thickness, $|\delta_s|$ the actual layer thickness, d_β the ion sound radius, γ the complex growth rate in a frame of reference that corotates with the electron fluid at the rational surface, and E_{ss} the tearing stability index.

	Low field	High field
B (T)	5.0	12.0
$\hat{Q}_*$	2.40×10^{-2}	5.77×10^{-2}
$\hat{P}_\varphi$	1.15	6.62
$\hat{P}_\perp$	3.77×10^{-1}	$2.17 \times 10^{+0}$
δ_s/d_β	$9.60 \times 10^{-1} - 1.45 \times 10^{-2}\,\mathrm{i}$	$1.44 \times 10^{+0} - 1.80 \times 10^{-2}\,\mathrm{i}$
$\|\delta_s\|$ (m)	4.69×10^{-3}	2.95×10^{-3}
γ (s^{-1})/E_{ss}	$1.93 \times 10^{-1} + 2.93 \times 10^{-3}\,\mathrm{i}$	$7.36 \times 10^{-1} + 9.16 \times 10^{-3}\,\mathrm{i}$

Figures 6.2 and 6.3 show values of $\mathrm{Re}(\delta_s/d_\beta)$ and $\mathrm{Im}(\delta_s/d_\beta)$, respectively, evaluated numerically as functions of $\hat{Q}_*$ and $\hat{P}$. In producing these figures, it is assumed that $\hat{P}_\varphi = \hat{P}_\perp = \hat{P}$. The fact that $\mathrm{Re}(\delta_s) > 0$ for all values of $\hat{Q}_*$ and $\hat{P}$

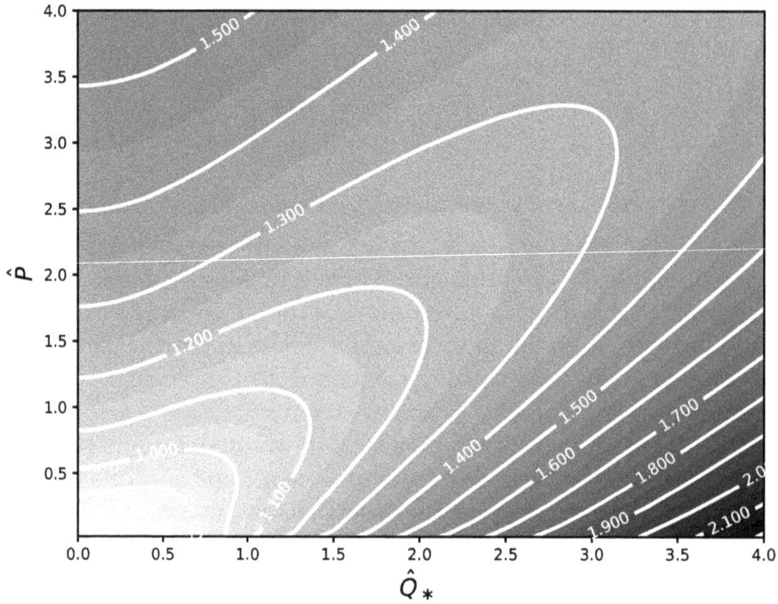

Figure 6.2. The real part of δ_s/d_β calculated as a function of $\hat{Q}_* \equiv Q_*/D^4$ and $\hat{P} \equiv P/D^6$.

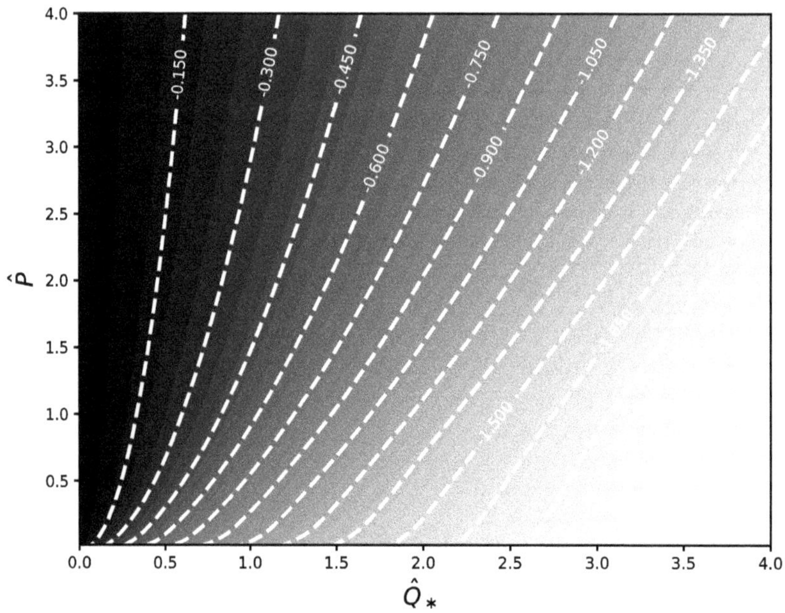

Figure 6.3. The imaginary part of δ_s/d_β calculated as a function of $\hat{Q}_* \equiv Q_*/D^4$ and $\hat{P} \equiv P/D^6$.

confirms that tearing modes are linearly unstable for $E_{ss} < 0$, but stable otherwise [4] (see equation (6.25)). It is clear that the layer width increases with increasing normalized diamagnetic frequency, $\hat{Q}_*$, and also with increasing normalized magnetic Prandtl number, $\hat{P}$.

References

[1] Ara G, Basu B, Coppi B, Laval G, Rosenbluth M N and Waddell B V 1978 Magnetic reconnection and $m = 1$ oscillations in current carrying plasmas *Ann. Phys., NY* **112** 443

[2] Brennan D P, Cole A J, Akcay C and Finn J M 2019 Two fluid stability of rotating tokamak plasmas using machine learning Bull *Bull. Am. Phys. Soc.* **64** 249 https://meetings.aps.org/Meeting/DPP19/Session/PP10.93

[3] Fitzpatrick R 2022 Influence of anomalous perpendicular transport on linear tearing mode dynamics in tokamak plasmas *Phys. Plasmas* **29** 032507

[4] Furth H P, Killeen J and Rosenbluth M N 1963 Finite-resistivity instabilities of a sheet pinch *Phys. Fluids* **6** 459

[5] Park J-K 2022 Parametric dependencies of resonant layer responses across linear, two-fluid, drift-MHD regimes *Phys. Plasmas* **29** 072506

[6] Wesson J A 1978 Hydrodynamic stability of tokamaks *Nucl. Fusion* **18** 87

Richard Fitzpatrick

Chapter 7

Error-field penetration in tokamak plasmas

7.1 Introduction

Tokamak plasmas are invariably subject to small-amplitude, static, magnetic perturbations—known as *error fields*—that are primarily generated by magnetic field-coil misalignments. A resonant error field can drive magnetic reconnection in a plasma that is intrinsically stable against tearing (known as a 'tearing-stable' plasma), resulting in the formation of a *locked* (i.e. nonrotating in the laboratory frame) magnetic island chain at the resonant magnetic flux surface (see section 5.16). Tokamak plasmas containing locked magnetic island chains often terminate in disruptions [14, 15]. Fortunately, error-field-driven magnetic reconnection is strongly suppressed by the naturally occurring rotation of the electron fluid at the resonant surface. However, when the error-field amplitude rises above a certain critical value, the electron fluid rotation at the resonant surface is suddenly arrested, and error-field-driven reconnection proceeds unhindered. This phenomenon is known as *error-field penetration* [4, 6, 7] and has been observed in many tokamak experiments [3, 8–13, 16, 17].

In this chapter, we shall calculate the critical error-field amplitude required to trigger penetration on the assumption that, prior to penetration, the rotational suppression of error-field-driven magnetic reconnection is sufficiently strong that the resonant plasma response is governed by linear layer physics [2, 4, 6, 7].

7.2 Asymptotic matching

Consider a tearing-stable, cylindrical, tokamak plasma of minor radius a and simulated major radius R_0. In accordance with the analysis given in chapter 3, suppose that the plasma is surrounded by a concentric, radially thin, rigid, resistive wall of minor radius $r_w > a$. Likewise, suppose that the wall is surrounded by a concentric, radially thin, magnetic field coil of minor radius $r_c > r_w$ that carries a nonrotating (in the laboratory frame) helical current of amplitude I_c. Let the current possess m periods in the poloidal direction and n periods in the toroidal direction.

doi:10.1088/978-0-7503-5367-0ch7

The static magnetic field generated by the field coil constitutes our error field. The error field resonates with the plasma at the rational magnetic flux surface, minor radius r_s, at which the safety factor (see section 3.2) takes the value m/n.

Setting $d/dt = 0$ (because the error field is static) in equations (3.187) and (3.188), we obtain

$$\Delta\hat{\Psi}_s = \Delta_{nw}\, \hat{\Psi}_s + \tilde{I}_c. \tag{7.1}$$

Here,

$$\hat{\Psi}_s = -i\, \frac{r_s}{R_0\, B_z}\left[\frac{\delta B_r}{m}\right]_{r_s}, \tag{7.2}$$

$$\Delta\hat{\Psi}_s = -\frac{r_s}{R_0\, B_z}\, [\delta B_\theta]_{r_{s-}}^{r_{s+}}, \tag{7.3}$$

where r, θ, z are conventional cylindrical coordinates, $\delta\mathbf{B}$ the perturbed magnetic field, and B_z the equilibrium toroidal magnetic field strength (see sections 3.3, 3.8, and 3.15). Clearly, $\hat{\Psi}_s$ is a measure of the reconnected magnetic flux driven at the rational surface by the error field, while $\Delta\hat{\Psi}_s$ is a measure of the helical current sheet that is induced at the surface. Moreover,

$$\Delta_{nw} = \Delta_{pw} + \frac{E_{sw}\, E_{ws}}{(-\tilde{E}_{ww})} \tag{7.4}$$

is the (real dimensionless) tearing stability index (of a tearing mode with poloidal mode number m and toroidal mode number n) in the absence of the wall,

$$\Delta_{pw} = E_{ss} \tag{7.5}$$

is the (real dimensionless) tearing stability index in the presence of a perfectly conducting wall at $r = r_w$ (see section 3.8), and

$$\tilde{I}_c = \left(\frac{\Delta_{nw} - \Delta_{pw}}{E_{ws}}\right)\left(\frac{r_w}{r_c}\right)^m\left(\frac{\mu_0\, I_c}{R_0\, B_z}\right). \tag{7.6}$$

Here, the (real dimensionless) plasma-wall coupling parameter, $E_{sw} > 0$, and the (real dimensionless) wall stability index, $\tilde{E}_{ww}$, are defined in sections 3.9 and 3.17, respectively. We expect $0 > \Delta_{nw} > \Delta_{pw}$, because the plasma is assumed to be tearing stable (see chapter 6), and $\tilde{E}_{ww} < 0$. In fact, if we make the approximation that the equilibrium plasma current external to the rational magnetic flux surface is negligible (i.e. $J_z' = 0$ in equation (3.77)), then it can easily be demonstrated that

$$\Delta_{nw} = \Delta_{pw} + \frac{2\, m\, (r_s/r_w)^{2\,m}}{1 - (r_s/r_w)^{2\,m}}, \tag{7.7}$$

$$\tilde{I}_c = \left(\frac{r_s}{r_c}\right)^m\left(\frac{\mu_0\, I_c}{R_0\, B_z}\right). \tag{7.8}$$

In the radially thin, resonant layer that surrounds the rational magnetic flux surface, equation (5.70) yields

$$\Delta\hat{\Psi}_s = S^{1/3}\,\hat{\Delta}\,\hat{\Psi}_s, \tag{7.9}$$

where S is the Lundquist number at the rational surface, and the complex layer-matching parameter, $\hat{\Delta}$, is defined in equation (5.69). Hence, asymptotic matching between the inner and the outer regions (see section 4.1) yields

$$\hat{\Psi}_s = \frac{\tilde{I}_c}{(-\Delta_{nw}) + S^{1/3}\,\hat{\Delta}}, \tag{7.10}$$

where use has been made of equations (7.1) and (7.9). The previous equation specifies the (normalized) reconnected magnetic flux, $\hat{\Psi}_s$, driven at the rational surface by the (normalized) error-field coil current, $\tilde{I}_c$. The complex layer parameter, $\hat{\Delta}$, specifies the strength of a shielding current that is induced in the resonant layer and acts to prevent driven magnetic reconnection. Note that the resistive wall has no influence on $\hat{\Psi}_s$ because no eddy currents are induced in the wall by a static error field.

7.3 The resonant-layer response

We saw in section 5.13 that the relevant linear resonant response regimes in a *low-field* tokamak fusion reactor are the *diffusive–resistive* regime and the *viscous–inertial* regime.

In the diffusive–resistive response regime (see section 5.9),

$$\hat{\Delta}_{DR} = C_{DR}\,e^{\,i\,\sigma\,\pi/2}\,\frac{|Q_E + Q_e|\,P_\perp^{1/4}}{(1 + 1/\tau)^{1/4}\,D^{1/2}}, \tag{7.11}$$

where $\sigma = \mathrm{sgn}(Q_E + Q_e)$, and

$$C_{DR} = \frac{2\pi\,\Gamma(3/4)}{\Gamma(1/4)}. \tag{7.12}$$

Here, $Q_E = S^{1/3}\,\omega_E\,\tau_H$ and $Q_e = S^{1/3}\,\omega_{*\,e}\,\tau_H$, where τ_H is the hydromagnetic timescale (see equation (5.43)), ω_E the E-cross-B frequency (see equation (5.44)), and $\omega_{*\,e}$ the electron diamagnetic frequency (see equation (5.45)). All quantities are evaluated at the rational surface. Moreover, $P_\perp = \tau_R/\tau_\perp$ and $D = S^{1/3}\,(1 + 1/\tau)^{-1/2}\,d_\beta/r_s$, where $\tau_R = S\,\tau_H$ is the resistive diffusion timescale (see equation (5.49)), $\tau_\perp$ the energy confinement timescale (see equation (5.52)), τ the ratio of the electron to the ion pressure gradient (see equation (4.5)), and d_β the ion sound radius (see equation (4.75)). In writing expression (7.11), we have made use of the fact that $Q \equiv S^{1/3}\,\omega\,\tau_H = 0$ for a static magnetic perturbation.

In the viscous–inertial response regime (see section 5.11),

$$\hat{\Delta}_{VI} = -C_{VI}\,e^{-i\,\sigma\,\pi/8}\,|Q_E + Q_e|^{-1/4}\,P_\varphi^{-1/4}, \tag{7.13}$$

where

$$C_{\rm VI} = \frac{\pi}{2}\,\frac{\Gamma(1/4)}{\Gamma(3/4)}. \tag{7.14}$$

Here, $P_\varphi = \tau_{\rm R}/\tau_\varphi$, where τ_φ is the momentum confinement timescale (see equation (5.50)).

Interpolating between the diffusive–resistive response regime and the viscous–inertial response regime, we can write

$$\hat{\Delta} = \frac{\hat{\Delta}_{\rm DR}\,\hat{\Delta}_{\rm VI}}{\hat{\Delta}_{\rm DR} + \hat{\Delta}_{\rm VI}}. \tag{7.15}$$

It follows that in a low-field tokamak fusion reactor,

$${\rm Re}(\hat{\Delta}) = -C_{\rm DR}\,\frac{|Q_{\rm E} + Q_{\rm e}|\,P_\perp^{1/4}}{(1 + 1/\tau)^{1/4}\,D^{1/2}}\,\frac{\cos(\pi/8)\,X_{\rm DI}}{1 + 2\,\sin(\pi/8)\,X_{\rm DI} + X_{\rm DI}^2}, \tag{7.16}$$

$${\rm Im}(\hat{\Delta}) = C_{\rm DR}\,\frac{(Q_{\rm E} + Q_{\rm e})\,P_\perp^{1/4}}{(1 + 1/\tau)^{1/4}\,D^{1/2}}\,\frac{1 + \sin(\pi/8)\,X_{\rm DI}}{1 + 2\,\sin(\pi/8)\,X_{\rm DI} + X_{\rm DI}^2}, \tag{7.17}$$

where

$$X_{\rm DI} = \frac{C_{\rm DR}}{C_{\rm VI}}\,\frac{|Q_{\rm E} + Q_{\rm e}|^{5/4}\,P_\perp^{1/4}\,P_\varphi^{1/4}}{(1 + 1/\tau)^{1/4}\,D^{1/2}}. \tag{7.18}$$

Figure 7.1 shows the real and imaginary components of $\hat{\Delta}$ as functions of the normalized E-cross-B frequency at the rational surface, $Q_{\rm E}$, calculated from equations (7.16)–(7.18)

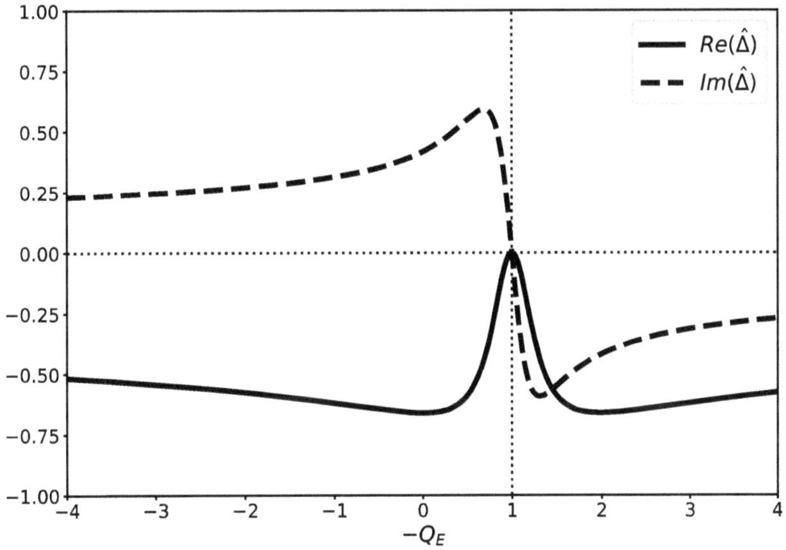

Figure 7.1. Complex layer-matching parameter, $\hat{\Delta}$, calculated from equations (7.16)–(7.18) for a low-field tokamak fusion reactor. The vertical dashed line corresponds to $Q_{\rm E} + Q_{\rm e} = 0$.

for a low-field tokamak fusion reactor. The calculation parameters are $\tau = 1, Q_e = 1.00$, $D = 3.02$, $P_\varphi = 874$, and $P_\perp = 287$ (see table 5.1). If we compare figure 7.1 with the exact numerical result shown in figure 5.5, then we can see that the analytic approximation (7.16)–(7.18) captures most of the salient features of the layer response. In particular, if we had tried to model the layer response using either the diffusive–resistive or the viscous–inertial response regime alone, then the agreement with the numerical results would have been very poor. The diffusive–resistive response regime predicts that $\hat{\Delta}$ is purely imaginary, and that $|\mathrm{Im}(\hat{\Delta})|$ is a monotonically increasing function of $|Q_E + Q_e|$. On the other hand, the viscous–inertial response regime does not predict a resonance (i.e. a point at which $\hat{\Delta} = 0$) at $Q_E + Q_e = 0$ (i.e. when the electron fluid at the rational surface is stationary). Neither of these predictions is consistent with the data shown in figure 5.5. Thus, it is clear that, in order to be reasonably accurate, an analytical approximation to the layer response must interpolate between a constant-ψ response regime (in this case, the diffusive–resistive regime) and a nonconstant-ψ response regime (in this case, the viscous–inertial regime).

We saw in section 5.13 that the relevant linear resonant response regimes in a *high-field* tokamak fusion reactor are the *viscous–resistive* regime and the *viscous–inertial* regime.

In the viscous–resistive response regime (see section 5.9),

$$\hat{\Delta}_{VR} = C_{VR}\, \mathrm{e}^{\,\mathrm{i}\,\sigma\,\pi/2}\, |Q_E + Q_e|\, P_\varphi^{1/6}, \tag{7.19}$$

where

$$C_{VR} = \frac{6^{2/3}\, \pi\, \Gamma(5/6)}{\Gamma(1/6)}. \tag{7.20}$$

Interpolating between the viscous–resistive response regime and the viscous–inertial response regime, we can write

$$\hat{\Delta} = \frac{\hat{\Delta}_{VR}\, \hat{\Delta}_{VI}}{\hat{\Delta}_{VR} + \hat{\Delta}_{VI}}. \tag{7.21}$$

It follows that in a high-field tokamak fusion reactor,

$$\mathrm{Re}(\hat{\Delta}) = -C_{VR}\, |Q_E + Q_e|\, P_\varphi^{1/6}\, \frac{\cos(\pi/8)\, X_{RI}}{1 + 2\sin(\pi/8)\, X_{RI} + X_{RI}^2}, \tag{7.22}$$

$$\mathrm{Im}(\hat{\Delta}) = C_{VR}\, (Q_E + Q_e)\, P_\varphi^{1/6}\, \frac{1 + \sin(\pi/8)\, X_{RI}}{1 + 2\sin(\pi/8)\, X_{RI} + X_{RI}^2}, \tag{7.23}$$

where

$$X_{RI} = \frac{C_{VR}}{C_{VI}}\, |Q_E + Q_e|^{5/4}\, P_\varphi^{5/12}. \tag{7.24}$$

Figure 7.2 shows the real and imaginary components of $\hat{\Delta}$ as functions of the normalized E-cross-B frequency at the rational surface, Q_E, calculated from

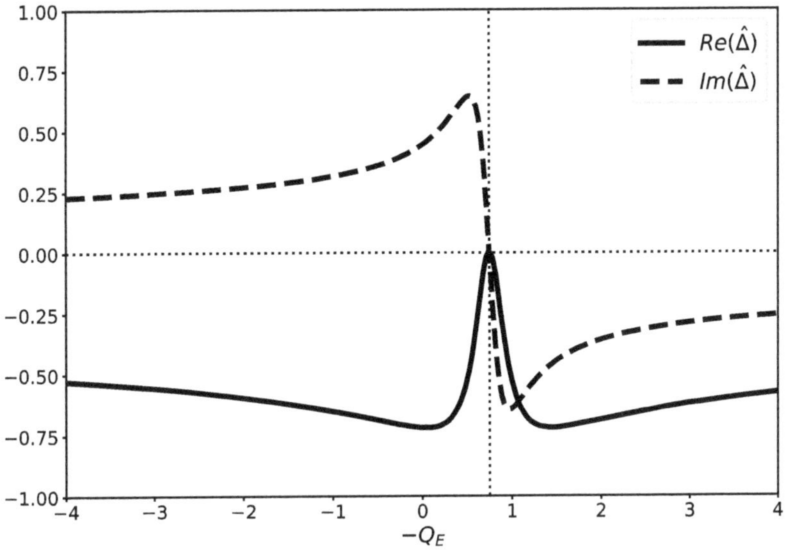

Figure 7.2. Complex layer-matching parameter, $\hat{\Delta}$, calculated from equations (7.22)–(7.24) for a high-field tokamak fusion reactor. The vertical dashed line corresponds to $Q_{\rm E} + Q_{\rm e} = 0$.

equations (7.22)–(7.24) for a high-field tokamak fusion reactor. The calculation parameters are $Q_{\rm e} = 0.75$ and $P_\varphi = 874$ (see table 5.1). As before, if we compare figure 7.2 with the exact numerical result shown in figure 5.6, then we can see that the analytic approximation (7.22)–(7.24) captures most of the salient features of the layer response.

7.4 Torque balance

According to equations (3.185) and (3.186),

$$\Delta\omega_{\rm E} = (\mathbf{k}\cdot\Delta\mathbf{V}_{\rm i})_{r=r_s} = -\sum_{p=1,\infty}(\alpha_{\rm p} + \beta_{\rm p}). \tag{7.25}$$

In other words, the shift in the E-cross-B frequency at the rational surface, $\Delta\omega_{\rm E}$, which develops in response to the electromagnetic torque exerted at the surface by the error field (see section 3.3), is mirrored by an equal shift in the ion fluid rotation frequency (as well as in the electron fluid rotation frequency). This is the essence of the *no-slip constraint*, (3.185), which follows from equations (2.321) and (2.322) because the torque modifies the E-cross-B velocity at the rational surface but does not affect the diamagnetic velocities.

Equations (3.190) and (3.191) yield

$$\left(\frac{1}{\tau_\theta} + \frac{j_{1\rm p}^{\,2}}{\tau_{\rm M}}\right)\alpha_{\rm p} = \frac{m^2\,[J_1(j_{1\rm p}\,\hat{r}_{\rm s})]^2}{\tau_{\rm A}^2\,\epsilon_{\rm s}^{\,2}\,[J_2(j_{1\rm p})]^2}\,{\rm Im}(\Delta\hat{\Psi}_{\rm s}\,\hat{\Psi}_{\rm s}^{\,*}), \tag{7.26}$$

$$\frac{j_{0p}^{\,2}}{\tau_M}\,\beta_p = \frac{n^2\,[J_0(j_{0p}\,\hat{r}_s)]^2}{\tau_A^2\,[J_1(j_{0p})]^2}\,\mathrm{Im}(\Delta\hat{\Psi}_s\,\hat{\Psi}_s^{\,*}), \tag{7.27}$$

where we have set $d/dt = 0$ (because the error field is static). Here,

$$\tau_\theta = \tau_{\theta\,i}(r_s) = \left[\frac{\tau_i}{\sqrt{2}\,\mu_{i\,11}\,f_t}\left(\frac{\epsilon}{q}\right)^2\right]_{r=r_s}, \tag{7.28}$$

$$\tau_M = \frac{\tau_\varphi}{\hat{r}_s^{\,2}}, \tag{7.29}$$

$$\tau_A = \epsilon_a\,n\,s_s\,\tau_H, \tag{7.30}$$

$$\epsilon(r) = \frac{r}{R_0}, \tag{7.31}$$

$\epsilon_s = \epsilon(r_s)$, $\epsilon_a = \epsilon(a)$, and $\hat{r}_s = r_s/a$. Furthermore, the dimensionless neoclassical viscosity parameter $\mu_{i\,11}$ is defined in equation (2.204), $\tau_{\theta\,i}$ is the ion poloidal flow-damping timescale, f_t the fraction of trapped particles (see equation (2.202)), τ_i the ion–ion collision time (see equation (2.21)), $q(r)$ the safety-factor profile (see section 3.2), and $s_s = s(r_s)$ the magnetic shear at the rational surface (see equation (5.28)). Moreover, $J_m(x)$ denotes a Bessel function, while j_{0p} and j_{1p} denote the pth zeros of the $J_0(x)$ and $J_1(x)$ Bessel functions, respectively.

Let

$$\Delta\omega_E = \omega_E - \omega_{E\,0}, \tag{7.32}$$

where $\omega_{E\,0}$ is the E-cross-B frequency at the rational surface in the absence of the error field. It is helpful to define $Q_{E\,0} = S^{1/3}\,\omega_{E\,0}\,\tau_H$. Equations (7.25)–(7.27) can be combined to give

$$\frac{1}{4\,(\epsilon_s\,s_s)^2}\left[\left(\frac{q_s}{\epsilon_s}\right)^2\left(\frac{\tau_\theta}{\tau_\varphi}\right)^{1/2} + 2\ln\left(\frac{a}{r_s}\right)\right]^{1/2}\frac{S^{4/3}}{P_\varphi}\,\mathrm{Im}(\Delta\hat{\Psi}_s\,\hat{\Psi}_s^{\,*}) = Q_{E\,0} - Q_E, \tag{7.33}$$

where $q_s = m/n$. Here, use has been made of the results [7]

$$\lim_{\epsilon\to 0}\sum_{p=1,\infty}\frac{\sqrt{\epsilon}\,[J_1(j_{1p}\,x)]^2}{[J_2(j_{1p})]^2\,(1+\epsilon\,j_{1p}^{\,2})} \simeq \frac{1}{4\,x}, \tag{7.34}$$

$$\sum_{p=1,\infty}\frac{[J_0(j_{0p}\,x)]^2}{[J_1(j_{0p})]^2\,j_{0p}^{\,2}} = \frac{1}{2}\ln\left(\frac{1}{x}\right). \tag{7.35}$$

Equations (7.9), (5.129), and (7.33) yield the torque balance equation [4, 6]

$$\frac{\hat{\Delta}_i(Q_E)\,\bar{I}_c^{\,2}}{[\zeta + \hat{\Delta}_r(Q_E)]^2 + [\hat{\Delta}_i(Q_E)]^2} = Q_{E\,0} - Q_E, \tag{7.36}$$

where

$$\hat{\Delta}_r = \mathrm{Re}(\hat{\Delta}), \tag{7.37}$$

$$\hat{\Delta}_i = \mathrm{Im}(\hat{\Delta}), \tag{7.38}$$

$$\zeta = S^{-1/3}\,(-\Delta_{nw}), \tag{7.39}$$

$$\bar{I}_c = \frac{1}{2\,\epsilon_s\,s_s}\left[\left(\frac{q_s}{\epsilon_s}\right)^2\left(\frac{\tau_\theta}{\tau_\varphi}\right)^{1/2} + 2\ln\left(\frac{a}{r_s}\right)\right]^{1/2}\left(\frac{S}{P_\varphi}\right)^{1/2}\tilde{I}_c. \tag{7.40}$$

The left-hand side of equation (7.36) represents an electromagnetic braking torque that develops at the rational surface and acts to halt the local electron fluid rotation. On the other hand, the right-hand side of the equation represents a viscous restoring torque that opposes any changes in the electron fluid rotation at the rational surface. Equation (7.36) can be rearranged to give

$$\bar{I}_c^{\,2} = F(Q_E), \tag{7.41}$$

where

$$F(Q_E) = (Q_{E\,0} - Q_E)\,\frac{[\zeta + \hat{\Delta}_r(Q_E)]^2 + [\hat{\Delta}_i(Q_E)]^2}{\hat{\Delta}_i(Q_E)}. \tag{7.42}$$

Figure 7.3 shows the torque balance function, $F(Q_E)$, calculated for a low-field tokamak fusion reactor with $Q_{E\,0} = 2.0$. The parameters used in the calculation are $\tau = 1$, $Q_e = 1.00$, $D = 3.02$, $P_\varphi = 874$, and $P_\perp = 287$, as well as $S = 1.35 \times 10^9$, $m = 2$, and $\Delta_{nw} = -2\,m$ (see tables 5.1 and 6.1). Note that $\hat{\Delta}_r \equiv \mathrm{Re}(\hat{\Delta})$ and $\hat{\Delta}_i \equiv \mathrm{Im}(\hat{\Delta})$ are specified in equations (7.16) and (7.17), respectively. The parameter ζ takes the value 3.62×10^{-3}. As indicated in the figure, when $\bar{I}_c^{\,2} = 2.0$ there are three possible values of the normalized E-cross-B frequency at the rational surface, Q_E, that satisfy equation (7.41). However, as is easily demonstrated, the middle solution is dynamically unstable [4]. We thus conclude that there are two dynamically stable branches of solutions to the torque balance equation, (7.36). On the *high-slip* branch (i.e. the rightmost solution in the figure), the electron fluid at the rational surface rotates with respect to the error field, generating a shielding current that suppresses driven magnetic reconnection. On the *low-slip* branch (i.e. the leftmost solution in the figure), the electron fluid rotation at the rational surface is arrested, there is no shielding current, and driven magnetic reconnection proceeds unhindered.

Figure 7.4 shows the torque balance function, $F(Q_E)$, calculated for a high-field tokamak fusion reactor with $Q_{E\,0} = 2.0$. The parameters used in the calculation are $Q_e = 0.75$ and $P_\varphi = 874$, as well as $S = 5.63 \times 10^8$, $m = 2$, and $\Delta_{nw} = -2\,m$ (see tables 5.1 and 6.1). Note that $\hat{\Delta}_r \equiv \mathrm{Re}(\hat{\Delta})$ and $\hat{\Delta}_i \equiv \mathrm{Im}(\hat{\Delta})$ are specified in equations

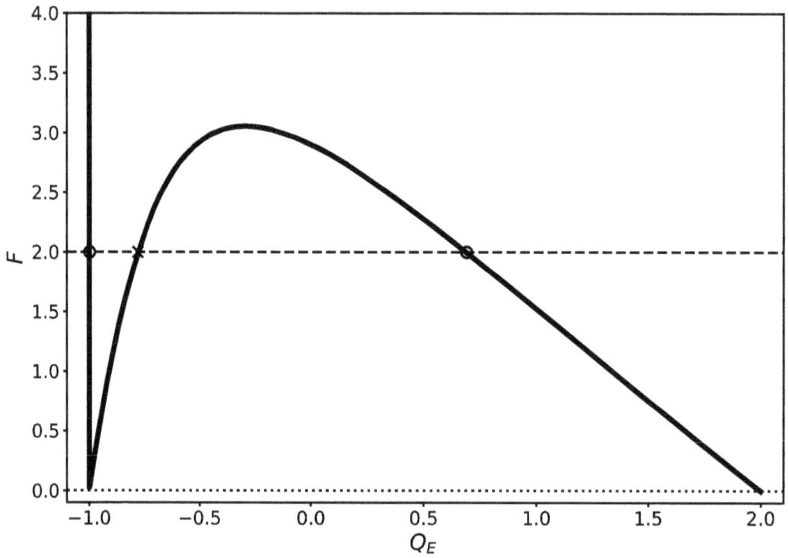

Figure 7.3. The torque balance curve in a low-field tokamak fusion reactor with $Q_{E\,0} = 2.0$. The intersection of the curve with the horizontal line of height 2.0 shows the possible steady-state values of $Q_{\rm E}$ when $\bar{I}_c^{\,2} = 2.0$. There are three solutions. Two (indicated by the circular markers) are dynamically stable, and one (indicated by the cross-shaped marker) is dynamically unstable.

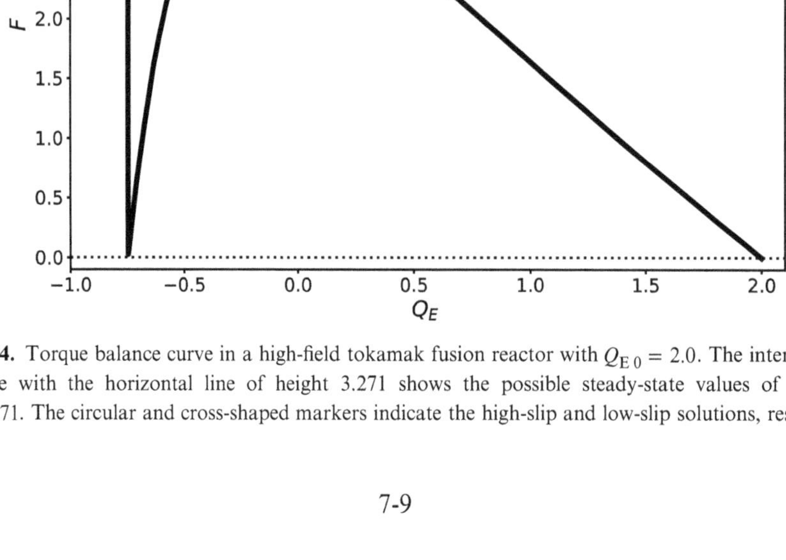

Figure 7.4. Torque balance curve in a high-field tokamak fusion reactor with $Q_{\rm E\,0} = 2.0$. The intersection of the curve with the horizontal line of height 3.271 shows the possible steady-state values of $Q_{\rm E}$ when $\bar{I}_c^{\,2} = 3.271$. The circular and cross-shaped markers indicate the high-slip and low-slip solutions, respectively.

(7.22) and (7.23), respectively. The parameter ζ takes the value 4.84×10^{-3}. As indicated in the figure, when $\bar{I}_c^{\,2}$ attains the critical value of 3.271, the high-slip and the dynamically unstable solutions of the torque balance equation merge together and annihilate one another. For $\bar{I}_c^{\,2} > 3.71$, only the low-slip solution branch exists. Thus, if the system is initially on the high-slip solution branch, and the error-field amplitude is raised such that $\bar{I}_c^{\,2}$ exceeds the critical value of 3.271, then there is a bifurcation from the high-slip to the low-slip solution branch. This bifurcation is associated with the sudden arrest of the electron fluid rotation at the rational surface, the collapse of the shielding current, and the onset of driven magnetic reconnection [4, 6]. Of course, the bifurcation corresponds to the error-field penetration phenomenon discussed in section 7.1.

7.5 Error-field penetration

If $\bar{I}_{\rm c\,pen}$ is the critical value of the normalized error-field coil current needed to trigger error-field penetration, then the critical radial magnetic field (in the absence of a shielding current) that must be generated at the rational surface in order to trigger penetration is

$$\left(\frac{b_r}{B_z}\right)_{\rm pen} = \frac{2\,m\,s_{\rm s}}{(-\Delta_{\rm nw})}\left[\left(\frac{q_{\rm s}}{\epsilon_{\rm s}}\right)^2\left(\frac{\tau_\theta}{\tau_\varphi}\right)^{1/2} + 2\ln\left(\frac{a}{r_{\rm s}}\right)\right]^{-1/2}\left(\frac{P_\varphi}{S}\right)^{1/2}\tilde{I}_{\rm c\,pen}. \tag{7.43}$$

Here, use has been made of equations (5.27), (5.129), and (7.40). Moreover, the *shielding factor* (i.e. the ratio of the magnetic flux that would be driven at the rational surface in the absence of a shielding current to that which is actually driven) takes the form

$$\Sigma(Q_{\rm E}) = \frac{\left([\zeta + \hat{\Delta}_{\rm r}(Q_{\rm E})]^2 + [\hat{\Delta}_{\rm i}(Q_{\rm E})]\right)^{1/2}}{\zeta}. \tag{7.44}$$

Finally, the reduced (by a factor of four—see equation (8.1)) width of the locked magnetic island chain driven at the rational surface is

$$w = \frac{W}{4} = \left(\frac{1}{n\,s_{\rm s}\,\epsilon_{\rm s}}\,\frac{b_r}{B_z}\,\frac{1}{\Sigma}\right)^{1/2} r_{\rm s}, \tag{7.45}$$

where use has been made of equations (5.27), (5.129), and (5.27). Figure 7.5 shows error-field penetration calculations made for a low-field tokamak fusion reactor. The calculation parameters are $\tau = 1$, $Q_{\rm e} = 1.00$, $D = 3.02$, $P_\varphi = 874$, and $P_\perp = 287$, as well as $S = 1.35 \times 10^9$, $m = 2$, and $\Delta_{\rm nw} = -2\,m$, and finally, $n = 1$, $s_{\rm s} = 1$, $\tau_{\rm i} = 1.36 \times 10^{-2}$ s, $\tau_\varphi = 1.60$ s, $R_0 = 7.59$ m, $a = R_0/3$, and $r_{\rm s} = a/2$ (see tables 2.1, 5.1, and 6.1). The top-left panel shows the critical radial magnetic field that must be induced at the rational surface (in the absence of shielding currents) in order to trigger error-field penetration as a function of the unperturbed electron fluid rotation frequency at the rational surface. It can be seen that diamagnetic levels of rotation

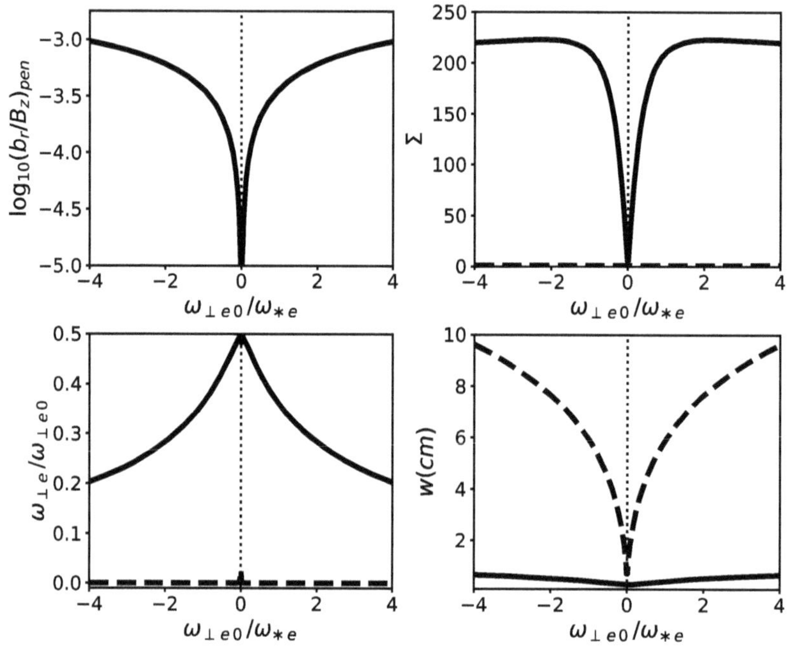

Figure 7.5. Analytically determined error-field penetration threshold in a low-field tokamak fusion reactor. The top-left panel shows the critical radial field needed to trigger penetration as a function of the unperturbed electron fluid rotation frequency. The bottom-left panel shows electron fluid rotation frequency before (solid curve) and after (dashed curve) penetration. The top-right panel shows the shielding factors before (solid curve) and after (dashed curve) penetration. The bottom-right panel shows the reduced width of the driven locked island chain before (solid curve) and after (dashed curve) penetration.

(i.e. $\omega_{\perp\,e} \sim \omega_{*\,e}$—see section 5.15) are sufficient to keep $(b_r/B_z)_{\rm pen}$ well above 10^{-4}, unless the unperturbed electron fluid rotation frequency at the rational surface, $\omega_{\perp\,e\,0} = \omega_{E\,0} + \omega_{*\,e}$, falls significantly below the electron diamagnetic frequency, $\omega_{*\,e}$. The bottom-left panel shows the electron fluid rotation frequency at the rational surface as a fraction of the unperturbed rotation frequency just before (solid curve) and just after (dashed curve) penetration. It can be seen that the electron fluid rotation frequency needs to be reduced by a substantial factor (i.e. at least, a factor of two) before penetration occurs. However, after penetration, the electron fluid rotation frequency is effectively reduced to zero. The top-right panel shows the shielding factors just before (solid curve) and just after (dashed curve) penetration. It can be seen that, prior to penetration, diamagnetic levels of rotation are sufficient to reduce the amount of driven magnetic reconnection at the rational surface by factors that exceed 200 (unless the unperturbed electron fluid rotation frequency at the resonant surface falls significantly below the electron diamagnetic frequency). However, after penetration, there is no shielding at all at the rational surface (i.e. $\Sigma \simeq 1$). Finally, the bottom-right panel shows the reduced width of the locked magnetic island chain driven at the rational surface just before (solid curve) and just after (dashed curve) penetration. It can be seen that, prior to penetration, the driven island width is of the order of 0.5 cm, which is about the same as the linear layer

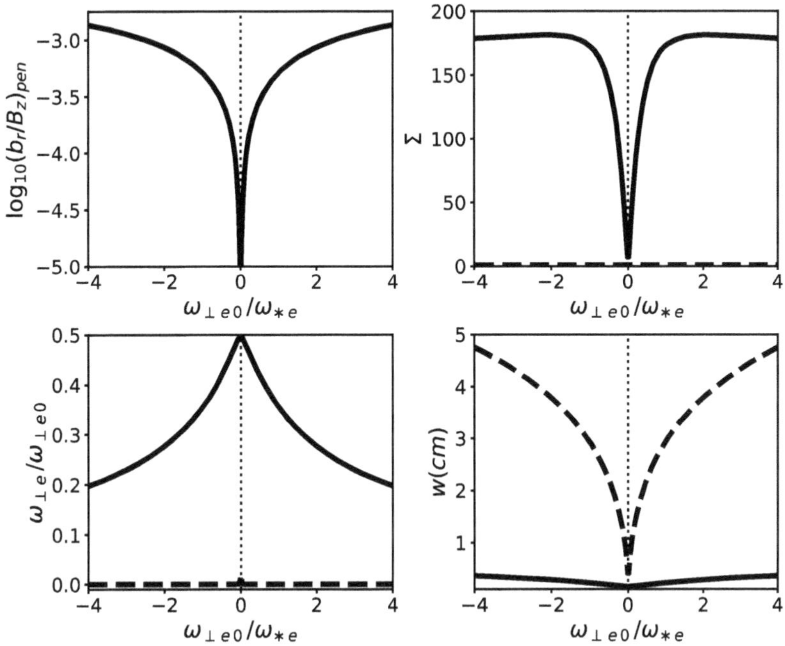

Figure 7.6. Analytically determined error-field penetration threshold in a high-field tokamak fusion reactor. The panels are the same as those described in figure 7.5.

width estimated in table 6.2. It follows that the use of linear theory to determine the error-field penetration threshold is a reasonable approximation (recall that linear theory would be completely invalidated were the island width to become much greater than the linear layer width—see section 5.16). On the other hand, after penetration, the driven island width typically exceeds 10 cm. Such a substantial island chain would significantly degrade the energy confinement properties of the plasma [1, 5] and might even trigger a disruption. It is also clear that, after penetration, the driven tearing perturbation is governed by nonlinear, rather than linear, physics. Figure 7.6 shows error-field penetration calculations made for a high-field tokamak fusion reactor. The calculation parameters are $Q_e = 0.75$ and $P_\varphi = 874$, as well as $S = 5.63 \times 10^8$, $m = 2$, and $\Delta_{nw} = -2\,m$, and finally, $n = 1$, $s_s = 1$, $\tau_i = 2.36 \times 10^{-3}$ s, $\tau_\varphi = 2.78 \times 10^{-1}$ s, $R_0 = 3.16$ m, $a = R_0/3$, and $r_s = a/2$ (see tables 2.1, 5.1, and 6.1). It can be seen, by comparison with figure 5.5, that error-field penetration in a high-field tokamak fusion reactor is fairly similar to that in a low-field reactor. The main differences are that the shielding factors and driven island widths are somewhat smaller in the high-field case compared to the low-field case.

7.6 The numerical solution of the layer equations

The low-field error-field penetration calculations shown in figure 7.5 use the analytical approximations (7.16) and (7.17) to determine the real and imaginary components of the layer-matching parameter, $\hat{\Delta}$. Likewise, the high-field error-field penetration calculations shown in figure 7.6 use the analytical approximations (7.22) and (7.23) to determine the layer-matching parameter. An alternative approach would be to determine the layer-matching parameter directly via numerical solution of the layer equations, as described in section 5.14. The latter approach is more accurate than relying on analytic approximations, but is also much more time-consuming.

Figure 7.7 shows error-field penetration calculations performed for a low-field tokamak fusion reactor using numerical solutions of the layer equations (with $Q = 0$, because the error field is static), rather than analytical approximations, to determine the layer-matching parameter. The calculation parameters are $\tau = 1$, $Q_e = 1.00$, $D = 3.02$, $P_\varphi = 874$, and $P_\perp = 287$, as well as $S = 1.35 \times 10^9$, $m = 2$, and $\Delta_{nw} = -2\,m$, and finally, $n = 1$, $s_s = 1$, $\tau_i = 1.36 \times 10^{-2}$ s, $\tau_\varphi = 1.60$ s, $R_0 = 7.59$ m, $a = R_0/3$, and $r_s = a/2$ (see tables 2.1, 5.1, and 6.1). It can be seen, by comparison with figure 7.5, that the analytical approximations (7.16) and (7.17) lead to a slight underestimate of the shielding factor prior to penetration. The calculations shown in figure 7.7 also exhibit a slight asymmetry between the positive and negative values of the unperturbed (by the error field) electron fluid rotation frequency, $\omega_{\perp\,e\,0}$, at the rational surface that is not captured by calculations that depend on analytical approximations. (If $\omega_{\perp\,e\,0} > 0$, then the tearing mode resonant at the rational surface would rotate in the electron diamagnetic direction were it naturally unstable. If $\omega_{\perp\,e\,0} < 0$, then the mode would rotate in the ion diamagnetic direction.) This asymmetry is due to ion diamagnetic flows. Nevertheless, the level of agreement

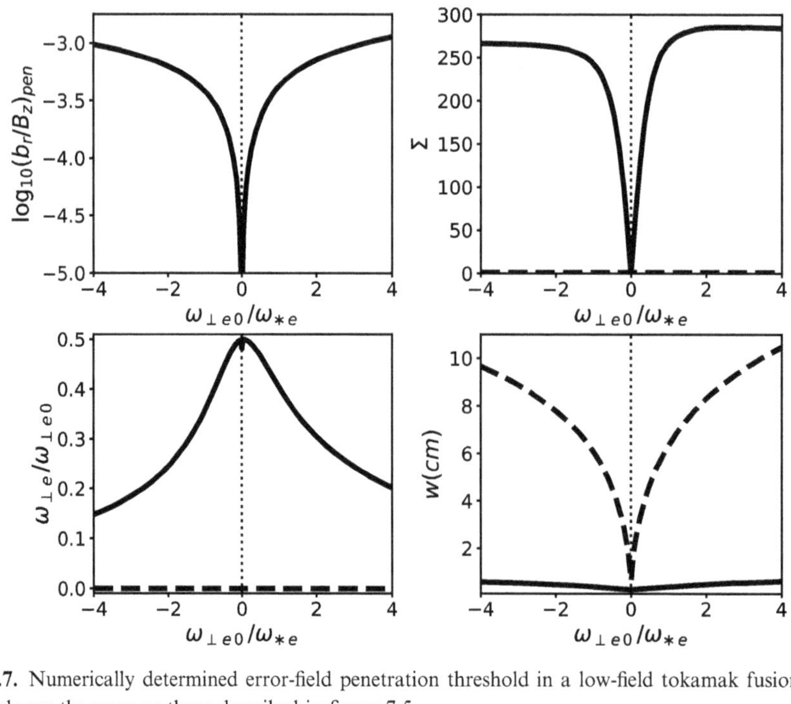

Figure 7.7. Numerically determined error-field penetration threshold in a low-field tokamak fusion reactor. The panels are the same as those described in figure 7.5.

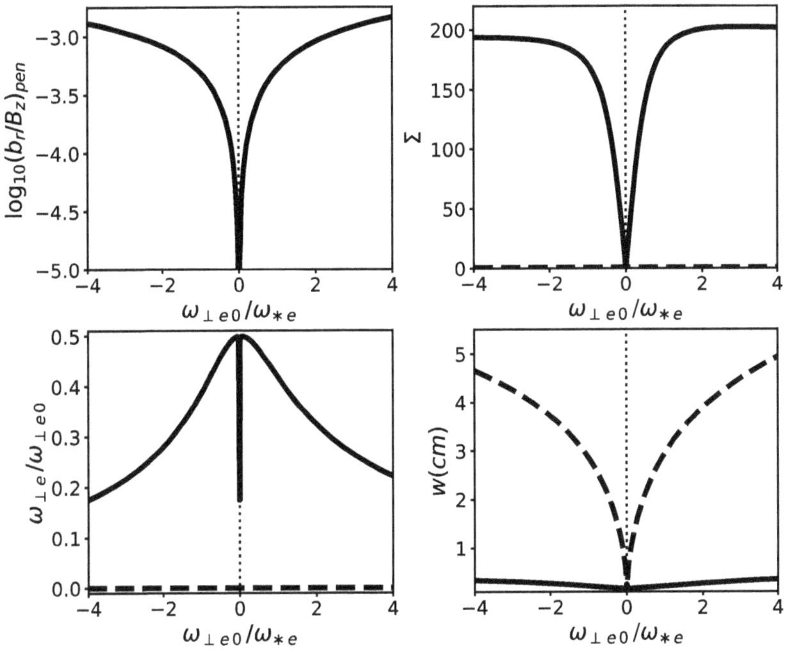

Figure 7.8. Numerically determined error-field penetration threshold in a high-field tokamak fusion reactor. The panels are the same as those described in figure 7.5.

between the calculations shown in figures 7.5 and 7.7 is sufficiently good to warrant the use of analytic approximations in calculations that do not require extreme accuracy.

Figure 7.8 shows error-field penetration calculations performed for a high-field tokamak fusion reactor using numerical solutions of the layer equations, rather than analytical approximations, to determine the layer-matching parameter. The calculation parameters are $\tau = 1$, $Q_e = 0.75$, $D = 2.26$, $P_\varphi = 874$, and $P_\perp = 287$, as well as $S = 5.63 \times 10^8$, $m = 2$, and $\Delta_{nw} = -2\,m$, and finally, $n = 1$, $s_s = 1$, $\tau_i = 2.36 \times 10^{-3}$ s, $\tau_\varphi = 2.78 \times 10^{-1}$ s, $R_0 = 3.16$ m, $a = R_0/3$, and $r_s = a/2$ (see tables 2.1, 5.1, and 6.1). Again, it can be seen, by comparison with figure 7.6, that calculations that rely on analytic approximations lead to a slight underestimate of the shielding factor prior to penetration and do not capture a slight asymmetry between the positive and negative values of the electron fluid rotation frequency at the rational surface.

References

[1] Chang Z and Callen J D 1990 Global energy confinement degradation due to macroscopic phenomena in tokamaks *Nucl. Fusion* **30** 219

[2] Cole A and Fitzpatrick R 2006 Drift-magnetohydrodynamical model of error-field penetration in tokamak plasmas *Phys. Plasmas* **13** 032503

[3] Fishpool G M and Haynes P S 1994 Field error instabilities in JET *Nucl. Fusion* **34** 109

[4] Fitzpatrick R 1993 Interaction of tearing modes with external structures in cylindrical geometry *Nucl. Fusion* **33** 1049

[5] Fitzpatrick R 1995 Helical temperature perturbations associated with tearing modes in tokamak plasmas *Phys. Plasmas* **2** 825
[6] Fitzpatrick R 1998 Bifurcated states of a rotating tokamak plasma in the presence of a static error-field *Phys. Plasmas* **5** 3325
[7] Fitzpatrick R 2022 Influence of anomalous perpendicular transport on linear tearing mode dynamics in tokamak plasmas *Phys. Plasmas* **29** 032507
[8] Hender T C *et al* 1992 Effect of resonant magnetic perturbations on COMPASS-C tokamak discharges *Nucl. Fusion* **32** 2091
[9] Howell D F, Hender T C and Cunningham G 2007 Locked mode thresholds on the MAST spherical tokamak *Nucl. Fusion* **47** 1336
[10] Menard J E *et al* 2010 Progress in understanding error-field physics in NSTX spherical torus plasmas *Nucl. Fusion* **50** 045008
[11] Scoville J T, La Haye R J, Kellman A K, Osborne T H, Stambaugh R D, Strait E J and Taylor T S 1991 Locked modes in DIII-D and a method for prevention of the low density mode *Nucl. Fusion* **31** 875
[12] Wang H-H *et al* 2018 Density scaling of $n = 1$ error field penetration in ohmically heated discharges in EAST *Nucl. Fusion* **58** 056024
[13] Wang N and the J-TEXT Team *et al* 2014 Study of the penetration of resonant magnetic perturbations in J-TEXT *Nucl. Fusion* **54** 064014
[14] Wesson J A 2011 *Tokamaks* 4th edn (Oxford: Oxford University Press) https://global.oup.com/academic/product/tokamaks-9780199592234
[15] Wesson J A *et al* 1989 Disruptions in JET *Nucl. Fusion* **29** 641
[16] Wolf R C and the TEXTOR Team *et al* 2005 Effect of the dynamic ergodic divertor in the TEXTOR tokamak on MHD stability *Nucl. Fusion* 1700
[17] Wolfe S M, Hutchinson I H, Granetz R S, Rice J and Hubbard A 2005 Nonaxisymmetric field effects on Alcator C-Mod *Phys. Plasmas* **12** 056110

Chapter 8

The nonlinear resonant response model

8.1 Introduction

As explained in section 5.1, when we describe the response of a tokamak plasma to a tearing-mode perturbation, the plasma can be divided into two regions [7]. In the *outer region*, which comprises most of the plasma, the tearing perturbation is governed by the equations of marginally stable ideal magnetohydrodynamics (MHD). However, these equations become singular at the *rational* magnetic flux surface where the safety factor (see section 3.2) takes the value m/n. Here, m and n are the poloidal and toroidal mode numbers, respectively, of the tearing mode. A radially thin region forms around the rational magnetic flux surface and constitutes the *inner region*. As discussed in section 5.16, the response of the plasma in the inner region to a rotating tearing perturbation in the outer region is accurately described by linear theory only when the width of the magnetic island chain that develops at the rational surface is less than the linear layer width. In the opposite limit, in which the island width exceeds the layer width, a nonlinear approach is required. The aim of this chapter is to employ the reduced drift-MHD model derived in chapter 4 to determine the *nonlinear* response of the plasma in the inner region to a rotating tearing perturbation in the outer region.

8.2 The reduced drift-MHD model

Following the approach given in section 5.2, it is convenient to set the normalization scale length, l, (see section 4.3) in our reduced drift-MHD model equal to the minor radius of the rational surface, r_s. The model evolves four scalar fields. These fields, which are defined in equations (5.1)–(5.4), are the (normalized) helical magnetic flux, ψ, the (normalized) perturbed total plasma pressure, N, the (normalized) MHD fluid stream function, ϕ, and the (normalized) ion parallel velocity, V. It is assumed that these four fields, as well as the auxiliary field, J, which is defined in equation (5.8), are only functions of the normalized radial coordinate $\hat{x} = (r - r_s)/r_s$, the helical angle $\zeta = m\,\theta - n\,\varphi$, and the normalized time $\hat{t} = t/(r_s/V_A)$. Here, r, θ, z are

doi:10.1088/978-0-7503-5367-0ch8

conventional cylindrical coordinates, $\varphi = z/R_0$ a simulated toroidal angle (see chapter 3), R_0 the simulated major radius of the plasma, and V_A the Alfvén speed (see equation (4.23)).

It is convenient to work in a frame of reference that *corotates* with the island chain. (This goal can be achieved by making the transformation $\phi \to \phi + (\hat{\omega}/m)\,\hat{x}$. See the following paragraph for the definitions of these quantities.) In the corotating reference frame, the normalized reconnected flux at the rational surface, $\hat{\Psi}_s(\hat{t})$ (see equation (3.184)), is assumed to be a positive real quantity. It is helpful to define the reduced (by a factor of four) radial width of the magnetic island chain that develops in the inner region:

$$w \equiv \frac{W}{4} = (L_s\, R_0\, \hat{\Psi}_s)^{1/2} \tag{8.1}$$

(see equation (5.129)). Here, L_s is the magnetic shear length at the rational surface (see equation (5.27)).

In the following, it is assumed that $\delta_s \ll w \ll r_s$, where δ_s is the linear layer width (see chapter 6). In other words, the width of the island chain is assumed to be much greater than the linear layer width but much less than the minor radius of the rational magnetic flux surface. Let $\hat{w} = w/r_s$. Reusing the analysis given in section 5.3, we find that

$$\psi(\hat{x}, \zeta, \hat{t}) \to \frac{\hat{x}^2}{2\,\hat{L}_s} + \hat{R}_0\, \hat{\Psi}_s(\hat{t}) \cos\zeta, \tag{8.2}$$

$$N(\hat{x}, \zeta, \hat{t}) \to -\hat{V}_*\, \hat{x}, \tag{8.3}$$

$$\phi(\hat{x}, \zeta, \hat{t}) \to \left[\frac{\hat{\omega}}{m} - \hat{V}_E(\hat{t})\right]\hat{x} - \frac{\varsigma}{4}\, \hat{V}_E'(\hat{t})\, \hat{x}^2, \tag{8.4}$$

$$V(\hat{x}, \zeta, \hat{t}) \to \tilde{V}_\parallel, \tag{8.5}$$

$$J(\hat{x}, \zeta, \hat{t}) \to -\left(\frac{2}{s_s} - 1\right)\frac{1}{\hat{L}_s}, \tag{8.6}$$

in the limit $|\hat{x}|/\hat{w} \gg 1$ (i.e. many island widths from the rational surface). Here, $\hat{L}_s = L_s/r_s$, $\hat{R}_0 = R_0/r_s$, $\hat{V}_E = V_E(r_s)/V_A$, $\hat{V}_* = V_*(r_s)/V_A$, $\tilde{V}_\parallel = \hat{d}_i\, V_{\parallel\,i}(r_s)/V_A$, $\hat{d}_i = d_i/r_s$, and $s_s = s(r_s)$, where $V_E(r)$ is the E-cross-B velocity profile in the outer region (see equation (5.21)), $V_*(r)$ the diamagnetic velocity profile (see equation (5.29)), $V_{\parallel\,i}(r)$ the ion parallel velocity profile (see equation (5.19)), $s(r)$ the magnetic shear profile (see equation (5.28)), and d_i the collisionless ion skin depth at the rational surface (see equation (4.24)). Moreover, $\omega = \hat{\omega}\, V_A/r_s$ is the rotation frequency of the tearing mode in the laboratory frame,

$$\varsigma = \mathrm{sgn}(\hat{x}), \tag{8.7}$$

and

$$\hat{V}_E' = \frac{1}{V_A}\left[r\,\frac{dV_E}{dr}\right]_{r_{s-}}^{r_{s+}}. \tag{8.8}$$

The parameter $\hat{V}_E'$ is introduced into the analysis in order to take account of the fact that the E-cross-B velocity profile in the outer region develops a gradient discontinuity at the rational surface in response to the localized electromagnetic torque that emerges at the surface. Note, finally, that in neglecting any dependence of $\hat{\Psi}_s$ on $\hat{x}$ in equation (8.2), we are making use of the so-called *constant-ψ approximation* [7, 8], which is valid as long as $|\Delta\hat{\Psi}_s|\,\hat{w}/\hat{\Psi}_s \ll 1$, where $\Delta\hat{\Psi}_s$ is defined in equations (3.73) and (3.183) [2, 4].

8.3 The rescaled reduced drift-MHD model

Let

$$X = \frac{\hat{x}}{\hat{w}}, \tag{8.9}$$

$$T = \omega_*\,t = \hat{\omega}_*\,\hat{t}, \tag{8.10}$$

where ω_* is the diamagnetic frequency at the rational surface (see equation (5.47)), and $\hat{\omega}_* = \omega_*/(V_A/r_s)$. It follows that $|X| \sim \mathcal{O}(1)$ in the immediate vicinity of the island chain.

It is helpful to define the rescaled fields $\Psi(X, \zeta, T)$, $\mathcal{N}(X, \zeta, T)$, $\Phi(X, \zeta, T)$, $\mathcal{V}(X, \zeta, T)$, and $\mathcal{J}(X, \zeta, T)$, where

$$\psi = \left(\frac{\hat{w}^{\,2}}{\hat{L}_s}\right)\Psi, \tag{8.11}$$

$$N = \left(\frac{\hat{\omega}_*\,\hat{w}}{m}\right)\mathcal{N}, \tag{8.12}$$

$$\phi = \left(\frac{\hat{\omega}_*\,\hat{w}}{m}\right)\Phi, \tag{8.13}$$

$$V = \tilde{V}_\parallel + \left(\frac{\hat{L}_s\,\hat{\omega}_*^{\,2}}{m^2\,c_\beta^{\,2}}\right)\mathcal{V}, \tag{8.14}$$

$$J = -\left(\frac{2}{s_s} - 1\right)\frac{1}{\hat{L}_s} + \left(\frac{\hat{L}_s\,\hat{\omega}_*^{\,2}}{m^2\,\hat{w}^{\,2}}\right)\mathcal{J}. \tag{8.15}$$

Here, c_β is a dimensionless measure of the plasma pressure at the rational surface (see equation (4.65)).

The reduced drift-MHD model specified in section 5.2 rescales to give

$$\frac{d(\ln \hat{w}^2)}{dT}\cos\zeta = \left\{\Phi - \left(\frac{\tau}{1+\tau}\right)\mathcal{N},\, \Psi\right\} + \epsilon_\beta\,\epsilon_c\,\epsilon_R\,\mathcal{J}, \tag{8.16}$$

$$\frac{\partial \mathcal{N}}{\partial T} = \{\Phi, \mathcal{N}\} + \{\mathcal{V}, \Psi\} + \epsilon_c\,\epsilon_p\,\{\mathcal{J}, \Psi\} + \epsilon_\parallel^{-1}\{\{\mathcal{N}, \Psi\}, \Psi\} + \epsilon_\perp\,\partial_X^2\mathcal{N}, \tag{8.17}$$

$$\frac{\partial(\partial_X^2\Phi)}{\partial T} = \partial_X\left\{\Phi + \frac{\mathcal{N}}{1+\tau},\, \partial_X\Phi\right\} + \{\mathcal{J}, \Psi\} + \epsilon_\varphi\,\partial_X^4\left(\Phi + \frac{\mathcal{N}}{1+\tau}\right), \tag{8.18}$$

$$\epsilon_c\,\frac{\partial \mathcal{V}}{\partial T} = \epsilon_c\,\{\Phi, \mathcal{V}\} + \{\mathcal{N}, \Psi\} + \epsilon_c\,\epsilon_\varphi\,\partial_X^2\mathcal{V}, \tag{8.19}$$

$$\partial_X^2\Psi = 1 + \epsilon_\beta\,\epsilon_c\,\mathcal{J}. \tag{8.20}$$

Here, $\partial_X \equiv \partial/\partial X$,

$$\{A, B\} \equiv \frac{\partial A}{\partial X}\frac{\partial B}{\partial \zeta} - \frac{\partial A}{\partial \zeta}\frac{\partial B}{\partial X}, \tag{8.21}$$

and we have set

$$\hat{E}_\parallel = \left(\frac{2}{s_s} - 1\right)\frac{\hat{\eta}_\parallel}{\hat{L}_s} \tag{8.22}$$

(see equation (5.32)), where $\hat{\eta}_\parallel$ is the normalized parallel plasma resistivity at the rational surface (see equation (5.14)). Furthermore,

$$\epsilon_\beta = c_\beta^{\,2}, \tag{8.23}$$

$$\epsilon_p = \left(\frac{L_p}{L_s}\right)^2, \tag{8.24}$$

$$\epsilon_c = \left(\frac{L_s}{L_p}\right)^2\left(\frac{d_\beta}{w}\right)^2 = \left(\frac{w_c}{w}\right)^2, \tag{8.25}$$

$$\epsilon_R = \frac{r_s^{\,2}}{\omega_*\,\tau_R\,w^2} = \left(\frac{w_R}{w}\right)^2, \tag{8.26}$$

$$\epsilon_\varphi = \frac{r_s^{\,2}}{\omega_*\,\tau_\varphi\,w^2} = \left(\frac{w_\varphi}{w}\right)^2, \tag{8.27}$$

$$\epsilon_\perp = \frac{r_s^{\,2}}{\omega_*\,\tau_\perp\,w^2} = \left(\frac{w_\perp}{w}\right)^2, \tag{8.28}$$

$$\epsilon_{\parallel} = \frac{\omega_{*}\,\tau_{\parallel}'\,r_s}{(\epsilon_s\,s_s\,n)^2\,w} = \frac{w_{\parallel}}{w}, \tag{8.29}$$

where

$$w_c = \frac{L_s\,d_\beta}{L_p}, \tag{8.30}$$

$$w_R = \frac{r_s}{\sqrt{\omega_{*}\,\tau_R}}, \tag{8.31}$$

$$w_\varphi = \frac{r_s}{\sqrt{\omega_{*}\,\tau_\varphi}}, \tag{8.32}$$

$$w_\perp = \frac{r_s}{\sqrt{\omega_{*}\,\tau_\perp}}, \tag{8.33}$$

$$w_{\parallel} = \frac{\omega_{*}\,\tau_{\parallel}'\,r_s}{(\epsilon_s\,s_s\,n)^2}. \tag{8.34}$$

Here, $d_\beta = c_\beta\,d_i$ is the ion sound radius, τ_R the resistive diffusion time (see equation (5.49)), τ_φ the toroidal momentum confinement time (see equation (5.50)), and $\tau_\perp$ the energy confinement time (see equation (5.52)). All of these quantities are evaluated at the rational surface. Moreover,

$$L_p = \frac{5}{3}\,(1 - c_\beta^2)\left[-\left(\frac{d\ln p}{dr}\right)_{r_s}\right]^{-1} \tag{8.35}$$

is the effective *pressure gradient scale length* at the rational surface, and $\epsilon_s = r_s/R_0$. Finally,

$$\tau_{\parallel}' = r_s^{\,2}\bigg/\left\{\frac{2}{3}\,(1 - c_\beta^2)\left[\left(\frac{\tau}{1+\tau}\right)\left(\frac{\eta_e}{1+\eta_e}\right)\chi_{\parallel\,e}'(r_s) + \left(\frac{1}{1+\tau}\right)\left(\frac{\eta_i}{1+\eta_i}\right)\chi_{\parallel\,i}'(r_s)\right]\right\}, \tag{8.36}$$

where

$$\chi_{\parallel\,e}' = \frac{v_{t\,e}\,L_s}{2\,\pi^{1/2}\,m}, \tag{8.37}$$

$$\chi_{\parallel\,i}' = \frac{v_{t\,i}\,L_s}{2\,\pi^{1/2}\,m}. \tag{8.38}$$

Here, $v_{t\,e}$ and $v_{t\,i}$ are the electron and ion thermal velocities, respectively (see equation (2.17)). Note that the parallel electron and ion energy diffusivities have been estimated from equations (2.319) and (2.320), respectively, on the assumption that $k_{\parallel} \simeq 2\,(m/L_s)\,\hat{w}$, which is the typical parallel wave number of the tearing perturbation at the edge of a magnetic island chain of reduced width w. Note that in writing equation (8.18), we have made use of the easily proved identity

$$\begin{aligned}&\{\Phi, \partial_X^2\Phi\} - \frac{1}{2\,(1+\tau)}(\partial_X^2\{\Phi, \mathcal{N}\} + \{\partial_X^2\Phi, \mathcal{N}\} + \{\partial_X^2\mathcal{N}, \Phi\})\\ &\equiv \partial_X\left\{\Phi + \frac{\mathcal{N}}{1+\tau}, \partial_X\Phi\right\}.\end{aligned} \tag{8.39}$$

Equations (8.16)–(8.20) must be solved subject to the boundary conditions (see equations (8.2)–(8.6) and (8.11)–(8.15))

$$\Psi(X, \zeta, T) \to \frac{X^2}{2} + \cos\zeta, \tag{8.40}$$

$$\mathcal{N}(X, \zeta, T) \to X, \tag{8.41}$$

$$\Phi(X, \zeta, T) \to v(T)\,X + \frac{\varsigma\, v'(T)\,X^2}{2}, \tag{8.42}$$

$$\mathcal{V}(X, \zeta, T) \to 0, \tag{8.43}$$

$$\mathcal{J}(X, \zeta, T) \to 0, \tag{8.44}$$

as $|X| \to \infty$. Here,

$$v(T) = \frac{\omega - \omega_E(r_s)}{\omega_*}, \tag{8.45}$$

$$v'(T) = -\frac{\hat{w}}{2\,\omega_*}\left[r\,\frac{d\omega_E}{dr}\right]_{r_{s-}}^{r_{s+}}, \tag{8.46}$$

where $\omega_E(r) = (m/r)\,V_E(r)$ is the E-cross-B frequency profile. Note that Ψ, $\mathcal{N}$, Φ, $\mathcal{V}$, and $\mathcal{J}$ are all $\mathcal{O}(1)$ quantities in the inner region. Note, further, that the boundary conditions (8.40)–(8.44), as well as the symmetry of the rescaled, reduced, drift-MHD equations, (8.16)–(8.20), ensure that Ψ, $\mathcal{V}$, and $\mathcal{J}$ are even functions of X, whereas $\mathcal{N}$ and Φ are odd functions.

8.4 The ordering scheme

Let us assume that

$$\epsilon_\beta \sim \epsilon_p \sim \epsilon_c \sim \epsilon_R \sim \epsilon_\varphi \sim \epsilon_\perp \sim \epsilon_\parallel \sim \epsilon, \tag{8.47}$$

where $\epsilon \ll 1$. Table 8.1 gives estimates for the values of the various parameters that characterize the rescaled reduced drift-MHD model, (8.16)–(8.20), in a low-field tokamak fusion reactor and a high-field tokamak fusion reactor (see chapter 1). These estimates are made using the following assumptions: $B = 5$ T (low field) or

Table 8.1. The parameters characterizing the rescaled reduced drift-MHD model for a low-field tokamak reactor and a high-field tokamak reactor (see equations (8.23)–(8.34) and (8.84)).

	Low-field	High-field
B (T)	5.0	12.0
ϵ_β	1.64×10^{-2}	1.64×10^{-2}
ϵ_p	7.47×10^{-2}	7.47×10^{-2}
w_c (m)	1.79×10^{-2}	7.46×10^{-3}
w_R (m)	8.09×10^{-4}	5.22×10^{-4}
w_φ (m)	2.39×10^{-2}	1.54×10^{-2}
$w_\perp$ (m)	1.37×10^{-2}	8.84×10^{-3}
$w_\parallel$ (m)	7.21×10^{-3}	3.01×10^{-3}
w_d (m)	6.97×10^{-3}	3.89×10^{-3}

$B = 12$ T (high field), $\beta = 0.02$, $T_e = T_i = 7$ keV, $m_i = (m_D + m_T)/2$ (where m_D and m_T are the deuteron and triton masses, respectively), $\Xi_{\perp\,i} = \chi_{\perp\,e} = \chi_{\perp\,i} = 1\ \mathrm{m^2\,s^{-1}}$, $m = 2$, $n = 1$, $r_s = a/2$ (where a is the minor radius of the plasma), $s_s = 1$, $\tau = 1$, $\eta_e = \eta_i = 1$, and $dp/dr = -p/a$. It is clear from table 8.1 and equations (8.25)–(8.34) that all of the ϵ parameters appearing in the rescaled, reduced, drift-MHD equations are less than unity, in accordance with equation (8.47), provided that the radial width of the magnetic island chain that develops in the inner region exceeds a few centimeters.

The orderings $\epsilon_\beta \ll 1$ and $\epsilon_p \ll 1$ are a direct consequence of the fact that conventional tokamak fusion reactors have large aspect ratios (i.e. $R_0 \gg r_s$) and confine low-β plasmas (i.e., $\beta \sim 0.02$). As will become apparent, the ordering $\epsilon_c \ll 1$ ensures that the island chain is sufficiently wide that ion acoustic waves propagating parallel to the magnetic field are able to smooth out any variations in the normalized plasma pressure, $\mathcal{N}$, around magnetic flux surfaces [9]. The orderings $\epsilon_R \ll 1$, $\epsilon_\varphi \ll 1$, and $\epsilon_\perp \ll 1$ ensure that the island chain is sufficiently wide that the perpendicular diffusions of magnetic flux, momentum, and energy are not dominant effects in the rescaled, reduced, drift-MHD equations. Finally, the ordering $\epsilon_\parallel\, \epsilon_\perp \ll 1$ ensures that the island chain is sufficiently wide that parallel energy transport smooths out any variations in the normalized plasma pressure, $\mathcal{N}$, around magnetic flux surfaces [3].

Finally, suppose that

$$\frac{\partial}{\partial T} \sim \mathcal{O}(\epsilon^3), \tag{8.48}$$

and let us expand the various fields in our rescaled model as follows:

$$\Psi = \Psi_0 + \epsilon^2\,\Psi_2 + \epsilon^3\,\Psi_3 + \cdots, \tag{8.49}$$

$$\Phi = \Phi_0 + \epsilon^2\,\Phi_2 + \epsilon^3\,\Phi_3 + \cdots, \tag{8.50}$$

$$\mathcal{N} = \mathcal{N}_0 + \epsilon^2\,\mathcal{N}_2 + \epsilon^3\,\mathcal{N}_3 + \cdots, \tag{8.51}$$

$$\mathcal{V} = \mathcal{V}_0 + \epsilon\,\mathcal{V}_1 + \cdots, \tag{8.52}$$

$$\mathcal{J} = \mathcal{J}_0 + \epsilon\,\mathcal{J}_1 + \cdots. \tag{8.53}$$

Here, Ψ_0, Ψ_1, etc. are assumed to be $\mathcal{O}(1)$ in the inner region.

8.5 The lowest-order solution

To the zeroth order in ϵ, equation (8.20) yields

$$\partial_X^2 \Psi_0 = 1. \tag{8.54}$$

Solving this equation subject to the boundary condition (8.40), we obtain

$$\Psi_0 = \Omega(X, \zeta) \equiv \frac{X^2}{2} + \cos\zeta. \tag{8.55}$$

Thus, we conclude that, to the lowest order in our expansion, the magnetic flux surfaces in the island region have the constant-ψ structure pictured in figure 5.7. The island O-points correspond to $\Omega = -1$ and $\zeta = (2k - 1)\,\pi$ (where k is an integer), the X-points correspond to $\Omega = +1$ and $\zeta = 2k\,\pi$, and the magnetic separatrix corresponds to $\Omega = +1$.

To the zeroth order in ϵ, equation (8.19) yields

$$\{\mathcal{N}_0, \Omega\} = 0, \tag{8.56}$$

where use has been made of equation (8.55). Given that $\mathcal{N}$ is an odd function of X, it follows that

$$\mathcal{N}_0(X, \zeta, T) = \varsigma\,\mathcal{N}_{(0)}(\Omega, T). \tag{8.57}$$

We conclude that parallel ion acoustic waves, whose dynamics are described by equation (8.19), smooth out any variations in the lowest-order normalized plasma pressure, $\mathcal{N}_0$, around magnetic flux surfaces [9]. By symmetry, $\mathcal{N}_0 = 0$ inside the magnetic separatrix of the island chain. In other words, the plasma pressure profile (and, by implication, the electron number density, electron temperature, and ion temperature profile—see section 4.1) is completely flattened inside the separatrix.

It is helpful to define

$$L(\Omega, T) = \frac{\partial \mathcal{N}_{(0)}}{\partial \Omega}. \tag{8.58}$$

It follows that $L(\Omega \leqslant 1, T) = 0$. Furthermore, equations (8.41) and (8.55) imply that

$$L(\Omega \to \infty, T) = \frac{1}{\sqrt{2\,\Omega}}. \tag{8.59}$$

To the zeroth order in ϵ, equation (8.16) yields

$$\{\Phi_0, \Omega\} = 0, \tag{8.60}$$

where use has been made of equations (8.55) and (8.57). Given that Φ is an odd function of X, it follows that

$$\Phi_0(X, \zeta, T) = \varsigma\, \Phi_{(0)}(\Omega, T). \tag{8.61}$$

We conclude that the lowest-order normalized MHD fluid stream function, Φ_0, is constant on magnetic flux surfaces. By symmetry, $\Phi_0 = 0$ inside the magnetic separatrix of the island chain. In other words, the stream-function profile is completely flattened inside the separatrix.

It is helpful to define

$$M(\Omega, T) = \frac{\partial \Phi_{(0)}}{\partial \Omega}. \tag{8.62}$$

It follows that $M(\Omega \leqslant 1, T) = 0$. Furthermore, equations (8.42) and (8.55) imply that

$$M(\Omega \to \infty, T) = \frac{v(T)}{\sqrt{2\,\Omega}} + v'(T). \tag{8.63}$$

To the zeroth order in ϵ, equation (8.17) yields

$$\{\mathcal{V}_0, \Omega\} = 0, \tag{8.64}$$

where use has been made of equations (8.55), (8.57), and (8.61). Given that $\mathcal{V}$ is an even function of X, we can write

$$\mathcal{V}_0(X, \zeta, T) = \mathcal{V}_{(0)}(\Omega, T). \tag{8.65}$$

In other words, the lowest-order normalized parallel ion velocity, $\mathcal{V}_0$, is also constant on magnetic flux surfaces.

Finally, to the zeroth order in ϵ, equation (8.18) yields

$$\{\mathcal{J}_0, \Omega\} = -\partial_X \left\{\Phi_0 + \frac{\mathcal{N}_0}{1+\tau}, \partial_X \Phi_0\right\} = \frac{1}{2}\left\{\partial_\Omega\left[M\left(M + \frac{L}{1+\tau}\right)\right] X^2, \Omega\right\}, \tag{8.66}$$

where use has been made of equations (8.55), (8.57), (8.58), (8.61), and (8.62). Moreover, $\partial_\Omega \equiv \partial/\partial\Omega$.

8.6 The flux-surface average operator

The *flux-surface average operator*, $\langle\cdots\rangle$, is defined as [8]

$$\langle A(\varsigma,\,\Omega,\,\zeta,\,T)\rangle \equiv \begin{cases} \displaystyle\int_{\zeta_0}^{2\pi-\zeta_0} \frac{A(\varsigma,\,\Omega\,\zeta,\,T) + A(-\varsigma,\,\Omega,\,\zeta,\,T)}{2\,[2\,(\Omega - \cos\zeta)]^{1/2}}\,\frac{d\zeta}{2\pi} & -1 \leqslant \Omega \leqslant 1 \\ \displaystyle\oint \frac{A(\varsigma,\,\Omega\,\zeta,\,T)}{[2\,(\Omega - \cos\zeta)]^{1/2}}\,\frac{d\zeta}{2\pi} & \Omega > 1 \end{cases}, \tag{8.67}$$

where $\zeta_0 = \cos^{-1}(\Omega)$ and $0 \leqslant \zeta_0 \leqslant \pi$. It follows that

$$\langle\{A,\,\Omega\}\rangle = 0 \tag{8.68}$$

for *any* $A(\varsigma,\,\Omega,\,\zeta,\,T)$. It is helpful to define

$$\tilde{A} \equiv A - \frac{\langle A\rangle}{\langle 1\rangle}. \tag{8.69}$$

It follows that

$$\langle\tilde{A}\rangle = 0 \tag{8.70}$$

for *any* $A(\varsigma,\,\Omega,\,\zeta,\,T)$. Equation (8.66) yields

$$\mathcal{J}_0(\Omega,\,\zeta,\,T) = \bar{\mathcal{J}}_0(\Omega,\,T) + \frac{1}{2}\,\partial_\Omega\left[M\left(M + \frac{L}{1+\tau}\right)\right]\tilde{X}^2, \tag{8.71}$$

where $\bar{\mathcal{J}}_0(\Omega,\,T)$ is an undetermined flux-surface function.

8.7 Fluid velocities

The flux-surface functions $M(\Omega,\,T)$ and $L(\Omega,\,T)$ are directly related to the lowest-order perpendicular velocities of the various plasma species in the island rest frame. In fact, as is clear from sections 4.2, 5.2, and 8.2,

$$V_{\perp\,\theta\,\mathrm{e}}(X,\,\zeta,\,T) = -\frac{r_\mathrm{s}\,\omega_*}{m}\,|X|\,Y_\mathrm{e}(\Omega,\,T), \tag{8.72}$$

$$V_{\perp\,\theta}(X,\,\zeta,\,T) = -\frac{r_\mathrm{s}\,\omega_*}{m}\,|X|\,M(\Omega,\,T), \tag{8.73}$$

$$V_{\perp\,\theta\,\mathrm{i}}(X,\,\zeta,\,T) = -\frac{r_\mathrm{s}\,\omega_*}{m}\,|X|\,Y_\mathrm{i}(\Omega,\,T), \tag{8.74}$$

where

$$Y_\mathrm{e}(\Omega,\,T) = M - \left(\frac{\tau}{1+\tau}\right)L, \tag{8.75}$$

$$Y_\mathrm{i}(\Omega,\,T) = M + \left(\frac{1}{1+\tau}\right)L, \tag{8.76}$$

Here, $V_{\perp\,\theta\,\mathrm{e}}$, $V_{\perp\,\theta}$, and $V_{\perp\,\theta\,\mathrm{i}}$ refer to the θ components of the perpendicular electron, MHD, and ion fluid velocities, respectively. Note that Y_e, M, and Y_i are all zero inside the magnetic separatrix of the island chain.

8.8 The need for a higher-order solution

To the lowest order in our expansion scheme, the rescaled, reduced, drift-MHD equations, (8.16)–(8.20), require the plasma pressure to be a magnetic flux-surface function. This should come as no surprise. In essence, a magnetic island chain is a helical magnetic equilibrium that evolves (in its local rest frame) on the very slow resistive timescale [8]. Hence, we would expect all of the results derived in section 2.25 to apply to the plasma in the immediate region of the island chain. Note, however, that, while solving the rescaled, reduced, drift-MHD equations to the lowest order tells us that $\mathcal{N} = \varsigma\, \mathcal{N}_{(0)}(\Omega, T)$, $\Phi = \varsigma\, \Phi_{(0)}(\Omega, T)$, and $\mathcal{V} = \mathcal{V}_{(0)}(\Omega, T)$, and that the lowest-order current density profile has the form (8.71), the lowest-order solution leaves the flux-surface functions $\mathcal{N}_{(0)}(\Omega, T)$, $\Phi_{(0)}(\Omega, T)$, $\mathcal{V}_{(0)}(\Omega, T)$, and $\mathcal{J}_0(\Omega, T)$ completely undetermined. In fact, in order to determine the forms of these four functions, it is necessary to solve the rescaled, reduced, drift-MHD equations to a higher order in our expansion scheme. In particular, we need to incorporate the terms that describe the perpendicular diffusion of magnetic flux, ion momentum, and energy into our analysis [5]. The essential point is that the magnetic island chain persists in the plasma for a sufficiently long time that the relatively small perpendicular diffusion terms are able to relax the lowest-order profiles across the island region (and, in fact, across the whole plasma).

8.9 The higher-order solution

In order to include the perpendicular transport term, we need to evaluate equation (8.16) to the third order in ϵ. By doing so, we obtain

$$\frac{d(\ln \hat{w}^2)}{dT}\cos\zeta = \{F_\mathrm{e}, \Omega\} + \epsilon_\beta\, \epsilon_\mathrm{c}\, \epsilon_R\, \mathcal{J}_0, \tag{8.77}$$

where

$$\begin{aligned} F_\mathrm{e}(\Omega, \zeta, T) = {}& \epsilon^2\,(\Phi_2 + \epsilon\, \Phi_3) - \epsilon^2\left(\frac{\tau}{1+\tau}\right)(\mathcal{N}_2 + \epsilon\, \mathcal{N}_3) \\ & - \epsilon^2\, \varsigma\, Y_\mathrm{e}\,(\Psi_2 + \epsilon\, \Psi_3), \end{aligned} \tag{8.78}$$

and use has been made of equations (8.49)–(8.53), (8.57), (8.58), (8.61), (8.62), and (8.75). The flux-surface average of equation (5.14) yields

$$\mathcal{J}_0(\Omega, T) = \frac{1}{\epsilon_\beta\, \epsilon_\mathrm{c}\, \epsilon_R}\frac{d(\ln \hat{w}^2)}{dT}\frac{\langle\cos\zeta\rangle}{\langle 1\rangle}, \tag{8.79}$$

where use has been made of equations (8.68), (8.70), and (8.71). Hence, we can write

$$\mathcal{J}_0(\Omega, \zeta, T) = \frac{1}{\epsilon_\beta\, \epsilon_c\, \epsilon_R}\frac{d(\ln \hat{w}^2)}{dT}\frac{\langle \cos\zeta\rangle}{\langle 1\rangle} + \frac{1}{2}\,\partial_\Omega\,(M\ \ Y_i)\,\tilde{X}^2, \tag{8.80}$$

where use has been made of equation (8.76). The first term on the right-hand side of the previous equation represents the parallel current driven inductively when the reconnected flux at the rational surface varies in time. The second term on the right-hand side of the previous equation represents the parallel return current driven by the perpendicular *polarization current* associated with the acceleration of the ion fluid around the magnetic separatrix of the island chain [10, 11]. In fact, it can easily be seen that if the ion fluid could pass freely through the separatrix (i.e. $\Phi_0 \propto X$ and $\mathcal{N}_0 \propto X$), then the polarization term would be zero.

In order to include the perpendicular transport term, we need to evaluate equation (8.17) to the first order in ϵ. By doing so, we obtain

$$0 = \epsilon\ \{\mathcal{V}_1, \Omega\} + \epsilon^2\, \epsilon_\parallel^{-1}\{\{\mathcal{N}_2, \Omega\}, \Omega\} + \epsilon_\perp\, \varsigma\, (X^2\, \partial_\Omega L + L), \tag{8.81}$$

where use has been made of equations (8.49)–(8.53), (8.55), (8.57), and (8.58). The flux-surface average of the previous equation yields

$$\langle X^2\rangle\, \partial_\Omega L + \langle 1\rangle\, L = \partial_\Omega\,(\langle X^2\rangle\, L) = 0, \tag{8.82}$$

where use has been made of equation (8.68). Here, we have made use of the easily proved result that $d_\Omega\langle X^n\rangle = (n-1)\,\langle X^{n-2}\rangle$, where $d_\Omega \equiv d/d\Omega$. We can solve the previous equation, subject to the boundary condition (8.59), to give

$$L(\Omega, T) = L(\Omega) = \begin{cases} 0 & -1 \leqslant \Omega \leqslant 1 \\ 1/\langle X^2\rangle & \Omega > 1 \end{cases}. \tag{8.83}$$

Here, we have taken into account the previously mentioned fact that $L = 0$ within the island separatrix.

Note that $L(\Omega)$ is discontinuous across the island separatrix, which implies that the pressure gradient—and, hence, the diamagnetic velocity—is also discontinuous across the separatrix. Of course, there is no real discontinuity. In fact, we would expect the discontinuity to be resolved in a layer on the separatrix whose width is obtained by balancing the parallel and perpendicular energy transport terms in equation (8.17) [3]. In other words, we expect the layer thickness to correspond to the island width at which $\epsilon_\parallel\, \epsilon_\perp \sim 1$. Hence, according to equations (8.28) and (8.29), the characteristic layer thickness is $4\, w_d$, where

$$w_d = \left(\frac{w_\parallel\, w_\perp^2}{4}\right)^{1/3}, \tag{8.84}$$

and $w_\parallel$ and $w_\perp$ are specified in equations (8.33) and (8.34), respectively. (The factor of four ensures that $W_d = 4\, w_d$ has the same value as the standard critical island width, W_c, defined in reference [3].) As is clear from table 8.1, w_d is of the order of half a centimeter in a tokamak fusion reactor.

In order to include the perpendicular transport term, we need to evaluate equation (8.18) to the first order in ϵ. By doing so, we obtain

$$0 = \epsilon \, \{\mathcal{J}_1, \Omega\} + \epsilon_\varphi \, \varsigma \, X \, \partial_\Omega \, [\partial_\Omega \, (X^3 \, \partial_\Omega Y_i)], \tag{8.85}$$

where use has been made of equations (8.49)–(8.53), (8.55), (8.57), (8.58), (8.61), (8.62), and (8.76). The flux-surface average of the previous equation yields

$$\partial_\Omega^2 \, (\langle X^4 \rangle \, d_\Omega Y_i) = 0. \tag{8.86}$$

We can solve the previous equation, subject to the boundary conditions (8.59) and (8.63), to give

$$Y_i(\Omega, T) = \frac{v'(T)}{\displaystyle\int_1^\infty \frac{d\Omega}{\langle X^4 \rangle}} \begin{cases} 0 & -1 \leqslant \Omega \leqslant 1 \\ \displaystyle\int_1^\Omega \frac{d\Omega'}{\langle X^4 \rangle} & \Omega > 1 \end{cases} \tag{8.87}$$

[5]. Here, we have rejected as unphysical the solution that blows up as $\Omega^{1/2}$ as $\Omega \to \infty$. We have also made use of the fact that $Y_i = 0$ within the magnetic separatrix of the island chain. Finally, we have demanded that the ion fluid velocity—and, hence, the function $Y_i(\Omega)$—be continuous across the separatrix, because the ion fluid possesses finite perpendicular viscosity. Note, however, that the discontinuity in the function $L(\Omega)$ across the island separatrix (see equation (8.83)) implies that the electron and MHD fluid velocities are discontinuous across the separatrix (see section 8.7). As previously mentioned, these discontinuities are resolved in a layer of characteristic thickness 4 w_d on the separatrix.

In order to include the perpendicular transport term, we need to evaluate equation (8.19) to the second order in ϵ. By doing so, we obtain

$$0 = \epsilon\{-\epsilon_c \, \varsigma \, M \, \mathcal{V}_1 - \epsilon \, \varsigma \, L \, \Psi_2 + \epsilon \, \mathcal{N}_2, \Omega\} + \epsilon_c \, \epsilon_\varphi \big(X^2 \, \partial_\Omega^2 \mathcal{V}_{(0)} + \partial_\Omega \mathcal{V}_{(0)}\big), \tag{8.88}$$

where use has been made of equations (8.49)–(8.53) (8.55), (8.57), (8.58), (8.61), (8.62), and (8.65). The flux-surface average of the previous equation yields

$$\langle X^2 \rangle \, \partial_\Omega^2 \mathcal{V}_{(0)} + \langle 1 \rangle \, \partial_\Omega \mathcal{V}_{(0)} = \partial_\Omega \left(\langle X^2 \rangle \, \partial_\Omega \mathcal{V}_{(0)}\right) = 0. \tag{8.89}$$

The previous equation can be solved, subject to the boundary condition (8.43), to give

$$\mathcal{V}_{(0)}(\Omega, T) = 0. \tag{8.90}$$

Hence, we conclude that the lowest-order ion parallel flow is unaffected by the presence of the island chain.

8.10 Asymptotic matching

Now that we have found the solution of our rescaled, reduced, drift-MHD equations in the immediate vicinity of the magnetic island chain, it is necessary to asymptotically match this solution to the solution in the outer region. Given that $\hat{\Psi}_s$ is real, it follows from equations (3.72), (3.73), (3.183), (3.184), and (8.9) that

$$\mathrm{Re}\left(\frac{\Delta\hat{\Psi}_s}{\hat{\Psi}_s}\right) = \frac{2\,r_s}{R_0\,B_z\,\hat{\Psi}_s}\int_{r_{s-}}^{r_{s+}}\oint\frac{\partial^2\psi}{\partial r^2}\cos\zeta\,\frac{d\zeta}{2\pi}\,dr = \frac{2}{\hat{w}\,\hat{R}_0\,\hat{\Psi}_s}\int_{-\infty}^{\infty}\oint\frac{\partial^2\psi}{\partial X^2}\cos\zeta\,\frac{d\zeta}{2\pi}\,dX, \tag{8.91}$$

$$\begin{aligned}\mathrm{Im}\left(\frac{\Delta\hat{\Psi}_s}{\hat{\Psi}_s}\right) &= -\frac{2\,r_s}{R_0\,B_z\,\hat{\Psi}_s}\int_{r_{s-}}^{r_{s+}}\oint\frac{\partial^2\psi}{\partial r^2}\sin\zeta\,\frac{d\zeta}{2\pi}\,dr\\ &= -\frac{2}{\hat{w}\,\hat{R}_0\,\hat{\Psi}_s}\int_{-\infty}^{\infty}\oint\frac{\partial^2\psi}{\partial X^2}\sin\zeta\,\frac{d\zeta}{2\pi}\,dX.\end{aligned} \tag{8.92}$$

Making use of equations (8.1) and (8.11), we obtain

$$\mathrm{Re}\left(\frac{\Delta\hat{\Psi}_s}{\hat{\Psi}_s}\right) = \frac{2}{\hat{w}}\int_{-\infty}^{\infty}\oint\partial_X^2\Psi\,\cos\zeta\,\frac{d\zeta}{2\pi}\,dX, \tag{8.93}$$

$$\mathrm{Im}\left(\frac{\Delta\hat{\Psi}_s}{\hat{\Psi}_s}\right) = -\frac{2}{\hat{w}}\int_{-\infty}^{\infty}\oint\partial_X^2\Psi\,\sin\zeta\,\frac{d\zeta}{2\pi}\,dX. \tag{8.94}$$

However, according to equations (8.20) and (8.53),

$$\partial_X^2\Psi = 1 + \epsilon_\beta\,\epsilon_c\,\mathcal{J}_0 + \epsilon\,\epsilon_\beta\,\epsilon_c\,\mathcal{J}_1. \tag{8.95}$$

Moreover, it is clear from equations (8.80) and (8.85) that $\mathcal{J}_0$ has the symmetry of $\cos\zeta$, whereas $\mathcal{J}_1$ has the symmetry of $\sin\zeta$. Hence, we deduce that

$$\mathrm{Re}\left(\frac{\Delta\hat{\Psi}_s}{\hat{\Psi}_s}\right) = \frac{2\,\epsilon_\beta\,\epsilon_c}{\hat{w}}\int_{-\infty}^{\infty}\oint\mathcal{J}_0\cos\zeta\,\frac{d\zeta}{2\pi}\,dX, \tag{8.96}$$

$$\mathrm{Im}\left(\frac{\Delta\hat{\Psi}_s}{\hat{\Psi}_s}\right) = -\frac{2\,\epsilon_\beta\,\epsilon_c}{\hat{w}}\int_{-\infty}^{\infty}\oint\epsilon\,\mathcal{J}_1\sin\zeta\,\frac{d\zeta}{2\pi}\,dX. \tag{8.97}$$

The previous two equations can also be written

$$\mathrm{Re}\left(\frac{\Delta\hat{\Psi}_s}{\hat{\Psi}_s}\right) = \frac{4\,\epsilon_\beta\,\epsilon_c}{\hat{w}}\int_{-1}^{\infty}\langle\mathcal{J}_0\cos\zeta\rangle\,d\Omega, \tag{8.98}$$

$$\mathrm{Im}\left(\frac{\Delta\hat{\Psi}_s}{\hat{\Psi}_s}\right) = -\frac{4\,\epsilon_\beta\,\epsilon_c}{\hat{w}}\int_{-1}^{\infty}\langle\epsilon\,\mathcal{J}_1\sin\zeta\rangle\,d\Omega = -\frac{4\,\epsilon_\beta\,\epsilon_c}{\hat{w}}\int_{-1}^{\infty}\langle X\,\epsilon\,\{\mathcal{J}_1,\,\Omega\}\rangle\,d\Omega. \tag{8.99}$$

Here, we have made use of the fact that $\mathcal{J}_0$ and $\mathcal{J}_1$ are both even functions of X, as well as the easily proved results $\partial(X,\,\zeta)/\partial(\Omega,\,\zeta) = 1/X$ and $\{\mathcal{J}_1,\,\Omega\} = -X\,(\partial\mathcal{J}_1/\partial\zeta)_\Omega$. Equations (8.85) and (8.99) can be combined to give [5]

$$\begin{aligned}
\mathrm{Im}\left(\frac{\Delta\hat{\Psi}_s}{\hat{\Psi}_s}\right) &= \frac{4\,\epsilon_\beta\,\epsilon_c\,\epsilon_\varphi}{\hat{w}}\int_1^\infty \langle X^2\,\partial_\Omega\,[\partial_\Omega(|X|^3\,\partial_\Omega Y_i)]\rangle d\Omega \\
&= \frac{4\,\epsilon_\beta\,\epsilon_c\,\epsilon_\varphi}{\hat{w}}\int_1^\infty \partial_\Omega\,(-\langle|X|\rangle\,Y_i + 2\,\langle|X|^3\rangle\,d_\Omega Y_i + \langle|X|^5\rangle\,d_\Omega^2 Y_i)d\Omega \\
&= \frac{4\,\epsilon_\beta\,\epsilon_c\,\epsilon_\varphi}{\hat{w}}\,\lim_{\Omega\to\infty}(-\langle|X|\rangle\,Y_i + 2\,\langle|X|^3\rangle\,d_\Omega Y_i + \langle|X|^5\rangle\,d_\Omega^2 Y_i),
\end{aligned} \tag{8.100}$$

where use has been made of the facts that Y_i is zero inside the island separatrix, Y_i is continuous across the separatrix, and $\partial_\Omega X^n = n\,X^{n-2}$. Combining the previous equation with equation (8.87), we obtain [5]

$$\mathrm{Im}\left(\frac{\Delta\hat{\Psi}_s}{\hat{\Psi}_s}\right) = -\frac{4\,\epsilon_\beta\,\epsilon_c\,\epsilon_\varphi\,v'}{\hat{w}}. \tag{8.101}$$

The previous equation yields

$$\left[r\,\frac{d\omega_E}{dr}\right]_{r_{s-}}^{r_{s+}} = \left(\frac{\tau_\varphi\,\hat{w}^4}{2\,\tau_H^2}\right)\mathrm{Im}\left(\frac{\Delta\hat{\Psi}_s}{\hat{\Psi}_s}\right), \tag{8.102}$$

where use has been made of equations (8.23), (8.25), (8.27), and (8.46), as well as

$$\frac{L_p}{L_s} = \frac{c_\beta}{\omega_*\,\tau_H}\,\frac{d_\beta}{r_s}. \tag{8.103}$$

Here, τ_H is the hydromagnetic time (see equation (5.43)). Equation (8.102) can also be obtained by integrating equation (3.165) across the rational surface, making use of equation (3.140), as well as the identification

$$\left[r\,\frac{d\omega_E}{dr}\right]_{r_{s-}}^{r_{s+}} = m\left[r\,\frac{\partial\Delta\Omega_\theta}{\partial r}\right]_{r_{s-}}^{r_{s+}}. \tag{8.104}$$

The previous identification merely states that the discontinuity in the MHD fluid velocity gradient that develops in the outer region at the rational surface is mirrored by an equal discontinuity in the ion fluid velocity gradient (because the discontinuity is ultimately due to a discontinuity in the E-cross-B velocity gradient, and there is no discontinuity in the diamagnetic velocity gradient). Equations (8.80) and (8.98) can be combined to give

$$\begin{aligned}
\mathrm{Re}\left(\frac{\Delta\hat{\Psi}_s}{\hat{\Psi}_s}\right) &= \frac{4}{\epsilon_R\,\hat{w}}\,\frac{d(\ln\hat{w}^2)}{dT}\int_{-1}^\infty \frac{\langle\cos\zeta\rangle^2}{\langle 1\rangle}\,d\Omega \\
&+ \frac{2\,\epsilon_\beta\,\epsilon_c}{\hat{w}}\int_{1+}^\infty \partial_\Omega(M\;Y_i)\langle\tilde{X}^2\cos\zeta\rangle d\Omega.
\end{aligned} \tag{8.105}$$

Making use of equations (8.76), (8.83), and (8.87), the previous equation yields

$$\mathrm{Re}\left(\frac{\Delta\hat{\Psi}_s}{\hat{\Psi}_s}\right) = \frac{4}{\epsilon_R\,\hat{w}}\,\frac{d(\ln\hat{w}^2)}{dT}\int_{-1}^{\infty}\frac{\langle\cos\zeta\rangle^2}{\langle 1\rangle}\,d\Omega - \frac{2\,\epsilon_\beta\,\epsilon_c\,v'}{(1+\tau)\,\hat{w}}\int_1^{\infty} d_\Omega\left(\frac{F_i}{\langle X^2\rangle}\right)\left\langle\tilde{X}^2\cos\zeta\right\rangle d\Omega$$
$$+\,\frac{2\,\epsilon_\beta\,\epsilon_c\,v'^{\,2}}{\hat{w}}\int_1^{\infty} d_\Omega\left(F_i^{\,2}\right)\left\langle\tilde{X}^2\cos\zeta\right\rangle d\Omega, \tag{8.106}$$

where

$$F_i(\Omega) = \int_1^{\Omega}\frac{d\Omega'}{\langle X^4\rangle(\Omega')}\bigg/\int_1^{\infty}\frac{d\Omega}{\langle X^4\rangle(\Omega)}. \tag{8.107}$$

Finally, equations (8.10), (8.23)–(8.27), (8.101), (8.103), and (8.106) give [6, 12]

$$\mathrm{Re}\left(\frac{\Delta\hat{\Psi}_s}{\hat{\Psi}_s}\right) = I_1\,\tau_R\,\frac{d}{dt}\left(\frac{4\,w}{r_s}\right) - I_2\left(\frac{c_\beta}{1+\tau}\right)\left(\frac{L_s}{L_p}\right)\left(\frac{d_\beta}{r_s}\right)\left(\frac{\tau_\varphi}{\tau_H}\right)\left(\frac{w}{r_s}\right)^2\mathrm{Im}\left(\frac{\Delta\hat{\Psi}_s}{\hat{\Psi}_s}\right)$$
$$- I_3\left(\frac{\tau_\varphi}{\tau_H}\right)^2\left(\frac{w}{r_s}\right)^7\left[\mathrm{Im}\left(\frac{\Delta\hat{\Psi}_s}{\hat{\Psi}_s}\right)\right]^2, \tag{8.108}$$

where

$$I_1 = 2\int_{-1}^{\infty}\frac{\langle\cos\zeta\rangle^2}{\langle 1\rangle}\,d\Omega, \tag{8.109}$$

$$I_2 = \frac{1}{4}\int_1^{\infty} d_\Omega\left(\frac{F_i}{\langle X^2\rangle}\right)\left(\langle X^4\rangle - \frac{\langle X^2\rangle^2}{\langle 1\rangle}\right)d\Omega, \tag{8.110}$$

$$I_3 = \frac{1}{16}\int_1^{\infty} d_\Omega\left(F_i^{\,2}\right)\left(\langle X^4\rangle - \frac{\langle X^2\rangle^2}{\langle 1\rangle}\right)d\Omega. \tag{8.111}$$

8.11 The evaluation of the integrals

In order to evaluate the integrals (8.109)–(8.111), it is helpful to define the new magnetic flux-surface label $k = [(1+\Omega)/2]^{1/2}$. It follows from equation (8.55) that $k = 0$ at the O-points of the magnetic island chain, and $k = 1$ on the magnetic separatrix. It can easily be demonstrated that

$$\mathcal{A}(k) \equiv 2\,k\,\langle 1\rangle = \frac{2}{\pi}\begin{cases} k\,K(k) & 0 \leqslant k \leqslant 1 \\ K(1/k) & k > 1 \end{cases}, \tag{8.112}$$

$$\mathcal{B}(k) \equiv \langle |X|\rangle = \begin{cases} (2/\pi)\,\sin^{-1}(k) & 0 \leqslant k \leqslant 1 \\ 1 & k > 1 \end{cases}, \tag{8.113}$$

$$\mathcal{C}(k) \equiv \frac{\langle X^2\rangle}{2\,k} = \frac{2}{\pi}\begin{cases}[E(k) + (k^2 - 1)\,K(k)]/k & 0 \leqslant k \leqslant 1\\ E(1/k) & k > 1\end{cases}, \tag{8.114}$$

$$\mathcal{D}(k) \equiv \frac{\langle |X|^3\rangle}{4\,k^2} = \begin{cases}(2/\pi)\,\sin^{-1}(k)\,[1 - 1/(2\,k^2)] & 0 \leqslant k \leqslant 1\\ 1 - 1/(2\,k^2) & k > 1\end{cases}, \tag{8.115}$$

$$\begin{aligned}\mathcal{E}(k) &\equiv \frac{\langle X^4\rangle}{8\,k^3}\\ &= \frac{2}{3\pi}\begin{cases}[2\,(2 - 1/k^2)\,E(k) + (3\,k^2 - 5 + 2/k^2)\,K(k)]/k & 0 \leqslant k \leqslant 1\\ [2\,(2 - 1/k^2)\,E(1/k) - (1 - 1/k^2)\,K(1/k)] & k > 1\end{cases},\end{aligned} \tag{8.116}$$

where

$$E(k) = \int_0^{\pi/2} (1 - k^2 \sin^2 u)^{1/2}\,du, \tag{8.117}$$

$$K(k) = \int_0^{\pi/2} (1 - k^2 \sin^2 u)^{-1/2}\,du \tag{8.118}$$

are standard elliptic integrals [1]. It follows that

$$I_n = \int_0^{\infty} \mathcal{I}_n(k)\,dk, \tag{8.119}$$

for $n = 1, 2, 3$, where

$$\mathcal{I}_1(k) = \frac{4[(2\,k^2 - 1)\,\mathcal{A} - 2\,k^2\,\mathcal{C}]^2}{\mathcal{A}}, \tag{8.120}$$

$$\mathcal{I}_2(k) = \begin{cases}0 & 0 \leqslant k < 1\\ (1 - k\,\mathcal{A}\,\mathcal{E}\,\mathcal{F}/\mathcal{C})[1/\mathcal{C} - \mathcal{C}/(\mathcal{A}\,\mathcal{E})]/\mathcal{F}(\infty) & k \geqslant 1\end{cases}, \tag{8.121}$$

$$\mathcal{I}_3(k) = \begin{cases}0 & 0 \leqslant k < 1\\ k\,\mathcal{F}[1 - \mathcal{C}^2/(\mathcal{A}\,\mathcal{E})]/\mathcal{F}^2(\infty) & k \geqslant 1\end{cases}, \tag{8.122}$$

$$\mathcal{F}(k) = \int_1^k \frac{dk'}{k'^2\,\mathcal{E}(k')}, \tag{8.123}$$

and use has been made of the easily proved result $d(k\,\mathcal{C})/dk = \mathcal{A}$.

The integrands $\mathcal{I}_1(k)$, $\mathcal{I}_2(k)$, and $\mathcal{I}_3(k)$ are shown in figure 8.1. Note that $\mathcal{I}_1(k)$ has a logarithmic (and therefore integrable) singularity at the island separatrix ($k = 1$). The values of the integrals themselves are

$$I_1 = 0.8227, \tag{8.124}$$

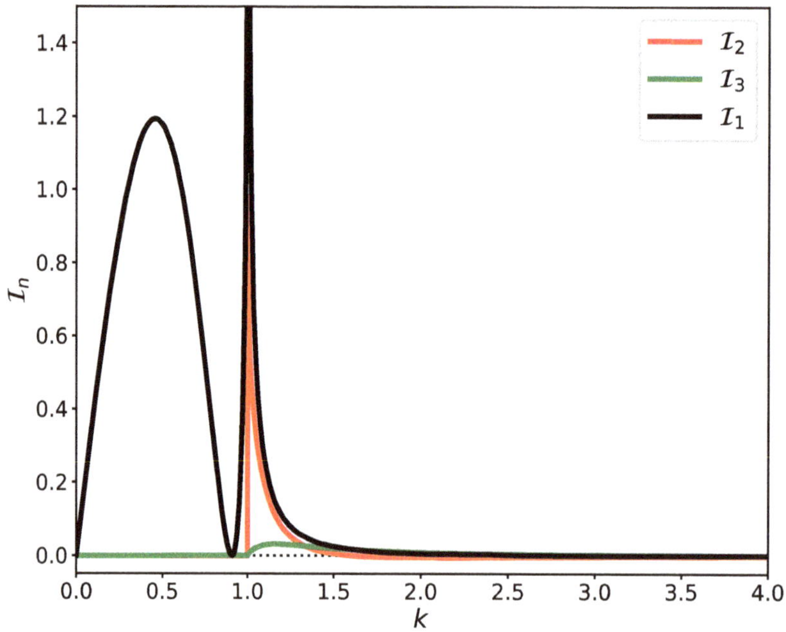

Figure 8.1. The integrands $\mathcal{I}_1(k)$, $\mathcal{I}_2(k)$, and $\mathcal{I}_3(k)$.

$$I_2 = 0.048\,89, \tag{8.125}$$

$$I_3 = 0.029\,44. \tag{8.126}$$

References

[1] Abramowitz M and Stegun I A (ed) 1964 *Handbook of Mathematical Functions with Formulas, Graphs, and Mathematical tables* (New York: Dover) https://store.doverpublications.com/0486612724.html

[2] Fitzpatrick R 1993 Interaction of tearing modes with external structures in cylindrical geometry *Nucl. Fusion* **33** 1049

[3] Fitzpatrick R 1995 Helical temperature perturbations associated with tearing modes in tokamak plasmas *Phys. Plasmas* **2** 825

[4] Fitzpatrick R 1998 Bifurcated states of a rotating tokamak plasma in the presence of a static error-field *Phys. Plasmas* **5** 3325

[5] Fitzpatrick R and Waelbroeck F L 2005 Two-fluid magnetic island dynamics in slab geometry. I. Isolated islands *Phys. Plasmas* **12** 022307

[6] Fitzpatrick R and Waelbroeck F L 2005 Two-fluid magnetic island dynamics in slab geometry. II. Islands interacting with resistive walls or resonant magnetic perturbations *Phys. Plasmas* **12** 022308

[7] Furth H P, Killeen J and Rosenbluth M N 1963 Finite-resistivity instabilities of a sheet pinch *Phys. Fluids* **6** 459

[8] Rutherford P H 1973 Nonlinear growth of the tearing mode *Phys. Fluids* **16** 1906

[9] Scott B D, Hassam A B and Drake J F 1985 Nonlinear evolution of drift-tearing modes *Phys. Fluids* **28** 275

[10] Smolyakov A I 1993 Nonlinear evolution of tearing modes in inhomogeneous plasmas *Plasma Phys. Control. Fusion* **35** 657

[11] Smolyakov A I, Hirose A, Lazzaro E, Re G B and Callen J D 1995 Rotating nonlinear magnetic islands in a tokamak plasma *Phys. Plasmas* **2** 1581

[12] Waelbroeck F L and Fitzpatrick R 1997 Rotation and locking of magnetic islands *Phys. Rev. Lett.* **78** 1703

IOP Publishing

Richard Fitzpatrick

Chapter 9

Nonlinear tearing-mode stability

9.1 Introduction

We saw in chapter 6 that a *linear* tearing mode is unstable when the (real dimensionless) tearing stability index, E_{ss} (see section 3.8), is positive, and is stable otherwise. In the linear regime, an unstable tearing mode grows *exponentially* in time on a timescale that is intermediate between the short hydromagnetic time, τ_H (see equation (5.43)), and the much longer resistive diffusion time, τ_R (see equation (5.49)). The mode reconnects magnetic flux at the so-called *rational* surface to produce a helical magnetic island chain (see section 5.16). The rational surface is the magnetic flux surface at which the tearing mode resonates with the equilibrium magnetic field (i.e. at which $\mathbf{k} \cdot \mathbf{B} = 0$, where $\mathbf{k}$ is the wave vector of the mode, and $\mathbf{B}$ is the equilibrium magnetic field). The tearing mode also *rotates* in the laboratory frame at the angular frequency

$$\omega_{\perp\,e} = \omega_E + \omega_{*\,e}, \tag{9.1}$$

where ω_E and $\omega_{*\,e}$ are the E-cross-B and electron diamagnetic frequencies, respectively, at the rational surface (see equations (5.21), (5.29), (5.44), and (5.45)). In fact, the previous equation implies that the tearing mode corotates with the *electron* fluid at the rational surface.

We saw in section 5.16 that linear tearing-mode theory breaks down as soon as the radial width of the magnetic island chain exceeds the linear layer width and must then be replaced by nonlinear theory. The aim of this chapter is to employ the analysis of chapter 8 to determine the *nonlinear* time evolution of an unstable tearing mode.

9.2 The Rutherford island-width evolution equation

Suppose that the rigid wall surrounding the plasma is perfectly conducting (i.e. $\tau_w \to \infty$, where τ_w is the time constant of the wall—see section 3.10) and that there is

doi:10.1088/978-0-7503-5367-0ch9

no current flowing through the external magnetic field coil (i.e. $\hat{I}_c = 0$, where $\hat{I}_c$ is the normalized coil current—see equation (3.194)). In this case, equations (3.187), (3.188), and (7.5), yield

$$\frac{\Delta\hat{\Psi}_s}{\hat{\Psi}_s} = \Delta_{pw}, \tag{9.2}$$

where $\Delta_{pw} = E_{ss}$ is the (real dimensionless) perfect-wall tearing stability index, $\hat{\Psi}_s$ the normalized reconnected helical magnetic flux at the rational surface (see equation (3.184)), and $\Delta\hat{\Psi}_s$ the normalized helical sheet current density at the rational surface (see equation (3.183)). It is clear from the previous equation that

$$\mathrm{Im}\left(\frac{\Delta\hat{\Psi}_s}{\hat{\Psi}_s}\right) = 0, \tag{9.3}$$

which implies that zero electromagnetic torque is exerted at the rational surface (see section 3.13).

Equation (9.2) yields

$$\mathrm{Re}\left(\frac{\Delta\hat{\Psi}_s}{\hat{\Psi}_s}\right) = \Delta_{pw}, \tag{9.4}$$

which can be combined with equations (8.1), (8.108), and (9.3) to give the so-called *Rutherford island-width evolution equation* [13]:

$$I_1\, \tau_R\, \frac{d}{dt}\left(\frac{W}{r_s}\right) = \Delta_{pw}. \tag{9.5}$$

Here, W is the full radial width of the magnetic island chain that develops at the rational surface, r_s the minor radius of the surface, and $I_1 = 0.8227$ (see equation (8.124)). We conclude that, in the nonlinear regime, the width of the magnetic island chain grows *algebraically* in time on the resistive diffusion timescale, τ_R. Obviously, this is a much slower time evolution than that predicted in the linear regime.

9.3 The composite linear/nonlinear model

According to equation (6.25), the time evolution of the reconnected magnetic flux (in a frame of reference that corotates with the magnetic island chain) in the linear regime is governed by

$$\frac{d\ln\hat{\Psi}_s}{dt}\,\tau_R\,\frac{\delta_s}{r_s} = \Delta_{pw}, \tag{9.6}$$

where δ_s is the linear layer width. The previous equation can be rearranged to give

$$\frac{\delta_s}{r_s}\,\tau_R\,\frac{d\hat{\Psi}_s}{dt} = \Delta_{pw}\,\hat{\Psi}_s. \tag{9.7}$$

Now, given that $W \propto \hat{\Psi}_s^{1/2}$ (see equation (8.1)), the Rutherford island-width evolution equation, (9.5), can be rewritten in the form

$$\frac{I_1\, W/2}{r_s}\,\tau_R\,\frac{d\hat{\Psi}_s}{dt} = \Delta_{pw}\,\hat{\Psi}_s. \tag{9.8}$$

It can be seen, via a comparison between the previous two equations, that a nonlinear magnetic island chain evolves in time in an analogous fashion to a linear layer whose width is $I_1\,W/2$. In other words, the essential nonlinearity in the nonlinear regime comes about because the effective layer width is amplitude dependent.

Equation (9.7) is valid when $\delta_s \gg W$ (i.e. when the linear layer width is much greater than the island width), whereas equation (9.8) is valid in the opposite limit. This observation allows us to formulate a composite time-evolution equation that encompasses both the linear regime and the nonlinear regime [3, 4, 8, 10]:

$$\frac{\delta_s + I_1\, W/2}{r_s}\,\tau_R\,\frac{d\hat{\Psi}_s}{dt} = \Delta_{pw}\,\hat{\Psi}_s. \tag{9.9}$$

The previous equation can also be written

$$\frac{2\,\delta_s + I_1\, W}{r_s}\,\tau_R\,\frac{dW}{dt} = \Delta_{pw}\,W. \tag{9.10}$$

Let

$$\hat{W} = \frac{I_1\, W}{2\,\delta_s}, \tag{9.11}$$

$$\hat{t} = \left(\frac{r_s\,\Delta_{pw}}{2\,\delta_s\,\tau_R}\right)t. \tag{9.12}$$

Equation (9.10) transforms into

$$\frac{d\hat{W}}{d\hat{t}} = \frac{\hat{W}}{1+\hat{W}}, \tag{9.13}$$

which can be solved to give

$$\hat{W} + \ln \hat{W} = \hat{t}, \tag{9.14}$$

assuming that $\hat{W} = 0$ at $\hat{t} = -\infty$. Figure 9.1 shows the time evolution of the magnetic island width predicted by the previous equation. It can be seen that the evolution makes a smooth transition from exponential growth when $\hat{W} \ll 1$ (i.e. when the island width is much less than the linear layer width) to algebraic growth when $\hat{W} \gg 1$ (i.e. when the island width is much greater than the linear layer width).

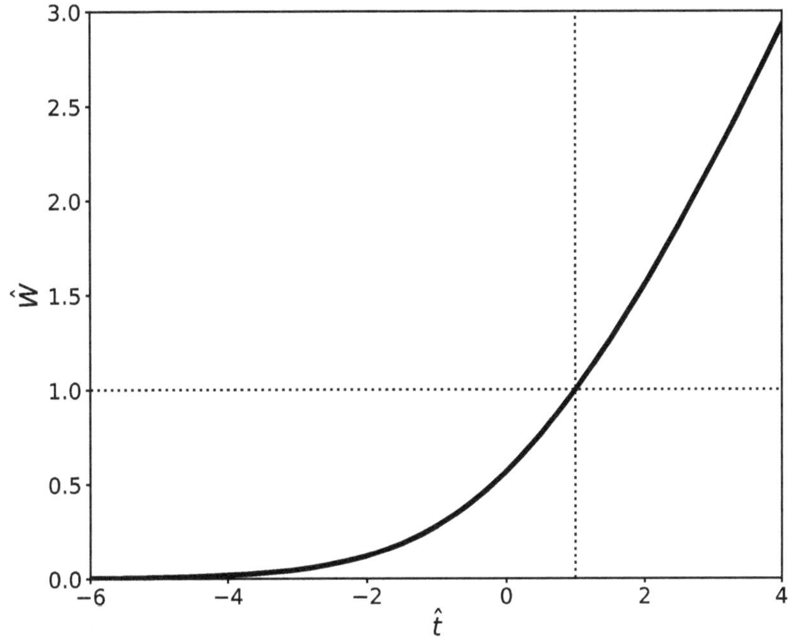

Figure 9.1. The time evolution of the magnetic island width predicted by the composite linear/nonlinear model.

9.4 Saturated island width

Equation (9.5) gives the impression that if $\Delta_{\rm pw} > 0$, then the width of the magnetic island chain grows without limit. In fact, this is not the case. Instead, the width of the island chain eventually stops growing, and the tearing mode attains a saturated steady state. In order to model this effect, it is necessary to perform asymptotic matching between the inner and outer regions (see sections 3.8 and 4.1) to a higher order, taking into account the finite width of the island chain [5, 12, 15, 18].

For the case of an island chain that is sufficiently wide to flatten the plasma pressure within its magnetic separatrix [6], the appropriate saturation theory is given in Hastie *et al* [9]. According to this theory, $\Delta_{\rm pw}$ in equation (9.5) must be replaced by

$$\Delta_{\rm pw}(0) - \left(0.8\,\alpha_s^{\,2} - 0.27\,\beta_s - 0.09\,\alpha_s\right)\frac{W}{r_s}, \tag{9.15}$$

where $\Delta_{\rm pw}(0)$ denotes the zero-island-width, perfect-wall, tearing stability index,

$$\alpha_s = -\left(\frac{r\,q\,J_z'}{s}\right)_{r=r_s}, \tag{9.16}$$

$$\beta_s = -\left(\frac{r\,q\,J_z''}{s}\right)_{r=r_s}, \tag{9.17}$$

$$s(r) = \frac{r\,q'}{q}, \tag{9.18}$$

$$J_z(r) = \frac{R_0\,\mu_0\,j_z(r)}{B_z}. \tag{9.19}$$

Here, r is a radial coordinate in cylindrical geometry, $q(r)$ the safety-factor profile (see section 3.2), $j_z(r)$ the equilibrium toroidal current density, R_0 the simulated major radius of the plasma, B_z the equilibrium toroidal magnetic field strength, and $'$ denotes d/dr.

Equations (9.5) and (9.15) can be combined to give

$$I_1\,\tau_R\,\frac{d}{dt}\left(\frac{W}{r_s}\right) = \Delta_{pw}(0) - \left(0.8\,\alpha_s^{\,2} - 0.27\,\beta_s - 0.09\,\alpha_s\right)\frac{W}{r_s}. \tag{9.20}$$

If we define

$$W_{pw} = \frac{\Delta_{pw}(0)\,r_s}{0.8\,\alpha_s^{\,2} - 0.27\,\beta_s - 0.09\,\alpha_s}, \tag{9.21}$$

$$\tau_{pw} = \frac{I_1\,\tau_R}{\Delta_{pw}(0)}\,\frac{W_{pw}}{r_s}, \tag{9.22}$$

then equation (9.20) reduces to

$$\tau_{pw}\,\frac{d}{dt}\left(\frac{W}{W_{pw}}\right) = 1 - \frac{W}{W_{pw}}. \tag{9.23}$$

The previous equation can be solved to give

$$W(t) = (1 - e^{-t/\tau_{pw}})\,W_{pw}, \tag{9.24}$$

assuming that the island width is zero at $t = 0$. It follows that the width of the island chain does not grow without limit but instead eventually attains the saturated value W_{pw}. Moreover, the time required to achieve saturation, τ_{pw}, is of the order of $\tau_R\,(W_{pw}/r_s)$.

Figure 9.2 shows the saturated island width of an $m = 2/n = 1$ tearing mode, which is generally the most unstable tearing mode in a conventional tokamak plasma [16], in a so-called 'Wesson' equilibrium [16] characterized by

$$J_z(r) = \frac{2}{q(0)}\left(1 - \frac{r^2}{a^2}\right)^{\nu}, \tag{9.25}$$

$$q(r) = \frac{q(a)\,(r/a)^2}{1 - (1 - r^2/a^2)^{1+\nu}}, \tag{9.26}$$

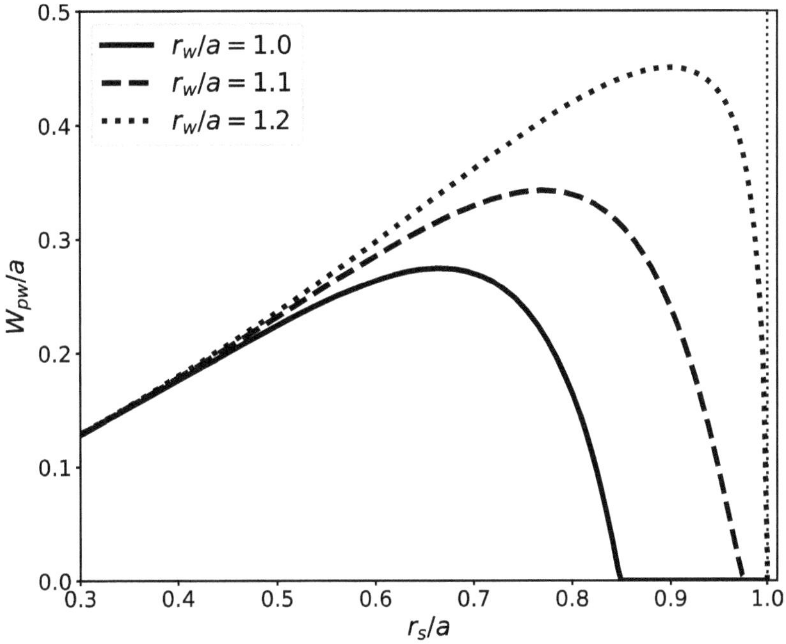

Figure 9.2. Saturated island width of the $m = 2/n = 1$ tearing mode for a Wesson equilibrium in which $q(0) = 0.8$ and $q(a)$ is varied. Here, r_w is the radius of the perfectly conducting wall that surrounds the plasma.

where $\nu = q(a)/q(0) - 1$. Here, $q(0)$ and $q(a)$ are the safety-factor values at the magnetic axis, $r = 0$, and the plasma boundary, $r = a$, respectively. In the calculation shown in figure 9.2, $\Delta_{\rm pw}(0)$ is determined by solving the cylindrical tearing-mode equation, subject to suitable boundary conditions, as described in section 3.8. Moreover, $q(0)$ is given the plausible value 0.8 [17], while $q(a)$ is varied in order to shift the position of the rational surface within the plasma. It can be seen that if the rational surface lies well inside the plasma boundary, then the 2/1 island chain saturates at a relatively small width (i.e. $W_{\rm pw}/a \ll 1$). On the other hand, if the rational surface lies close to the plasma boundary, then the island chain saturates at a much higher width. However, the presence of a close-fitting perfectly conducting wall can mitigate this effect to some extent.

Given that the minor radius, a, of a tokamak fusion reactor is a few meters (see table 1.2), while the typical linear layer width is a few millimeters (see table 6.2), it is clear from figure 9.2 that the saturated island width of an unstable tearing mode in a tokamak fusion reactor is bound to be very much larger than the linear layer width. Thus, we conclude that the time evolution of an unstable tearing mode in such a reactor will only be governed by linear theory for a comparatively short period of time after its onset, its subsequent time evolution being governed by nonlinear theory. Furthermore, it takes a time on the order of the resistive diffusion time, $\tau_{\rm R}$, for the mode to attain its final saturated amplitude.

Note, finally, that a saturated magnetic island chain degrades the overall energy confinement properties of a tokamak plasma because it flattens the pressure profile within its magnetic separatrix (see chapter 8 and section 14.18) [6]. Moreover, the degree of degradation is roughly proportional to $W_{\rm pw}/a$ [2].

9.5 The island rotation frequency

Equations (5.45) and (9.3) imply that

$$v' = 0 \tag{9.27}$$

for an isolated magnetic island chain (i.e. a chain that is not interacting with a resistive wall or an error field). It follows from equations (8.75), (8.76), (8.83), and (8.87) that the stream functions for the electron, magnetohydrodynamic (MHD), and ion fluids take the respective forms

$$Y_{\rm e}(\Omega) = \begin{cases} 0 & -1 \leqslant \Omega \leqslant 1 \\ -1/\langle X^2\rangle & \Omega > 1 \end{cases}, \tag{9.28}$$

$$M(\Omega) = \left(\frac{1}{1+\tau}\right)\begin{cases} 0 & -1 \leqslant \Omega \leqslant 1 \\ -1/\langle X^2\rangle & \Omega > 1 \end{cases}, \tag{9.29}$$

$$Y_{\rm i}(\Omega) = 0, \tag{9.30}$$

in the vicinity of the island chain. Here, Ω is a magnetic flux-surface label that takes a value of zero at the island O-points, a value of unity on the magnetic separatrix, and a value much larger than unity far from the island chain (see equation (8.55)). Moreover, τ is ratio of the electron and ion pressure gradients at the rational surface (see equation (4.5)). The function $\langle X^2\rangle(\Omega)$ is specified in section 8.11. Finally, the poloidal components of the perpendicular velocities of the three fluids in the island rest frame are related to their stream functions according to equations (8.72)–(8.74).

Figure 9.3 shows typical normalized perpendicular velocity profiles of the electron, MHD, and ion fluids across the O-points and the X-points of an isolated magnetic island chain in the island rest frame, as determined from equations (8.72)–(8.74) and (9.28)–(9.30) [7]. Here, $\hat{V}_\theta = (m/r_{\rm s})V_{\perp\,\theta}/\omega_*$, where m is the poloidal mode number of the tearing mode, and ω_* is the diamagnetic frequency at the rational surface (see equation (5.47)). It can be seen that all three fluids corotate with the island chain inside the magnetic separatrix. The ion fluid velocity profile is completely unaffected by the presence of the chain. On the other hand, the electron and MHD fluid velocity profiles are modified by the island chain. In particular, both profiles are discontinuous across the magnetic separatrix. As explained in section 8.9, these discontinuities are resolved in a thin layer of thickness $4\,w_{\rm d}$ on the separatrix (see equation (8.84) and table 8.1). It is clear from the figure that the island chain rotates in the ion diamagnetic (i.e. negative) direction with respect to the unperturbed E-cross-B frame at the rational surface [7].

Equations (8.45), (8.63), (9.27), and (9.29) yield $v = -1/(1+\tau)$, or

$$\omega = \omega_{\perp\,{\rm i}}, \tag{9.31}$$

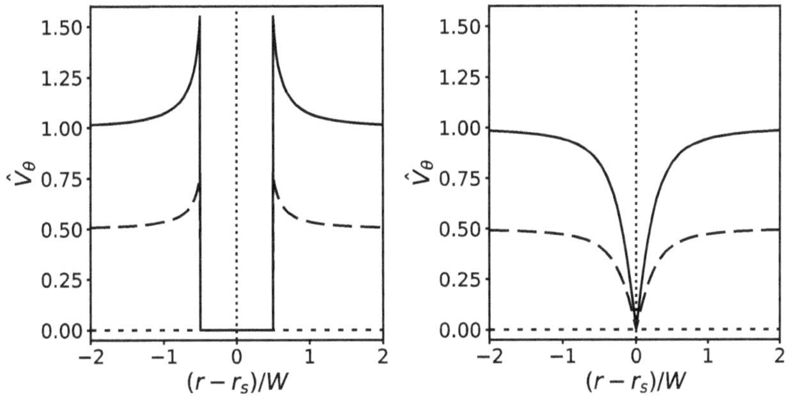

Figure 9.3. Normalized velocity profiles across the O-points (left-hand panel) and the X-points (right-hand panel) of an isolated magnetic island chain in the island rest frame. The solid, dashed, and dotted lines correspond to the electron, MHD, and ion fluids, respectively. Here, $\tau = 1$.

where

$$\omega_{\perp\,\mathrm{i}} = \omega_{\mathrm{E}} + \omega_{*\,\mathrm{i}}, \tag{9.32}$$

and

$$\omega_{*\,\mathrm{i}} = -\left(\frac{1}{1+\tau}\right)\omega_{*} = -\frac{\omega_{*\,\mathrm{e}}}{\tau} \tag{9.33}$$

is the ion diamagnetic frequency at the rational surface (see equations (5.45) and (5.46)). We conclude that, unlike a linear tearing mode, which is convected by the electron fluid at the rational surface (see section 6.2), a nonlinear tearing mode is convected by the *ion* fluid [7]. The reason for this difference is discussed in section 1.12. Incidentally, there is clear experimental evidence that wide magnetic island chains in tokamak plasmas rotate in the ion diamagnetic direction relative to the E-cross-B frame at the rational surface [1, 11].

Finally, it is reasonable to suppose that the critical island width that must be exceeded before a magnetic island chain switches from rotating in the electron direction (with respect to the unperturbed local E-cross-B frame) to rotating in the ion direction is that above which the pressure profile flattens within the separatrix [14].

References

[1] Buratti P, Alessi E, Baruzzo M, Casolari A, Giovannozzi E, Giroud C, Hawkes N, Menmuir S and Pucella G 2016 Diagnostic application of magnetic islands rotation in JET *Nucl. Fusion* **56** 076004

[2] Chang Z and Callen J D 1990 Global energy confinement degradation due to macroscopic phenomena in tokamaks *Nucl. Fusion* **30** 219

[3] Chudnovskiy A N and the T-10 Team *et al* 2003 MHD-mode locking by controlled halo-current in the T-10 tokamak *Nucl. Fusion* **43** 681

[4] Eliseev L G, Ivanov N V, Kakurin A M, Melnikov A V and Perfilov S V 2015 Magnetic island and plasma rotation under external resonant magnetic perturbation in the T-10 tokamak *Phys. Plasmas* **22** 052504
[5] Escande D F and Ottaviani M 2004 Simple and rigorous solution for the nonlinear tearing mode *Phys. Lett.* A **323** 278
[6] Fitzpatrick R 1995 Helical temperature perturbations associated with tearing modes in tokamak plasmas *Phys. Plasmas* **2** 825
[7] Fitzpatrick R and Waelbroeck F L 2005 Two-fluid magnetic island dynamics in slab geometry. I. Isolated islands *Phys. Plasmas* **12** 022307
[8] Fitzpatrick R 2021 Further modeling of q_{95} windows for the suppression of edge localized modes by resonant magnetic perturbations in the DIII-D tokamak *Phys. Plasmas* **28** 022503
[9] Hastie R J, Militello F and Porcelli F 2005 Nonlinear saturation of tearing mode islands *Phys. Rev. Lett.* **95** 065001
[10] Huang W and Zhu P 2020 Analytical model of plasma response to external magnetic perturbation in absence of no-slip condition *Phys. Plasmas* **27** 022514
[11] La Haye R J, Petty C C, Strait E J, Waelbroeck F L and Wilson H R 2003 Propagation of magnetic islands in the $E_r = 0$ frame of co-injected neutral beam driven discharges in the DIII-D tokamak *Phys. Plasmas* **10** 3644
[12] Militello F and Porcelli F 2004 Simple analysis of the nonlinear saturation of the tearing mode *Phys. Plasmas* **11** L13
[13] Rutherford P H 1973 Nonlinear growth of the tearing mode *Phys. Fluids* **16** 1906
[14] Scott B D, Hassam A B and Drake J F 1985 Nonlinear evolution of drift-tearing modes *Phys. Fluids* **28** 275
[15] Thyagaraja A 1981 Perturbation analysis of a simple model of magnetic island structures *Phys. Fluids* **24** 1716
[16] Wesson J A 1978 Hydrodynamic stability of tokamaks *Nucl. Fusion* **18** 87
[17] Wesson J A 2011 *Tokamaks* 4th edn (Oxford: Oxford University Press) https://global.oup.com/academic/product/tokamaks-9780199592234
[18] White R A, Monticello D A, Rosenbluth M N and Waddell B V 1977 Saturation of the tearing mode *Phys. Fluids* **20** 800

IOP Publishing

Richard Fitzpatrick

Chapter 10

Rotation braking in tokamak plasmas

10.1 Introduction

An unstable tearing mode in a tokamak plasma reconnects magnetic flux at the so-called *rational* magnetic flux surface, where the mode resonates with the equilibrium magnetic field (see section 3.7), to produce a *rotating* (in the laboratory frame) magnetic island chain (see sections 5.16 and 9.5).

Mode locking is a process by which the rotation of a slowly growing magnetic island chain in a tokamak plasma is braked due to electromagnetic interaction with a rigid electrically conducting wall surrounding the plasma (see sections 3.9 and 3.10), causing the chain to eventually *lock* (i.e. become stationary in the laboratory frame) to a static error field (see section 7.1) [2–4, 6, 9–14]. Locked magnetic island chains are strongly correlated with disruptions (i.e. sudden, catastrophic losses of thermal and magnetic energy) in tokamak plasmas [5]. In most tokamaks, the role of the conducting wall is played by the metallic vacuum vessel that surrounds the plasma.

As we saw in the previous chapter, the growth of a magnetic island chain in a tokamak fusion reactor is only governed by linear theory for a comparatively short period of time after its onset, its subsequent time evolution being governed by nonlinear theory. Hence, it is reasonable to suppose that the slowing down and locking of a magnetic island chain in such a reactor is governed by *nonlinear* theory. The aim of this chapter is to employ the nonlinear resonant response model derived in chapter 8 to investigate rotation braking in tokamak fusion reactors. It is assumed that, once the rotation frequency of the island chain has been reduced to a small value by rotation braking, even the smallest residual error field would be sufficient to completely arrest the chain's rotation.

10.2 Rotation braking by a thin conducting wall

Let us, first of all, investigate the rotation braking of a nonlinear magnetic island chain due to electromagnetic interaction with a *thin* conducting wall (in the absence

doi:10.1088/978-0-7503-5367-0ch10

of an error field). Note that, by a 'thin' wall, we mean one in which the skin depth in the wall material is greater than the wall's radial thickness.

Assuming that the chain possesses the rotation frequency ω in the laboratory frame, so that $d/dt \equiv -\mathrm{i}\,\omega$, and that $\hat{I}_c = 0$ (i.e. there is no error field), equation (3.188) yields

$$\hat{\Psi}_w = \frac{E_{sw}/(-\tilde{E}_{ww})\,\hat{\Psi}_s}{1 - \mathrm{i}\,\tau_{LR}\,\omega}. \tag{10.1}$$

Here, $\hat{\Psi}_s$ is the normalized reconnected magnetic flux at the rational surface (see equations (3.72) and (3.184)), $\hat{\Psi}_w$ is the normalized helical magnetic flux that penetrates the wall (see equations (3.82) and (3.192)), the real dimensionless parameters E_{sw} and $\tilde{E}_{ww}$ are defined in equations (3.87) and (3.195), respectively,

$$\tau_{LR} = \frac{\tau_w}{(-\tilde{E}_{ww})} \tag{10.2}$$

is the effective L/R time of the wall, and τ_w the time constant of the wall (see equation (3.103)). Note that equation (10.1) is only valid in the so-called *thin-wall limit*,

$$\frac{\delta_w}{r_w} \ll \frac{1}{|\omega|\,\tau_w} \tag{10.3}$$

(see equation (3.104)), where δ_w is the radial thickness of the wall and r_w its minor radius. The previous inequality ensures that the perturbed radial magnetic field only exhibits weak radial variation across the wall.

Equations (3.187) and (10.1) yield

$$\frac{\Delta\hat{\Psi}_s}{\hat{\Psi}_s} = \Delta_{pw} + (\Delta_{nw} - \Delta_{pw})\left(\frac{1 + \mathrm{i}\,\tau_{LR}\,\omega}{1 + \tau_{LR}^2\,\omega^2}\right), \tag{10.4}$$

where use has been made of equations (7.4) and (7.5). Here, Δ_{pw} is the (real, dimensionless) tearing stability index when the wall is perfectly conductive (i.e. $\tau_w \to \infty$), whereas Δ_{nw} is the (real, dimensionless) tearing stability index when there is no wall (i.e. $\tau_w \to 0$). We expect $\Delta_{nw} - \Delta_{pw} > 0$. In other words, we expect the magnetic island chain to be stabilized by the presence of a perfectly conducting wall [3, 10].

It follows from the previous equation that

$$\mathrm{Re}\left(\frac{\Delta\hat{\Psi}_s}{\hat{\Psi}_s}\right) = \Delta_{pw} + (\Delta_{nw} - \Delta_{pw})\left(\frac{1}{1 + \tau_{LR}^2\,\omega^2}\right), \tag{10.5}$$

$$\mathrm{Im}\left(\frac{\Delta\hat{\Psi}_s}{\hat{\Psi}_s}\right) = (\Delta_{nw} - \Delta_{pw})\left(\frac{\tau_{LR}\,\omega}{1 + \tau_{LR}^2\,\omega^2}\right). \tag{10.6}$$

Finally, making use of the island saturation theory presented in section 9.4, we obtain

$$\mathrm{Re}\left(\frac{\Delta\hat{\Psi}_s}{\hat{\Psi}_s}\right) = \Delta_{pw}(0)\left(1 - \frac{w}{w_{pw}}\right) + \left[\Delta_{nw}(0) - \Delta_{pw}(0)\right]\left(\frac{1}{1 + \tau_{LR}^2\,\omega^2}\right), \tag{10.7}$$

$$\mathrm{Im}\left(\frac{\Delta\hat{\Psi}_s}{\hat{\Psi}_s}\right) = \left[\Delta_{nw}(0) - \Delta_{pw}(0)\right]\left(\frac{\tau_{LR}\,\omega}{1 + \tau_{LR}^2\,\omega^2}\right). \tag{10.8}$$

Here, $\Delta_{pw}(0)$ is the perfect-wall tearing stability index when the width of the magnetic island chain at the rational surface is zero. Likewise, $\Delta_{nw}(0)$ is the no-wall tearing stability index when the width of the magnetic island chain at the rational surface is zero. Finally, $w_{pw} = W_{pw}/4$, where W_{pw} is the saturated radial magnetic island width when the wall is perfectly conducting (see equation (9.21)).

Equations (8.108), (10.7), and (10.8) yield the following modified Rutherford island-width evolution equation [13]:

$$\begin{aligned} I_1\,\tau_R\,\frac{d}{dt}\left(\frac{4\,w}{r_s}\right) = {} & \Delta_{pw}(0)\left(1 - \frac{w}{w_{pw}}\right) + \left[\Delta_{nw}(0) - \Delta_{pw}(0)\right]\left(\frac{1}{1 + \tau_{LR}^2\,\omega^2}\right) \\ & + I_2\left[\Delta_{nw}(0) - \Delta_{pw}(0)\right]\left(\frac{c_\beta}{1+\tau}\right)\left(\frac{L_s}{L_p}\right)\left(\frac{d_\beta}{r_s}\right)\left(\frac{\tau_\varphi}{\tau_H}\right)\left(\frac{w}{r_s}\right)^2\left(\frac{\tau_{LR}\,\omega}{1 + \tau_{LR}^2\,\omega^2}\right) \\ & + I_3\left[\Delta_{nw}(0) - \Delta_{pw}(0)\right]^2\left(\frac{\tau_\varphi}{\tau_H}\right)^2\left(\frac{w}{r_s}\right)^7\left(\frac{\tau_{LR}\,\omega}{1 + \tau_{LR}^2\,\omega^2}\right)^2. \end{aligned} \tag{10.9}$$

Here, r_s is the minor radius of the rational surface, τ_R the resistive evolution time (see equation (5.49)), τ_H the hydromagnetic time (see equation (5.43)), τ_φ the toroidal momentum confinement time (see equation (5.50)), c_β a dimensionless measure of the plasma pressure at the rational surface (see equation (4.65)), τ the ratio of the electron and ion pressure gradients at the rational surface (see equation (4.5)), $4\,w$ the radial width of the magnetic island chain, $d_\beta = c_\beta\,d_i$ the ion sound radius, d_i the collisionless ion skin depth at the rational surface (see equation (4.24)), L_s the magnetic shear length at the rational surface (see equation (5.27)), and L_p the effective pressure gradient scale length at the rational surface (see equation (8.35)). Furthermore, $I_1 = 0.8227$, $I_2 = 0.048\,89$, and $I_3 = 0.029\,44$ (see section 8.11).

The first term on the right-hand side of the previous equation governs the growth and saturation of the magnetic island chain when its rotation frequency is sufficiently large that the wall acts as a perfect conductor. The second term describes the loss of wall stabilization when the chain's rotation frequency is sufficiently small that the perturbed magnetic field can penetrate through the wall [3, 10]. The third and fourth terms represent the destabilizing effect of the ion polarization current induced in the vicinity of the rational surface when the ion fluid is diverted around the island chain's magnetic separatrix [7, 13]. Unlike the case of an isolated island chain (see chapter 9), the polarization effect is nonzero because, according to equations (8.74), (8.87), (8.101), and (10.8), the electromagnetic braking torque exerted on the plasma in the immediate vicinity of the rational surface due to interaction with the conducting wall generates a finite ion flow in the island rest frame.

Assuming that the island chain corotates with the ion fluid at the rational surface (see section 9.5), equations (3.189) and (9.31) imply that

$$\omega = \omega_{\perp\,\mathrm{i}} - \sum_{p=1,\infty} (\alpha_\mathrm{p} + \beta_\mathrm{p}), \tag{10.10}$$

where $\omega_{\perp\,\mathrm{i}}$, which is defined in equation (9.32), is the unperturbed (by any electromagnetic torques that develop at the rational surface) rotation frequency. Here, we are treating α_p and β_p as constants because the electromagnetic torque that develops at the rational surface has no explicit time dependence (assuming that the width of the island chain grows on a timescale that is much greater than τ_LR and τ_φ). Equations (3.180), (3.190), (3.191), (4.23), (5.27), (5.43), (5.50), (7.28), and (7.34)–(7.35) can be combined to give

$$\sum_{p=1,\infty} (\alpha_\mathrm{p} + \beta_\mathrm{p}) = \left(\frac{\epsilon_\mathrm{s}}{q_\mathrm{s}}\right)^2 \left(\frac{\tau_\mathrm{V}}{\tau_\mathrm{H}^2}\right) \left(\frac{w}{r_\mathrm{s}}\right)^4 \mathrm{Im}\left(\frac{\Delta\hat{\Psi}_\mathrm{s}}{\hat{\Psi}_\mathrm{s}}\right), \tag{10.11}$$

where

$$\tau_\mathrm{V} = \frac{1}{4}\left[\left(\frac{q_\mathrm{s}}{\epsilon_\mathrm{s}}\right)^2 \left(\frac{\tau_\theta}{\tau_\varphi}\right)^{1/2} + 2\ln\left(\frac{a}{r_\mathrm{s}}\right)\right]\tau_\varphi. \tag{10.12}$$

Here, τ_θ is the poloidal flow-damping time (see equation (7.28)), $\epsilon_\mathrm{s} = r_\mathrm{s}/R_0$, R_0 is the simulated major radius of the plasma, a the minor radius of the plasma, $q_\mathrm{s} = m/n$, m is the poloidal mode number of the island chain, and n the toroidal mode number of the island chain. Equations (10.8), (10.10), and (10.11) can be combined to give the torque balance equation [6]

$$\omega_{\perp\,\mathrm{i}} - \omega = \left[\Delta_\mathrm{nw}(0) - \Delta_\mathrm{pw}(0)\right]\left(\frac{\epsilon_\mathrm{s}}{q_\mathrm{s}}\right)^2 \left(\frac{\tau_V}{\tau_\mathrm{H}^2}\right) \left(\frac{w}{r_\mathrm{s}}\right)^4 \left(\frac{\tau_\mathrm{LR}\,\omega}{1 + \tau_\mathrm{LR}^2\,\omega^2}\right). \tag{10.13}$$

The left-hand side of the previous equation represents the viscous restoring torque that acts to prevent changes in the plasma rotation at the rational surface, while the right-hand side represents the electromagnetic braking torque acting on the plasma in the vicinity of the rational surface due to the eddy current induced in the conducting wall.

Let

$$x = \frac{w}{w_\mathrm{pw}}, \tag{10.14}$$

$$y = \frac{\omega}{\omega_{\perp\,\mathrm{i}}}, \tag{10.15}$$

$$T = \frac{t}{\tau_{pw}}, \tag{10.16}$$

where τ_{pw} is defined in equation (9.22). Thus, x is the width of the magnetic island chain relative to its saturated width when the wall is perfectly conducting, y is the island rotation frequency relative to its value when there is no interaction with the wall, and T is time normalized to the typical time required for the island chain to attain is final saturated width. The modified Rutherford equation, (10.9), and the torque balance equation, (10.13), can be written in the non-dimensional forms [13]

$$\frac{dx}{dT} = 1 - x + \beta_1\left(\frac{1}{1 + \alpha_1^{\,2}\, y^2}\right) + \gamma_1\left(\frac{x^2\, y}{1 + \alpha_1^{\,2}\, y^2}\right) + \delta_1\left[\frac{x^7\, y^2}{(1 + \alpha_1^{\,2}\, y^2)^2}\right], \tag{10.17}$$

$$\zeta\,\frac{dy}{dT} = 1 - y - \epsilon_1\left(\frac{x^4\, y}{1 + \alpha_1^{\,2}\, y^2}\right), \tag{10.18}$$

respectively, where

$$\alpha_1 = \tau_{LR}\,|\omega_{\perp\,i}|, \tag{10.19}$$

$$\beta_1 = \frac{\Delta_{nw}(0) - \Delta_{pw}(0)}{\Delta_{pw}(0)}, \tag{10.20}$$

$$\gamma_1 = I_2\,\alpha_1\,\beta_1\left(\frac{c_\beta}{1 + \tau}\right)\left(\frac{L_s}{L_p}\right)\left(\frac{d_\beta}{r_s}\right)\left(\frac{\tau_\varphi}{\tau_H}\right)\left(\frac{w_{pw}}{r_s}\right)^2, \tag{10.21}$$

$$\delta_1 = I_3\,\Delta_{pw}(0)\,(\alpha_1\,\beta_1)^2\left(\frac{\tau_\varphi}{\tau_H}\right)^2\left(\frac{w_{pw}}{r_s}\right)^7, \tag{10.22}$$

$$\epsilon_1 = \left[\Delta_{nw}(0) - \Delta_{pw}(0)\right]\left(\frac{\epsilon_s}{q_s}\right)^2\left(\frac{\tau_{LR}\,\tau_V}{\tau_H^2}\right)\left(\frac{w_{pw}}{r_s}\right)^4, \tag{10.23}$$

and $0 < \zeta \ll 1$. Here, we have added artificial plasma inertia (i.e. the $\zeta\, dy/dT$ term) into the torque balance equation, (10.18), in order to distinguish between dynamically stable and dynamically unstable solutions [6].

The steady-state solutions of the torque balance equation, (10.18), correspond to the roots of the cubic polynomial

$$\alpha_1^{\,2}\, y^3 - \alpha_1^{\,2}\, y^2 + (1 + \epsilon_1\, x^4)y - 1 = 0. \tag{10.24}$$

If $\alpha_1 < \sqrt{27}$, then the previous equation only possesses one real root. In this situation, the normalized island rotation frequency, y, slows down in a smooth reversible fashion as the normalized island width, x, increases [6]. On the other hand, if $\alpha_1 > \sqrt{27}$, then the previous equation possesses three real roots. However, the intermediate root is dynamically unstable. In this situation, the torque balance equation possesses two branches of dynamically stable solutions [6]. The *high-rotation solution branch* is characterized by $1/3 \leqslant y \leqslant 1$. On the other hand, the *low-rotation solution branch* is characterized by $0 \leqslant y \leqslant 1/3$. The two solution branches are separated by a *forbidden band* of island rotation frequencies. The existence of this forbidden band has been verified experimentally [8]. The high-rotation solution branch ceases to exist when the normalized island width exceeds a critical value, and there is a bifurcation to the low-rotation solution branch. Likewise, the low-rotation solution branch ceases to exist when the normalized island width falls below a different critical value, and there is a bifurcation to the high-rotation solution branch. Thus, if $\alpha_{\rm l} > \sqrt{27}$, then the slowing down of the island rotation frequency with increasing island width is neither smooth nor reversible [6]. The steady-state solutions of the torque balance equation are illustrated in figure 10.1.

In the limit $\alpha_{\rm l} \gg \sqrt{27}$, we can find approximate solutions of the torque balance equation, (10.18). The high-rotation solution branch is characterized by $1/2 \leqslant y \leqslant 1$ and is such that

$$y \simeq \frac{1}{2}\left(1 + \left[1 - \left(\frac{x}{x_{\rm h}}\right)^4\right]^{1/2}\right), \tag{10.25}$$

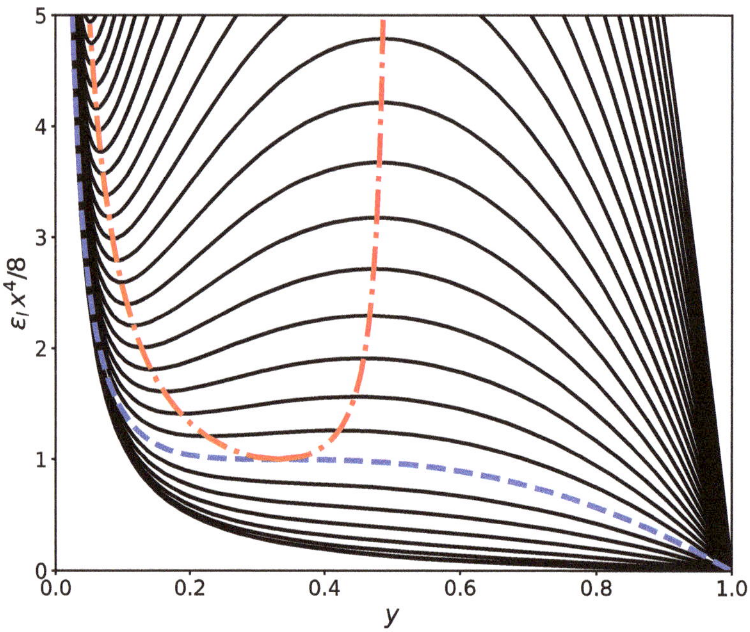

Figure 10.1. Steady-state solutions of the torque balance equation, (10.18). The solid lines show solution curves for various different values of $\alpha_{\rm l}$. The dashed line shows the critical solution curve, $\alpha_{\rm l} = \sqrt{27}$. The dash-dotted line shows the boundary of the forbidden band of island rotation frequencies. The high-rotation solution branch lies above the critical solution curve, to the right of the forbidden band. The low-rotation solution branch lies above the critical solution curve, to the left of the forbidden band. Within the forbidden band, all of the solutions of the torque balance equation are dynamically unstable. When a high-rotation/low-rotation solution reaches the boundary of the forbidden band, a bifurcation takes place to a low-rotation/high-rotation solution with the same value of $\epsilon_{\rm l}\, x^4/8$.

where

$$x_h = \left(\frac{\alpha_l^2}{4\,\epsilon_l}\right)^{1/4}. \tag{10.26}$$

The high-rotation solution branch ceases to exist when the normalized island width, x, exceeds the critical value, x_h—at which point, the island rotation frequency has been reduced to half of its original value—and there is a bifurcation to the low-rotation solution branch. The low-rotation solution branch is characterized by $0 \leqslant y \leqslant 1/\alpha_l$ and is such that

$$y \simeq \frac{1}{\alpha_l}\left\{\left(\frac{x}{x_l}\right)^4 - \left[\left(\frac{x}{x_l}\right)^8 - 1\right]^{1/2}\right\}, \tag{10.27}$$

where

$$x_l = \left(\frac{2\,\alpha_l}{\epsilon_l}\right)^{1/4}. \tag{10.28}$$

The low-rotation solution branch ceases to exist when the normalized island width falls below the critical value x_l, and there is a bifurcation to the high-rotation solution branch. Note that $x_h \gg x_l$, which indicates that once the width of the island chain has grown sufficiently wide to trigger a high-rotation to low-rotation bifurcation, the chain is unlikely to ever attain a high-rotation state again (because its width would have to shrink by a considerable factor to trigger the reverse bifurcation). In the limit $\alpha_l \gg \sqrt{27}$, the forbidden band of normalized island rotation frequencies corresponds to $1/\alpha_l < y < 1/2$ (see figure 10.1).

It is clear from equation (10.17) that the slowing down of the island chain's rotation due to the eddy current induced in the conducting wall has a *destabilizing* effect on the chain. In fact, if there is no substantial slowing down, then the normalized modified Rutherford equation (10.17) reduces to

$$\frac{dx}{dT} \simeq 1 - x, \tag{10.29}$$

assuming that $\alpha_l \gg \beta_l^{1/2}, \gamma_l^{1/2}, \delta_l^{1/4}$, indicating that the normalized island width, x, saturates at its perfect-wall value, unity. On the other hand, on the low-rotation solution branch, assuming that $\alpha_l \gg \sqrt{27}$ and $x \gg x_l$, equation (10.17) reduces to

$$\frac{dx}{dT} \simeq 1 - x + \beta_l + \frac{\gamma_l}{\epsilon_l\,x^2} + \frac{\delta_l}{\epsilon_l^{\,2}\,x}, \tag{10.30}$$

indicating that the normalized island width saturates at a value that is greater than unity (because β_l, γ_l, δ_l, and ϵ_l are all positive quantities). The island chain is destabilized by the loss of wall stabilization and also by the ion polarization effect associated with rotation braking.

Figure 10.2 shows numerical solutions of equations (10.17) and (10.18) obtained for a low-field tokamak fusion reactor and a high-field tokamak fusion reactor (see chapter 1). The simulation parameters are determined using the following assumptions: $B = 5$ T (low field) or $B = 12$ T (high field), $\beta = 0.02$, $T_e = T_i = 7$ keV, $m_i = (m_D + m_T)/2$ (where m_D and m_T are the deuteron and triton masses, respectively), $\Xi_{\perp\,i} = \chi_{\perp\,e} = \chi_{\perp\,i} = 1\ \mathrm{m^2\,s^{-1}}$, $\tau = 1$, $\eta_e = \eta_i = 1$, $dp/dr = -p/a$, $\omega_{\perp\,i} = 4\,\omega_{*\,e}$. Here, $\omega_{*\,e}$ is the electron diamagnetic frequency (see equation (5.45)). The poloidal and toroidal mode numbers of the magnetic island chain are $m = 2$ and $n = 1$, respectively. The wall parameters are $\tau_w = 10^{-1}$ s and $r_w/a = 1.1$, which correspond to a moderately conducting, close-fitting wall. The plasma equilibrium is assumed to be of the Wesson type (see section 9.4), with $q(0) = 0.8$ and $q(a) = 6.0$. It follows that $r_s = 0.560\,a$. The perfect-wall saturated island width is $W_{pw}/a = 0.264$. Finally, the various dimensionless parameters appearing in equations (10.17) and (10.18) take the values $\alpha_1 = 7.21 \times 10^1$ (low field) or $\alpha_1 = 1.73 \times 10^2$ (high field), $\beta_1 = 0.104$, $\gamma_1 = 7.98$, $\delta_1 = 8.61 \times 10^6$, $\epsilon_1 = 2.05 \times 10^4$, and $\zeta = 0.1$.

It can be seen from figure 10.2 that as the normalized width, x, of the island chain grows over time, the chain's normalized rotation frequency, y, is gradually reduced until it has been reduced to about half of its original value, at which point there is a sudden collapse in the rotation frequency to a very low value. The collapse in the rotation frequency causes the chain to be further destabilized due to the loss of wall stabilization and ion polarization effects. Consequently, the final saturated width of the island chain is greater than the perfect-wall saturated island width (i.e. $x > 1$). Note that a low-field tokamak fusion reactor is more susceptible to rotation braking than a high-field fusion reactor because of its lower diamagnetic frequency (see table 6.1) and consequent lower ion fluid rotation. The slowing-down curves shown in the figure are similar in form to those observed experimentally when a wide magnetic island chain interacts electromagnetically with a resistive wall in a toroidal confinement device [4].

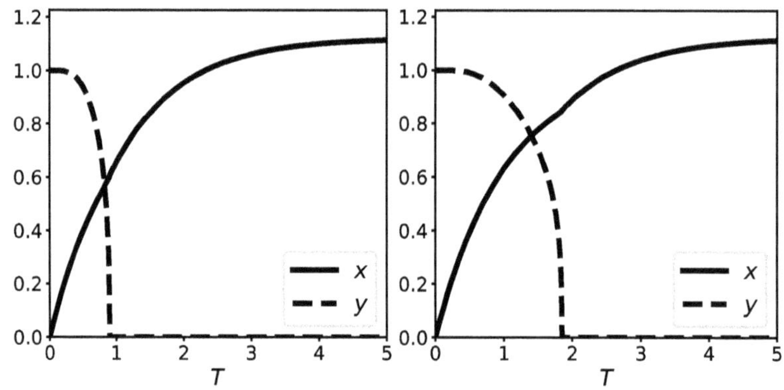

Figure 10.2. Simulations of island rotation braking in a low-field tokamak fusion reactor (left panel) and a high-field tokamak fusion reactor (right panel), both of which have thin walls. Here, x and y are the normalized island width and the rotation frequency, respectively.

Figure 10.3 displays the results of a series of simulations of the type shown in figure 10.2 in which the radius of the rational surface is scanned over a range of values (by changing $q(a)$ while keeping $q(0) = 0.8$ fixed). The wall parameters are $r_w/a = 1.2$ and $\tau_w = 5 \times 10^{-2}$ s. It can be seen that if the rational surface lies well inside the plasma boundary, then the island chain attains its final, perfect-wall saturated width without a collapse in its rotation frequency. On the other hand, if the rational surface lies closer to the plasma boundary, then a collapse in the rotation frequency is triggered before the chain attains its final saturated width. The critical island width at which the rotation collapse is triggered lies between 20% and 10% of the plasma's minor radius. Moreover, the final saturated width of the island chain exceeds the perfect-wall value due to the loss of wall stabilization and ion polarization effects. Finally, it is again clear that a low-field tokamak fusion reactor is more susceptible to rotation braking than a high-field fusion reactor.

Figure 10.4 shows the results of a series of simulations of the type shown in figure 10.2 in which the unperturbed island chain rotation frequency is scanned over a range of values for various different wall time constants. The plasma parameters are $q(0) = 0.8$ and $q(a) = 7.0$, which correspond to $r_s = 0.517\,a$, while the wall radius is $r_w = 1.2\,a$. It is clear that if the wall is very resistive (i.e. $\tau_w \lesssim 10^{-3}$ s), then the critical island width above which the chain's rotation frequency collapses is only weakly dependent on the unperturbed rotation frequency and is about 10% of the plasma's minor radius. On the other hand, for the case of a highly conducting wall (i.e. $\tau_w \lesssim 1$ s), the diamagnetic levels of plasma rotation are sufficient to completely suppress the collapse in the island rotation frequency, unless the unperturbed rotation frequency lies close to zero. As before, it is clear that a low-field tokamak fusion reactor is more susceptible to rotation braking than a high-field fusion reactor.

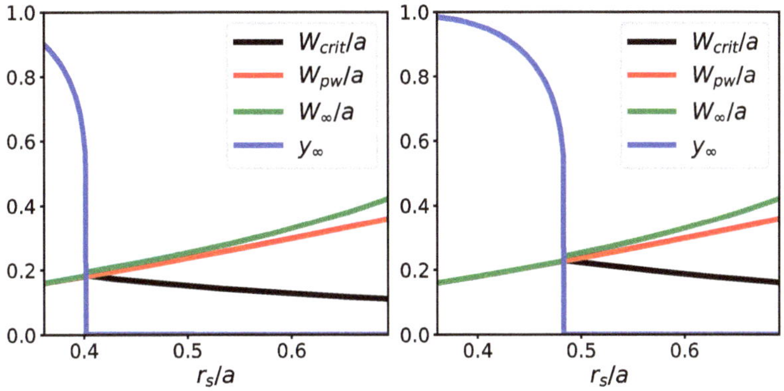

Figure 10.3. The results of a series of simulations of the type shown in figure 10.2 in which the radius of the rational surface is scanned over a range of values (by changing the edge safety factor). W_{crit} is the critical island chain width above which the chain's rotation frequency collapses. W_{pw} is the perfect-wall saturated island width. W_∞ is the final saturated island width. y_∞ is the final normalized island rotation frequency. The left-hand/right-hand panels correspond to a low-field tokamak fusion reactor and a high-field tokamak fusion reactor, respectively.

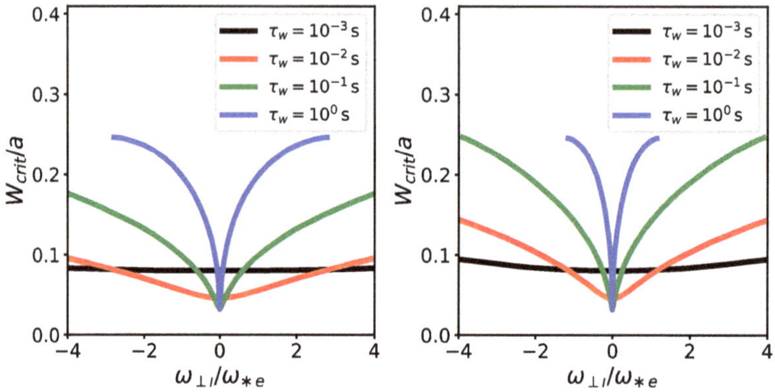

Figure 10.4. The results of a series of simulations of the type shown in figure 10.2 in which the unperturbed island rotation frequency is scanned over a range of values for various different wall time constants. $W_{\rm crit}$ is the critical island chain width above which the chain's rotation frequency collapses. The left-hand/right-hand panels correspond to a low-field tokamak fusion reactor and a high-field tokamak fusion reactor, respectively.

10.3 Rotation braking by a thick conducting wall

The calculation presented in the previous section has a number of deficiencies. The first and most important deficiency is that the criterion (10.3) for the validity of the thin-wall approximation (i.e. that the radial variation of the perturbed magnetic flux across the wall is relatively weak) is not always satisfied in practice. For instance, in a low-field tokamak fusion reactor (see section 1.5) with diamagnetic levels of ion fluid rotation (i.e. $\omega_{\perp\,\rm i} \simeq \omega_{*\,\rm e}$), the critical wall thickness above which the thin-wall approximation fails is $\delta_{\rm w\,max} \simeq 9.7 \times 10^{-4}/\tau_{\rm w}({\rm s})$ m. In the case of a high-field reactor, the critical wall thickness is $\delta_{\rm w\,max} \simeq 1.7 \times 10^{-4}/\tau_{\rm w}({\rm s})$ m. It follows that the thin-wall approximation holds for resistive walls (i.e. $\tau_{\rm w} < 10^{-1}$ s) but not for highly conducting walls (i.e. $\tau_{\rm w} > 10^{-1}$ s), because the latter type of wall would have to be impossibly thin.

A second deficiency in our previous calculation is that it neglects plasma inertia. This neglect is reasonable when the magnetic island chain is rotating steadily but not when the chain's rotation frequency collapses, because such a collapse is associated with a rapid deceleration of the chain and a consequent rapid deceleration of the plasma in the vicinity of the resonant surface.

A third deficiency in our previous calculation is that it assumes that the eddy current excited in the wall has a simple $e^{-i\,\omega\,t}$ time dependence, where ω is the instantaneous rotation frequency of the island chain. This assumption is reasonable when the island chain is rotating steadily, but not when its rotation frequency collapses, in which case we expect the rapid deceleration of the chain to excite a *transient* eddy current in the wall [4].

The aim of this section is to generalize the analysis given in the previous section in order to take into account thick walls, plasma inertia, and a transient component of the wall eddy current. Note that, by a 'thick' wall, we mean one in which the skin depth in the wall material is less than the wall's radial thickness. It is still reasonable to assume that the wall's thickness, $\delta_{\rm w}$, is much less than its minor radius, $r_{\rm w}$.

Suppose that the wall extends from $r = r_w$ to $r = r_w + \delta_w$, where $\delta_w \ll r_w$. Here, r is a conventional cylindrical coordinate. Let $\delta\psi(r, t) = R_0\, B_z\, \delta\hat{\psi}(r, t)$ be the perturbed magnetic flux within the wall (see equation (3.20)). Here, B_z is the equilibrium toroidal magnetic field strength. Ohm's law inside the wall yields (see equation (3.101))

$$\frac{\partial^2 \delta\hat{\psi}}{\partial r^2} \simeq \frac{\mu_0}{\eta_w}\frac{\partial \delta\hat{\psi}}{\partial t}, \tag{10.31}$$

where η_w is the electrical resistivity of the wall material. The previous equation must be solved subject to the boundary conditions

$$\delta\hat{\psi}(r_w, t) = \hat{\Psi}_w(t), \tag{10.32}$$

$$\frac{\partial \ln \delta\hat{\psi}(r_w + \delta_w, t)}{\partial \ln r} = -m. \tag{10.33}$$

The first boundary condition follows from equations (3.82) and (3.192). Note that we have effectively redefined $\hat{\Psi}_w$ to be the normalized helical magnetic flux that penetrates the *inner* (in r) boundary of the wall. The second boundary condition follows because, in the vacuum region outside the wall, a well-behaved solution of the cylindrical tearing-mode equation, (3.60), varies as r^{-m}.

Let

$$\rho = \frac{r - r_w}{\delta_w}, \tag{10.34}$$

$$\delta\hat{\psi}(r, t) = \hat{\Psi}_w(t)\, F(\rho, t). \tag{10.35}$$

Equations (10.31)–(10.33) yield

$$\frac{\partial^2 F}{\partial \rho^2} = \tau_w\, \zeta_w\left(\gamma\, F + \frac{\partial F}{\partial t}\right), \tag{10.36}$$

$$F(0, t) = 1, \tag{10.37}$$

$$\frac{\partial \ln F(1, t)}{\partial \rho} \simeq -m\, \zeta_w, \tag{10.38}$$

where

$$\gamma = \frac{d \ln \hat{\Psi}_w}{dt}, \tag{10.39}$$

$$\tau_w = \frac{\mu_0\, r_w\, \delta_w}{\eta_w}, \tag{10.40}$$

$$\zeta_w = \frac{\delta_w}{r_w} \ll 1. \tag{10.41}$$

Let us first search for a solution of

$$\frac{d^2F_0}{d\rho^2} = \gamma\,\tau_w\,\zeta_w\,F_0, \tag{10.42}$$

subject to the boundary conditions

$$F_0(0) = 1, \tag{10.43}$$

$$\frac{d\ln F_0(1)}{d\rho} = -m\,\zeta_w. \tag{10.44}$$

We obtain

$$F_0(\rho) = \frac{\alpha_w\cosh[\alpha_w\,(\rho-1)] - m\,\zeta_w\sinh[\alpha_w\,(\rho-1)]}{\alpha_w\cosh\alpha_w + m\,\zeta_w\sinh\alpha_w}, \tag{10.45}$$

where

$$\alpha_w = \sqrt{\gamma\,\tau_w\,\zeta_w}\,. \tag{10.46}$$

However, in the physically relevant limit,

$$|\gamma|\,\tau_w \gg \zeta_w, \tag{10.47}$$

expression (10.45) simplifies to give

$$F_0(\rho) \simeq \frac{\cosh[\alpha_w\,(\rho-1)]}{\cosh\alpha_w}. \tag{10.48}$$

Incidentally, it is clear from equation (10.48) that if the inequality (10.47) is satisfied, then the boundary condition (10.44) effectively reduces to

$$\frac{dF_0(1)}{d\rho} \simeq 0. \tag{10.49}$$

Let us write

$$F(\rho, t) = F_0(\rho) + F_1(\rho, t), \tag{10.50}$$

where

$$F_1(0, t) = 0, \tag{10.51}$$

$$\frac{\partial\ln F_1(1, t)}{\partial\rho} = -m\,\zeta_w. \tag{10.52}$$

Expression (10.50) automatically satisfies the boundary conditions (10.37) and (10.38). It is clear, by analogy with equation (10.49), that if the inequality (10.47) is satisfied, then the boundary condition (10.52) effectively reduces to

$$\frac{\partial F_1(1, t)}{\partial \rho} \simeq 0. \tag{10.53}$$

Let us write

$$F_1(\rho, t) = \sum_{j=1,\infty} f_j(t)\, \frac{\sin(\beta_{\mathrm{w}\,j}\,\rho)}{\beta_{\mathrm{w}\,j}}, \tag{10.54}$$

where

$$\beta_{\mathrm{w}\,j} = \left(j - \frac{1}{2}\right)\pi. \tag{10.55}$$

Expression (10.54) automatically satisfies the boundary conditions (10.51) and (10.53). Equations (10.36) and (10.50) can be combined to give

$$\begin{aligned}\sum_{k=1,\infty} \frac{df_k}{dt}\, \frac{\sin(\beta_{\mathrm{w}\,k}\,\rho)}{\beta_{\mathrm{w}\,k}} &= -\sum_{k=1,\infty} (\gamma + \lambda_k)\, f_k\, \frac{\sin(\beta_{\mathrm{w}\,k}\,\rho)}{\beta_{\mathrm{w}\,k}} \\ &- \frac{\alpha_{\mathrm{w}}}{2\,\gamma}\, \frac{d\gamma}{dt}\left\{\frac{(\rho - 1)\,\sinh[\alpha_{\mathrm{w}}\,(\rho - 1)]}{\cosh \alpha_{\mathrm{w}}} - \frac{\cosh[\alpha_{\mathrm{w}}\,(\rho - 1)]\,\sinh \alpha_{\mathrm{w}}}{\cosh^2 \alpha_{\mathrm{w}}}\right\} \\ &\sin(\beta_{\mathrm{w}\,k}\,\rho),\end{aligned} \tag{10.56}$$

where

$$\lambda_j = \frac{(j - 1/2)^2\, \pi^2}{\tau_{\mathrm{w}}\, \zeta_{\mathrm{w}}}. \tag{10.57}$$

Finally, multiplying equation (10.56) by $2\,\sin(\beta_{\mathrm{w}\,j}\,\rho)$ and integrating from $\rho = 0$ to $\rho = 1$, we obtain [4]

$$\frac{df_j}{dt} + (\gamma + \lambda_j)\, f_j = \sigma_j\, \frac{d\gamma}{dt}, \tag{10.58}$$

where

$$\sigma_j = \frac{2\,\lambda_j}{(\gamma + \lambda_j)^2}. \tag{10.59}$$

Here, use has been made of the easily demonstrated results:

$$\int_0^1 2\,\sin(\beta_{\mathrm{w}\,j}\,\rho)\,\sin(\beta_{\mathrm{w}\,k}\,\rho)\,d\rho = \delta_{jk}, \tag{10.60}$$

$$\int_0^1 \left\{ \frac{(\rho - 1)\sinh[\alpha_w(\rho - 1)]}{\cosh\alpha_w} - \frac{\cosh[\alpha_w(\rho - 1)]\sinh\alpha_w}{\cosh^2\alpha_w} \right\} \sin(\beta_{wj}\,\rho)\,d\rho$$
$$= -\frac{2\,\alpha_w\,\beta_{wj}}{(\alpha_w^2 + \beta_{wj}^2)^2}. \tag{10.61}$$

Equation (3.83) generalizes to give

$$\Delta\hat{\Psi}_w = \left[r\,\frac{\partial\delta\hat{\psi}}{\partial r} \right]_{r=r_w}^{r=r_w+\delta_w} \simeq \frac{\hat{\Psi}_w}{\zeta_w}\left[\frac{\partial F}{\partial \rho} \right]_{\rho=0}^{\rho=1}, \tag{10.62}$$

where use has been made of equations (3.192), (3.193), (10.34), (10.35), and (10.41). Note that $\Delta\hat{\Psi}_w$ is a measure of the normalized net eddy current induced in the wall. It follows from equations (10.48), (10.46), (10.50), and (10.54) that [4]

$$\frac{\Delta\hat{\Psi}_w}{\hat{\Psi}_w} = G(t), \tag{10.63}$$

where

$$G(t) = \sqrt{\frac{\gamma\,\tau_w}{\zeta_w}}\,\tanh\!\left(\sqrt{\gamma\,\tau_w\,\zeta_w}\right) - \frac{1}{\zeta_w}\sum_{j=1,\infty} f_j(t). \tag{10.64}$$

Clearly, equation (10.63) specifies the relation between the net eddy current induced in the wall and the helical magnetic flux that penetrates the inner boundary of the wall.

The first term on the right-hand side of equation (10.64),

$$G_0(\gamma) = \sqrt{\frac{\gamma\,\tau_w}{\zeta_w}}\,\tanh\!\left(\sqrt{\gamma\,\tau_w\,\zeta_w}\right), \tag{10.65}$$

specifies the net eddy current induced in the wall by a steadily rotating island chain. In the thin-wall limit (see equations (3.104), (10.39), (10.41), and (10.47)) [4],

$$\zeta_w \ll |\gamma|\,\tau_w \ll \frac{1}{\zeta_w}, \tag{10.66}$$

the steady-state wall response reduces to (see equation (3.102))

$$G_0(\gamma) \simeq \gamma\,\tau_w, \tag{10.67}$$

which is consistent with the analysis employed in the previous section. On the other hand, in the *thick-wall limit* [4],

$$|\gamma|\,\tau_w \gg \frac{1}{\zeta_w}, \tag{10.68}$$

we obtain

$$G_0(\gamma) \simeq \sqrt{\frac{\gamma\,\tau_w}{\zeta_w}}. \tag{10.69}$$

The previous expression represents a steady-state wall response in which the eddy current only penetrates a distance of the order of the skin depth into the wall from its inner boundary.

The second term on the right-hand side of equation (10.64),

$$G_1(t) = -\frac{1}{\zeta_w}\sum_{j=1,\infty} f_j(t), \tag{10.70}$$

specifies the transient eddy current excited in the wall. As is clear from equation (10.58), a transient eddy current is excited when the island rotation frequency (which is proportional to γ) changes in time.

Equation (3.188) (with $\hat{I}_c = 0$, because there is no error field) and equation (10.63) yield

$$\hat{\Psi}_w = \frac{E_{sw}\,\hat{\Psi}_s}{G + (-\tilde{E}_{ww})}. \tag{10.71}$$

Writing

$$\hat{\Psi}_s(t) = |\hat{\Psi}_s|(t)\exp\left(-{\rm i}\int_0^t \omega(t')\,dt'\right), \tag{10.72}$$

where $\omega(t)$ is the instantaneous island rotation frequency, equations (10.39) and (10.71) imply that

$$\gamma(t) = -{\rm i}\,\omega + 2\,\frac{d\ln w}{dt} - \frac{dG/dt}{G + (-\tilde{E}_{ww})}. \tag{10.73}$$

Here, we have made use of the fact that $w \propto |\hat{\Psi}_s|^{1/2}$ (see equation (8.1)). Equations (3.187) and (10.71) give

$$\frac{\Delta\hat{\Psi}_s}{\hat{\Psi}_s} = \Delta_{pw}(0)\left(1 - \frac{w}{w_{pw}}\right) + \left[\Delta_{nw}(0) - \Delta_{pw}(0)\right]\left[\frac{(-\tilde{E}_{ww})}{G + (-\tilde{E}_{ww})}\right], \tag{10.74}$$

where use has been made of equations (7.4) and (7.5) as well as the island saturation theory presented in section 9.4. Hence,

$$\mathrm{Re}\left(\frac{\Delta\hat{\Psi}_s}{\hat{\Psi}_s}\right) = \Delta_{pw}(0)\left(1 - \frac{w}{w_{pw}}\right) + \left[\Delta_{nw}(0) - \Delta_{pw}(0)\right]\left(\frac{\mathcal{R}}{\mathcal{R}^2 + \mathcal{I}^2}\right), \tag{10.75}$$

$$\mathrm{Im}\left(\frac{\Delta\hat{\Psi}_s}{\hat{\Psi}_s}\right) = \left[\Delta_{nw}(0) - \Delta_{pw}(0)\right]\left(\frac{\mathcal{I}}{\mathcal{R}^2 + \mathcal{I}^2}\right), \tag{10.76}$$

where

$$\mathcal{R} = 1 + \frac{\mathrm{Re}(G)}{(-\tilde{E}_{\mathrm{ww}})}, \tag{10.77}$$

$$\mathcal{I} = -\frac{\mathrm{Im}(G)}{(-\tilde{E}_{\mathrm{ww}})}. \tag{10.78}$$

Equations (8.108), (10.75), and (10.76) yield the following modified Rutherford island-width evolution equation, which is a generalization of equation (10.9):

$$\begin{aligned} I_1\,\tau_{\mathrm{R}}\,\frac{d}{dt}\left(\frac{4\,w}{r_{\mathrm{s}}}\right) &= \Delta_{\mathrm{pw}}(0)\left(1-\frac{w}{w_{\mathrm{pw}}}\right) + \left[\Delta_{\mathrm{nw}}(0)-\Delta_{\mathrm{pw}}(0)\right]\left(\frac{\mathcal{R}}{\mathcal{R}^2+\mathcal{I}^2}\right) \\ &+ I_2\left[\Delta_{\mathrm{nw}}(0)-\Delta_{\mathrm{pw}}(0)\right]\left(\frac{c_\beta}{1+\tau}\right)\left(\frac{L_{\mathrm{s}}}{L_{\mathrm{p}}}\right)\left(\frac{d_\beta}{r_{\mathrm{s}}}\right)\left(\frac{\tau_\varphi}{\tau_{\mathrm{H}}}\right)\left(\frac{w}{r_{\mathrm{s}}}\right)^2\left(\frac{\mathcal{I}}{\mathcal{R}^2+\mathcal{I}^2}\right) \\ &+ I_3\left[\Delta_{\mathrm{nw}}(0)-\Delta_{\mathrm{pw}}(0)\right]^2\left(\frac{\tau_\varphi}{\tau_{\mathrm{H}}}\right)^2\left(\frac{w}{r_{\mathrm{s}}}\right)^7\left(\frac{\mathcal{I}}{\mathcal{R}^2+\mathcal{I}^2}\right)^2. \end{aligned} \tag{10.79}$$

All of the parameters appearing in this equation are defined in the previous section.

As in the previous section, the instantaneous island rotation frequency can be written (see equation (10.10))

$$\omega = \omega_{\perp\,\mathrm{i}} - \sum_{p=1,\infty}(\alpha_{\mathrm{p}}+\beta_{\mathrm{p}}). \tag{10.80}$$

However, according to equations (3.190) and (3.191), the time evolution of the quantities α_{p} and β_{p} is specified by

$$\tau_M\,\frac{d\alpha_{\mathrm{p}}}{dt} + \left(\zeta_\theta + j_{1\mathrm{p}}^{\,2}\right)\alpha_{\mathrm{p}} = g_{\theta\,\mathrm{p}}\left(\frac{\tau_\varphi}{\tau_{\mathrm{H}}^2}\right)\left(\frac{w}{r_{\mathrm{s}}}\right)^4 \mathrm{Im}\left(\frac{\Delta\hat{\Psi}_{\mathrm{s}}}{\hat{\Psi}_{\mathrm{s}}}\right), \tag{10.81}$$

$$\tau_M\,\frac{d\beta_{\mathrm{p}}}{dt} + j_{0\mathrm{p}}^{\,2}\,\beta_{\mathrm{p}} = g_{\varphi\,\mathrm{p}}\left(\frac{\epsilon_{\mathrm{s}}}{q_{\mathrm{s}}}\right)^2\left(\frac{\tau_\varphi}{\tau_{\mathrm{H}}^2}\right)\left(\frac{w}{r_{\mathrm{s}}}\right)^4 \mathrm{Im}\left(\frac{\Delta\hat{\Psi}_{\mathrm{s}}}{\hat{\Psi}_{\mathrm{s}}}\right), \tag{10.82}$$

where all of the parameters appearing in the previous two equations are defined in the previous section, except for

$$\tau_{\mathrm{M}} = \left(\frac{a}{r_{\mathrm{s}}}\right)^2\tau_\varphi, \tag{10.83}$$

$$\zeta_\theta = \frac{\tau_{\mathrm{M}}}{\tau_\theta}, \tag{10.84}$$

$$g_{\theta p} = \left[\frac{J_1(j_{1p}\, r_s/a)}{J_2(j_{1p})}\right]^2, \tag{10.85}$$

$$g_{\varphi p} = \left[\frac{J_0(j_{0p}\, r_s/a)}{J_1(j_{0p})}\right]^2. \tag{10.86}$$

Here, $J_0(z)$ and $J_1(z)$ are standard Bessel functions, and j_{np} denotes the pth zero of the $J_n(z)$ Bessel function [1]. Note that the terms involving d/dt in equations (10.81) and (10.82) represent plasma inertia.

Let $x = w/w_{pw}$, $y = \omega/|\omega_{\perp\, i}|$, and $T = t\, |\omega_{\perp\, i}|$. Thus, x is the width of the magnetic island chain relative to its saturated width when the wall is perfectly conducting, y is the island rotation frequency relative to the magnitude of its value when there is no interaction with the wall, and T is time normalized to the typical time required for the island chain complete a full rotation. Equations (10.73) and (10.75)–(10.82) can be converted into the following closed set of normalized equations that govern the time evolution of the island chain's rotation frequency:

$$y = \mathrm{sgn}(\omega_{\perp\, i}) - \sum_{p=1,\infty} (\hat{\alpha}_p + \hat{\beta}_p), \tag{10.87}$$

where

$$\hat{\tau}_{pw}\, \frac{dx}{dT} = 1 - x + \beta_1\left(\frac{\mathcal{R}}{\mathcal{R}^2 + \mathcal{I}^2}\right) + \gamma_1'\, x^2\left(\frac{\mathcal{I}}{\mathcal{R}^2 + \mathcal{I}^2}\right) + \delta_1'\, x^7\left(\frac{\mathcal{I}}{\mathcal{R}^2 + \mathcal{I}^2}\right)^2, \tag{10.88}$$

$$\hat{\tau}_M\, \frac{d\hat{\alpha}_p}{dT} = -\left(\zeta_\theta + j_{1p}^{\,2}\right)\hat{\alpha}_p + g_{\theta\, p}\, \epsilon_1'\, x^4\left(\frac{\mathcal{I}}{\mathcal{R}^2 + \mathcal{I}^2}\right), \tag{10.89}$$

$$\hat{\tau}_M\, \frac{d\hat{\beta}_p}{dT} = -j_{0p}^{\,2}\, \hat{\beta}_p + g_{\varphi\, p}\, \theta_1\, x^4\left(\frac{\mathcal{I}}{\mathcal{R}^2 + \mathcal{I}^2}\right). \tag{10.90}$$

Here,

$$\mathcal{R} = \mathrm{Re}(\mathcal{G}), \tag{10.91}$$

$$\mathcal{I} = -\mathrm{Im}(\mathcal{G}), \tag{10.92}$$

where

$$\mathcal{G} = 1 + \sqrt{\frac{\hat{\gamma}\, \hat{\tau}_w}{\hat{\zeta}_w}}\, \tanh\left(\sqrt{\hat{\gamma}\, \hat{\tau}_w\, \hat{\zeta}_w}\right) - \sum_{j=1,\infty} \hat{f}_j. \tag{10.93}$$

Moreover,

$$\hat{\gamma} = -\mathrm{i}\, y + 2\,\frac{d\ln x}{dT} - \frac{d\ln \mathcal{G}}{dT}, \tag{10.94}$$

$$\frac{d\hat{f}_j}{dT} = -\left(\hat{\gamma} + \hat{\lambda}_j\right)\hat{f}_j + \hat{\sigma}_j\,\frac{d\hat{\gamma}}{dT}, \tag{10.95}$$

$$\hat{\lambda}_j = \frac{(j - 1/2)^2\,\pi^2}{\hat{\tau}_{\rm w}\,\hat{\zeta}_{\rm w}}, \tag{10.96}$$

$$\hat{\sigma}_j = \frac{2\,\hat{\lambda}_j}{\hat{\zeta}_{\rm w}\,(\hat{\gamma} + \hat{\lambda}_j)^2}. \tag{10.97}$$

Finally, β_1 is specified in equation (10.20), ζ_θ is specified in equation (10.84), and

$$\hat{\tau}_{\rm w} = \frac{\tau_{\rm w}\,|\omega_{\perp\,{\rm i}}|}{(-\tilde{E}_{\rm ww})}, \tag{10.98}$$

$$\hat{\tau}_{\rm M} = \tau_{\rm M}\,|\omega_{\perp\,{\rm i}}|, \tag{10.99}$$

$$\hat{\tau}_{\rm pw} = \tau_{\rm pw}\,|\omega_{\perp\,{\rm i}}|, \tag{10.100}$$

$$\hat{\zeta}_{\rm w} = \frac{\delta_{\rm w}\,(-\tilde{E}_{\rm ww})}{r_{\rm w}}, \tag{10.101}$$

$$\gamma_1{}' = I_2\,\beta_1\!\left(\frac{c_\beta}{1+\tau}\right)\!\left(\frac{L_{\rm s}}{L_{\rm p}}\right)\!\left(\frac{d_\beta}{r_{\rm s}}\right)\!\left(\frac{\tau_\varphi}{\tau_{\rm H}}\right)\!\left(\frac{W_{\rm pw}}{r_{\rm s}}\right)^2, \tag{10.102}$$

$$\delta_1{}' = I_3\,\Delta_{\rm pw}(0)\,\beta_1{}^2\!\left(\frac{\tau_\varphi}{\tau_{\rm H}}\right)^2\!\left(\frac{W_{\rm pw}}{r_{\rm s}}\right)^7, \tag{10.103}$$

$$\epsilon_1{}' = \left[\Delta_{\rm nw}(0) - \Delta_{\rm pw}(0)\right]\!\left(\frac{\tau_\varphi}{\tau_{\rm H}^2\,|\omega_{\perp\,{\rm i}}|}\right)\!\left(\frac{W_{\rm pw}}{r_{\rm s}}\right)^4, \tag{10.104}$$

$$\theta_1 = \left(\frac{\epsilon_{\rm s}}{q_{\rm s}}\right)^2 \epsilon_1{}'. \tag{10.105}$$

Note that x, y, $\hat{\alpha}_{\rm p}$, and $\hat{\beta}_{\rm p}$ are real quantities, whereas $\hat{\gamma}$ and f_j are complex.

The type of rotation braking calculation discussed in this section is far more computationally intensive than the type discussed in the previous section, because the former involves the solution of a great many more differential equations than the latter. However, the new calculation is an improvement on the previous one because it allows us to determine the timescale on which rotating braking occurs. Our previous calculation is unable to achieve this goal because it neglects plasma inertia.

Let us investigate a specific example. Consider a high-field tokamak fusion reactor (see chapter 1) characterized by $B = 12$ T, $\beta = 0.02$, $T_e = T_i = 7$ keV, $m_i = (m_D + m_T)/2$ (where m_D and m_T are the deuteron and triton masses, respectively), $\Xi_{\perp\,i} = \chi_{\perp\,e} = \chi_{\perp\,i} = 1\ \mathrm{m^2\,s^{-1}}$, $\tau = 1$, $\eta_e = \eta_i = 1$, $dp/dr = -p/a$, and $\omega_{\perp\,i} = 4\,\omega_{*\,e}$. The wall parameters are $\eta_w = 6.9 \times 10^{-7}\ \Omega\, m$ (which is the electrical resistivity of stainless steel), $r_w = 1.2\,a$, and $\delta_w = 0.1\,a$. The plasma equilibrium is assumed to be of the Wesson type (see section 9.4), with $q(0) = 0.8$ and $q(a) = 6.0$. The poloidal and toroidal mode numbers of the tearing mode are $m = 2$ and $n = 1$, respectively. It follows that $r_s = 0.560\,a$. The perfect-wall saturated island width is $W_{pw}/a = 0.272$, the poloidal flow-damping time is $\tau_\theta = 4.59 \times 10^{-5}$ s, the wall time constant is $\tau_w = 0.24$ s, the momentum confinement time is $\tau_M = 1.1$ s (see equation (3.180)), and the typical type required for the magnetic island to attain its final saturated width is $\tau_{pw} = 6.6 \times 10^1$ s. The normalized parameters that characterize our model take the values $\hat{\tau}_w = 431$, $\hat{\tau}_M = 8.33 \times 10^3$, $\hat{\tau}_{pw} = 4.95 \times 10^5$, $\zeta_\theta = 2.42 \times 10^4$, $\hat{\zeta}_w = 0.352$, $\beta_1 = 6.95 \times 10^{-2}$, $\gamma_1' = 3.28 \times 10^{-2}$, $\delta_1' = 1.66 \times 10^2$, $\epsilon_1' = 1.72 \times 10^4$, and $\theta_1 = 1.50 \times 10^2$. We conclude that the effective L/R time of the wall is about 500 times larger than the typical time required for the unperturbed magnetic island chain to complete a full rotation (i.e. $\hat{\tau}_w \sim 500$), the momentum confinement time is about 10^4 times larger than the island rotation time (i.e. $\hat{\tau}_M \sim 10^4$), and the island saturation time is about 5×10^5 times larger than the island rotation time (i.e., $\hat{\tau}_{pw} \sim 5 \times 10^5$).

It turns out that 100 poloidal and toroidal velocity harmonics are sufficient to describe the time evolution of the plasma poloidal and toroidal rotation profiles in a reasonably accurate manner. Consequently, we shall neglect all α_p and β_p variables with $p > p_{max} = 100$ in our calculation. In order to compensate for the truncation of the sum in equation (10.87), we shall replace this equation by

$$y = \mathrm{sgn}(\omega_{\perp\,i}) - \sum_{p=1,p_{max}} \left(\frac{\hat{\alpha}_p}{S_\theta} + \frac{\beta_p}{S_\varphi}\right), \tag{10.106}$$

where

$$S_\theta = 4\left(\frac{r_s}{a}\right)\sqrt{\zeta_\theta}\sum_{p=1,p_{max}} \frac{g_{\theta\,p}}{\zeta_\theta + j_{1p}^{\,2}}, \tag{10.107}$$

$$S_\varphi = \frac{2}{\ln(a/r_s)}\sum_{p=1,p_{max}} \frac{g_{\varphi\,p}}{j_{0p}^{\,2}}. \tag{10.108}$$

(See equations (7.34) and (7.35).) For the calculation in hand, $S_\theta = 0.708$ and $S_\varphi = 0.997$.

For the transient wall harmonics, an examination of equations (10.95) and (10.97) implies that, roughly speaking, all harmonics in the range $1 \leqslant j < 2\,j_{crit}$, where

$\hat{\lambda}_{j_{\rm crit}} \sim |\hat{\gamma}|$, are important in the calculation. Hence, given that $|\hat{\gamma}| \leqslant 1$, we deduce from equation (10.96) that $j_{\rm crit} \sim (\hat{\tau}_{\rm w}\,\hat{\zeta}_{\rm w})^{1/2}/\pi \sim 7$. This result merely implies that when the island chain is rotating at its unperturbed rotation frequency, the transient eddy current induced in the wall only penetrates radially into about the inner 7th part of the wall. Hence, we need to retain all transient wall harmonics up to those of order $j = 14$ in order to resolve this relatively thin current distribution. In the following, we shall keep all transient wall harmonics up to $j = 20$ (i.e. we shall neglect $\hat{f}_j$ variables with $j > j_{\rm max} = 20$ in our calculation), so as to ensure that all important transient wall harmonics are retained in the calculation. It follows that the final set of coupled, first-order, ordinary differential equations that makes up our model consists of 243 real equations.

Figure 10.5 shows the numerical solution of our set of differential equations. The solution is qualitatively similar to that obtained for a thin wall (see figure 10.2). As before, it can be seen that as the normalized width, x, of the island chain grows in time, the chain's normalized rotation frequency, y, is gradually reduced, until it has been reduced to about half of its original value, at which point there is a sudden collapse in the rotation frequency to a very low value. The rotation collapse occurs when $x = 0.69$, which corresponds to $W/a = 0.19$. It is clear from the right-hand panel of figure 10.5 that the rotation collapse takes place over a time interval of about 100 normalized time units, which corresponds to about 15 ms. This timescale is similar to the hybrid timescale $(\tau_\theta\,\tau_{\rm M})^{1/2} = 7$ ms. Hence, it is plausible that the timescale for the rotation collapse is determined by a combination of poloidal flow damping and perpendicular viscosity.

We can construct a torque balance diagnostic:

$$\mathcal{T}(T) = y - 1 + \xi_1\,x^4\left(\frac{\mathcal{I}}{\mathcal{R}^2 + \mathcal{I}^2}\right), \tag{10.109}$$

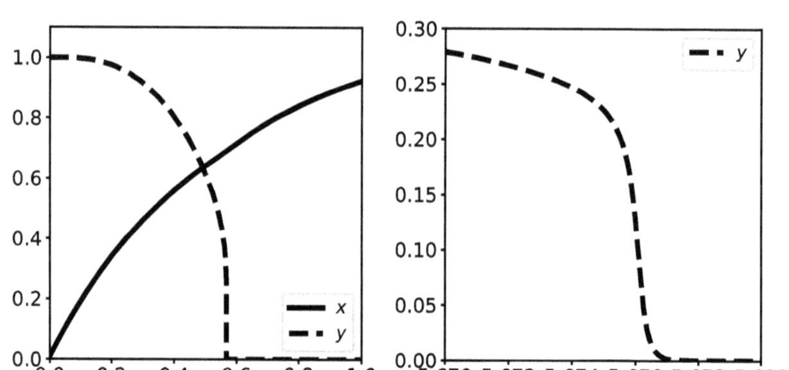

Figure 10.5. A simulation of island rotation braking in a high-field tokamak fusion reactor with a thick wall. Here, x and y are the normalized island width and the rotation frequency, respectively.

where

$$\xi_1 = \left[\Delta_{\rm nw}(0) - \Delta_{\rm pw}(0)\right]\left(\frac{\epsilon_{\rm s}}{q_{\rm s}}\right)^2\left(\frac{\tau_V}{\tau_{\rm H}^2\,|\omega_{\perp\,{\rm i}}|}\right)\left(\frac{w_{\rm pw}}{r_{\rm s}}\right)^4, \tag{10.110}$$

and τ_V is specified in equation (10.12). The quantity $\mathcal{T}$ takes the value zero when the plasma is in torque balance—in other words, when the electromagnetic braking torque exerted at the rational surface is exactly balanced by the viscous restoring torque. Obviously, plasma inertia plays no role in the rotation braking process when the plasma is in torque balance. On the other hand, if $\mathcal{T}$ is nonzero, then the electromagnetic braking torque is not balanced by the viscous restoring torque, which indicates that plasma inertia is playing a role in the braking process. Figure 10.6 displays the time evolution of the torque balance diagnostic in the rotation braking simulation shown in figure 10.5. It can be seen that the plasma is in torque balance to a very good approximation both before and after the rotation collapse. However, during the rotation collapse, the plasma is clearly not in torque balance, indicating that plasma inertia plays an important role in the rotation collapse. According to the right-hand panel of figure 10.6, after torque balance breaks down during the rotation collapse, it takes a time interval of the order of 4000 normalized time units, which corresponds to about 0.53 s, for torque balance to be reestablished. This timescale is similar to the momentum confinement timescale, $\tau_{\rm M} = 1.1$ s. Hence, it is plausible that torque balance is reestablished by plasma viscosity.

Figure 10.7 shows the time evolution of the quantities

$$\hat{V}_\theta(T) = \sum_{p=1,p_{\max}} \frac{\hat{\alpha}_{\rm p}}{S_\theta}, \tag{10.111}$$

$$\hat{V}_\varphi(T) = \sum_{p=1,p_{\max}} \frac{\hat{\beta}_{\rm p}}{S_\varphi}, \tag{10.112}$$

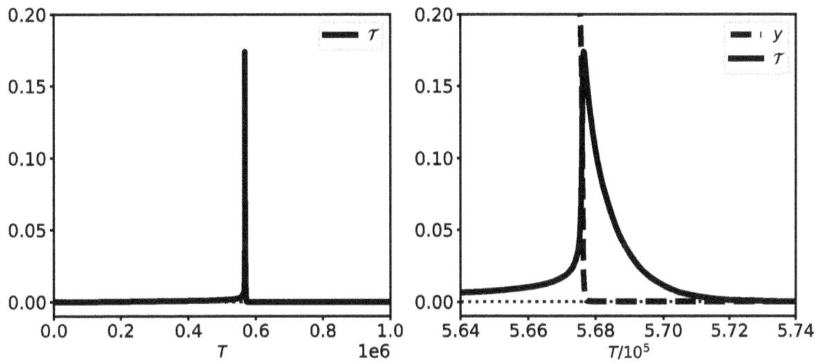

Figure 10.6. A simulation of island rotation braking in a high-field tokamak fusion reactor with a thick wall. Here, y and $\mathcal{T}$ are the normalized rotation frequency and a torque balance diagnostic, respectively.

in the rotation braking simulation shown in figure 10.5. The rotation braking process causes the normalized rotation frequency of the island chain, y, to decrease from unity (assuming that $\omega_{\perp\,\mathrm{i}} > 0$) to a value that is very much smaller than unity. In other words, $\Delta y \simeq -1$. Hence, it is clear from equations (10.106), (10.111), and (10.112) that $\Delta \hat{V}_\theta + \Delta \hat{V}_\varphi = 1$. Here, $\Delta \hat{V}_\theta$ is the fraction of the decrease in the island rotation frequency that is due to a shift in the poloidal ion fluid angular velocity at the rational surface, while $\Delta \hat{V}_\varphi$ is the fraction of the decrease that is due to a shift in the toroidal ion fluid angular velocity at the rational surface. In fact, it is apparent from figure 10.7 that about 53% of the decrease in the rotation frequency is due to a poloidal velocity shift, the remaining 47% being due to a toroidal velocity shift. It is also apparent from the figure's right-hand panel that the rotation collapse, which takes place on a timescale of about 100 normalized time units (i.e. $(\tau_\theta\, \tau_\mathrm{M})^{1/2}$), is due to a sudden shift in the *poloidal* angular velocity at the rational surface. In fact, this sudden shift is responsible for the loss of torque balance during the rotation collapse. The corresponding shift in the toroidal angular velocity at the rational surface takes place on a timescale of 4000 normalized time units (i.e. τ_M). Note that after the sudden shift that is associated with rotation collapse, the poloidal velocity subsequently readjusts to its final value on the τ_M timescale.

Figure 10.8 shows the time evolution of the transient wall harmonics in the rotation braking simulation shown in figure 10.5. It can be seen that the transient wall harmonics are only important (i.e. $|f_j| \sim O(1)$) during the rotation collapse. Note that $|f_{j_\mathrm{max}}| \ll |f_1|, |f_{j_\mathrm{crit}}|$ at all times, indicating that our calculation has included all of the important transient wall harmonics. Prior to the rotation collapse, $|f_{j_\mathrm{crit}}| \gg |f_1|$, indicating that the (very small) transient eddy current induced in the wall is localized to within a skin depth of the inner boundary of the wall. However, during the rotation collapse, the low-j transient wall harmonics become dominant, indicating that the transient eddy current has penetrated to the outer boundary of the wall. It can been seen from the right-hand panel of figure 10.8 that the longest-wavelength f_1 transient wall harmonic excited by the rotation collapse decays away

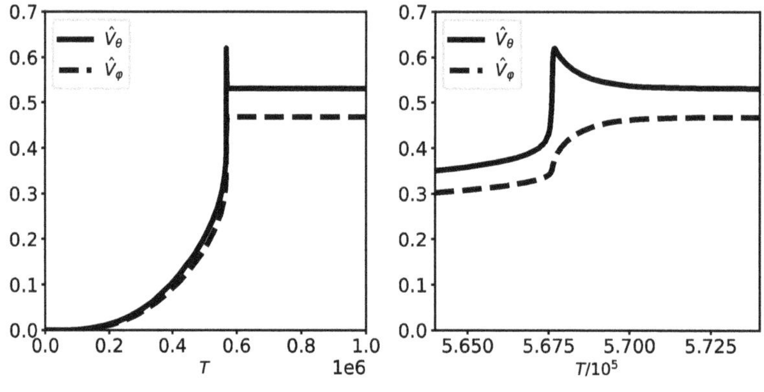

Figure 10.7. A simulation of island rotation braking in a high-field tokamak fusion reactor with a thick wall. Here, $\hat{V}_\theta$ and $\hat{V}_\varphi$ are measures of the poloidal and toroidal ion fluid velocities, respectively, at the rational surface.

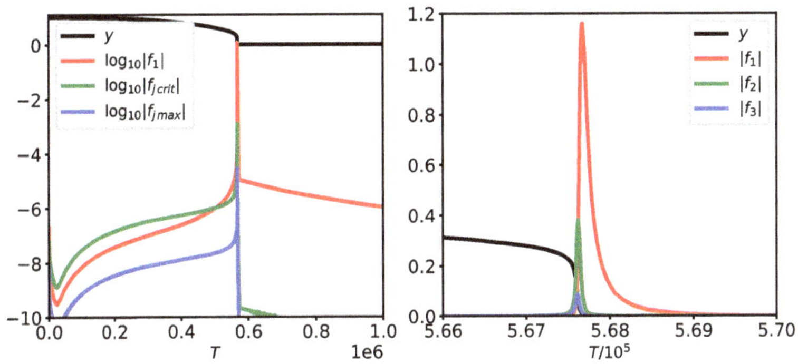

Figure 10.8. A simulation of island rotation braking in a high-field tokamak fusion reactor with a thick wall. Here, y and f_j are the normalized rotation frequency and wall transient harmonics, respectively.

after a time interval of about 1000 normalized time units, which corresponds to 0.13 s. This timescale is similar to the time constant of the wall, $\tau_w = 0.24$ s. Hence, it is plausible that the transient eddy current induced by the rotation collapse decays away after a time interval of the order of the wall time constant.

10.4 An improved torque balance model

It is clear from the calculation presented in the previous section that plasma inertia and the transient wall eddy current are only important during the rotation collapse. At all other times, it should be possible to accurately model the wall response using equation (10.65), which specifies the response of a wall of arbitrary thickness to a steadily rotating island chain, with $\gamma = -\mathrm{i}\,\omega$. This observation leads us to formulate the following generalization of the torque balance model presented in section 10.2:

$$\frac{dx}{dT} = 1 - x + \beta_1\left(\frac{\mathcal{R}}{\mathcal{R}^2 + \mathcal{I}^2}\right) + \gamma_1'\,x^2\left(\frac{\mathcal{I}}{\mathcal{R}^2 + \mathcal{I}^2}\right) + \delta_1'\,x^7\left(\frac{\mathcal{I}}{\mathcal{R}^2 + \mathcal{I}^2}\right)^2, \quad (10.113)$$

$$\zeta\,\frac{dy}{dT} = 1 - y - \xi_1\,x^4\left(\frac{\mathcal{I}}{\mathcal{R}^2 + \mathcal{I}^2}\right), \quad (10.114)$$

where

$$\mathcal{G} = 1 + \sqrt{\frac{-\mathrm{i}\,y\,\hat{\tau}_w}{\hat{\zeta}_w}}\,\tanh\sqrt{-\mathrm{i}\,y\,\hat{\tau}_w\,\hat{\zeta}_w}\,. \quad (10.115)$$

Here, x is the width of the magnetic island chain relative to its saturated width when the wall is perfectly conducting, y is the island rotation frequency relative to its value when there is no interaction with the wall, and T is time normalized to the typical time required for the island chain to attain its final saturated width. Moreover, β_1, γ_1', δ_1', ξ_1, $\mathcal{R}$, $\mathcal{I}$, $\hat{\tau}_w$, and $\hat{\zeta}_w$ are defined in equations (10.20), (10.102), (10.103), (10.110), (10.91), (10.92), (10.98), and (10.101), respectively. The improved torque balance model should be capable of accurately predicting the critical island width that must be exceeded in order to trigger a collapse in the island rotation frequency, even when the thin-wall approximation breaks down, despite not being able to accurately predict the timescale on which the rotation collapse occurs.

Figure 10.9 shows the critical island width required to trigger a rotation collapse calculated as a function of the wall thickness for various different unperturbed island rotation frequencies using the improved torque balance model. Separate calculations are performed for a low-field tokamak fusion reactor and a high-field tokamak fusion reactor (see chapter 1). The calculation parameters are determined using the following assumptions: $B = 5\,\mathrm{T}$ (low field) or $B = 12\,\mathrm{T}$ (high field), $\beta = 0.02$, $T_e = T_i = 7\,\mathrm{keV}$, $m_i = (m_D + m_T)/2$ (where m_D and m_T are the deuteron and triton masses, respectively), $\Xi_{\perp\,i} = \chi_{\perp\,e} = \chi_{\perp\,i} = 1\,\mathrm{m^2\,s^{-1}}$, $\tau = 1$, $\eta_e = \eta_i = 1$, and $dp/dr = -p/a$. The plasma equilibrium is assumed to be of the Wesson type (see section 9.4), with $q(0) = 0.8$ and $q(a) = 6.0$. The poloidal and toroidal mode numbers of the tearing mode are $m = 2$ and $n = 1$, respectively. It follows that $r_s = 0.560\,a$. The perfect-wall saturated island width is $W_{pw}/a = 0.272$. The wall radius and resistivity are assumed to be $r_w = 1.2\,a$ and $\eta_w = 6.9 \times 10^{-7}\,\Omega\,m$ (which is the electrical resistivity of stainless steel), respectively.

As is clear from figure 10.9, the critical island width required to trigger rotation collapse increases with increasing wall thickness (because the wall becomes less electrically resistive) until a critical thickness is reached above which the critical island width becomes independent of the wall thickness. Of course, the critical wall thickness is that above which the thin-wall approximation breaks down. The eddy current induced by the rotating island chain in a wall whose thickness is greater than the critical thickness (which corresponds to a skin depth in the wall material) only penetrates a skin depth into the wall from its inner boundary, which implies that the effective thickness of the wall becomes the skin depth, rather than its actual thickness. According to the figure, for a plasma with diamagnetic levels of ion fluid rotation (i.e. $\omega_{\perp\,i} \sim \omega_{*\,e}$), the critical island width is less than 10% of the plasma's

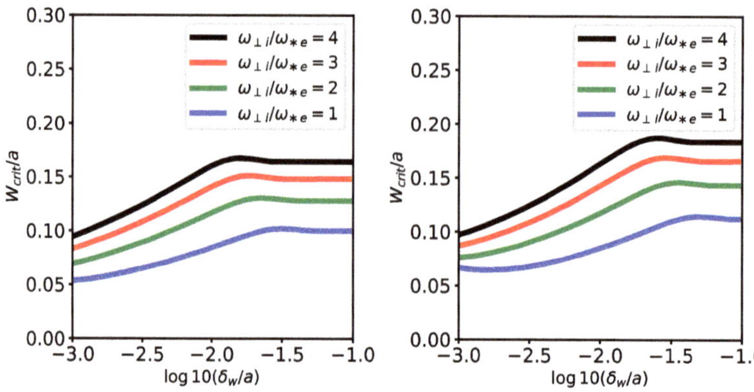

Figure 10.9. The critical island width required to trigger rotation collapse calculated as a function of the wall thickness for various unperturbed island rotation frequencies using the improved torque balance model. The left-hand/right-hand panels correspond to a low-field tokamak fusion reactor and a high-field tokamak fusion reactor, respectively.

minor radius for thin (i.e. $\delta_w/a \sim 10^{-3}$) resistive walls. On the other hand, the critical island width is about twice this value for thick (i.e. $\delta_w/a \sim 10^{-1}$) conducting walls. As before, it is apparent that a low-field tokamak fusion reactor is more susceptible to rotation braking than a high-field fusion reactor because of its lower diamagnetic frequency and consequent lower ion fluid rotation.

References

[1] Abramowitz M and Stegun I A (ed) 1964 *Handbook of Mathematical Functions with Formulas, Graphs, and Mathematical tables* (New York: Dover) https://store.doverpublications.com/0486612724.html

[2] Berge G, Sandal L K and Wesson J A 1989 Rotation and mode locking in tokamaks *Phys. Scr.* **40** 173

[3] Bondeson A and Persson M 1988 Stabilization by resistive walls and q-limit disruptions in tokamaks *Nucl. Fusion* **28** 1887

[4] Chapman B E, Fitzpatrick R, Craig D, Martin P and Spizzo G 2004 Observation of tearing mode deceleration and locking due to eddy currents induced in a conducting shell *Phys. Plasmas* **11** 2156

[5] de Vries P C, Johnson M F, Alper B, Buratti P, Hender T C, Koslowski H R, Riccardo V and the JET-EFDA Contributors 2011 Survey of disruption causes at JET *Nucl. Fusion* **51** 053018

[6] Fitzpatrick R 1993 Interaction of tearing modes with external structures in cylindrical geometry *Nucl. Fusion* **33** 1049

[7] Fitzpatrick R and Waelbroeck F L 2005 Two-fluid magnetic island dynamics in slab geometry. II. Islands interacting with resistive walls or resonant magnetic perturbations *Phys. Plasmas* **12** 022308

[8] Gates D A and Hender T C 1996 Resistive wall induced forbidden bands of mode rotation frequency on the COMPASS-D tokamak *Nucl. Fusion* **36** 273

[9] Hender T C, Gimblett C G and Robinson D C 1989 Effects of a resistive wall on magnetohydrodynamic instabilities *Nucl. Fusion* **29** 1279

[10] Jensen T H and Chu M S 1983 A linear model for the tearing mode of a tokamak plasma with flow and a resistive wall boundary condition *J. Plasma Phys.* **30** 57

[11] Nave M F F and Wesson J A 1990 Mode locking in tokamaks *Nucl. Fusion* **30** 2575

[12] Persson M and Bondeson A 1989 Wall locking and density limit disruptions *Nucl. Fusion* **29** 989

[13] Waelbroeck F L and Fitzpatrick R 1997 Rotation and locking of magnetic islands *Phys. Rev. Lett.* **78** 1703

[14] Zohm H, Kallenbach A, Bruhns H, Fussmann G and Klüber O 1990 Plasma angular-momentum loss by MHD mode locking *Europhys. Lett.* **11** 745

Chapter 11

The nonlinear neoclassical resonant response model

11.1 Introduction

In chapters 4 and 8, we derived a reduced model that describes the nonlinear response of a tokamak plasma to a tearing perturbation in the *inner region*, which is situated in the immediate vicinity of the so-called *rational* magnetic flux surface at which the perturbation resonates with the equilibrium magnetic field (see chapter 3). Our derivation employs the simplifying approximation of ignoring the specifically neoclassical terms (e.g. the terms associated with ion poloidal flow damping [22] and the bootstrap current [1]) in the neoclassical fluid equations, (2.370)–(2.374). Our justification for this approximation is that the terms in question have their origin in friction between trapped and passing particles. However, trapped ions make radial excursions from magnetic flux surfaces that are of the order of the ion banana width (see section 2.7). This width is a few centimeters in a tokamak fusion reactor (see table 2.4). Hence, if the radial width of the inner region is much less than the ion banana width, then we would not expect ion neoclassical effects to contribute to the plasma response in the inner region (because the trapped ions would average over the spatial structure of the inner region). The electron banana width is of the order of a few tenths of a centimeter in a tokamak fusion reactor (see table 2.4). Thus, we would not expect electron neoclassical effects to contribute to the plasma response in the radially thin resistive layers characteristic of the linear response regime (see chapter 6). On the other hand, in the nonlinear response regime, a tearing mode generates a magnetic island chain at the rational surface whose radial width (which constitutes the effective width of the inner region) can easily exceed the ion banana width (see section 5.16). Under these circumstances, there is no justification for neglecting the specifically neoclassical terms in the neoclassical fluid equations. The aim of this chapter is to repeat the analyses given in chapters 4 and 8 while including

doi:10.1088/978-0-7503-5367-0ch11

the neoclassical terms in equations (2.370)–(2.374), in order to produce a reduced model that is suitable for analyzing the dynamics of wide magnetic island chains.

11.2 The neoclassical drift-magnetohydrodynamic equations

As before, we adopt definitions (4.1)–(4.6) for the constants n_0, p_0, η_e, η_i, τ, and τ' that parameterize the equilibrium plasma density, pressure, and temperature and their gradients at the rational surface. We also make the simplifying assumption that perturbed electron and ion temperature profiles in the inner region are functions of the perturbed electron number density profile. Our neoclassical fluid equations, (2.370)–(2.374), reduce to the following set of *neoclassical drift-magnetohydrodynamic (MHD) fluid equations*:

$$\begin{aligned} &m_i\, n_0\left[\frac{\partial \mathbf{V}}{\partial t} + (\mathbf{V}\cdot\nabla)\mathbf{V} + \frac{1}{1+\tau}(\mathbf{V}_*\cdot\nabla)\mathbf{V}_E\right] + \nabla p - \mathbf{j}\times\mathbf{B} \\ &-\frac{m_i\, n_0}{\tau_\theta}\left(\mathbf{V}_i - \alpha_\theta\,\frac{1}{1+\tau}\,\frac{\eta_i}{1+\eta_i}\,\mathbf{V}_*\right)\cdot\mathbf{e}_\theta\,\mathbf{e}_\theta \\ &-m_i\, n_0\, \Xi_\perp \nabla\cdot\left(\nabla\mathbf{V}_i + \nabla\mathbf{V}_i^\dagger - \frac{2}{3}\nabla\cdot\mathbf{V}_i\,\mathbf{I}\right) = 0, \end{aligned} \tag{11.1}$$

$$\begin{aligned} &\mathbf{E} + \mathbf{V}\times\mathbf{B} + \frac{1}{e\, n_0}\left[\nabla p - \frac{1}{1+\tau}(\mathbf{b}\cdot\nabla p) - \mathbf{j}\times\mathbf{B}\right] \\ &= \eta_\perp\left(\mathbf{j}_\perp - \frac{\mathbf{b}\times\nabla p}{B_z}\right) + \eta_\parallel\,(j_\parallel - j_{bs})\,\mathbf{b}, \end{aligned} \tag{11.2}$$

$$\begin{aligned} &\frac{3}{2}\frac{\partial p}{\partial t} + \frac{3}{2}\mathbf{V}\cdot\nabla p + \frac{5}{2}p_0\,\nabla\cdot\mathbf{V} + \frac{5}{2}\lambda\, p_0\nabla\cdot\mathbf{V}_* \\ &-\frac{5}{2}\frac{\tau'}{1+\tau'}\frac{p_0}{e\, n_0}\nabla\cdot\left[\left(j_{bs} + \frac{E_\parallel}{\eta_\parallel}\right)\mathbf{b}\right] \\ &-\frac{\tau'}{1+\tau'}\frac{p_0}{e\, n_0}\nabla\cdot\left[\left(j_{nc\,e} + \frac{E_\parallel}{\eta_{\parallel\,e}}\right)\mathbf{b}\right] - \frac{5}{2}\frac{1}{1+\tau'}\frac{p_0}{e\, n_0}\nabla\cdot(j_{nc\,i}\,\mathbf{b}) \\ &-\chi_\parallel\,\mathbf{b}\cdot\nabla(\mathbf{b}\cdot\nabla p) - \chi_\perp\,\nabla^2 p = 0. \end{aligned} \tag{11.3}$$

Here, the E-cross-B, diamagnetic, MHD, and ion fluid velocities, $\mathbf{V}_E$, $\mathbf{V}_*$, $\mathbf{V}$, and $\mathbf{V}_i$, respectively, are defined in equations (4.10)–(4.13). Moreover, the dimensionless parameter λ, the ion perpendicular momentum diffusivity $\Xi_\perp$, the parallel energy diffusivity $\chi_\parallel$, and the perpendicular energy diffusivity $\chi_\perp$, are defined in equations (4.16)–(4.19). Finally,

$$\tau_\theta = \tau_{\theta\,i}(r_s), \tag{11.4}$$

$$\alpha_\theta = \alpha_1\,(1 - \alpha_2\,f_{\rm t\,s}), \tag{11.5}$$

$$f_{\rm t\,s} = f_{\rm t}(r_{\rm s}), \tag{11.6}$$

$$j_{\rm bs} = -\alpha_{\rm bs}\,\frac{1}{B_z}\,\frac{\partial p}{\partial r} \tag{11.7}$$

$$j_{\rm nc\,e} = -\alpha_{\rm nc\,e}\,\frac{1}{B_z}\,\frac{\partial p}{\partial r}, \tag{11.8}$$

$$j_{\rm nc\,i} = -\alpha_{\rm nc\,i}\,\frac{1}{B_z}\,\frac{\partial p}{\partial r}, \tag{11.9}$$

$$\alpha_{\rm bs} = f_{\rm t\,s}\,\frac{q_{\rm s}}{\epsilon_{\rm s}}\left[\beta_{11}\left(1 - \alpha_1\,\frac{1}{1+\tau}\,\frac{\eta_{\rm i}}{1+\eta_{\rm i}}\right) - \beta_{12}\,\frac{\tau}{1+\tau}\frac{\eta_{\rm e}}{1+\eta_{\rm e}}\right], \tag{11.10}$$

$$\alpha_{\rm nc\,e} = f_{\rm t\,s}\,\frac{q_{\rm s}}{\epsilon_{\rm s}}\left[\epsilon_1\left(1 - \alpha_1\,\frac{1}{1+\tau}\,\frac{\eta_{\rm i}}{1+\eta_{\rm i}}\right) - \epsilon_2\,\frac{\tau}{1+\tau}\frac{\eta_{\rm e}}{1+\eta_{\rm e}}\right], \tag{11.11}$$

$$\alpha_{\rm nc\,i} = f_{\rm t\,s}\,\frac{q_{\rm s}}{\epsilon_{\rm s}}\,(-\alpha_2), \tag{11.12}$$

$$\eta_\perp = \mu_{\rm e\,11}\,f_{\rm t\,s}\left(\frac{q_{\rm s}}{\epsilon_{\rm s}}\right)^2\frac{1}{\sigma_\perp(r_{\rm s})}, \tag{11.13}$$

$$\eta_\parallel = \frac{1}{\gamma_{11} - \delta_{11}\,f_{\rm t\,s}}\,\frac{1}{\sigma_\perp(r_{\rm s})}, \tag{11.14}$$

$$\eta_{\parallel\,\rm e} = \frac{1}{\epsilon_3 - \epsilon_4\,f_{\rm t\,s}}\,\frac{1}{\sigma_\perp(r_{\rm s})}. \tag{11.15}$$

Here, r is a radial cylindrical coordinate, $\mathbf{e}_\theta$ is a unit vector in the poloidal direction (see section 3.2), $r_{\rm s}$ is the minor radius of the rational surface, $q_{\rm s} = q(r_{\rm s})$ is the safety factor at the rational surface (see equation (3.2)), $\epsilon_{\rm s} = \epsilon(r_{\rm s})$ is the inverse aspect ratio at the rational surface (see equation (3.18)), the ion neoclassical poloidal flow-damping time $\tau_{\theta\,\rm i}(r)$ is defined in equation (2.332), the fraction of trapped particles $f_{\rm t}(r)$ is defined in equation (2.202), the dimensionless neoclassical parameters $\mu_{\rm e\,11}$, α_1, α_2, β_{11}, β_{12}, γ_{11}, δ_{11}, ϵ_3, and ϵ_4 are defined in equations (2.209), (2.217), (2.218), (2.243), (2.244), (2.247), (2.251), (2.347), and (2.348), respectively, and the plasma perpendicular electrical conductivity $\sigma_\perp(r)$ is defined in equation (2.41). As before, e is the magnitude of the electron charge, $m_{\rm i}$ the ion mass, n_0 the equilibrium electron number density at the rational surface, p_0 the equilibrium total plasma pressure at the rational surface, $\mathbf{E}$ the electric field strength, $\mathbf{B}$ the magnetic field strength, $\mathbf{j}$ the current density, p the total plasma pressure, and $\mathbf{b} = \mathbf{B}/B_z$. Finally, $E_\parallel = \mathbf{E} \cdot \mathbf{b}$, $j_\parallel = \mathbf{j} \cdot \mathbf{b}$, and $\mathbf{j}_\perp = \mathbf{j} - j_\parallel\, \mathbf{b}$.

If we compare equation (11.1) with its non-neoclassical concomitant, (4.7), then we can see that the former equation contains an additional term that is due to the action of ion neoclassical poloidal flow damping. The term in question involves τ_θ, which is the ion neoclassical poloidal flow-damping time at the rational surface. If we compare equation (11.2) with its non-neoclassical counterpart, (4.8), then we can see that the former equation contains an additional term that is due to the action of the bootstrap current. The term in question involves $j_{\rm bs}$, which is the parallel bootstrap current density at the rational surface. Finally, if we compare equation (11.3) with its non-neoclassical counterpart, (4.9), then we can see that the former equation contains many additional terms that are due to neoclassical parallel momentum and heats flows.

11.3 The reduced neoclasssical drift-MHD model

The Alfvén speed, V_A, and the collisionless ion skin depth, $d_{\rm i}$, are defined in equations (4.23) and (4.24), respectively. Let l be a typical variational length scale in the inner region. It is convenient to adopt the following normalization scheme that renders all quantities in the neoclassical drift-MHD fluid equations dimensionless: $\hat{\nabla} = l\,\nabla$, $\hat{t} = t/(l/V_{\rm A})$, $\hat{d}_{\rm i} = d_{\rm i}/l$, $\hat{\mathbf{E}} = \mathbf{E}/(B_z\, V_{\rm A})$, $\hat{\mathbf{j}} = \mathbf{j}/(B_z/\mu_0\, l)$, $\hat{j}_{\rm bs} = j_{\rm bs}/(B_z/\mu_0\, l)$, $\hat{j}_{\rm nc\,e,i} = j_{\rm nc\,e,i}/(B_z/\mu_0\, l)$, $\hat{p} = p/(B_z^{\,2}/\mu_0)$, $\hat{p}_0 = p_0/(B_z^{\,2}/\mu_0)$, $\hat{\mathbf{V}} = \mathbf{V}/V_{\rm A}$, $\hat{\mathbf{V}}_{*,{\rm i,E}} = \mathbf{V}_{*,{\rm i,E}}/V_{\rm A}$, $\hat{V}_\parallel = V_{\parallel\,{\rm i}}/V_{\rm A}$, $\hat{x} = (r - r_{\rm s})/l$, $\hat{\Xi}_\theta = l/(V_{\rm A}\,\tau_\theta)$, $\hat{\Xi}_\perp = \Xi_\perp/(l\,V_{\rm A})$, $\hat{\eta}_{\parallel,\perp,\parallel\,{\rm e}} = \eta_{\parallel,\perp,\parallel\,{\rm e}}/(\mu_0\,l\,V_{\rm A})$, $\hat{\chi}_{\parallel,\perp} = \chi_{\parallel,\perp}/(l\,V_{\rm A})$. Here, $V_{\parallel\,{\rm i}}$ is the ion parallel fluid velocity (see equation (2.321)). Equations (11.1)–(11.3) yield the following set of normalized neoclassical drift-MHD fluid equations:

$$
\begin{aligned}
&\frac{\partial \hat{\mathbf{V}}}{\partial \hat{t}} + \hat{\nabla}\left(\frac{\hat{V}^2}{2}\right) - \hat{\mathbf{V}} \times (\hat{\nabla} \times \hat{\mathbf{V}}) \\
&\quad + \frac{1}{1+\tau}\,\frac{1}{2}[\hat{\nabla}(\hat{\mathbf{V}}_{\rm E} \cdot \hat{\mathbf{V}}_*) - \hat{\mathbf{V}}_{\rm E} \times (\hat{\nabla} \times \hat{\mathbf{V}}_*) - \hat{\mathbf{V}}_* \times (\hat{\nabla} \times \hat{\mathbf{V}}_{\rm E}) \\
&\quad + (\hat{\nabla} \cdot \hat{\mathbf{V}}_{\rm E})\,\hat{\mathbf{V}}_* - (\hat{\nabla} \cdot \hat{\mathbf{V}}_*)\,\hat{\mathbf{V}}_{\rm E} + \hat{\nabla} \times (\hat{\mathbf{V}}_{\rm E} \times \hat{\mathbf{V}}_*)] + \hat{\nabla}\hat{p} - \hat{\mathbf{j}} \times \mathbf{b} \\
&\quad + \hat{\Xi}_\theta\left(\hat{\mathbf{V}}_{\rm i} - \alpha_\theta\,\frac{1}{1+\tau}\,\frac{\eta_{\rm i}}{1+\eta_{\rm i}}\,\hat{\mathbf{V}}_*\right) \cdot \mathbf{e}_\theta\,\mathbf{e}_\theta - \hat{\Xi}_\perp \hat{\nabla} \cdot \left(\hat{\nabla}\hat{\mathbf{V}}_{\rm i} + \hat{\nabla}\hat{\mathbf{V}}_{\rm i}^\dagger - \frac{2}{3}\,\hat{\nabla} \cdot \hat{\mathbf{V}}_{\rm i}\,\mathbf{I}\right) = 0,
\end{aligned}
\tag{11.16}
$$

$$
\begin{aligned}
&\hat{\mathbf{E}} + \hat{\mathbf{V}} \times \mathbf{b} + \hat{d}_{\rm i}\left[\hat{\nabla}\hat{p} - \frac{1}{1+\tau}\,(\mathbf{b} \cdot \hat{\nabla}\hat{p})\,\mathbf{b} - \hat{\mathbf{j}} \times \mathbf{b}\right] \\
&\quad = \hat{\eta}_\perp\,(\hat{\mathbf{j}}_\perp - \mathbf{b} \times \hat{\nabla}\hat{p}) + \hat{\eta}_\parallel\,(\hat{j}_\parallel - \hat{j}_{\rm bs})\,\mathbf{b},
\end{aligned}
\tag{11.17}
$$

$$\frac{3}{2}\frac{\partial \hat{p}}{\partial \hat{t}} + \frac{3}{2}\hat{\mathbf{V}}\cdot\hat{\nabla}\hat{p} + \frac{5}{2}\hat{p}_0\,\hat{\nabla}\cdot\hat{\mathbf{V}} + \frac{5}{2}\lambda\,\hat{p}_0\hat{\nabla}\cdot\hat{\mathbf{V}}_*$$
$$-\frac{5}{2}\frac{\tau'}{1+\tau'}\hat{d}_{\rm i}\,\hat{p}_0\,\hat{\nabla}\cdot\left[\left(\hat{j}_{\rm bs}+\frac{\hat{E}_\parallel}{\hat{\eta}_\parallel}\right)\mathbf{b}\right]$$
$$-\frac{\tau'}{1+\tau'}\hat{d}_{\rm i}\,\hat{p}_0\,\hat{\nabla}\cdot\left[\left(\hat{j}_{\rm nc\,e}+\frac{\hat{E}_\parallel}{\hat{\eta}_{\parallel\,\rm e}}\right)\mathbf{b}\right]-\frac{5}{2}\frac{\tau'}{1+\tau'}\hat{d}_{\rm i}\,\hat{p}_0\hat{\nabla}\cdot(\hat{j}_{\rm nc\,i}\,\mathbf{b}) \tag{11.18}$$
$$-\hat{\chi}_\parallel\,\mathbf{b}\cdot\hat{\nabla}(\mathbf{b}\cdot\hat{\nabla}\hat{p})-\hat{\chi}_\perp\,\hat{\nabla}^2\hat{p}=0,$$

where

$$\hat{j}_{\rm bs} = -\alpha_{\rm bs}\,\frac{\partial \hat{p}}{\partial \hat{x}}, \tag{11.19}$$

$$\hat{j}_{\rm nc\,e,i} = -\alpha_{\rm nc\,e,i}\,\frac{\partial \hat{p}}{\partial \hat{x}}, \tag{11.20}$$

$$\hat{\mathbf{V}}_{\rm E} = \hat{\mathbf{E}}\times\mathbf{b}, \tag{11.21}$$

$$\hat{\mathbf{V}}_* = \hat{d}_{\rm i}\,\mathbf{b}\times\hat{\nabla}\hat{p}, \tag{11.22}$$

$$\hat{\mathbf{V}} = \hat{\mathbf{V}}_{\rm E} + \hat{V}_\parallel\,\mathbf{b}, \tag{11.23}$$

$$\hat{\mathbf{V}}_{\rm i} = \hat{\mathbf{V}} + \frac{1}{1+\tau}\,\hat{\mathbf{V}}_*. \tag{11.24}$$

Finally, Maxwell's equations yield

$$\hat{\nabla}\cdot\mathbf{b} = 0, \tag{11.25}$$

$$\hat{\nabla}\times\hat{\mathbf{E}} = -\frac{\partial \mathbf{b}}{\partial \hat{t}}, \tag{11.26}$$

$$\hat{\mathbf{j}} = \hat{\nabla}\times\mathbf{b}. \tag{11.27}$$

As before (see section 4.4), all quantities in the inner region are assumed to be functions of $\hat{x}$, $\zeta = m\,\theta - n\,\varphi$, and $\hat{t}$ only. Here, θ and φ are the poloidal and toroidal angles, respectively, while m and n are the poloidal and toroidal mode numbers, respectively, of the tearing mode (see sections 3.2 and 3.3). We can write

$$\hat{p} = \hat{p}_0 + \delta p^{(1)}, \tag{11.28}$$

$$\mathbf{b} = (1+\delta b^{(1)})\mathbf{n} + \hat{\nabla}\times(\psi^{(1)}\,\mathbf{n}), \tag{11.29}$$

$$\hat{\mathbf{E}} = \hat{\nabla}\phi^{(1)} + \left(\hat{E}_\parallel^{(2)} - \frac{\partial^{(1)}\psi^{(1)}}{\partial \hat{t}}\right)\mathbf{n} + \hat{\nabla}\times\left(\frac{\partial^{(1)}\chi^{(1)}}{\partial \hat{t}}\,\mathbf{n}\right), \tag{11.30}$$

$$\hat{\mathbf{V}}_E = \hat{\nabla}\phi^{(1)} \times \mathbf{b} - \hat{\nabla}\left(\frac{\partial^{(1)}\chi^{(1)}}{\partial \hat{t}}\right), \tag{11.31}$$

$$\hat{\mathbf{V}}_* = \hat{d}_i\, \mathbf{b} \times \hat{\nabla}\delta p^{(1)}, \tag{11.32}$$

$$\hat{\mathbf{V}} = \hat{\mathbf{V}}_E + \hat{V}_\parallel^{(1)}\, \mathbf{b} + \hat{\nabla}\Upsilon^{(2)}, \tag{11.33}$$

$$\hat{\mathbf{V}}_i = \hat{\mathbf{V}} + \frac{1}{1+\tau}\, \hat{\mathbf{V}}_*, \tag{11.34}$$

where $\mathbf{n}$ is defined in equation (4.37), $\hat{E}_\parallel^{(2)}$ is the (normalized) constant inductive component of the parallel electric field that maintains the equilibrium parallel current density in the inner region against ohmic decay, and $\hat{\nabla}^2\chi^{(1)} = \delta b^{(1)}$. The superscript (1) indicates a quantity that is first order in our ordering scheme. Zeroth-order terms are left without superscripts, while second-order terms are given the superscript (2).

Evaluating the normalized neoclassical drift-MHD fluid equations, (11.16)–(11.18), up to the second order, we obtain

$$\begin{aligned}
&\hat{\nabla}(\delta p^{(1)} + \delta b^{(1)}) + \left(\frac{\partial^{(1)}\hat{V}_\parallel^{(1)}}{\partial \hat{t}} + \left[\hat{V}_\parallel^{(1)}, \phi^{(1)}\right] + [\psi^{(1)}, \delta b^{(1)}]\right)\mathbf{n} \\
&+ \hat{\nabla}\left(\frac{\partial^{(1)}\phi^{(1)}}{\partial \hat{t}} - \frac{\hat{d}_i[\phi^{(1)}, \delta p^{(1)}]}{2\,(1+\tau)}\right) \times \mathbf{n} \\
&+ \frac{1}{2}\,\hat{\nabla}\left(\hat{\mathbf{V}}_\perp^{(1)} \cdot \hat{\mathbf{V}}_\perp^{(1)} + \frac{\hat{\mathbf{V}}_E^{(1)} \cdot \hat{\mathbf{V}}_*^{(1)}}{1+\tau} + \delta b^{(1)}\,\delta b^{(1)} + \frac{2\,\epsilon^{(1)}\,\epsilon^{(1)}}{q_s^2}\right) \\
&- \hat{\nabla}^2\phi^{(1)}\,\hat{\nabla}\phi^{(1)} + \frac{\hat{d}_i}{2\,(1+\tau)}\left(\hat{\nabla}^2\phi^{(1)}\,\hat{\nabla}\delta p^{(1)} + \hat{\nabla}^2\delta p^{(1)}\,\hat{\nabla}\phi^{(1)}\right) + J^{(1)}\,\hat{\nabla}\psi^{(1)} \\
&+ \hat{\Xi}_\theta^{(1)}\left(\frac{\epsilon^{(1)}}{q_s}\,\hat{V}_\parallel^{(1)} - \frac{\partial}{\partial \hat{x}}\left[\phi^{(1)} - \frac{\hat{d}_i}{1+\tau}\left(1 - \alpha_\theta\,\frac{\eta_i}{1+\eta_i}\right)\delta p^{(1)}\right]\right)\mathbf{e}_\theta - \Xi_\perp^{(1)}\,\hat{\nabla}^2\mathbf{V}_i^{(1)} = 0,
\end{aligned} \tag{11.35}$$

$$\begin{aligned}
&\hat{d}_i\,\hat{\nabla}(\delta p^{(1)} + \delta b^{(1)}) + \left(\hat{E}_\parallel^{(2)} - \frac{\partial^{(1)}\psi^{(1)}}{\partial \hat{t}} - [\psi^{(1)}, \phi^{(1)}] + \frac{\hat{d}_i}{1+\tau}[\psi^{(1)}, \delta p^{(1)}]\right. \\
&\left. + \hat{d}_i[\psi^{(1)}, \delta b^{(1)}] + \hat{\eta}_\parallel^{(1)}\,J^{(1)} - \hat{\eta}_\parallel^{(1)}\,\alpha_{bs}\,\frac{\partial \delta p^{(1)}}{\partial \hat{x}}\right)\mathbf{n} \\
&+ \hat{\nabla}\left[\Upsilon^{(2)} - \hat{\eta}_\perp^{(1)}\,(\delta b^{(1)} - \delta p^{(1)})\right] \times \mathbf{n} + \frac{\hat{d}_i}{2}\,\hat{\nabla}\left(\delta b^{(1)}\,\delta b^{(1)} + \frac{2\,\epsilon^{(1)}\,\epsilon^{(1)}}{q_s^2}\right) \\
&+ \hat{d}_i\,J^{(1)}\,\hat{\nabla}\psi^{(1)} - 2\,\delta b^{(1)}\,\hat{\nabla}\phi^{(1)} = 0,
\end{aligned} \tag{11.36}$$

$$\begin{aligned}
&\frac{3}{2}\,\frac{\partial^{(1)}\delta p^{(1)}}{\partial \hat{t}} + \frac{3}{2}\,[\delta p^{(1)},\,\phi^{(1)}] \\
&\quad + \frac{5}{2}\,\hat{p}_0\left(-\frac{\partial^{(1)}\delta b^{(1)}}{\partial \hat{t}} + [\delta b^{(1)},\,\delta\phi^{(1)}] + \left[\hat{V}_{\parallel}^{(1)},\,\psi^{(1)}\right] + \hat{\nabla}^2\Upsilon^{(2)}\right) \\
&\quad + \frac{5}{2}\,\lambda\,\hat{p}_0\,\hat{d}_{\rm i}\,[\delta p^{(1)},\,\delta b^{(1)}] + \frac{5}{2}\,\hat{d}_{\rm i}\,\hat{p}_0\,\alpha_{\rm nc}\left[\frac{\partial\delta p^{(1)}}{\partial\hat{x}},\,\psi^{(1)}\right] \\
&\quad + \frac{5}{2}\,\frac{\hat{d}_{\rm i}\,\hat{p}_0\,\alpha_{\parallel}}{\hat{\eta}_{\parallel}^{(1)}}\left[\frac{\partial^{(1)}\psi^{(1)}}{\partial\hat{t}},\,\psi^{(1)}\right] \\
&\quad - \hat{\chi}_{\parallel}^{(-1)}[[\delta p^{(1)},\,\psi^{(1)}],\,\psi^{(1)}] - \hat{\chi}_{\perp}^{(1)}\,\hat{\nabla}^2\delta p^{(1)} = 0,
\end{aligned} \tag{11.37}$$

where $J^{(1)}$ is defined in equation (2.332), $[A,\,B] \equiv \hat{\nabla}A \times \hat{\nabla}B\cdot\mathbf{n}$, and

$$\alpha_{\rm nc} = \frac{\tau'}{1+\tau'}\,\alpha_{\rm bs} + \frac{2}{5}\,\frac{\tau'}{1+\tau'}\,\alpha_{\rm nc\,e} + \frac{1}{1+\tau'}\,\alpha_{\rm nc\,i}, \tag{11.38}$$

$$\alpha_{\parallel} = \frac{\tau'}{1+\tau'}\left(1 + \frac{2}{5}\,\frac{\hat{\eta}_{\parallel}}{\hat{\eta}_{\parallel\,\rm e}}\right). \tag{11.39}$$

To the first order, equations (11.35) and (11.36) again give the equilibrium force balance constraint

$$\delta b^{(1)} = -\delta p^{(1)}. \tag{11.40}$$

(see section 4.4). The scalar product of equation (11.35) with $\mathbf{n}$ yields

$$\begin{aligned}
\frac{\partial^{(1)}\hat{V}_{\parallel}^{(1)}}{\partial\hat{t}} &= \left[\phi^{(1)},\,\hat{V}_{\parallel}^{(1)}\right] - [\delta p^{(1)},\,\psi^{(1)}] \\
&\quad - \hat{\Xi}_{\theta}^{(1)}\,\frac{\epsilon^{(1)}}{q_{\rm s}}\left(\frac{\epsilon^{(1)}}{q_{\rm s}}\,\hat{V}_{\parallel}^{(1)} - \frac{\partial}{\partial\hat{x}}\left[\phi^{(1)} - \frac{\hat{d}_{\rm i}}{1+\tau}\left(1 - \alpha_{\theta}\,\frac{\eta_{\rm i}}{1+\eta_{\rm i}}\right)\delta p^{(1)}\right]\right) \\
&\quad + \hat{\Xi}_{\perp}^{(1)}\,\hat{\nabla}^2\,\hat{V}_{\parallel}^{(1)}.
\end{aligned} \tag{11.41}$$

The scalar product of equation (11.36) with $\mathbf{n}$ gives

$$\begin{aligned}
\frac{\partial^{(1)}\psi^{(1)}}{\partial\hat{t}} &= [\phi^{(1)},\,\psi^{(1)}] + \hat{d}_{\rm i}\,\frac{\tau}{1+\tau}[\delta p^{(1)},\,\psi^{(1)}] \\
&\quad + \hat{\eta}_{\parallel}^{(1)}\left(J^{(1)} - \alpha_{\rm bs}\,\frac{\delta p^{(1)}}{\partial\hat{x}}\right) + \hat{E}_{\parallel}^{(2)}.
\end{aligned} \tag{11.42}$$

The scalar product of the curl of equation (11.35) with $\mathbf{n}$ yields

$$\begin{aligned}\frac{\partial^{(1)}\hat{\nabla}^2\phi^{(1)}}{\partial\hat{t}} &= \left[\phi^{(1)},\,\hat{\nabla}^2\phi^{(1)}\right]\\
&+\frac{\hat{d}_i}{2\,(1+\tau)}\left(\hat{\nabla}^2[\phi^{(1)},\,\delta p^{(1)}]+\left[\hat{\nabla}^2\phi^{(1)},\,\delta p^{(1)}\right]+\left[\hat{\nabla}^2\delta p^{(1)},\,\phi^{(1)}\right]\right)\\
&+[J^{(1)},\,\psi^{(1)}]+\hat{\Xi}_\theta^{(1)}\,\frac{\partial}{\partial\hat{x}}\left(\frac{\epsilon^{(1)}}{q_s}\,\hat{V}_\parallel^{(1)}-\frac{\partial}{\partial\hat{x}}\left[\phi^{(1)}-\frac{\hat{d}_i}{1+\tau}\left(1-\alpha_\theta\,\frac{\eta_i}{1+\eta_i}\right)\delta p^{(1)}\right]\right)\\
&+\hat{\Xi}_\perp^{(1)}\,\hat{\nabla}^4\left(\phi^{(1)}-\frac{\hat{d}_i}{1+\tau}\,\delta p^{(1)}\right).\end{aligned}\tag{11.43}$$

Finally, the scalar product of the curl of equation (11.36) with $\mathbf{n}$ gives

$$\hat{\nabla}^2\Upsilon^{(2)}=2[\delta p^{(1)},\,\phi^{(1)}]+\hat{d}_i[J^{(1)},\,\psi^{(1)}].\tag{11.44}$$

The previous equation can be combined with equations (11.37) and (11.40) to produce

$$\begin{aligned}\frac{\partial^{(1)}\delta p^{(1)}}{\partial\hat{t}} &= [\phi^{(1)},\,\delta p^{(1)}]-c_\beta^2\left[\hat{V}_\parallel^{(1)},\,\psi^{(1)}\right]\\
&-c_\beta^2\,\hat{d}_i[J^{(1)},\,\psi^{(1)}]-c_\beta^2\,\hat{d}_i\left[\alpha_{nc}\,\frac{\partial\delta p^{(1)}}{\partial\hat{x}},\,\psi^{(1)}\right]\\
&-c_\beta^2\,\hat{d}_i\left[\frac{\alpha_\parallel}{\hat{\eta}_\parallel^{(1)}}\,\frac{\partial^{(1)}\psi^{(1)}}{\partial\hat{t}},\,\psi^{(1)}\right]+\frac{2}{3}\,(1-c_\beta^2)\,\hat{\chi}_\parallel^{(-1)}[[\delta p^{(1)},\,\psi^{(1)}],\,\psi^{(1)}]\\
&+\frac{2}{3}\,(1-c_\beta^2)\,\hat{\chi}_\perp^{(1)}\,\hat{\nabla}^2\delta p^{(1)},\end{aligned}\tag{11.45}$$

where c_β is defined in equations (4.65) and (4.66).

Our final reduced neoclassical drift-MHD model takes the form [7, 8, 10]

$$\frac{\partial\psi}{\partial\hat{t}}=[\phi,\,\psi]+\hat{d}_i\,\frac{\tau}{1+\tau}[\delta p,\,\psi]+\hat{\eta}_\parallel\left(J-\alpha_{bs}\,\frac{\partial\delta p}{\partial\hat{x}}\right)+\hat{E}_\parallel,\tag{11.46}$$

$$\begin{aligned}\frac{\partial\delta p}{\partial\hat{t}} &= [\phi,\,\delta p]-c_\beta^2\left[\hat{V}_\parallel,\,\psi\right]-c_\beta^2\,\hat{d}_i[J,\,\psi]\\
&-c_\beta^2\,\hat{d}_i\left[\alpha_{nc}\,\frac{\partial\delta p}{\partial\hat{x}},\,\psi\right]-c_\beta^2\,\hat{d}_i\left[\frac{\alpha_\parallel}{\hat{\eta}_\parallel}\,\frac{\partial\psi}{\partial\hat{t}},\,\psi\right]\\
&+\frac{2}{3}\,(1-c_\beta^2)\,\hat{\chi}_\parallel[[\delta p,\,\psi],\,\psi]+\frac{2}{3}\,(1-c_\beta^2)\,\hat{\chi}_\perp\,\hat{\nabla}^2\delta p,\end{aligned}\tag{11.47}$$

$$\begin{aligned}\frac{\partial U}{\partial\hat{t}} &= [\phi,\,U]+\frac{\hat{d}_i}{2\,(1+\tau)}\left(\hat{\nabla}^2[\phi,\,\delta p]+[U,\,\delta p]+\left[\hat{\nabla}^2\delta p,\,\phi\right]\right)\\
&+[J,\,\psi]+\hat{\Xi}_\theta\,\frac{\partial}{\partial\hat{x}}\left(\frac{\epsilon_s}{q_s}\,\hat{V}_\parallel-\frac{\partial}{\partial\hat{x}}\left[\phi-\frac{\hat{d}_i}{1+\tau}\left(1-\alpha_\theta\,\frac{\eta_i}{1+\eta_i}\right)\delta p\right]\right)\\
&+\hat{\Xi}_\perp\,\hat{\nabla}^4\left(\phi-\frac{\hat{d}_i}{1+\tau}\,\delta p\right),\end{aligned}\tag{11.48}$$

$$\frac{\partial \hat{V}_\parallel}{\partial \hat{t}} = \left[\phi,\, \hat{V}_\parallel\right] - [\delta p,\, \psi] - \hat{\Xi}_\theta\, \frac{\epsilon_s}{q_s}\left(\frac{\epsilon_s}{q_s}\, \hat{V}_\parallel - \frac{\partial}{\partial \hat{x}}\left[\phi - \frac{\hat{d}_i}{1+\tau}\left(1 - \alpha_\theta\, \frac{\eta_i}{1+\eta_i}\right)\delta p\right]\right) + \hat{\Xi}_\perp\, \hat{\nabla}^2 \hat{V}_\parallel, \tag{11.49}$$

where

$$J = -\frac{2\,\epsilon_s}{q_s} + \hat{\nabla}^2\psi, \tag{11.50}$$

$$U = \hat{\nabla}^2\phi. \tag{11.51}$$

Here, we have suppressed the ordering superscripts. If we compare equations (11.46)–(11.51) to our previous reduced non-neoclassical drift-MHD equations, (4.67)–(4.74), then we can see that the former set of equations contains many additional terms. The additional term involving the parameter α_{bs} in equation (11.46) is due to the bootstrap current. The additional terms involving the parameters α_{nc} and $\alpha_\parallel$ in equation (11.47) are due to neoclassical parallel momentum and heat fluxes. Finally, the additional terms involving the parameter Ξ_θ in equations (11.48) and (11.49) are due to neoclassical poloidal flow damping.

11.4 Magnetic field-line curvature

As mentioned in section 1.14, the bootstrap current has a destabilizing effect on wide magnetic island chains [2, 5, 16]. It turns out that the mean curvature of magnetic field lines in the inner region has a stabilizing effect on such chains that is similar in magnitude to the destabilizing effect of the bootstrap current [11, 15]. Hence, it is not consistent to include the bootstrap current in our analysis without also including curvature effects. The appropriate curvature terms are derived in reference [13]. We shall simply incorporate them into our model, which generalizes to give:

$$\frac{\partial \psi}{\partial \hat{t}} = [\phi,\, \psi] + \hat{d}_i\, \frac{\tau}{1+\tau}[\delta p,\, \psi] + \hat{\eta}_\parallel\left(J - \alpha_{bs}\, \frac{\partial \delta p}{\partial \hat{x}}\right) + \hat{E}_\parallel, \tag{11.52}$$

$$\begin{aligned} \frac{\partial \delta p}{\partial \hat{t}} = {} & [\phi,\, \delta p] - c_\beta^{\,2}\left[\hat{V}_\parallel,\, \psi\right] - c_\beta^{\,2}\, \hat{d}_i[J,\, \psi] \\ & - c_\beta^{\,2}\, \hat{d}_i\left[\alpha_{nc}\, \frac{\partial \delta p}{\partial \hat{x}},\, \psi\right] - c_\beta^{\,2}\, \hat{d}_i\left[\frac{\alpha_\parallel}{\hat{\eta}_\parallel}\, \frac{\partial \psi}{\partial \hat{t}},\, \psi\right] \\ & - c_\beta^{\,2}\left[\phi + \hat{d}_i\, \frac{\tau}{1+\tau}\, \delta p,\, H\right] + \frac{2}{3}\,(1 - c_\beta^2)\, \hat{\chi}_\parallel[[\delta p,\, \psi],\, \psi] \\ & + \frac{2}{3}\,(1 - c_\beta^2)\, \hat{\chi}_\perp\, \hat{\nabla}^2\delta p, \end{aligned} \tag{11.53}$$

$$
\begin{aligned}
\frac{\partial U}{\partial \hat{t}} = {} & [\phi,\, U] + \frac{\hat{d}_\mathrm{i}}{2\,(1+\tau)}\left(\hat{\nabla}^2[\phi,\, \delta p] + [U,\, \delta p] + [\hat{\nabla}^2\delta p,\, \phi]\right) \\
& + [J,\, \psi] + [\delta p,\, H] \\
& + \hat{\Xi}_\theta\,\frac{\partial}{\partial \hat{x}}\left(\frac{\epsilon_\mathrm{s}}{q_\mathrm{s}}\,\hat{V}_\parallel - \frac{\partial}{\partial \hat{x}}\left[\phi - \frac{\hat{d}_\mathrm{i}}{1+\tau}\left(1 - \alpha_\theta\,\frac{\eta_\mathrm{i}}{1+\eta_\mathrm{i}}\right)\delta p\right]\right) \\
& + \hat{\Xi}_\perp\,\hat{\nabla}^4\left(\phi - \frac{\hat{d}_\mathrm{i}}{1+\tau}\,\delta p\right),
\end{aligned}
\tag{11.54}
$$

$$
\begin{aligned}
\frac{\partial \hat{V}_\parallel}{\partial \hat{t}} = {} & \left[\phi,\, \hat{V}_\parallel\right] - [\delta p,\, \psi] + \frac{c_\beta^{\,2}\,\hat{d}_\mathrm{i}}{1+\tau}\,[\hat{V}_\parallel,\, 2\,H - \delta p] \\
& - \hat{\Xi}_\theta\,\frac{\epsilon_\mathrm{s}}{q_\mathrm{s}}\left(\frac{\epsilon_\mathrm{s}}{q_\mathrm{s}}\,\hat{V}_\parallel - \frac{\partial}{\partial \hat{x}}\left[\phi - \frac{\hat{d}_\mathrm{i}}{1+\tau}\left(1 - \alpha_\theta\,\frac{\eta_\mathrm{i}}{1+\eta_\mathrm{i}}\right)\delta p\right]\right) + \hat{\Xi}_\perp\,\hat{\nabla}^2\,\hat{V}_\parallel.
\end{aligned}
\tag{11.55}
$$

Here,

$$H = \frac{2\,\hat{x}}{\hat{L}_\mathrm{c}}, \tag{11.56}$$

and $\hat{L}_\mathrm{c} = L_\mathrm{c}/l$, where

$$L_\mathrm{c} = \frac{R_0^{\,2}}{r_\mathrm{s}\,(1 - 1/q_\mathrm{s}^{\,2})} \tag{11.57}$$

is the *magnetic curvature length* (i.e. the mean radius of curvature of magnetic field lines at the rational surface) [12, 17]. Note that the previous expression for L_c is only valid in a large-aspect-ratio, low-β plasma with circular magnetic flux surfaces. A more general expression is given in section A.8.

11.5 The rescaled reduced neoclassical drift-MHD model

Following the approach adopted in sections 5.2 and 8.2, it is convenient to set the normalization scale length, l, in our reduced neoclassical drift-MHD model equal to the minor radius of the rational surface, r_s. It is also convenient to work in a frame of reference that *corotates* with the magnetic island chain that develops in the inner region. This goal can be achieved by making the transformations $\phi \to \phi + (\hat{\omega}/m)\,\hat{x}$ and $\hat{V}_\parallel \to \hat{V}_\parallel + (q_\mathrm{s}/\epsilon_\mathrm{s})\,(\hat{\omega}/m)$, where $\omega = \hat{\omega}\,V_\mathrm{A}/r_\mathrm{s}$ is the rotation frequency of the tearing mode in the laboratory frame. In the corotating reference frame, the normalized reconnected flux at the rational surface, $\hat{\Psi}_\mathrm{s}(\hat{t})$ (see equations (3.72) and (3.184)), is assumed to be a positive real quantity. It is helpful to define the reduced (by a factor of four) radial width of the magnetic island chain: $w = (L_\mathrm{s}\,R_0\,\hat{\Psi}_\mathrm{s})^{1/2}$ (see equation (5.129)). Here, R_0 is the plasma major radius (see section 3.2), and L_s is the magnetic shear length at the rational surface (see equation (5.27)).

As before (see section 8.2), it is assumed that $\delta_s \ll w \ll r_s$, where δ_s is the linear layer width (see chapter 6). In other words, the width of the island chain is assumed to be much greater than the linear layer width but much less than the minor radius of the rational magnetic flux surface. Let $\hat{w} = w/r_s$. Reusing the analyses given in sections 5.3 and 8.2, we find that

$$\psi(\hat{x}, \zeta, \hat{t}) \to \frac{\hat{x}^2}{2\,\hat{L}_s} + \hat{R}_0\, \hat{\Psi}_s(\hat{t}) \cos\zeta, \tag{11.58}$$

$$\delta p(\hat{x}, \zeta, \hat{t}) \to \frac{\hat{V}_*}{\hat{d}_i}\, \hat{x}, \tag{11.59}$$

$$\phi(\hat{x}, \zeta, \hat{t}) \to \left[\frac{\hat{\omega}}{m} - \hat{V}_E(\hat{t})\right]\hat{x} - \frac{\varsigma}{4}\, \hat{V}_E{}'(\hat{t})\, \hat{x}^2, \tag{11.60}$$

$$\hat{V}_\parallel(\hat{x}, \zeta, \hat{t}) \to \frac{q_s}{\epsilon_s}\left[\frac{\hat{\omega}}{m} - \hat{V}_E(\hat{t}) - \frac{\hat{V}_*}{1+\tau}\left(1 - \alpha_\theta\,\frac{\eta_i}{1+\eta_i}\right) - \frac{\varsigma}{2}\, \hat{V}_E{}'(\hat{t})\, \hat{x}\right], \tag{11.61}$$

$$J(\hat{x}, \zeta, \hat{t}) \to -\left(\frac{2}{s_s} - 1\right)\frac{1}{\hat{L}_s}, \tag{11.62}$$

in the limit $|\hat{x}|/\hat{w} \gg 1$ (i.e. many island widths from the rational surface). Here, $\hat{L}_s = L_s/r_s$, $\hat{R}_0 = R_0/r_s$, $\hat{V}_E = V_E(r_s)/V_A$, $\hat{V}_* = V_*(r_s)/V_A$, and $s_s = s(r_s)$, where $V_E(r)$ is the E-cross-B velocity profile in the outer region (see equation (5.21)), $V_*(r)$ the diamagnetic velocity profile (see equation (5.29)), and $s(r)$ the magnetic shear profile (see equation (5.28)). Moreover, $\varsigma = \mathrm{sgn}(\hat{x})$, and $\hat{V}_E{}' = [r\, dV_E/dr]_{r_{s-}}^{r_{s+}}/V_A$. The parameter $\hat{V}_E{}'$ is introduced into the analysis in order to take account of the fact that the E-cross-B velocity profile in the outer region (i.e. everywhere in the plasma apart from the immediate vicinity of the magnetic island chain) develops a gradient discontinuity at the rational surface in response to the localized electromagnetic torque that emerges at the surface (see section 8.2). Note, finally, that in neglecting any dependance of $\hat{\Psi}_s$ on $\hat{x}$ in equation (11.58), we are making use of the so-called *constant-ψ approximation* [9, 19], which is valid as long as $|\Delta\hat{\Psi}_s|\, \hat{w}/\hat{\Psi}_s \ll 1$ [4, 6]. Here, $\Delta\hat{\Psi}_s$ is defined in equations (3.73) and (3.183).

Equations (11.58)–(11.62) are analogous to the boundary conditions in our previous reduced non-neoclassical drift-MHD model, (8.2)–(8.6), apart from equation (11.61). The latter equation is derived on the assumption that ion neoclassical poloidal flow damping relaxes the ion poloidal velocity profile to its neoclassical value (see section 2.18) many island widths away from the rational surface.

Let $X = \hat{x}/\hat{w}$ and $T = \omega_*\, t = \hat{\omega}_*\, \hat{t}$, where ω_* is the diamagnetic frequency at the rational surface (see equation (5.47)), and $\hat{\omega}_* = \omega_*/(V_A/r_s)$. It follows that $|X| \sim \mathcal{O}(1)$ in the immediate vicinity of the island chain. It is helpful to define the rescaled fields $\Psi(X, \zeta, T)$, $\mathcal{N}(X, \zeta, T)$, $\Phi(X, \zeta, T)$, $\mathcal{V}(X, \zeta, T)$, and $\mathcal{J}(X, \zeta, T)$, where

$$\psi = \left(\frac{\hat{w}^{\,2}}{\hat{L}_{\rm s}}\right)\Psi, \tag{11.63}$$

$$\delta p = -\left(\frac{\hat{\omega}_*\,\hat{w}}{m\,\hat{d}_{\rm i}}\right)\mathcal{N}, \tag{11.64}$$

$$\phi = \left(\frac{\hat{\omega}_*\,\hat{w}}{m}\right)\Phi, \tag{11.65}$$

$$\hat{V}_\parallel = \left(\frac{\hat{L}_{\rm s}\,\hat{\omega}_*^{\,2}}{m^2\,c_\beta^{\,2}\,\hat{d}_{\rm i}}\right)\mathcal{V}, \tag{11.66}$$

$$J = -\left(\frac{2}{s_{\rm s}} - 1\right)\frac{1}{\hat{L}_{\rm s}} + \left(\frac{\hat{L}_{\rm s}\,\hat{\omega}_*^{\,2}}{m^2\,\hat{w}^{\,2}}\right)\mathcal{J}. \tag{11.67}$$

These fields are (essentially) the same as those used in our previous rescaled reduced non-neoclassical drift-MHD model (see section 8.2).

Equations (11.50)–(11.56) rescale to give

$$\frac{d(\ln \hat{w}^2)}{dT}\cos\zeta = \left\{\Phi - \frac{\tau}{1+\tau}\,\mathcal{N},\,\Psi\right\} + \epsilon_\beta\,\epsilon_{\rm c}\,\epsilon_{\rm R}\,\mathcal{J} + \zeta_{\rm bs}\,\epsilon_\beta\,\epsilon_{\rm c}\,\epsilon_{\rm R}\,(\partial_X\mathcal{N} - 1), \tag{11.68}$$

$$\begin{aligned}\frac{\partial\mathcal{N}}{\partial T} &= \{\Phi,\,\mathcal{N}\} + \{\mathcal{V},\,\Psi\} + \epsilon_{\rm p}\,\epsilon_{\rm c}\,\{\mathcal{J},\,\Psi\}\\ &\quad + \zeta_{\rm g}\,\epsilon_{\rm p}\,\epsilon_{\rm c}\left\{\Phi - \frac{\tau}{1+\tau}\,\mathcal{N},\,X\right\}\\ &\quad - \zeta_{\rm nc}\,\epsilon_{\rm p}\,\epsilon_{\rm c}\,\{\partial_X\mathcal{N},\,\Psi\} + \epsilon_\parallel^{\,-1}\{\{\mathcal{N},\,\Psi\},\,\Psi\} + \epsilon_\perp\,\partial_X^2\mathcal{N},\end{aligned} \tag{11.69}$$

$$\begin{aligned}\frac{\partial(\partial_X^2\,\Phi)}{\partial T} &= \partial_X\left\{\Phi + \frac{\mathcal{N}}{1+\tau},\,\partial_X\Phi\right\} + \{\mathcal{J},\,\Psi\} - \zeta_{\rm g}\,\{\mathcal{N},\,X\}\\ &\quad + \frac{\epsilon_\theta}{\epsilon_{\rm q}}\,\partial_X\left(\xi^{-1}\,\mathcal{V} - \partial_X\left[\Phi + \frac{1}{1+\tau}\left(1 - \alpha_\theta\,\frac{\eta_{\rm i}}{1+\eta_{\rm i}}\right)\mathcal{N}\right]\right)\\ &\quad + \epsilon_\varphi\,\partial_X^4\left(\Phi + \frac{\mathcal{N}}{1+\tau}\right),\end{aligned} \tag{11.70}$$

$$\begin{aligned}\epsilon_{\rm c}\,\frac{\partial\mathcal{V}}{\partial T} &= \epsilon_{\rm c}\,\{\Phi,\,\mathcal{V}\} + \{\mathcal{N},\,\Psi\} + \frac{1}{1+\tau}\,\{\mathcal{V},\,2\,\zeta_{\rm g}\,\epsilon_{\rm p}\,\epsilon_{\rm c}^{\,2}\,X + \epsilon_{\rm c}\,\epsilon_\beta\,\mathcal{N}\}\\ &\quad - \epsilon_{\rm c}\,\epsilon_\theta\left(\mathcal{V} - \xi\,\partial_X\left[\Phi + \frac{1}{1+\tau}\left(1 - \alpha_\theta\,\frac{\eta_{\rm i}}{1+\eta_{\rm i}}\right)\mathcal{N}\right]\right) + \epsilon_{\rm c}\,\epsilon_\varphi\,\partial_X^2\mathcal{V},\end{aligned} \tag{11.71}$$

$$\partial_X^2 \Psi = 1 + \epsilon_\beta\, \epsilon_c\, \mathcal{J}. \tag{11.72}$$

Here, $\partial_X \equiv \partial/\partial X$, $\{A, B\} \equiv (\partial A/\partial X)\,(\partial B/\partial \zeta) - (\partial A/\partial \zeta)\,(\partial B/\partial X)$, and we have set

$$\hat{E}_\parallel = \left[\left(\frac{2}{s_s} - 1\right)\frac{1}{\hat{L}_s} + \alpha_{bs}\,\frac{\hat{V}_*}{\hat{d}_i}\right]\hat{\eta}_\parallel. \tag{11.73}$$

We have also made use of the identity (8.39). Furthermore,

$$\epsilon_q = \left(\frac{\epsilon_s}{q_s}\right)^2, \tag{11.74}$$

$$\epsilon_\theta = \left(\frac{\epsilon_s}{q_s}\right)^2 \frac{1}{\omega_*\,\tau_\theta}, \tag{11.75}$$

and

$$\xi = \sqrt{\frac{\epsilon_p}{\epsilon_q}} = \frac{L_p}{L_s}\,\frac{q_s}{\epsilon_s}, \tag{11.76}$$

$$\zeta_{bs} = f_{t\,s}\,\xi\left[\beta_{11}\left(1 - \alpha_1\,\frac{1}{1+\tau}\,\frac{\eta_i}{1+\eta_i}\right) - \beta_{12}\,\frac{\tau}{1+\tau}\,\frac{\eta_e}{1+\eta_e}\right]\left(\frac{w}{d_\beta}\right)^2 = \left(\frac{w}{w_{bs}}\right)^2, \tag{11.77}$$

$$\begin{aligned}\zeta_{nc} = f_{t\,s}\,\xi&\left[\left(\beta_{11} + \frac{2}{5}\,\epsilon_1\right)\left(1 - \alpha_1\,\frac{1}{1+\tau}\,\frac{\eta_i}{1+\eta_i}\right)\right.\\ &\left.- \left(\beta_{12} + \frac{2}{5}\,\epsilon_2\right)\frac{\tau}{1+\tau}\,\frac{\eta_e}{1+\eta_e} - \frac{\alpha_2}{\tau'}\right]\left(\frac{w}{d_\beta}\right)^2\\ = \left(\frac{w}{w_{nc}}\right)^2,&\end{aligned} \tag{11.78}$$

$$\zeta_g = 2\,\frac{L_p}{L_c}\left(\frac{w}{d_\beta}\right)^2 = \left(\frac{w}{w_g}\right)^2, \tag{11.79}$$

where

$$w_{bs} = d_\beta\left\{f_{t\,s}\,\xi\left[\beta_{11}\left(1 - \alpha_1\,\frac{1}{1+\tau}\,\frac{\eta_i}{1+\eta_i}\right) - \beta_{12}\,\frac{\tau}{1+\tau}\,\frac{\eta_e}{1+\eta_e}\right]\right\}^{-1/2}, \tag{11.80}$$

$$w_{nc} = d_\beta\left\{f_{t\,s}\,\xi\left[\left(\beta_{11} + \frac{2}{5}\,\epsilon_1\right)\left(1 - \alpha_1\,\frac{1}{1+\tau}\,\frac{\eta_i}{1+\eta_i}\right) - \left(\beta_{12} + \frac{2}{5}\,\epsilon_2\right)\frac{\tau}{1+\tau}\,\frac{\eta_e}{1+\eta_e} - \frac{\alpha_2}{\tau'}\right]\right\}^{-1/2}, \tag{11.81}$$

$$w_{\rm g} = d_\beta\left(\frac{L_{\rm c}}{2\,L_{\rm p}}\right)^{1/2}. \tag{11.82}$$

Here, $L_{\rm p}$ is the effective pressure gradient scale length at the rational surface (see equation (8.35)), d_β is the ion sound radius (see equation (4.75)), and the dimensionless quantities ϵ_β, $\epsilon_{\rm p}$, $\epsilon_{\rm c}$, $\epsilon_{\rm R}$, ϵ_φ, $\epsilon_\perp$, and $\epsilon_\parallel$ are defined in section 8.3.

Equations (11.68)–(11.72) must be solved subject to the boundary conditions (see equations (11.58)–(11.62) and (11.63)–(11.67))

$$\Psi(X, \zeta, T) \rightarrow \frac{X^2}{2} + \cos\zeta, \tag{11.83}$$

$$\mathcal{N}(X, \zeta, T) \rightarrow X, \tag{11.84}$$

$$\Phi(X, \zeta, T) \rightarrow v(T)\,X + \frac{\varsigma\, v'(T)\,X^2}{2}, \tag{11.85}$$

$$\mathcal{V}(X, \zeta, T) \rightarrow \xi\left[v(T) + \frac{1}{1+\tau}\left(1 - \alpha_\theta\,\frac{\eta_{\rm i}}{1+\eta_{\rm i}}\right) + \varsigma\, v'(T)\,X\right], \tag{11.86}$$

$$\mathcal{J}(X, \zeta, T) \rightarrow 0, \tag{11.87}$$

as $|X| \rightarrow \infty$. Here,

$$v(T) = \frac{\omega - \omega_{\rm E}}{\omega_*}, \tag{11.88}$$

$$v'(T) = -\frac{\hat{w}}{2\,\omega_*}\left[r\,\frac{d\omega_{\rm E}}{dr}\right]_{r_{s-}}^{r_{s+}}, \tag{11.89}$$

where $\omega_{\rm E} = (m/r_{\rm s})\,V_{\rm E}(r_{\rm s})$ is the E-cross-**B** frequency at the rational surface. Note that Ψ, $\mathcal{N}$, Φ, $\mathcal{V}$, and $\mathcal{J}$ are all $\mathcal{O}(1)$ quantities in the inner region. Note, further, that the boundary conditions (11.83)–(11.87), as well as the symmetry of the rescaled reduced neoclassical drift-MHD equations, (11.68)–(11.72), ensure that Ψ, $\mathcal{V}$, and $\mathcal{J}$ are even functions of X, whereas $\mathcal{N}$ and Φ are odd functions.

Finally, asymptotic matching between the inner region and the surrounding plasma yields (see section 8.10)

$${\rm Re}\left(\frac{\Delta\hat{\Psi}_{\rm s}}{\hat{\Psi}_{\rm s}}\right) = \frac{2\,\epsilon_\beta\,\epsilon_{\rm c}}{\hat{w}}\int_{-\infty}^{\infty}\oint \mathcal{J}\cos\zeta\,\frac{d\zeta}{2\pi}\,dX, \tag{11.90}$$

$${\rm Im}\left(\frac{\Delta\hat{\Psi}_{\rm s}}{\hat{\Psi}_{\rm s}}\right) = -\frac{2\,\epsilon_\beta\,\epsilon_{\rm c}}{\hat{w}}\int_{-\infty}^{\infty}\oint \mathcal{J}\sin\zeta\,\frac{d\zeta}{2\pi}\,dX. \tag{11.91}$$

11.6 The ordering scheme

Tables 8.1 and 11.1 give estimates for the values of the various parameters that characterize our rescaled reduced neoclassical drift-MHD model, (11.68)–(11.72), in a low-field tokamak fusion reactor and a high-field tokamak fusion reactor (see chapter 1). These estimates are made using the following assumptions: $B = 5$ T (low field) or $B = 12$ T (high field), $\beta = 0.02$, $T_e = T_i = 7$ keV, $m_i = (m_D + m_T)/2$ (where m_D and m_T are the deuteron and triton masses, respectively), $\Xi_{\perp\,i} = \chi_{\perp\,e} = \chi_{\perp\,i} = 1\ \mathrm{m^2\,s^{-1}}$, $m = 2$, $n = 1$, $r_s = a/2$ (where a is the minor radius of the plasma), $s_s = 1$, $\tau = 1$, $\tau' = 1$, $\eta_e = \eta_i = 1$, and $dp/dr = -p/a$.

Let us adopt the orderings

$$\epsilon_\beta \sim \epsilon_p \sim \epsilon_q \sim \epsilon_\theta \sim \epsilon_c \sim \epsilon_R \sim \epsilon_\perp \sim \epsilon_\parallel \sim \epsilon, \qquad (11.92)$$

where $\epsilon \ll 1$. As is clear from tables 8.1 and 11.1, these orderings are reasonable provided that the radial width of the magnetic island chain that develops in the inner region exceeds a few centimeters.

Let us also adopt the orderings

$$\xi \sim \zeta_{bs} \sim \zeta_{nc} \sim \zeta_g \sim 1. \qquad (11.93)$$

As is clear from tables 8.1 and 11.1, the ordering $\xi \sim 1$ is consistent with the orderings (11.92). Unfortunately, the orderings $\zeta_{bs} \sim \zeta_{nc} \sim \zeta_g \sim 1$ are not consistent with the orderings (11.92). (It would actually be more consistent to adopt the orderings $\zeta_{bs} \sim \zeta_{nc} \sim \zeta_g \gg 1$.) Nevertheless, we are forced to adopt the orderings $\zeta_{bs} \sim \zeta_{nc} \sim \zeta_g \sim 1$ in order to make further progress. (By adopting these orderings, we are essentially ensuring that the bootstrap current and curvature terms in our analysis are not so large that they lead to a breakdown of the constant-ψ approximation.)

Table 11.1. The parameters that characterize the rescaled reduced neoclassical drift-MHD equations for a low-field tokamak reactor and a high-field tokamak reactor (see equations (11.6), (11.74), (11.75), and (11.80)–(11.82)).

	Low field	High field
B (T)	5.0	12.0
$f_{t\,s}$	0.565	0.565
ϵ_q	6.94×10^{-4}	6.94×10^{-3}
ϵ_θ	1.79×10^{-2}	4.30×10^{-2}
w_{bs} (m)	3.89×10^{-3}	1.62×10^{-3}
w_{nc} (m)	3.42×10^{-3}	1.42×10^{-3}
w_g (m)	1.32×10^{-2}	5.51×10^{-3}

Finally, suppose that

$$\frac{\partial}{\partial T} \sim \mathcal{O}(\epsilon^3), \tag{11.94}$$

and let us expand the various fields in our rescaled model as follows:

$$\Psi = \Psi_0 + \epsilon^2\,\Psi_2 + \epsilon^3\,\Psi_3 + \cdots, \tag{11.95}$$

$$\Phi = \Phi_0 + \epsilon^2\,\Phi_2 + \epsilon^3\,\Phi_3 + \cdots, \tag{11.96}$$

$$\mathcal{N} = \mathcal{N}_0 + \epsilon^2\,\mathcal{N}_2 + \epsilon^3\,\mathcal{N}_3 + \cdots, \tag{11.97}$$

$$\mathcal{V} = \mathcal{V}_0 + \epsilon\,\mathcal{V}_1 + \cdots, \tag{11.98}$$

$$\mathcal{J} = \mathcal{J}_0 + \epsilon\,\mathcal{J}_1 + \cdots. \tag{11.99}$$

Here, Ψ_0, Ψ_1, etc. are assumed to be $\mathcal{O}(1)$ in the inner region.

11.7 The zeroth-order solution

To the zeroth order in ϵ, equation (11.72) yields

$$\partial_X^2 \Psi_0 = 1. \tag{11.100}$$

Solving this equation subject to the boundary condition (11.83), we obtain

$$\Psi_0 = \Omega(X, \zeta) \equiv \frac{X^2}{2} + \cos\zeta. \tag{11.101}$$

Thus, we again conclude that, to the lowest order in our expansion, the magnetic flux surfaces in the island region have the constant-ψ structure pictured in figure 5.7. The island O-points correspond to $\Omega = -1$ and $\zeta = (2k - 1)\,\pi$ (where k is an integer), the X-points correspond to $\Omega = +1$ and $\zeta = 2k\,\pi$, and the magnetic separatrix corresponds to $\Omega = +1$.

To the zeroth order in ϵ, equation (11.71) yields

$$\{\mathcal{N}_0, \Omega\} = 0, \tag{11.102}$$

where use has been made of equation (11.101). Given that $\mathcal{N}$ is an odd function of X, it follows that

$$\mathcal{N}_0(X, \zeta, T) = \varsigma\,\mathcal{N}_{(0)}(\Omega, T). \tag{11.103}$$

By symmetry, $\mathcal{N}_0 = 0$ inside the magnetic separatrix of the island chain. Let

$$L(\Omega, T) = \frac{\partial \mathcal{N}_{(0)}}{\partial \Omega}. \tag{11.104}$$

Note that $L(\Omega \leqslant 1, T) = 0$. Equations (11.84) and (11.101) imply that

$$L(\Omega \to \infty, T) = \frac{1}{\sqrt{2\,\Omega}}. \tag{11.105}$$

To the zeroth order in ϵ, equation (11.68) yields

$$\{\Phi_0, \Omega\} = 0, \tag{11.106}$$

where use has been made of equations (11.101) and (11.103). Given that Φ is an odd function of X, it follows that

$$\Phi_0(X, \zeta, T) = \varsigma\, \Phi_{(0)}(\Omega, T). \tag{11.107}$$

By symmetry, $\Phi_0 = 0$ inside the magnetic separatrix of the island chain. Let

$$M(\Omega, T) = \frac{\partial \Phi_{(0)}}{\partial \Omega}. \tag{11.108}$$

Note that $M(\Omega \leqslant 1, T) = 0$. Equations (11.85) and (11.101) imply that

$$M(\Omega \to \infty, T) = \frac{v(T)}{\sqrt{2\,\Omega}} + v'(T). \tag{11.109}$$

To the zeroth order in ϵ, equation (11.69) yields

$$\{\mathcal{V}_0, \Omega\} = 0, \tag{11.110}$$

where use has been made of equations (11.101), (11.103), and (11.107). Given that $\mathcal{V}$ is an even function of X, we can write

$$\mathcal{V}_0(X, \zeta, T) = \mathcal{V}_{(0)}(\Omega, T). \tag{11.111}$$

Finally, to the zeroth order in ϵ, equation (11.70) gives

$$\begin{aligned}\{\mathcal{J}_0, \Omega\} = &- \zeta_g\, \{L\,|X|, \Omega\} + \frac{1}{2}\left\{\partial_\Omega\left[M\left(M + \frac{L}{1+\tau}\right)\right]X^2, \Omega\right\} \\ &- \frac{\epsilon_\theta}{\epsilon_q}\, X\, \partial_\Omega\left(\xi^{-1}\,\mathcal{V}_{(0)} - |X|\,F\right),\end{aligned} \tag{11.112}$$

where

$$F(\Omega, T) = M(\Omega, T) + \frac{1}{1+\tau}\left(1 - \alpha_\theta\,\frac{\eta_i}{1+\eta_i}\right)L(\Omega, T), \tag{11.113}$$

and use has been made of equations (11.101), (11.103), (11.104), (11.107), and (11.108). Moreover, $\partial_\Omega \equiv \partial/\partial\Omega$. Note that $F(\Omega \leqslant 1, T) = 0$.

Let us write

$$\mathcal{J}_0(\Omega, \zeta, T) = \mathcal{J}_0^{(c)}(\Omega, \zeta, T) + \mathcal{J}_0^{(s)}(\Omega, \zeta, T), \tag{11.114}$$

where $\mathcal{J}_0^{(c)}$ has the symmetry of $\cos\zeta$, whereas $\mathcal{J}_0^{(s)}$ has the symmetry of $\sin\zeta$. It follows from equations (11.112) and (11.114) that

$$\{\mathcal{J}_0^{(c)}, \Omega\} = -\zeta_g \{L\,|X|, \Omega\} + \frac{1}{2}\left\{\partial_\Omega\left[M\left(M + \frac{L}{1+\tau}\right)\right]X^2, \Omega\right\}, \tag{11.115}$$

which implies that

$$\mathcal{J}_0^{(c)}(\Omega, \zeta, T) = \bar{\mathcal{J}}_0(\Omega, T) - \zeta_g\, L\, \widetilde{|X|} + \frac{1}{2}\,\partial_\Omega\left[M\left(M + \frac{L}{1+\tau}\right)\right]\widetilde{X^2}, \tag{11.116}$$

where the $\widetilde{X}$ operator is defined in section 8.6, and $\bar{\mathcal{J}}_0(\Omega, T)$ is an undetermined flux-surface function.

Equations (11.112) and (11.114) also yield

$$\{\mathcal{J}_0^{(s)}, \Omega\} = -\frac{\epsilon_\theta}{\epsilon_q}\, X\,\partial_\Omega\left(\xi^{-1}\,\mathcal{V}_{(0)} - |X|\,F\right). \tag{11.117}$$

The flux-surface average (see section 8.6) of the previous equation gives

$$\left\langle X\,\partial_\Omega(\xi^{-1}\,\mathcal{V}_{(0)} - |X|\,F)\right\rangle = 0. \tag{11.118}$$

Inside the separatrix of the magnetic island chain, recalling that $F(\Omega \leqslant 1, T) = 0$, the previous equation reduces to

$$\langle X\rangle\partial_\Omega\mathcal{V}_{(0)} = 0, \tag{11.119}$$

whereas outside the separatrix, it gives

$$\partial_\Omega\left(\xi^{-1}\,\langle|X|\rangle\,\mathcal{V}_{(0)} - \langle X^2\rangle F\right) = 0, \tag{11.120}$$

where use has been made of the easily proved identity $\langle X\,\partial_\Omega G\rangle \equiv \partial_\Omega\langle X\,G\rangle$. Thus, we conclude that

$$\mathcal{V}_{(0)}(\Omega, T) = c_0(T) \tag{11.121}$$

inside the separatrix, and

$$\mathcal{V}_{(0)}(\Omega, T) = \xi\,\langle X^2\rangle\,F(\Omega, T) + c_1(T) \tag{11.122}$$

outside the separatrix. Here, use has been made of $\langle|X|\rangle = 1$ outside the separatrix. However, the boundary condition (11.86), combined with equations (11.101), (11.105), (11.109), and (11.113), implies that $c_1(T) = 0$. Finally, equations (11.117), (11.121), and (11.122) yield

$$\{\mathcal{J}^{(s)}, \Omega\} = -\frac{\epsilon_\theta}{\epsilon_q}\, X\,\partial_\Omega\,[(\langle X^2\rangle - |X|)F]. \tag{11.123}$$

11.8 The higher-order solution

Our zeroth-order solution is characterized by three undetermined flux-surface functions: $\bar{\mathcal{J}}_0(\Omega, T)$, $L(\Omega, T)$, and $F(\Omega, T)$. In order to determine the forms of these functions, we need to expand our model to a higher order (see section 8.8).

Expanding equation (11.68) to the third order in ϵ, we obtain

$$\frac{d(\ln \hat{w}^2)}{dT}\cos\zeta = \{F_e, \Omega\} + \epsilon_\beta\,\epsilon_c\,\epsilon_R\,\mathcal{J}_0 + \zeta_{bs}\,\epsilon_\beta\,\epsilon_c\,\epsilon_R(|X|\,L - 1), \qquad (11.124)$$

where

$$F_e(\Omega, \zeta, T) = \epsilon^2\,(\Phi_2 + \epsilon\,\Phi_3) - \epsilon^2\left(\frac{\tau}{1+\tau}\right)(\mathcal{N}_2 + \epsilon\,\mathcal{N}_3) - \epsilon^2\,\varsigma\,Y_e\,(\Psi_2 + \epsilon\,\Psi_3), \qquad (11.125)$$

and use has been made of equations (8.75), (11.95)–(11.99), (11.103), (11.104), (11.107), and (11.108). The flux-surface average of equation (11.124) yields

$$\bar{\mathcal{J}}_0(\Omega, T) = \frac{1}{\epsilon_\beta\,\epsilon_c\,\epsilon_R}\,\frac{d(\ln \hat{w}^2)}{dT}\,\frac{\langle\cos\zeta\rangle}{\langle 1\rangle} - \zeta_{bs}\left(\frac{L}{\langle 1\rangle} - 1\right), \qquad (11.126)$$

where use has been made of equations (8.68), (8.70), (11.114), and (11.116). Note that $\langle \mathcal{J}_0^{(s)}\rangle = 0$, by symmetry. Hence,

$$\mathcal{J}_0^{(c)}(\Omega, \zeta, T) = \frac{1}{\epsilon_\beta\,\epsilon_c\,\epsilon_R}\,\frac{d(\ln \hat{w}^2)}{dT}\,\frac{\langle\cos\zeta\rangle}{\langle 1\rangle} - \zeta_{bs}\left(\frac{L}{\langle 1\rangle} - 1\right) - \zeta_g\,L\,\widetilde{|X|} + \frac{1}{2}\,\partial_\Omega\left[M\left(M + \frac{L}{1+\tau}\right)\right]\widetilde{X^2}, \qquad (11.127)$$

where use has been made of equation (11.116).

Expanding equation (11.69) to the first order in ϵ, we obtain

$$0 = \epsilon\,\{\mathcal{V}_1, \Omega\} + \epsilon^2\,\epsilon_\parallel^{-1}\{\{\mathcal{N}_2, \Omega\}, \Omega\} + \epsilon_\perp\,\varsigma\,(X^2\,\partial_\Omega L + L), \qquad (11.128)$$

where use has been made of equations (11.95)–(11.99), (11.101), (11.103), and (11.104). The flux-surface average of the previous equation yields

$$\partial_\Omega\,(\langle X^2\rangle\,L) = 0. \qquad (11.129)$$

We can solve the previous equation, subject to the boundary condition (11.105), to give

$$L(\Omega, T) = L(\Omega) = \begin{cases} 0 & -1 \leqslant \Omega \leqslant 1 \\ 1/\langle X^2\rangle & \Omega > 1 \end{cases}. \qquad (11.130)$$

Here, we have taken account of the previously mentioned fact that $L = 0$ within the island separatrix. Note that $L(\Omega)$ is discontinuous across the island separatrix, which implies that the pressure gradient is also discontinuous across the separatrix. As discussed in section 8.9, we expect this discontinuity to be resolved in a layer of characteristic thickness $4\,w_d$ on the separatrix (see equation (8.84) and table 8.1).

Expanding equation (11.71) to the second order in ϵ, we obtain

$$0 = \epsilon\{-\epsilon_c\,\varsigma\,M\,\mathcal{V}_1 - \epsilon\,\varsigma\,L\,\Psi_2 + \epsilon\,\mathcal{N}_2, \Omega\} - \epsilon_c\,\epsilon_\theta\,(\mathcal{V}_{(0)} - \xi\,|X|\,F) + \epsilon_c\,\epsilon_\varphi\left(X^2\,\partial_\Omega^2\mathcal{V}_{(0)} + \partial_\Omega\mathcal{V}_{(0)}\right), \qquad (11.131)$$

where use has been made of equations (11.95)–(11.99), (11.101), (11.103), (11.104), (11.107), (11.108), (11.111), and (11.113). The flux-surface average of the previous equation yields

$$\epsilon_\varphi\, \partial_\Omega\left(\left\langle X^{\,2}\right\rangle \partial_\Omega \mathcal{V}_{(0)}\right) = \epsilon_\theta\left(\left\langle 1\right\rangle \mathcal{V}_{(0)} - \xi\, F\right). \tag{11.132}$$

According to equation (11.121), recalling that $F(\Omega \leqslant 1,\, T) = 0$, we get

$$0 = \epsilon_\theta\, \langle 1\rangle\, c_0(T) \tag{11.133}$$

inside the separatrix, which implies that $c_0(T) = 0$. Hence, given that we have already concluded that $c_1(T) = 0$, equations (11.121) and (11.122) yield

$$\mathcal{V}_{(0)}(\Omega,\, T) = \xi\, \left\langle X^{\,2}\right\rangle\, F(\Omega,\, T). \tag{11.134}$$

The previous equation can be combined with equation (11.132) to produce

$$\epsilon_\varphi\, \partial_\Omega\, [\langle X^{\,2}\rangle\, \partial_\Omega\, (\langle X^{\,2}\rangle\, F)] = \epsilon_\theta(\langle X^{\,2}\rangle\, \langle 1\rangle - 1)F. \tag{11.135}$$

It is apparent from the previous equation that there is no discontinuity in F across the separatrix. Thus, given that $F(\Omega \leqslant 1,\, T) = 0$, it follows that $F(1_+,\, T) = 0$. It is also clear from equations (11.105), (11.109), and (11.113) that $F(\Omega \to \infty,\, T) = v'(T)$. Thus, we can write

$$F(\Omega,\, T) = v'(T)\begin{cases} 0 & -1 \leqslant \Omega \leqslant 1 \\ \mathcal{F}(\Omega) & \Omega > 1 \end{cases}, \tag{11.136}$$

where

$$\epsilon_\varphi\, d_\Omega\, [\langle X^{\,2}\rangle\, d_\Omega\, (\langle X^{\,2}\rangle\, \mathcal{F})] = \epsilon_\theta(\langle X^{\,2}\rangle\, \langle 1\rangle - 1)\mathcal{F}, \tag{11.137}$$

and

$$\mathcal{F}(1) = 0, \tag{11.138}$$

$$\mathcal{F}(\Omega \to \infty) = 1. \tag{11.139}$$

Here, $d_\Omega \equiv d/d\Omega$.

11.9 Asymptotic matching

The asymptotic matching relations (11.90) and (11.91) can be re-expressed in the forms (see section 8.10),

$$\mathrm{Re}\left(\frac{\Delta\hat{\Psi}_{\rm s}}{\hat{\Psi}_{\rm s}}\right) = \frac{4\,\epsilon_\beta\,\epsilon_{\rm c}}{\hat{w}}\int_{-1}^{\infty}\langle \mathcal{J}^{(c)}\cos\zeta\rangle\, d\Omega, \tag{11.140}$$

$$\mathrm{Im}\left(\frac{\Delta\hat{\Psi}_{\rm s}}{\hat{\Psi}_{\rm s}}\right) = -\frac{4\,\epsilon_\beta\,\epsilon_{\rm c}}{\hat{w}}\int_{-1}^{\infty}\langle \mathcal{J}^{(s)}\sin\zeta\rangle\, d\Omega = -\frac{4\,\epsilon_\beta\,\epsilon_{\rm c}}{\hat{w}}\int_{-1}^{\infty}\langle X\,\{\mathcal{J}^{(s)},\,\Omega\}\rangle\, d\Omega. \tag{11.141}$$

Equations (11.123), (11.136), and (11.141) yield

$$\mathrm{Im}\left(\frac{\Delta\hat{\Psi}_s}{\hat{\Psi}_s}\right) = \frac{4\,\epsilon_\beta\,\epsilon_c\,\epsilon_\theta\,v'}{\hat{w}\,\epsilon_q}\int_1^\infty \langle X^2\,\partial_\Omega\,[(\langle X^2\rangle - |X|)\mathcal{F}]\rangle d\Omega. \tag{11.142}$$

Employing the easily proved identity $\langle X^2\,\partial_\Omega G\rangle \equiv \partial_\Omega\langle X^2\,G\rangle - \langle G\rangle$, we obtain

$$\begin{aligned}\mathrm{Im}\left(\frac{\Delta\hat{\Psi}_s}{\hat{\Psi}_s}\right) &= \frac{4\,\epsilon_\beta\,\epsilon_c\,\epsilon_\theta\,v'}{\hat{w}\,\epsilon_q}[(\langle X^2\rangle^2 - \langle |X|^3\rangle)\mathcal{F}]_1^\infty - \frac{4\,\epsilon_\beta\,\epsilon_c\,\epsilon_\theta\,v'}{\hat{w}\,\epsilon_q}\int_1^\infty (\langle X^2\rangle\,\langle 1\rangle - 1)\mathcal{F}\,d\Omega\\ &= -\frac{4\,\epsilon_\beta\,\epsilon_c\,\epsilon_\theta\,v'}{\hat{w}\,\epsilon_q}\int_1^\infty (\langle X^2\rangle\,\langle 1\rangle - 1)\mathcal{F}\,d\Omega,\end{aligned} \tag{11.143}$$

where use has been made of equations (11.138), (11.139), as well as the fact that $\langle X^2\rangle^2 - \langle |X|^3\rangle \sim \Omega^{-1}$ as $\Omega \to \infty$. However, according to equation (11.137),

$$\mathrm{Im}\left(\frac{\Delta\hat{\Psi}_s}{\hat{\Psi}_s}\right) = -\frac{4\,\epsilon_\beta\,\epsilon_c\,\epsilon_\varphi\,v'}{\hat{w}\,\epsilon_q}[\langle X^2\rangle\,d_\Omega(\langle X^2\rangle\,\mathcal{F})]_1^\infty = -\frac{4\,\epsilon_\beta\,\epsilon_c\,\epsilon_\varphi\,v'}{\hat{w}\,\epsilon_q}, \tag{11.144}$$

where use has been made of equations (11.138) and (11.139). The previous equation yields

$$\left[r\,\frac{d\omega_E}{dr}\right]_{r_{s-}}^{r_{s+}} = \left(\frac{\epsilon_s}{q_s}\right)^2\left(\frac{\tau_\varphi\,\hat{w}^4}{2\,\tau_H^2}\right)\mathrm{Im}\left(\frac{\Delta\hat{\Psi}_s}{\hat{\Psi}_s}\right), \tag{11.145}$$

where use has been made of equations (8.23), (8.25), (8.27), and (8.46), and (8.103). Here, τ_H is the hydromagnetic time (see equation (5.43)), and τ_φ is the toroidal momentum confinement time (see equation (5.43)). Equation (11.145) can also be obtained by integrating equation (3.165) across the rational surface, making use of equation (3.165) as well as the identification

$$\left[r\,\frac{d\omega_E}{dr}\right]_{r_{s-}}^{r_{s+}} = -n\left[r\,\frac{\partial\Delta\Omega_z}{\partial r}\right]_{r_{s-}}^{r_{s+}}. \tag{11.146}$$

If we compare equation (11.145) with equation (8.102), then we can see that the discontinuity in the MHD fluid velocity gradient that develops at the rational surface is a factor of $(\epsilon_s/q_s)^2$ smaller in our neoclassical drift-MHD model than the corresponding discontinuity in our non-neoclassical drift-MHD model. The reason for this reduction is that the strong poloidal flow damping present in the former model prevents the poloidal plasma rotation profile from being modified by the localized electromagnetic torque produced at the rational surface. Instead, only the toroidal plasma rotation profile is modified by the electromagnetic torque [4, 18].

It follows from equations (11.113), (11.127), (11.130), (11.136), and (11.140) that

$$
\begin{aligned}
\mathrm{Re}\left(\frac{\Delta\hat{\Psi}_s}{\hat{\Psi}_s}\right) = {} & \frac{4}{\epsilon_R\,\hat{w}}\,\frac{d(\ln\hat{w}^2)}{dT}\int_{-1}^{\infty}\frac{\langle\cos\zeta\rangle^2}{\langle 1\rangle}\,d\Omega - \frac{4\,\epsilon_\beta\,\epsilon_c\,\zeta_{bs}}{\hat{w}}\int_{1}^{\infty}\frac{\langle\cos\zeta\rangle}{\langle 1\rangle\,\langle X^2\rangle}\,d\Omega\\
& - \frac{4\,\epsilon_\beta\,\epsilon_c\,\zeta_g}{\hat{w}}\int_{1}^{\infty}\frac{\langle\widetilde{|X|}\cos\zeta\rangle}{\langle X^2\rangle}\,d\Omega\\
& - \frac{2\,\epsilon_\beta\,\epsilon_c}{(1+\tau)^2\,\hat{w}}\left(\alpha_\theta\,\frac{\eta_i}{1+\eta_i}\right)\left(1-\alpha_\theta\,\frac{\eta_i}{1+\eta_i}\right)\int_{1-}^{\infty} d_\Omega\left(\frac{\mathcal{H}}{\langle X^2\rangle^2}\right)\left\langle\widetilde{X^2}\cos\zeta\right\rangle d\Omega \quad (11.147)\\
& - \frac{2\,\epsilon_\beta\,\epsilon_c\,v'}{(1+\tau)\,\hat{w}}\left(1-2\,\alpha_\theta\,\frac{\eta_i}{1+\eta_i}\right)\int_{1}^{\infty} d_\Omega\left(\frac{\mathcal{F}}{\langle X^2\rangle}\right)\left\langle\widetilde{X^2}\cos\zeta\right\rangle d\Omega\\
& + \frac{2\,\epsilon_\beta\,\epsilon_c\,v'^{\,2}}{\hat{w}}\int_{1}^{\infty} d_\Omega\,(\mathcal{F}^2)\left\langle\widetilde{X^2}\cos\zeta\right\rangle d\Omega.
\end{aligned}
$$

Here,

$$
\mathcal{H}(\Omega) = \begin{cases} 0 & -1 \leqslant \Omega \leqslant 1\\ 1 & \Omega > 1 \end{cases}, \tag{11.148}
$$

and use has been made of the fact that $\mathcal{F}(\Omega)$ is continuous across the separatrix. The previous equation can be combined with equations (8.10), (8.23)–(8.27), (8.103), (11.77), (11.79), and (11.144) to give [2, 3, 7, 14, 15, 20, 21, 23]

$$
\begin{aligned}
\mathrm{Re}\left(\frac{\Delta\hat{\Psi}_s}{\hat{\Psi}_s}\right) = {} & I_1\,\tau_R\,\frac{d}{dt}\left(\frac{4\,w}{r_s}\right)\\
& - I_2\left[\beta_{11}\left(1-\alpha_1\,\frac{1}{1+\tau}\,\frac{\eta_i}{1+\eta_i}\right) - \beta_{12}\,\frac{\tau}{1+\tau}\frac{\eta_e}{1+\eta_e}\right] f_{t\,s}\left(\frac{q_s}{\epsilon_s}\right) c_\beta^2\left(\frac{L_s}{L_p}\right)\left(\frac{r_s}{w}\right)\\
& + I_2\,c_\beta^2\left(\frac{2\,L_s^{\,2}}{L_p\,L_c}\right)\left(\frac{r_s}{w}\right)\\
& + I_3\left(\alpha_\theta\,\frac{\eta_i}{1+\eta_i}\right)\left(1-\alpha_\theta\,\frac{\eta_i}{1+\eta_i}\right)\left(\frac{c_\beta}{1+\tau}\right)^2\left(\frac{L_s}{L_p}\right)^2\left(\frac{d_\beta}{r_s}\right)^2\left(\frac{r_s}{w}\right)^3 \quad (11.149)\\
& - I_4\left(1-2\,\alpha_\theta\,\frac{\eta_i}{1+\eta_i}\right)\left(\frac{\epsilon_s}{q_s}\right)^2\left(\frac{c_\beta}{1+\tau}\right)\left(\frac{L_s}{L_p}\right)\left(\frac{d_\beta}{r_s}\right)\left(\frac{\tau_\varphi}{\tau_H}\right)\left(\frac{w}{r_s}\right)^2\mathrm{Im}\left(\frac{\Delta\hat{\Psi}_s}{\hat{\Psi}_s}\right)\\
& - I_5\left(\frac{\epsilon_s}{q_s}\right)^4\left(\frac{\tau_\varphi}{\tau_H}\right)^2\left(\frac{w}{r_s}\right)^7\left[\mathrm{Im}\left(\frac{\Delta\hat{\Psi}_s}{\hat{\Psi}_s}\right)\right]^2,
\end{aligned}
$$

where

$$
I_1 = 2\int_{-1}^{\infty}\frac{\langle\cos\zeta\rangle^2}{\langle 1\rangle}\,d\Omega, \tag{11.150}
$$

$$
I_2 = 2\int_{1}^{\infty}\left(\frac{\langle|X|^3\rangle}{\langle X^2\rangle} - \frac{1}{\langle 1\rangle}\right)d\Omega, \tag{11.151}
$$

$$I_3 = \int_{1-}^{\infty} d_\Omega \left(\frac{\mathcal{H}}{\langle X^2\rangle^2}\right)\left(\langle X^4\rangle - \frac{\langle X^2\rangle^2}{\langle 1\rangle}\right) d\Omega, \tag{11.152}$$

$$I_4 = \frac{1}{4}\int_{1}^{\infty} d_\Omega \left(\frac{\mathcal{F}}{\langle X^2\rangle}\right)\left(\langle X^4\rangle - \frac{\langle X^2\rangle^2}{\langle 1\rangle}\right) d\Omega, \tag{11.153}$$

$$I_5 = \frac{1}{16}\int_{1}^{\infty} d_\Omega \,(\mathcal{F}^{\,2})\left(\langle X^4\rangle - \frac{\langle X^2\rangle^2}{\langle 1\rangle}\right) d\Omega. \tag{11.154}$$

11.10 The evaluation of the integrals

It is helpful to define the new magnetic flux-surface label $k = [(1 + \Omega)/2]^{1/2}$ (see section 8.11). It follows from equation (11.101) that $k = 0$ at the O-points of the magnetic island chain, and $k = 1$ on the magnetic separatrix. Hence, we can write

$$I_1 = \int_{0}^{\infty} \frac{4\,[(2\,k^2 - 1)\,\mathcal{A} - 2\,k^2\,\mathcal{C}]^2}{\mathcal{A}}\, dk, \tag{11.155}$$

$$I_2 = 16\int_{1}^{\infty}\left(\frac{\mathcal{D}}{\mathcal{C}} - \frac{1}{\mathcal{A}}\right) k^2\, dk, \tag{11.156}$$

$$I_3 = \frac{2\pi}{3} - \int_{1}^{\infty} \frac{4}{\mathcal{C}}\left(\frac{\mathcal{E}\,\mathcal{A}}{\mathcal{C}^2} - 1\right) dk, \tag{11.157}$$

$$I_4 = \int_{1}^{\infty} (k\,\mathcal{C}\,d_k\mathcal{F} - \mathcal{A}\,\mathcal{F})\left(\frac{\mathcal{E}}{\mathcal{C}^2} - \frac{1}{\mathcal{A}}\right) k\, dk, \tag{11.158}$$

$$I_5 = \int_{1}^{\infty} \mathcal{F}\,d_k\mathcal{F}\left(\mathcal{E} - \frac{\mathcal{C}^2}{\mathcal{A}}\right) k^3\, dk. \tag{11.159}$$

where

$$d_k\,[\mathcal{C}\,d_k(k\,\mathcal{C}\,\mathcal{F})] = \lambda_{\theta\varphi}\,(\mathcal{A}\,\mathcal{C} - 1)\,k\,\mathcal{F}, \tag{11.160}$$

$d_k \equiv d/dk$,

$$\lambda_{\theta\varphi} = \frac{4\,\epsilon_\theta}{\epsilon_\varphi}, \tag{11.161}$$

$$\mathcal{F}(1) = 0, \tag{11.162}$$

$$\mathcal{F}(\infty) = 1. \tag{11.163}$$

Here, the functions $\mathcal{A}(k)$, $\mathcal{C}(k)$, $\mathcal{D}(k)$, and $\mathcal{E}(k)$ are defined in section 8.11. Note that the factor $2\pi/3$ in equation (11.157) is generated by the discontinuity in the function

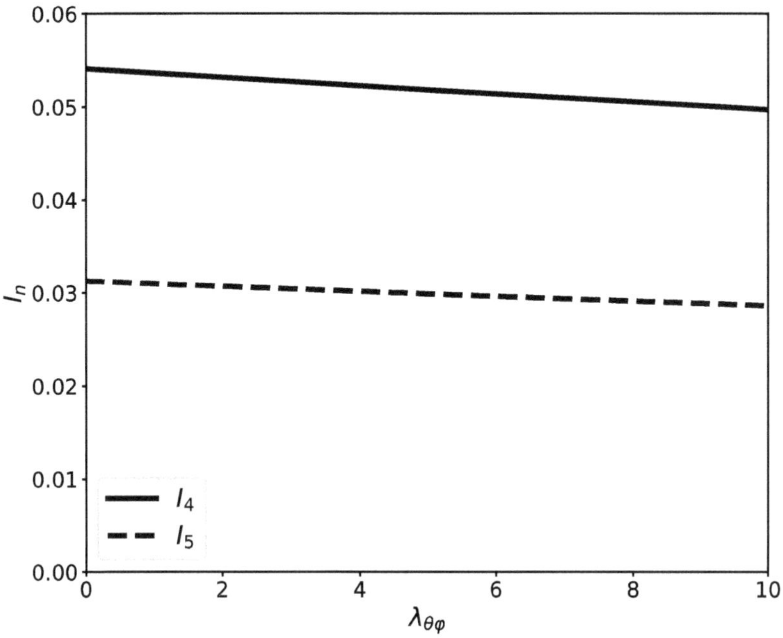

Figure 11.1. The integrals I_4 and I_5 evaluated as functions of $\lambda_{\theta\varphi}$.

$L(\Omega)$ (i.e. the discontinuity in the pressure gradient) across the separatrix of the magnetic island chain [23]. If this contribution is omitted, then the sign of the integral I_3 is reversed.

The values of the first three integrals are

$$I_1 = 0.8227, \tag{11.164}$$

$$I_2 = 1.5835, \tag{11.165}$$

$$I_3 = 1.3814. \tag{11.166}$$

Figure 11.1 shows the values of the integrals I_4 and I_5 as functions of $\lambda_{\theta\varphi}$. Note that I_4 and I_5 are both only very weak functions of $\lambda_{\theta\varphi}$.

References

[1] Bickerton R J, Connor J W and Taylor J B 1971 Diffusion driven plasma currents and bootstrap tokamak *Nat. Phys. Sci.* **229** 110

[2] Carrera R, Hazeltine R D and Kotschenreuther M 1986 Island bootstrap current modification of the nonlinear dynamics of the tearing mode *Phys. Fluids* **29** 899

[3] Connor J W, Waelbroeck F L and Wilson H R 2001 The role of polarization current in magnetic island evolution *Phys. Plasmas* **8** 2835

[4] Fitzpatrick R 1993 Interaction of tearing modes with external structures in cylindrical geometry *Nucl. Fusion* **33** 1049

[5] Fitzpatrick R 1995 Helical temperature perturbations associated with tearing modes in tokamak plasmas *Phys. Plasmas* **2** 825
[6] Fitzpatrick R 1998 Bifurcated states of a rotating tokamak plasma in the presence of a static error-field *Phys. Plasmas* **5** 3325
[7] Fitzpatrick R 2016 An improved neoclassical drift-magnetohydrodynamical fluid model of helical magnetic island equilibria in tokamak plasmas *Phys. Plasmas* **23** 052506
[8] Fitzpatrick R and Waelbroeck F L 2009 Effect of flow damping on drift-tearing magnetic islands in tokamak plasmas *Phys. Plasmas* **16** 072507
[9] Furth H P, Killeen J and Rosenbluth M N 1963 Finite-resistivity instabilities of a sheet pinch *Phys. Fluids* **6** 459
[10] Furuya A, Yagi M and Itoh S-I 2003 Linear analysis of neoclassical tearing mode based on the four-field reduced neoclassical MHD equation *J. Phys. Soc. Japan* **72** 313
[11] Glasser A H, Greene J M and Johnson J L 1975 Resistive instabilities in general toroidal plasma configurations *Phys. Fluids* **18** 875
[12] Glasser A H, Greene J M and Johnson J L 1976 Resistive instabilities in a tokamak *Phys. Fluids* **19** 567
[13] Hazeltine R D, Kotschenreuther M and Morrison P G 1985 A four-field model for tokamak plasma dynamics *Phys. Fluids* **28** 2466
[14] Hegna C C 1999 Nonlinear dynamics of pressure driven magnetic islands in low aspect ratio tokamaks *Phys. Plasmas* **6** 3980
[15] Kotschenreuther M, Hazeltine R D and Morrison P J 1985 Nonlinear dynamics of magnetic islands with curvature and pressure *Phys. Fluids* **28** 294
[16] La Haye R J 2006 Neoclassical tearing modes and their control *Phys. Plasmas* **13** 055501
[17] Lütjens H, Luciani J-F and Garbet X 2001 Curvature effects on the dynamics of tearing modes in tokamaks *Phys. Plasmas* **8** 4267
[18] Nave M F F and Wesson J A 1990 Mode locking in tokamaks *Nucl. Fusion* **30** 2575
[19] Rutherford P H 1973 Nonlinear growth of the tearing mode *Phys. Fluids* **16** 1906
[20] Smolyakov A I 1993 Nonlinear evolution of tearing modes in inhomogeneous plasmas *Plasma Phys. Control. Fusion* **35** 657
[21] Smolyakov A I, Hirose A, Lazzaro E, Re G B and Callen J D 1995 Rotating nonlinear magnetic islands in a tokamak plasma *Phys. Plasmas* **2** 1581
[22] Stix T H 1973 Decay of poloidal rotation in a tokamak plasma *Phys. Fluids* **16** 1260
[23] Waelbroeck F L and Fitzpatrick R 1997 Rotation and locking of magnetic islands *Phys. Rev. Lett.* **78** 1703

IOP Publishing

Richard Fitzpatrick

Chapter 12

Neoclassical tearing modes

12.1 Introduction

A nonlinear tearing mode in a tokamak plasma reconnects magnetic flux (see section 5.16), leading to the generation of a helical magnetic island chain at the so-called *rational* magnetic flux surface at which the mode resonates with the equilibrium magnetic field. As we saw in chapters 8 and 11, if the full radial width of the island chain exceeds the critical value of $4\, w_{\rm d}$ (see equation (8.84) and table 8.1), then the plasma pressure profile is flattened in the region lying within the chain's magnetic separatrix [1]. The flattening of the pressure profile causes a reduction in the bootstrap current [2] inside the separatrix that has a destabilizing effect on the chain [3]. A tearing mode that is driven into instability by this mechanism, rather than the usual free energy sources for a tearing mode (i.e. global current and pressure gradients), is known as a *neoclassical tearing mode* [4].

Neoclassical tearing modes were originally identified experimentally in the Tokamak Fusion Test Reactor (TFTR) [5] and have subsequently been observed in many other tokamaks [6–10]. The flattening of the pressure profile within the magnetic separatrix of a neoclassical tearing mode leads to a degradation of the energy confinement properties of the plasma [11] that limits the maximum attainable β value (see equation (1.23)) [9]. Consequently, neoclassical tearing modes are nowadays regarded as the main obstacle to obtaining β values in tokamak plasmas that are adequate for the achievement of thermonuclear fusion [4, 12, 13].

The fact that a magnetic island chain can only locally flatten the plasma pressure profile (thereby generating a destabilizing local reduction in the bootstrap current) when its radial width exceeds the critical value $4\, w_{\rm d}$ leads to the conclusion that neoclassical tearing modes are actually *metastable* [1]. In other words, some sort of seed perturbation must be applied to the relevant rational magnetic flux surface in order to trigger a neoclassical tearing mode. In practice, the seed perturbation usually takes the form of a transient magnetic perturbation that is resonant at the rational surface [14]. Such perturbations arise naturally in tokamak plasmas as a

doi:10.1088/978-0-7503-5367-0ch12

consequence of plasma instabilities such as internal kink modes and edge localized modes [15].

Neoclassical tearing modes can be stabilized by driving a parallel (to the magnetic field) current in the vicinity of the rational surface by means of radio-frequency (rf) electromagnetic waves injected into the plasma; the idea is to replace the missing bootstrap current within the island chain's magnetic separatrix [7, 16–20].

The aim of this chapter is to use the nonlinear neoclassical resonant response model derived in the previous chapter to investigate the physics of neoclassical tearing modes.

12.2 The isolated magnetic island chain

Suppose that the rigid wall surrounding the plasma is perfectly conducting (i.e. $\tau_w \to \infty$, where τ_w is the time constant of the wall—see section 3.10) and that there is no current flowing through the external magnetic field coil (i.e. $\hat{I}_c = 0$, where $\hat{I}_c$ is the normalized coil current—see equation (3.194)). In this situation, the magnetic island chain that develops at the rational surface is said to be *isolated* (because it is not subject to any outside influence). For an isolated island chain, equations (3.187), (3.188), (7.5), as well as the analysis of section 9.4, lead to the relation

$$\frac{\Delta\hat{\Psi}_s}{\hat{\Psi}_s} = \Delta - \frac{W}{W_s}, \tag{12.1}$$

where $\Delta \equiv \Delta_{pw}(0)$ is the (real dimensionless) perfect-wall tearing stability index evaluated at zero island width, $\hat{\Psi}_s$ the normalized reconnected helical magnetic flux at the rational surface (see equation (3.184)), $\Delta\hat{\Psi}_s$ the normalized helical sheet current density at the rational surface (see equation (3.183)), $W = 4\,w$ the full radial width of the magnetic island chain, and

$$W_s = \frac{r_s}{0.8\,\alpha_s^{\,2} - 0.27\,\beta_s - 0.09\,\alpha_s}. \tag{12.2}$$

Here, r_s is the minor radius of the rational surface, and the dimensionless parameters α_s and β_s are defined in equations (9.16) and (9.17), respectively. It is clear from the previous equation that

$$\mathrm{Im}\left(\frac{\Delta\hat{\Psi}_s}{\hat{\Psi}_s}\right) = 0, \tag{12.3}$$

which implies that zero electromagnetic torque is exerted at the rational surface (see section 3.13).

12.3 The island rotation frequency

Equations (11.144) and (12.3) yield

$$v' = 0. \tag{12.4}$$

It follows from equation (11.136) that $F(\Omega) = 0$. Hence, equations (11.113) and (11.130) imply that the magnetohydrodynamic (MHD) stream function of an isolated magnetic island chain takes the form

$$M(\Omega) = -\frac{1}{1+\tau}\left(1 - \alpha_\theta\,\frac{\eta_{\rm i}}{1+\eta_{\rm i}}\right)\begin{cases} 0 & -1 \leqslant \Omega \leqslant 1 \\ 1/\langle X^2\rangle & \Omega > 1 \end{cases}. \tag{12.5}$$

Here, Ω is a magnetic flux-surface label that takes the value zero at the island O-points, unity on the magnetic separatrix, and a value much larger than unity far from the island chain (see equation (11.101)). Moreover, $\eta_{\rm i}$ is a dimensionless measure of the ion temperature gradient at the rational surface (see equation (4.4)), and τ is the ratio of the electron and ion pressure gradients at the rational surface (see equation (4.5)). The quantity

$$\alpha_\theta = \alpha_1\,(1 - \alpha_2\,f_{\rm t\,s}) = 1.17\,(1 - 0.461\,f_{\rm t\,s}), \tag{12.6}$$

where $f_{\rm t\,s}$ is the fraction of trapped particles at the rational surface (see equation (2.202)), parameterizes the influence of ion neoclassical poloidal flow damping (see equations (2.217), (2.218), and (11.5)). Finally, the function $\langle X^2\rangle(\Omega)$ is specified in section 8.11. Note that $\langle X^2\rangle(\Omega) \to \sqrt{2\,\Omega}$ as $\Omega \to \infty$. It follows from equation (12.5) that

$$M(\Omega \to \infty) = -\frac{1}{1+\tau}\left(1 - \alpha_\theta\,\frac{\eta_{\rm i}}{1+\eta_{\rm i}}\right)\frac{1}{\sqrt{2\,\Omega}}. \tag{12.7}$$

However, equations (11.109) and (12.4) yield

$$M(\Omega \to \infty) = \frac{v}{\sqrt{2\,\Omega}}, \tag{12.8}$$

where (see equation (11.88))

$$v = \frac{\omega - \omega_{\rm E}}{\omega_*}. \tag{12.9}$$

Here, ω is the rotation frequency of magnetic island chain in the laboratory frame (see section 5.4), $\omega_{\rm E}$ the E-cross-B frequency at the rational surface (see equation (5.44)), and ω_* the diamagnetic frequency at the rational surface (see equation (5.47)). The previous three equations give

$$v = -\frac{1}{1+\tau}\left(1 - \alpha_\theta\,\frac{\eta_{\rm i}}{1+\eta_{\rm i}}\right), \tag{12.10}$$

which implies that

$$\omega = \omega_{\rm E} + \left(1 - \alpha_\theta\,\frac{\eta_{\rm i}}{1+\eta_{\rm i}}\right)\omega_{*\,{\rm i}}, \tag{12.11}$$

where

$$\omega_{*\,\mathrm{i}} = -\left(\frac{1}{1+\tau}\right)\omega_* \qquad (12.12)$$

is the ion diamagnetic frequency at the rational surface (see equation (5.46)). We conclude that, as a result of ion neoclassical poloidal flow damping (i.e. $\alpha_\theta \neq 0$), an isolated magnetic island chain does not exactly corotate with the unperturbed ion fluid at the rational surface (i.e. $\omega \neq \omega_{\mathrm{E}} + \omega_{*\,\mathrm{i}}$). (Recall that non-neoclassical theory leads to the prediction that a wide magnetic island chain does exactly corotate with the unperturbed ion fluid at the rational surface. See section 9.5.) Instead, assuming that $0 < \eta_{\mathrm{i}} < \infty$, and given that $0 < \alpha_\theta < 1$ under most circumstances (see equation (12.6)), we deduce that a wide magnetic island chain propagates in the electron diamagnetic direction with respect to the unperturbed ion fluid at the rational surface, but in the ion diamagnetic direction with respect to the unperturbed MHD fluid [21–23]. This behavior should be contrasted with that of a linear tearing mode, which propagates in the electron diamagnetic direction with respect to the unperturbed MHD fluid (see chapter 6).

According to equations (8.75), (8.76), (11.130), and (12.5), the electron and ion fluid stream functions take the respective forms

$$Y_{\mathrm{e}}(\Omega) = -1 + \alpha_\theta\,\frac{1}{1+\tau}\,\frac{\eta_{\mathrm{i}}}{1+\eta_{\mathrm{i}}}\,\frac{1}{\langle X^2\rangle}, \qquad (12.13)$$

$$Y_{\mathrm{i}}(\Omega) = \alpha_\theta\,\frac{1}{1+\tau}\,\frac{\eta_{\mathrm{i}}}{1+\eta_{\mathrm{i}}}\,\frac{1}{\langle X^2\rangle}. \qquad (12.14)$$

The poloidal components of the perpendicular velocities of the electron, MHD, and ion fluids in the island rest frame are related to their respective stream functions, Y_{e}, M, and Y_{i}, according to equations (8.72)–(8.74).

Figure 12.1 shows typical normalized perpendicular velocity profiles of the electron, MHD, and ion fluids across the O-points and the X-points of an isolated magnetic island chain in the island rest frame, as determined from equations (8.72)–(8.74), (12.5), (12.13), and (12.14). Here, r is the magnetic flux-surface minor radius, $\hat{V}_\theta = (m/r_{\mathrm{s}})\,V_{\perp\,\theta}/\omega_*$, and m is the poloidal mode number of the tearing mode. It can be seen that all three fluids corotate with the island chain inside the magnetic separatrix. However, unlike the non-neoclassical case (see figure 9.3), ion neoclassical poloidal flow damping (i.e. $\alpha_\theta \neq 0$) causes the ion fluid velocity profile to be modified by the presence of the island chain. In fact, all three velocity profiles are discontinuous across the magnetic separatrix. As explained in section 8.9, these discontinuities are resolved in a thin layer of thickness $4\,w_{\mathrm{d}}$ on the separatrix. It is clear from the figure that the island chain rotates in the ion diamagnetic (i.e. negative) direction with respect to the unperturbed E-cross-B frame (i.e. the unperturbed MHD fluid) at the rational surface, but in the electron diamagnetic (i.e. positive) direction with respect to the unperturbed ion fluid.

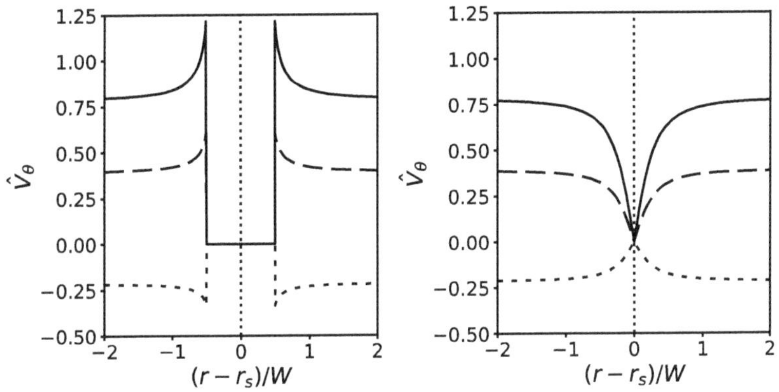

Figure 12.1. Normalized perpendicular velocity profiles across the O-points (left-hand panel) and the X-points (right-hand panel) of an isolated magnetic island chain in the island rest frame. The solid, dashed, and dotted lines correspond to the electron, MHD, and ion fluids, respectively. Here, $\tau = 1$, $\eta_i = 1$, and $f_{t\,s} = 0.565$ (see table 11.1).

12.4 The generalized Rutherford equation

Equations (11.149), (12.1), and (12.3) lead to the so-called *generalized Rutherford equation* [4] which governs the time evolution of the width of an isolated magnetic island chain:

$$
\begin{aligned}
I_1\,\tau_R\,\frac{d}{dt}\left(\frac{W}{r_s}\right) = \Delta - \frac{W}{W_s} &+ c_{bs}\,c_\beta^2\left(\frac{q_s}{\epsilon_s}\right)\left(\frac{L_s}{L_p}\right)\left(\frac{r_s}{W}\right)\\
&- c_{curv}\,c_\beta^2\left(\frac{L_s}{L_c}\right)\left(\frac{L_s}{L_p}\right)\left(\frac{r_s}{W}\right)\\
&- c_{polz}\,c_\beta^2\left(\frac{L_s}{L_p}\right)^2\left(\frac{d_\beta}{r_s}\right)^2\left(\frac{r_s}{W}\right)^3,
\end{aligned}
\tag{12.15}
$$

where

$$c_{bs} = 4\,I_2 f_{t\,s}\left[\beta_{11}\left(1 - \alpha_1\,\frac{1}{1+\tau}\,\frac{\eta_i}{1+\eta_i}\right) - \beta_{12}\,\frac{\tau}{1+\tau}\,\frac{\eta_e}{1+\eta_e}\right], \tag{12.16}$$

$$c_{curv} = 8\,I_2, \tag{12.17}$$

$$c_{polz} = 64\,I_3\left(\frac{1}{1+\tau}\right)^2\left(\alpha_\theta\,\frac{\eta_i}{1+\eta_i}\right)\left(1 - \alpha_\theta\,\frac{\eta_i}{1+\eta_i}\right). \tag{12.18}$$

Here, $I_1 = 0.8227$, $I_2 = 1.5835$, $I_3 = 1.3814$ (see section 11.10), $\alpha_1 = 1.17$ (see equation (2.217)), $\beta_{11} = 1.64$ (see equation (2.243)), and $\beta_{12} = 1.23$ (see equation (2.244)). Moreover, τ_R is the resistive diffusion time (see equation (5.49)), η_e a dimensionless measure of the electron temperature gradient at the rational surface (see equation (4.3)),

$q_s = q(r_s)$ the safety factor at the rational surface (see equation (3.2)), $\epsilon_s = \epsilon(r_s)$ the inverse aspect ratio at the rational surface (see equation (3.18)), $c_\beta^{\,2}$ a dimensionless measure of the plasma pressure at the rational surface (see equations (4.65) and (4.66)), L_s the magnetic shear length at the rational surface (see equation (5.27)), L_p the effective pressure gradient scale length at the rational surface (see equation (8.35)), L_c the magnetic curvature length at the rational surface (see equation (11.57)), and d_β the ion sound radius (see equation (4.75)).

The first term on the right-hand side of equation (11.149) is the linear tearing stability index, Δ [24]. A classical (i.e. non-neoclassical) tearing mode is unstable when $\Delta > 0$ and stable otherwise [25]. The free energy that drives a classical tearing mode is derived from equilibrium current gradients within the plasma (see chapter 3).

The second term on the right-hand side of equation (11.149) is a stabilizing (i.e. negative) saturation term that prevents an unstable classical tearing mode from growing indefinitely [26].

The third term on the right-hand side of equation (11.149) is a destabilizing (i.e. positive) term that is due to the loss of the bootstrap current within the magnetic separatrix of the island chain consequent on the flattening of the plasma pressure profile in this region [3].

The fourth term on the right-hand side of equation (11.149) is a stabilizing term due to magnetic field-line curvature [27, 28]. Note that this term also crucially depends on the flattening of the pressure profile within the magnetic separatrix.

The final term on the right-hand side of equation (11.149) represents the stabilizing effect of the ion polarization current induced in the vicinity of the rational surface when the ion fluid is diverted around the island chain's magnetic separatrix [29, 23, 30, 31]. Note that the polarization term is stabilizing because it is generally the case that

$$0 < \alpha_\theta \frac{\eta_i}{1 + \eta_i} < 1. \tag{12.19}$$

In other words, it is generally the case that the magnetic island chain rotates in the ion diamagnetic direction with respect to the unperturbed MHD fluid at the rational surface but in the electron diamagnetic direction with respect to the unperturbed ion fluid (see section 12.3). Note, further, that the dominant contribution to the polarization term comes from a discontinuity in the ion fluid velocity across the magnetic separatrix [31] (see figure 12.1). This discontinuity only occurs because ion neoclassical poloidal flow damping prevents the island chain from exactly corotating with the ion fluid (see section 12.3). Furthermore, the discontinuity is resolved in a layer of thickness 4 w_d on the separatrix, indicating that the polarization term is unlikely to survive if the island width falls below 4 w_d. Note, finally, that our polarization term does not exhibit the so-called *neoclassical enhancement of inertia* [30, 32] (according to which it would be a factor $(\epsilon_s/q_s)^2$ larger). As shown in reference [21], the neoclassical enhancement of inertia is only operative when the neoclassical ion poloidal flow-damping term is the dominant term in the ion equation of parallel motion. However, as is clear from the estimates made in the previous chapter, this is not likely to be the case in a tokamak fusion reactor.

Given that the plasma pressure profile is only flattened within the separatrix of the magnetic island chain when $W > 4\, w_{\rm d}$ and that the bootstrap, curvature, and polarization terms depend crucially on this flattening, it is conventional to modify equation (12.15) as follows [1, 4, 33]:

$$
\begin{aligned}
I_1\, \tau_{\rm R}\, \frac{d}{dt}\left(\frac{W}{r_{\rm s}}\right) = \Delta - \frac{W}{W_{\rm s}} &+ c_{\rm bs}\, c_\beta^2\left(\frac{q_{\rm s}}{\epsilon_{\rm s}}\right)\left(\frac{L_{\rm s}}{L_{\rm p}}\right)\left(\frac{r_{\rm s}}{W}\right)\left(\frac{W^2}{W^2 + W_{\rm d}^2}\right)\\
&- c_{\rm curv}\, c_\beta^2\left(\frac{L_{\rm s}}{L_{\rm c}}\right)\left(\frac{L_{\rm s}}{L_{\rm p}}\right)\left(\frac{r_{\rm s}}{\sqrt{W^2 + W_{\rm d}^2}}\right)\\
&- c_{\rm polz}\, c_\beta^2\left(\frac{L_{\rm s}}{L_{\rm p}}\right)^2\left(\frac{d_\beta}{r_{\rm s}}\right)^2\left(\frac{r_{\rm s}}{\sqrt{W^2 + W_{\rm d}^2}}\right)^3,
\end{aligned}
\qquad (12.20)
$$

where $W_{\rm d} = 4\, w_{\rm d}$, and $w_{\rm d}$ is defined in equation (8.84). The fact that the bootstrap term scales as W^1 at small island widths, while the curvature term scales as W^0, can be established analytically [1, 33]. The small-island-width behavior of the polarization term is a guess (that is irrelevant because the term turns out to be negligible). Other modifications are possible. For instance, if the island width were to fall below the ion banana width, then we would expect the ion contribution to the destabilizing bootstrap term to disappear (because the trapped ions would average over the flattening of the pressure profile within the separatrix, so there would be no reduction in the ion contribution to the bootstrap current in this region) [34].

We can write the generalized Rutherford equation in the form

$$
I_1\, \tau_{\rm R}\, \frac{a}{r_{\rm s}}\frac{d\hat{W}}{dt} = F(\hat{W}), \qquad (12.21)
$$

where

$$
F(\hat{W}) = \Delta - f_{\rm sat}\, \hat{W} + f_{\rm bs}\, \frac{\hat{W}}{\hat{W}^2 + \hat{W}_{\rm d}^2} - \frac{f_{\rm curv}}{\left(\hat{W}^2 + \hat{W}_{\rm d}^2\right)^{1/2}} - \frac{f_{\rm polz}}{\left(\hat{W}^2 + \hat{W}_{\rm d}^2\right)^{3/2}}, \qquad (12.22)
$$

and $\hat{W} = W/a$, $\hat{W}_{\rm d} = W_{\rm d}/a$,

$$
f_{\rm sat} = \frac{a}{W_{\rm s}}, \qquad (12.23)
$$

$$
f_{\rm bs} = c_{\rm bs}\, c_\beta^2\left(\frac{q_{\rm s}}{\epsilon_{\rm s}}\right)\left(\frac{L_{\rm s}}{L_{\rm p}}\right)\left(\frac{r_{\rm s}}{a}\right), \qquad (12.24)
$$

$$f_{\rm curv} = c_{\rm curv}\, c_\beta^2 \left(\frac{L_s}{L_c}\right)\left(\frac{L_s}{L_p}\right)\left(\frac{r_s}{a}\right), \tag{12.25}$$

$$f_{\rm polz} = c_{\rm polz}\, c_\beta^2 \left(\frac{L_s}{L_p}\right)^2\left(\frac{d_\beta}{r_s}\right)^2\left(\frac{r_s}{a}\right)^3. \tag{12.26}$$

Here, a is the plasma minor radius.

Table 12.1 shows estimates for the various dimensionless parameters that characterize the right-hand side of the generalized Rutherford equation, (12.22), for the cases of a low-field tokamak fusion reactor and a high-field tokamak fusion reactor (see chapter 1). These estimates are made using the following assumptions: $B = 5$ T (low field) or $B = 12$ T (high field), $\beta = 0.02$, $T_e = T_i = 7$ keV, $m_i = (m_D + m_T)/2$ (where m_D and m_T are the deuteron and triton masses, respectively), $\Xi_{\perp\,i} = \chi_{\perp\,e} = \chi_{\perp\,i} = 1\ {\rm m^2\,s^{-1}}$, $m = 2$, $n = 1$, $\tau = 1$, $\eta_e = \eta_i = 1$, and $dp/dr = -p/a$. The plasma equilibrium is assumed to be of the Wesson type with $q(0) = 0.8$, $r_s/a = 0.5$, and $r_w/a = 1.2$ (see section 9.4). It can be seen from the table that the parameters in the generalized Rutherford equation all take the same values for low-field reactors and high-field reactors, apart from the critical normalized island width above which the plasma pressure is flattened within the island chain's magnetic separatrix, $\hat{W}_d$, which is smaller in the former reactor type.

Figure 12.2 shows the various terms on the right-hand side of the generalized Rutherford equation, (12.22), plotted as functions of the normalized island width, $\hat{W}$, for a low-field tokamak fusion reactor and a high-field tokamak fusion reactor (using the parameters given in table 12.1). It can be seen that the destabilizing bootstrap term is the dominant term. Moreover, the stabilizing curvature term is much smaller than the bootstrap term, but is important at small island widths. Furthermore, the stabilizing saturation term is only important at large island widths. Finally, the stabilizing polarization term is completely negligible (because there is no neoclassical enhancement of inertia). It is clear that a high-field tokamak fusion

Table 12.1. Parameters characterizing the generalized Rutherford equation in a low-field tokamak fusion reactor and a high-field tokamak fusion reactor.

	Low field	High field
B (T)	5.0	12.0
$f_{\rm sat}$	7.47	7.47
$f_{\rm bs}$	0.712	0.712
$f_{\rm curv}$	3.99×10^{-2}	3.99×10^{-2}
$f_{\rm polz}$	9.34×10^{-7}	9.34×10^{-7}
$\hat{W}_d$	1.27×10^{-2}	1.70×10^{-2}

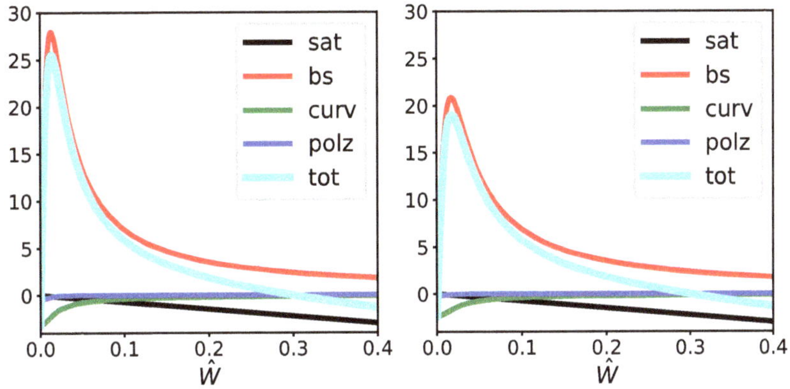

Figure 12.2. The various terms on the right-hand side of the generalized Rutherford equation calculated for a low-field tokamak fusion reactor (left-hand panel) and a high-field tokamak fusion reactor (right-hand panel) as functions of the normalized island width. Here, 'sat', 'bs', 'curv', and 'polz', refer to the saturation, bootstrap, curvature, and polarization terms, respectively, while 'tot' refers to the sum of all of these terms.

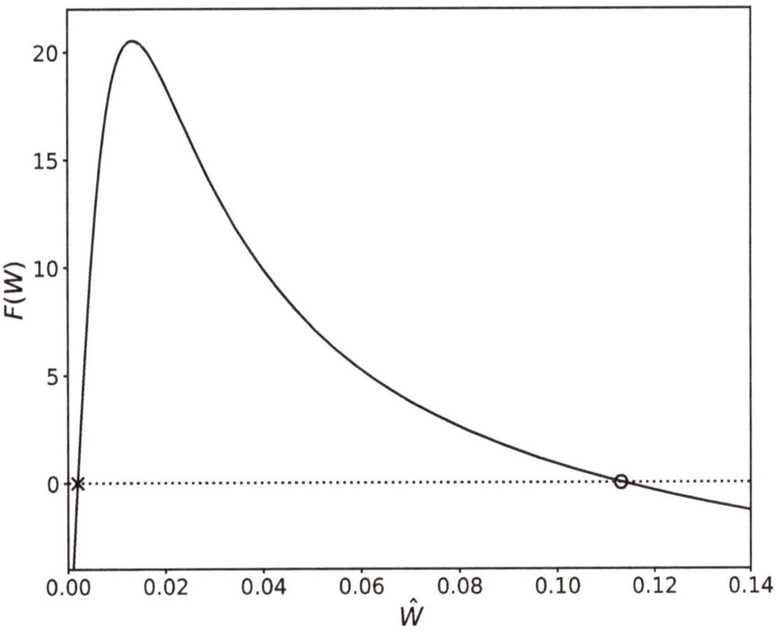

Figure 12.3. Right-hand side of the generalized Rutherford equation plotted as a function of the normalized island width for an $m = 2/n = 1$ neoclassical tearing mode in a low-field tokamak fusion reactor. The linear tearing stability index is assumed to take the value $\Delta = -5$.

reactor is slightly less susceptible to a neoclassical tearing mode than a low-field reactor (because the destabilizing bootstrap term is smaller in the former case).

Figure 12.3 shows the function $F(\hat{W})$ calculated for an $m = 2/n = 1$ tearing mode (m is the poloidal mode number, while n is the toroidal mode number) in a low-field tokamak fusion reactor (using the parameters given in the left-hand column of table 12.1) on the assumption that the linear tearing stability index takes the value $\Delta = -5$. The fact that $\Delta < 0$ implies that the classical 2/1 tearing mode is stable. On the other hand, as is clear from the figure, the neoclassical 2/1 tearing mode is potentially unstable because there exists a range of island widths for which $F(\hat{W}) > 0$. According to equation (12.21), the roots of $F(\hat{W}) = 0$ correspond to possible steady-state island widths. There are two such roots shown in the figure, marked with an X and an O. The smaller root (marked with an X) corresponds to a dynamically unstable equilibrium (because $dF/d\hat{W} > 0$), whereas the larger root (marked with an O) corresponds to a dynamically stable equilibrium (because $dF/d\hat{W} < 0$). In fact, the smaller root specifies a 'seed' island width that must be exceeded in order to trigger the neoclassical tearing mode, whereas the larger root corresponds to the mode's final saturated island width [4]. It can be seen that, in the example shown in the figure, the seed island width is about 0.2% of the plasma's minor radius, while the final saturated island width is about 11% of the minor radius.

12.5 Stabilization via rf-driven current

When rf electromagnetic waves (e.g. electron cyclotron waves [35]) are injected into a tokamak plasma and are absorbed in the vicinity of the rational surface of an unstable neoclassical tearing mode, they can potentially stabilize the mode by one of three different mechanisms. First, the current driven by the absorbed waves [36] can locally modify the equilibrium toroidal current profile, leading to a decrease in the linear tearing stability index, Δ [37]. Second, the heating of the plasma due to the absorbed waves can locally decrease the plasma resistivity, leading to a local enhancement of the inductive toroidal plasma current that offsets the destabilizing reduction in the bootstrap current consequent on the flattening of the pressure profile within the mode's magnetic separatrix [7]. Third, the current driven by the absorbed waves can directly fill in the hole in the bootstrap current profile left by the flattening of the pressure profile within the separatrix [7, 19]. We shall concentrate on the third stabilization mechanism, which is the only practical one in a tokamak fusion reactor.

In the presence of rf-driven current, equation (11.2) generalizes such that

$$\eta_{\|}\,(j_{\|} - j_{bs})\,\mathbf{b} \rightarrow \eta_{\|}\,(j_{\|} - j_{bs} - j_{rf})\,\mathbf{b}. \tag{12.27}$$

Here, j_{rf} is the parallel current density generated by absorbed rf waves. If $\hat{j}_{rf} = j_{rf}/(B_z/\mu_0\, r_s)$, where B_z is the equilibrium toroidal magnetic field strength, then it is convenient to write

$$\hat{j}_{rf} = \zeta_{rf}\,\alpha_{bs}\,c_\beta^{2}\,\frac{r_s}{L_p}\,F_{rf}(\hat{x},\,\zeta). \tag{12.28}$$

Here, α_{bs} is defined in equation (11.10); $\hat{x} = (r - r_s)/r_s$, where r is a radial cylindrical coordinate; and $\zeta = m\,\theta - n\,\varphi$, where θ and φ are poloidal and toroidal angles, respectively. Moreover, $F_{rf}(\hat{x},\,\zeta)$ is the rf-driven current density profile, normalized such that the maximum value of F_{rf} is unity. Finally, ζ_{rf} is the ratio of the maximum rf-driven current density to the equilibrium bootstrap current density at the rational surface.

In the presence of rf current drive, equation (11.68) generalizes to give

$$\frac{d(\ln \hat{w}^2)}{dT}\cos\zeta = \left\{\Phi - \frac{\tau}{1+\tau}\mathcal{N},\,\Psi\right\} + \epsilon_\beta\,\epsilon_c\,\epsilon_R\,\mathcal{J} + \zeta_{bs}\,\epsilon_\beta\,\epsilon_c\,\epsilon_R\,(\partial_X\mathcal{N} - 1 + \zeta_{rf}\,F_{rf}). \tag{12.29}$$

Hence, equation (11.126) becomes

$$\bar{\mathcal{J}}_0(\Omega,\,T) = \frac{1}{\epsilon_\beta\,\epsilon_c\,\epsilon_R}\,\frac{d(\ln \hat{w}^2)}{dT}\,\frac{\langle\cos\zeta\rangle}{\langle 1\rangle} - \zeta_{bs}\left(\frac{L}{\langle 1\rangle} - 1 + \zeta_{rf}\,\frac{\langle F_{rf}\rangle}{\langle 1\rangle}\right), \tag{12.30}$$

and, after some straightforward analysis, the right-hand side of the generalized Rutherford equation, (12.22), is modified such that

$$F(\hat{W}) = \Delta - f_{sat}\,\hat{W} + \frac{f_{bs}}{\hat{W}}\left(\frac{\hat{W}^2}{\hat{W}^2 + \hat{W}_d^{\,2}} - \eta_{rf}\,\zeta_{rf}\right) - \frac{f_{curv}}{\left(\hat{W}^2 + \hat{W}_d^{\,2}\right)^{1/2}} - \frac{f_{polz}}{\left(\hat{W}^2 + \hat{W}_d^{\,2}\right)^{3/2}}, \tag{12.31}$$

where

$$\eta_{rf} = -\frac{4}{I_2}\int_{-1}^{\infty}\frac{\langle F_{rf}\rangle\,\langle\cos\zeta\rangle}{\langle 1\rangle}\,d\Omega = -\frac{16}{I_2}\int_0^{\infty}\langle F_{rf}\rangle\left(2\,k^2 - 1 - \frac{2\,k^2\,\mathcal{C}}{\mathcal{A}}\right)k\,dk. \tag{12.32}$$

Here, the flux-surface average operator $\langle\cdots\rangle$ is defined in section 8.6, $k = [(1+\Omega)/2]^{1/2}$, and the functions $\mathcal{A}(k)$ and $\mathcal{C}(k)$ are defined in section 8.11.

We can think of the dimensionless parameter η_{rf} as a sort of rf current drive efficiency. If $\eta_{rf} = 1$, then the destabilizing bootstrap term (i.e. the third term) on the right-hand side of the modified generalized Rutherford equation, (12.31), can be completely suppressed (assuming that $\hat{W} \gg \hat{W}_d$) by driving a current density at the rational surface that matches the missing bootstrap current density (i.e. $\zeta_{rf} = 1$). In this case, the rf current drive is 100% efficient at stabilizing the neoclassical tearing mode. On the other hand, if $\eta_{rf} = 0.5$, then the destabilizing bootstrap term in the generalized Rutherford equation can only be completely eliminated by driving a current density at the rational surface that is twice the missing bootstrap current density (i.e. $\zeta_{rf} = 2$). In this case, the rf current drive is only 50% efficient.

Suppose that the normalized rf-driven current density profile is such that

$$F_{rf}(X,\,\zeta) = \exp\left[-\frac{(X-\delta_{rf})^2}{\Delta_{rf}^2}\right] \times \begin{cases} 1 & \pi\,(1-\tau_{rf}) < \zeta < \pi\,(1+\tau_{rf}) \\ 0 & \text{otherwise} \end{cases}, \tag{12.33}$$

where $X = 4\,(r - r_s)/W$. In other words, the profile is a Gaussian whose full width at half maximum (in r) is $0.416\,\Delta_{rf}\,W$; it is centered on the magnetic flux surface whose minor radius is offset from that of the rational surface by $0.25\,\delta_{rf}\,W$. Moreover, the driven current is concentrated at the island chain O-points. The degree of concentration is measured by the so-called duty cycle parameter, τ_{rf}. ($\tau_{rf} = 1$ corresponds to no concentration, whereas $\tau_{rf} \to 0$ corresponds to complete concentration.) In practice, the concentration is achieved by modulating the rf power such that it is turned on each time an island O-point rotates past the source and turned off each time an island X-point rotates past the source. The duty cycle parameter, τ_{rf}, is the fraction of time that the rf power is turned on.

It can easily be demonstrated that

$$\langle F_{rf}\rangle(k) = \frac{1}{\pi}\int_0^{\tau_{rf}\,\pi/2} \frac{\mathcal{F}_{rf}\left(2\sqrt{k^2 - \sin^2\varphi}\right)}{\sqrt{k^2 - \sin^2\varphi}}\,d\varphi \tag{12.34}$$

for $k > 1$, and

$$\langle F_{rf}\rangle(k) = \frac{1}{\pi}\int_0^{\varphi_{max}} \frac{\mathcal{F}_{rf}(2\,k\,\cos\varphi)}{\sqrt{1 - k^2\,\sin^2\varphi}}\,d\varphi \tag{12.35}$$

for $0 \leqslant k < 1$, where

$$\mathcal{F}_{rf}(X) = \frac{1}{2}\left(\exp\left[-\frac{(X - \delta_{rf})^2}{\Delta_{rf}^2}\right] + \exp\left[-\frac{(X + \delta_{rf})^2}{\Delta_{rf}^2}\right]\right), \tag{12.36}$$

$$\varphi_{max} = \begin{cases} \sin^{-1}[\sin(\tau_{rf}\,\pi/2)/k] & \sin(\tau_{rf}\,\pi/2)/k < 1 \\ \pi/2 & \text{otherwise} \end{cases}. \tag{12.37}$$

Figures 12.4 and 12.5 illustrate how the current drive efficiency parameter, η_{rf}, varies with the current drive profile width parameter, Δ_{rf}, and the duty cycle parameter, τ_{rf}, in cases in which there is no misalignment between the current drive profile and the rational surface (i.e. $\delta_{rf} = 0$). It can be seen that η_{rf} attains a maximum value of about 0.94 when $\Delta_{rf} \simeq 2$ and $\tau_{rf} \simeq 0.6$ [38]. Thus, the optimal full width at half maximum of the current drive profile is about 83% of the width of the magnetic island chain. It can also be seen that optimal modulation of the driven current increases the maximum achievable current drive efficiency by about 50%, while reducing the required rf power by 40% [38].

Figure 12.6 illustrates what happens when the current drive profile is misaligned with the rational surface (i.e. when the peak driven current density does not lie on the rational surface). It can be seen that misalignment decreases the current drive efficiency. Indeed, if the misalignment becomes too great, then the efficiency becomes negative (indicating that the current drive destabilizes, rather than stabilizes, the neoclassical tearing mode) [38]. When $\Delta_{rf} \simeq 2$, the maximum tolerable misalignment is about $\delta_{rf} \simeq 2.5$, which corresponds to a misalignment in minor radius of about 63% of the island width [38].

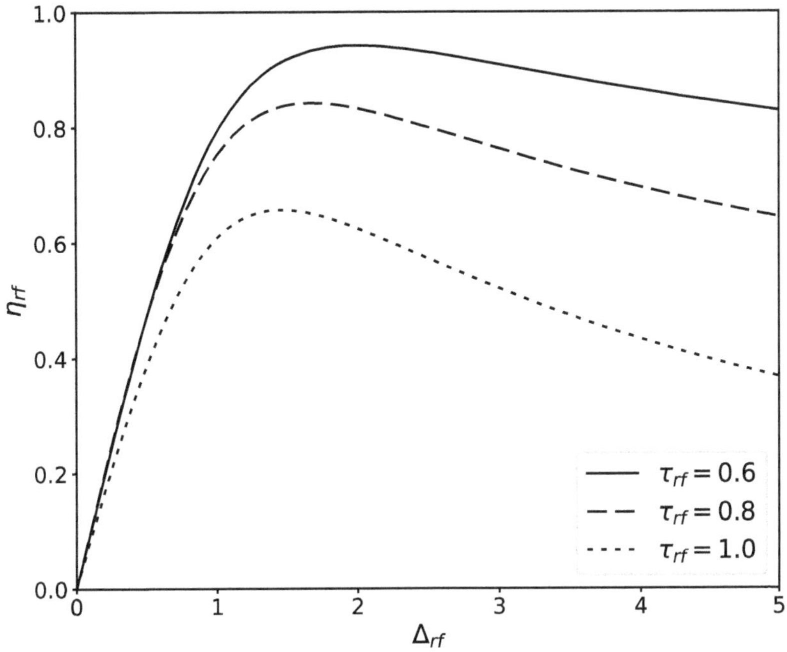

Figure 12.4. Rf current drive efficiency parameter, $\eta_{\rm rf}$, calculated as a function of the current drive profile width parameter, $\Delta_{\rm rf}$, for various different values of the duty cycle parameter, $\tau_{\rm rf}$. Here, $\delta_{\rm rf} = 0$.

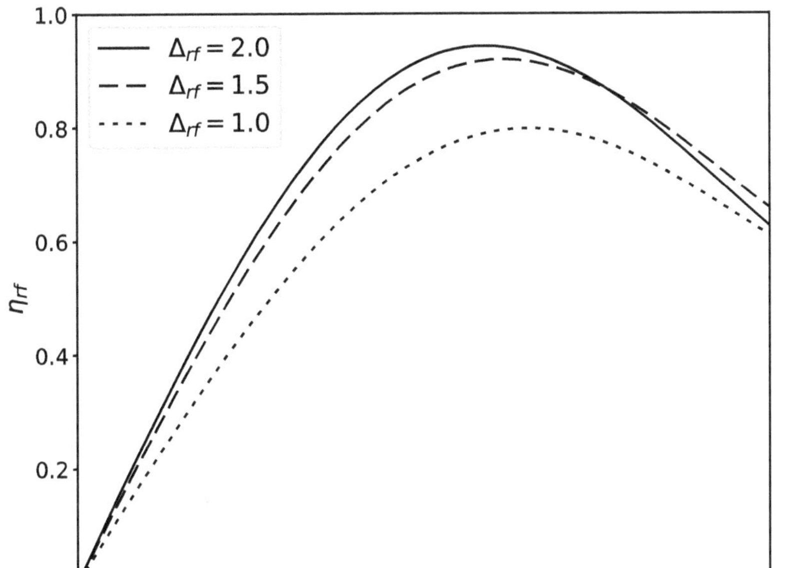

Figure 12.5. Rf current drive efficiency parameter, $\eta_{\rm rf}$, calculated as a function of the duty cycle parameter, $\tau_{\rm rf}$, for various different values of the current drive profile width parameter, $\Delta_{\rm rf}$. Here, $\delta_{\rm rf} = 0$.

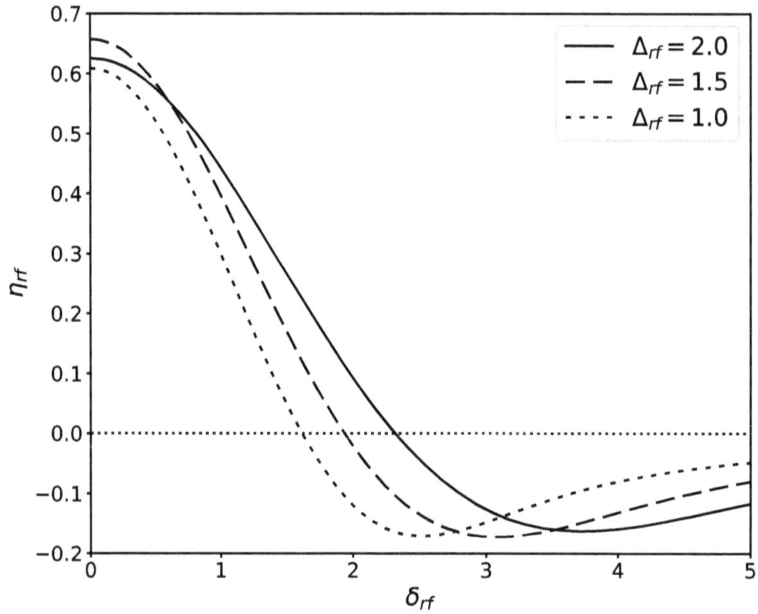

Figure 12.6. Rf current drive efficiency parameter, $\eta_{\rm rf}$, calculated as a function of the current drive profile misalignment parameter, $\delta_{\rm rf}$, for various different values of the current drive profile width parameter, $\Delta_{\rm rf}$. Here, $\tau_{\rm rf} = 1$.

The broad conclusion that can be drawn from figures 12.4–12.5 is that the maximum current drive efficiency is achieved when the driven current profile matches the profile of the missing bootstrap current (i.e. when it is centered on the rational surface, has the radial extent of about an island width, and is slightly concentrated at the island O-points).

Consider a neoclassical tearing mode for which the right-hand side of the generalized Rutherford equation case is specified in figure 12.3. Suppose that the magnetic island chain has achieved its final steady-state width, $\hat{W}_0 \simeq 0.11$. Let us apply an rf current drive profile whose width and misalignment parameters are $\Delta_{\rm rf\,0}$ and $\delta_{\rm rf\,0}$, respectively. The applied current drive causes the island chain to shrink in width. Assuming that the current drive profile remains constant, the effective width and misalignment parameters when the normalized island width has shrunk to the value $\hat{W}$ are

$$\Delta_{\rm rf} = \frac{\Delta_{\rm rf\,0}\,\hat{W}_0}{\hat{W}}, \tag{12.38}$$

$$\delta_{\rm rf} = \frac{\delta_{\rm rf\,0}\,\hat{W}_0}{\hat{W}}, \tag{12.39}$$

respectively.

Figure 12.7 shows how the right-hand side of the generalized Rutherford equation specified in figure 12.3 is modified by an rf current drive profile with optimal width,

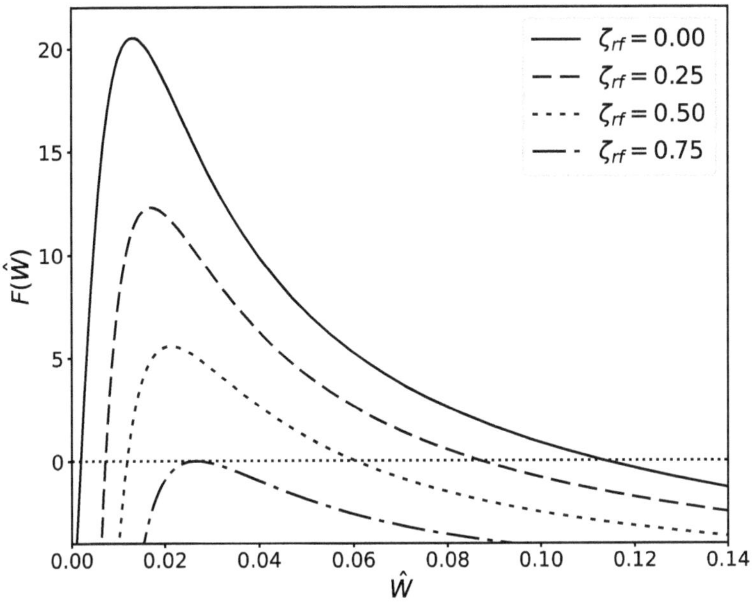

Figure 12.7. The right-hand side of the generalized Rutherford equation, plotted as a function of the normalized island width for an $m = 2/n = 1$ neoclassical tearing mode in a low-field tokamak fusion reactor and evaluated for various different current drive amplitudes. The linear tearing stability index is assumed to take the value $\Delta = -5$. Moreover, $\Delta_{\rm rf\,0} = 2.0$, $\delta_{\rm rf\,0} = 0.0$, and $\tau_{\rm rf} = 0.6$.

$\Delta_{\rm rf\,0} = 2.0$ (i.e. a full width at half maximum of 9.2% of the plasma's minor radius), optimal alignment, $\delta_{\rm rf\,0} = 0.0$, and optimal duty cycle, $\tau_{\rm rf} = 0.6$. It can be seen that the neoclassical tearing mode is completely stabilized (i.e. $F(\hat{W}) < 0$ for all $\hat{W}$) once the peak amplitude of the driven current density exceeds 75% of the equilibrium bootstrap current density at the rational surface (i.e. $\zeta_{\rm rf} > 0.75$). Of course, as soon as the island width has shrunk to a value less than the seed island width shown in figure 12.3 (i.e. $\hat{W} = 0.02$), the rf power can be switched off, and the neoclassical tearing mode will not return (unless it is triggered again by a transient resonant magnetic perturbation).

Figure 12.8 shows how the right-hand side of the generalized Rutherford equation specified in figure 12.3 is modified by an unmodulated (i.e. $\tau_{\rm rf} = 1$) rf current drive profile with optimal width, $\Delta_{\rm rf\,0} = 2.0$, and optimal alignment, $\delta_{\rm rf\,0} = 0.0$. It can be seen that the neoclassical tearing mode is completely stabilized once the peak amplitude of the driven current density exceeds 291% of the equilibrium bootstrap current density at the rational surface (i.e. $\zeta_{\rm rf} > 2.91$). The large increase in the requisite driven current density, relative to the modulated case shown in figure 12.7, comes about because unmodulated current drive becomes very inefficient at stabilizing a neoclassical tearing mode once the island width shrinks to such an extent that it is much less than the current drive profile width. In other words, $\eta_{\rm rf} \to 0$ as $\Delta_{\rm rf} \to \infty$ for unmodulated current drive, whereas $\eta_{\rm rf}$ remains finite as $\Delta_{\rm rf} \to \infty$ for modulated current drive [4, 39]. To be more specific, in the case of

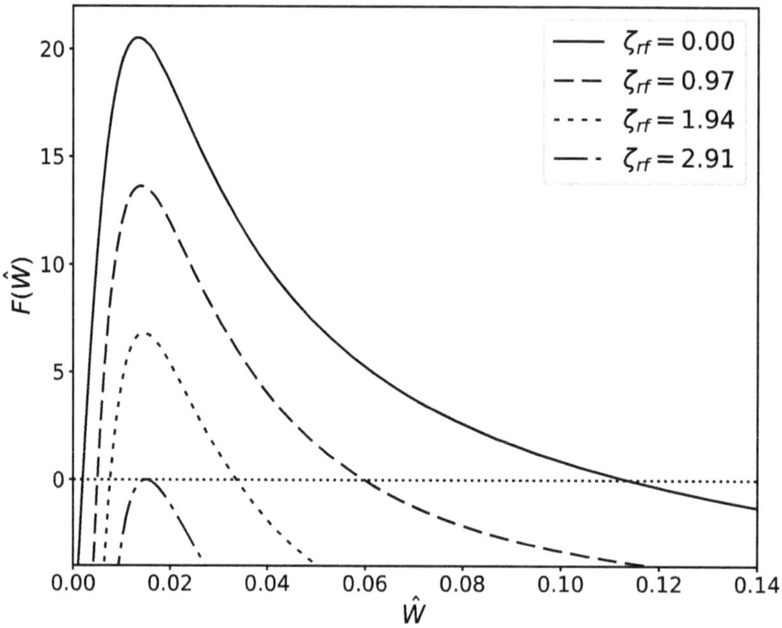

Figure 12.8. The right-hand side of the generalized Rutherford equation, plotted as a function of the normalized island width for an $m = 2/n = 1$ neoclassical tearing mode in a low-field tokamak fusion reactor and evaluated for various different current drive amplitudes. The linear tearing stability index is assumed to take the value $\Delta = -5$. Moreover, $\Delta_{\rm rf\,0} = 2.0$, $\delta_{\rm rf\,0} = 0.0$, and $\tau_{\rm rf} = 1.0$.

unmodulated current drive, the stabilizing effect of the current driven at the island O-points is partially canceled by the destabilizing effect of the current driven at the island X-points [18]. This cancellation becomes almost exact when the island width is much less than the current drive profile width [4]. Of course, there is no such cancellation in the case of modulated current drive because no current is driven at the X-points.

Figure 12.9 shows how the right-hand side of the generalized Rutherford equation specified in figure 12.3 is modified by an rf current drive profile with optimal width, $\Delta_{\rm rf\,0} = 2.0$, non-optimal alignment, $\delta_{\rm rf\,0} = 2.0$, and optimal duty cycle, $\tau_{\rm rf} = 0.6$. It can be seen that the misalignment of the current drive profile causes the critical peak current density required to stabilize the neoclassical tearing mode to rise to 261% of the equilibrium bootstrap current density at the rational surface.

The broad conclusion that can be drawn from figures 12.7–12.9 is that a neoclassical tearing mode can be stabilized by a properly aligned, modulated, current drive in which the driven current density profile has a similar width to the magnetic island chain and a similar peak magnitude to the equilibrium bootstrap current density at the rational surface.

Finally, the successful stabilization of neoclassical tearing modes via electron cyclotron wave current drive has been achieved in many tokamak experiments, including ASDEX Upgrade [20], DIII-D [16], and JT-60U [17].

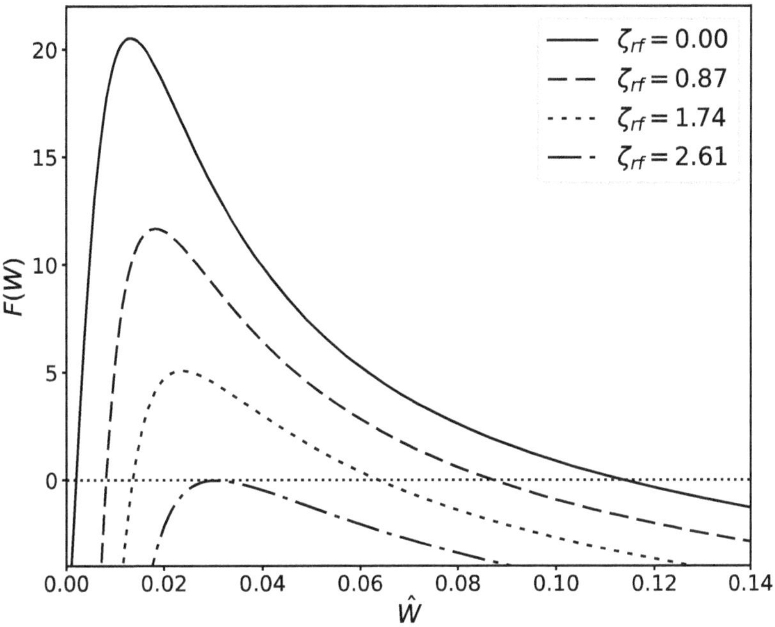

Figure 12.9. The right-hand side of the generalized Rutherford equation, plotted as a function of the normalized island width for an $m = 2/n = 1$ neoclassical tearing mode in a low-field tokamak fusion reactor and evaluated for various different current drive amplitudes. The linear tearing stability index is assumed to take the value $\Delta = -5$. Moreover, $\Delta_{\mathrm{rf}\,0} = 2.0$, $\delta_{\mathrm{rf}\,0} = 2.0$, and $\tau_{\mathrm{rf}} = 1.0$.

References

[1] Fitzpatrick R 1995 Helical temperature perturbations associated with tearing modes in tokamak plasmas *Phys. Plasmas* **2** 825

[2] Bickerton R J, Connor J W and Taylor J B 1971 Diffusion driven plasma currents and bootstrap tokamak *Nat. Phys. Sci.* **229** 110

[3] Carrera R, Hazeltine R D and Kotschenreuther M 1986 Island bootstrap current modification of the nonlinear dynamics of the tearing mode *Phys. Fluids* **29** 899

[4] La Haye R J 2006 Neoclassical tearing modes and their control *Phys. Plasmas* **13** 055501

[5] Chang Z, Callen J D, Fredrickson E D, Budny R V, Hegna C C, McGuire K M, Zarnstorff M C and the TFTR Group 1995 Observation of nonlinear neoclassical pressure-gradient-driven tearing modes in TFTR *Phys. Rev. Lett.* **74** 4663

[6] Gates D A, Lloyd B, Morris A W, McArdle G, O'Brien M, Valovic M, Warrick C D and Wilson H R and the COMPASS-D Team and ECRH Team 1997 Neoclassical islands on COMPASS-D *Nucl. Fusion* **37** 1593

[7] Hegna C C and Callen J D 1977 On the stabilization of neoclassical magnetohydrodynamic tearing modes using localized current drive or heating *Phys. Plasmas* **4** 2940

[8] Isayama A, Kamada Y, Ozeki T and Isei N 1999 Measurement of magnetic island width in long-pulse, high-discharges in JT-60U *Plasma Phys. Control. Fusion* **41** 35

[9] La Haye R J, Lao L L, Strait E J and Taylor T S 1997 High beta tokamak operation in DIII-D limited at low density/collisionality by resistive tearing modes *Nucl. Fusion* **37** 397

[10] Zohm H, Gates D A, Wilson H R, Gantenbein G, Gruber O, Günter S, Martaschek M, Morris A W, Sokoli M, Wagner D and the ASDEX Upgrade Team, COMPASS-D Team and ECRH Team 1977 Neoclassical MHD in ASDEX Upgrade and COMPASS-D *Plasma Phys. Control. Fusion* **39** B237

[11] Chang Z and Callen J D 1990 Global energy confinement degradation due to macroscopic phenomena in tokamaks *Nucl. Fusion* **30** 219

[12] Buttery R J *et al* 2000 Neoclassical tearing modes *Plasma Phys. Control. Fusion* **42** B61

[13] Sauter O *et al* 1997 Beta limits in long-pulse tokamak discharges *Phys. Plasmas* **4** 1654

[14] Hegna C C, Callen J D and La Haye R J 1999 Dynamics of seed magnetic island formation due to geometrically coupled perturbations *Phys. Plasmas* **6** 130

[15] Wesson J A 2011 *Tokamaks* 4th edn (Oxford: Oxford University Press) https://global.oup.com/academic/product/tokamaks-9780199592234

[16] La Haye R J, Günter S, Humphreys D A, Lohr J, Luce T C, Maraschek M E, Petty C C, Prater R, Scoville J T and Strait E J 2002 Control of neoclassical tearing modes in DIII-D *Phys. Plasmas* **9** 2051

[17] Nagasaki K, Isayama A, Ide S and the JT-60 Team 2003 Stabilization effect of early ECCD on a neoclassical tearing mode in the JT-60U tokamak *Nucl. Fusion* **43** L7

[18] Reiman A H 1983 Suppression of magnetic islands by RF-driven currents *Phys. Fluids* **26** 1338

[19] Zohm H 1997 Stabilization of neoclassical tearing modes by electron cyclotron current drive *Phys. Plasmas* **4** 3433

[20] Zohm H, Gantenbein G, Gude A, Günter S, Leuterer F, Maraschek M, Meskat J, Suttrop W, Yu Q, ASDEX Upgrade Team and ECRH-Group (AUG) 2001 Neoclassical tearing modes and their stabilization by electron cyclotron current drive in ASDEX upgrade *Phys. Plasmas* **8** 2009

[21] Fitzpatrick R 2016 An improved neoclassical drift-magnetohydrodynamical fluid model of helical magnetic island equilibria in tokamak plasmas *Phys. Plasmas* **23** 052506

[22] Fitzpatrick R and Waelbroeck F L 2009 Effect of flow damping on drift-tearing magnetic islands in tokamak plasmas *Phys. Plasmas* **16** 072507

[23] Smolyakov A I 1993 Nonlinear evolution of tearing modes in inhomogeneous plasmas *Plasma Phys. Control. Fusion* **35** 657

[24] Furth H P, Killeen J and Rosenbluth M N 1963 Finite-resistivity instabilities of a sheet pinch *Phys. Fluids* **6** 459

[25] Rutherford P H 1973 Nonlinear growth of the tearing mode *Phys. Fluids* **16** 1906

[26] Hastie R J, Militello F and Porcelli F 2005 Nonlinear saturation of tearing mode islands *Phys. Rev. Lett.* **95** 065001

[27] Hegna C C 1999 Nonlinear dynamics of pressure driven magnetic islands in low aspect ratio tokamaks *Phys. Plasmas* **6** 3980

[28] Kotschenreuther M, Hazeltine R D and Morrison P J 1985 Nonlinear dynamics of magnetic islands with curvature and pressure *Phys. Fluids* **28** 294

[29] Connor J W, Waelbroeck F L and Wilson H R 2001 The role of polarization current in magnetic island evolution *Phys. Plasmas* **8** 2835

[30] Smolyakov A I, Hirose A, Lazzaro E, Re G B and Callen J D 1995 Rotating nonlinear magnetic islands in a tokamak plasma *Phys. Plasmas* **2** 1581

[31] Waelbroeck F L and Fitzpatrick R 1997 Rotation and locking of magnetic islands *Phys. Rev. Lett.* **78** 1703

[32] Hinton F L and Robertson J A 1984 Neoclassical dielectric property of a tokamak plasma *Phys. Fluids* **27** 1243
[33] Lütjens H, Luciani J-F and Garbet X 2001 Curvature effects on the dynamics of tearing modes in tokamaks *Phys. Plasmas* **8** 4267
[34] Poli E, Peeters A G, Bergmann A, Günter S and Pinches S D 2002 Reduction of the ion drive and ρ_θ^* scaling of the neoclassical tearing mode *Phys. Rev. Lett.* **88** 075001
[35] Prater R 2004 Heating and current drive by electron cyclotron waves *Phys. Plasmas* **11** 2349
[36] Fisch N J 1978 Confining a tokamak plasma with RF-driven currents *Phys. Rev. Lett.* **41** 873
[37] Westerhof E 1990 Tearing mode stabilization by local current density perturbations *Nucl. Fusion* **30** 1143
[38] Prater R, La Haye R J, Lohr J, Luce T C, Petty C C, Ferron J R, Humphreys D A, Strait E J, Perkins F W and Harvey R W 2003 Discharge improvement through control of neoclassical tearing modes by localized ECCD in DIII-D *Nucl. Fusion* **43** 1128
[39] Sauter O 2004 On the contribution of local current density to neoclassical tearing mode stabilization *Phys. Plasmas* **11** 4808
[40] JET Team, prepared by Huysmans G T A 1999 Observation of neoclassical tearing modes in JET *Nucl. Fusion* **39** 1965

IOP Publishing

Tearing Mode Dynamics in Tokamak Plasmas

Richard Fitzpatrick

Chapter 13

Mode locking in tokamak plasmas

13.1 Introduction

A tearing mode in a tokamak plasma reconnects magnetic flux at the so-called *rational* magnetic flux surface, where the mode resonates with the equilibrium magnetic field, to produce a *rotating* (in the laboratory frame) helical magnetic island chain.

As previously mentioned in chapter 10, *mode locking* is a process by which the rotation of a slowly growing magnetic island chain in a tokamak plasma is braked due to electromagnetic interaction with a rigid, electrically conducting wall surrounding the plasma, causing the chain to eventually *lock* (i.e. become stationary in the laboratory frame) to a static resonant magnetic perturbation (e.g. an error field) [2, 3, 5, 6, 9, 14, 15, 17, 20, 22, 24, 27, 28]. Locked magnetic island chains are strongly correlated with disruptions (i.e. sudden, catastrophic losses of thermal and magnetic energy) in tokamak plasmas [8].

In chapter 10, we investigated the interaction of a rotating magnetic island chain with a conducting wall in the absence of a magnetic perturbation. In this chapter, we shall concentrate on the interaction of a rotating magnetic island chain with a static resonant magnetic perturbation in the absence of a wall. The generalization to the case in which there is both a wall and a magnetic perturbation is straightforward [28].

13.2 Asymptotic matching

Let us assume that the helical magnetic island chain possesses m periods in the poloidal direction and n periods in the toroidal direction. Let r_s be the minor radius of the rational surface.

Let us reuse the analysis given in sections 3.17, 7.2, and 9.4. In the absence of a wall (i.e. $\tau_w = 0$) but in the presence of a static magnetic perturbation generated by a steady helical current, I_c, with m periods in the poloidal direction and n periods in the toroidal direction, running through an external magnetic field coil located outside the plasma at minor radius r_c, we find that [9]

$$\frac{\Delta\hat{\Psi}_s}{\hat{\Psi}_s} = \Delta - \frac{W}{W_s} + \left(\frac{W_c}{W}\right)^2 e^{\,i\varphi}, \tag{13.1}$$

where

$$\varphi(t) = \int_0^t \omega(t')\,dt', \tag{13.2}$$

$$W_c = 4\left(\frac{r_s}{r_c}\right)^{m/2}\left(\frac{\mu_0\, I_c\, L_s}{B_z}\right)^{1/2}. \tag{13.3}$$

Here, $\Delta \equiv \Delta_{nw}(0)$ is the (real dimensionless) no-wall tearing stability index evaluated at zero island width (see equation (7.4)), $\hat{\Psi}_s$ the normalized reconnected helical magnetic flux at the rational surface (see equation (3.184)), $\Delta\hat{\Psi}_s$ the normalized helical sheet current density at the rational surface (see equation (3.183)), W the full radial width of the magnetic island chain (see equation (5.129)), φ the helical phase of the island chain measured with respect to that of the static resonant magnetic perturbation, ω the instantaneous island rotation frequency in the laboratory frame (see equation (3.189)), L_s the magnetic shear length at the rational surface (see equation (5.29)), and B_z the equilibrium toroidal magnetic field strength. Moreover, W_s is specified in equation (12.2), and W_c is a measure of the width of the locked magnetic island chain that the resonant magnetic perturbation would induce at the rational surface in the absence of localized shielding currents. Finally, we have made the simplifying approximation that the equilibrium plasma current external to the rational surface is negligible (see section 7.2).

13.3 The Rutherford island width evolution equation

Equations (11.149) and (13.1) yield the generalized Rutherford equation for a rotating magnetic island chain interacting with a static resonant magnetic perturbation (see section 12.4):

$$\begin{aligned} I_1\,\tau_R\,\frac{d}{dt}\left(\frac{W}{r_s}\right) &= \Delta - \frac{W}{W_s} + c_{bs}\,c_\beta^2\left(\frac{q_s}{\epsilon_s}\right)\left(\frac{L_s}{L_p}\right)\left(\frac{r_s}{W}\right) - c_{curv}\,c_\beta^2\left(\frac{L_s}{L_c}\right)\left(\frac{L_s}{L_p}\right)\left(\frac{r_s}{W}\right) \\ &\quad - c_{polz}\,c_\beta^2\left(\frac{L_s}{L_p}\right)^2\left(\frac{d_\beta}{r_s}\right)^2\left(\frac{r_s}{W}\right)^3 + c_{polz1}\,c_\beta\left(\frac{\epsilon_s}{q_s}\right)^2\left(\frac{L_s}{L_p}\right)\left(\frac{d_\beta}{r_s}\right)\left(\frac{\tau_\varphi}{\tau_H}\right)\left(\frac{W_c}{r_s}\right)^2\sin\varphi \\ &\quad + c_{polz2}\left(\frac{\epsilon_s}{q_s}\right)^4\left(\frac{\tau_\varphi}{\tau_H}\right)^2\left(\frac{W_c}{r_s}\right)^4\left(\frac{W}{r_s}\right)^3\sin^2\varphi + \left(\frac{W_c}{W}\right)^2\cos\varphi. \end{aligned} \tag{13.4}$$

Here, c_{bs}, c_{curv}, and c_{polz} are defined in equations (12.16), (12.17), and (12.18), respectively, while

$$c_{polz1} = \frac{I_4}{2^4}\left(\frac{1}{1+\tau}\right)\left(1 - 2\,\alpha_\theta\,\frac{\eta_i}{1+\eta_i}\right), \tag{13.5}$$

$$c_{\rm polz2} = \frac{I_5}{2^{14}}. \tag{13.6}$$

Moreover, $I_1 = 0.8227$, $\tau_{\rm R}$ is the resistive diffusion time (see equation (5.49)), $q_s = q(r_s) = m/n$ the safety factor at the rational surface (see equation (3.2)), $\epsilon_s = \epsilon(r_s)$ the inverse aspect ratio at the rational surface (see equation (3.18)), $c_\beta^{\,2}$ a dimensionless measure of the plasma pressure at the rational surface (see equations (4.65) and (4.66)), $L_{\rm p}$ the effective pressure gradient scale length at the rational surface (see equation (8.35)), $L_{\rm c}$ the magnetic curvature length at the rational surface (see equation (11.57)), d_β the ion sound radius (see equation (4.75)), $\tau_{\rm H}$ the hydromagnetic time (see equation (5.43)), τ_φ the toroidal momentum confinement time (see equation (5.50)), τ the ratio of the electron and ion pressure gradients at the rational surface (see equation (4.5)), and $\eta_{\rm i}$ a dimensionless measure of the ion temperature gradient at the rational surface (see equation (4.3)). Finally, the dimensionless ion neoclassical poloidal flow-damping parameter α_θ is defined in equation (12.6), and the dimensionless integrals I_4 and I_5 are specified in figure 11.1. Note that, in writing equation (13.4), we have assumed that $W \gg W_{\rm d}$ (see equation (12.20)) (i.e. that the island chain is sufficiently wide that the plasma pressure profile is flattened within its magnetic separatrix).

The first term on the right-hand side of equation (13.4) is the linear tearing stability index, Δ [11]. The second term is a stabilizing (i.e. negative) saturation term [12]. The third term is a destabilizing (i.e. positive) term that is due to the loss of the bootstrap current within the magnetic separatrix of the island chain consequent on the flattening of the plasma pressure profile in this region [4]. The fourth term is a stabilizing term due to magnetic field-line curvature [13, 18]. The fifth term represents the stabilizing effect of the ion polarization current induced in the vicinity of the rational surface when the ion fluid is diverted around the island chain's magnetic separatrix [7, 25–27]. The sixth and seventh terms are additional ion polarization terms that arise from the modified ion flow in the immediate vicinity of the island chain when the chain is subject to an externally generated electromagnetic torque [10, 27]. The final term is due to the direct interaction between the island chain and the resonant magnetic perturbation [9].

It is convenient to rewrite equation (13.4) in the form

$$\tilde{\tau}_{\rm R}\,\frac{d\hat{W}}{dt} = \Delta - f_{\rm sat}\,\hat{W} + \frac{f_{\rm bs}}{\hat{W}} - \frac{f_{\rm curv}}{\hat{W}} - \frac{f_{\rm polz}}{\hat{W}^3} + f_{\rm polz1}\,\hat{W}_{\rm c}^{\,2}\,\sin\varphi + f_{\rm polz2}\,\hat{W}_{\rm c}^{\,4}\,\hat{W}^3\,\sin^2\varphi$$
$$+\,\frac{\hat{W}_{\rm c}^{\,2}}{\hat{W}^2}\,\cos\varphi. \tag{13.7}$$

Here, $\tilde{\tau}_{\rm R} = I_1\,\tau_{\rm R}\,(a/r_s)$, $\hat{W} = W/a$, and $\hat{W}_{\rm c} = W_{\rm c}/a$, where a is the plasma's minor radius. Moreover, $f_{\rm sat}$, $f_{\rm bs}$, $f_{\rm curv}$, and $f_{\rm polz}$ are defined in equations (12.23)–(12.26), respectively, while

$$f_{\rm polz1} = c_{\rm polz1}\,c_\beta\left(\frac{\epsilon_s}{q_s}\right)^2\left(\frac{L_s}{L_{\rm p}}\right)\left(\frac{d_\beta}{r_s}\right)\left(\frac{\tau_\varphi}{\tau_{\rm H}}\right)\left(\frac{a}{r_s}\right)^2, \tag{13.8}$$

$$f_{\rm polz2} = c_{\rm polz2}\left(\frac{\epsilon_{\rm s}}{q_{\rm s}}\right)^4\left(\frac{\tau_\varphi}{\tau_{\rm H}}\right)^2\left(\frac{a}{r_{\rm s}}\right)^7. \tag{13.9}$$

Table 13.1 gives estimates for the various dimensionless parameters that characterize the right-hand side of the generalized Rutherford equation, (13.7), for the cases of a low-field tokamak fusion reactor and a high-field tokamak fusion reactor (see chapter 1). These estimates are made using the following assumptions: $B = 5\,{\rm T}$ (low field) or $B = 12\,{\rm T}$ (high field), $\beta = 0.02$, $T_{\rm e} = T_{\rm i} = 7\,{\rm keV}$, $m_{\rm i} = (m_{\rm D} + m_{\rm T})/2$ (where $m_{\rm D}$ and $m_{\rm T}$ are the deuteron and triton masses, respectively), $\Xi_{\perp\,{\rm i}} = \chi_{\perp\,{\rm e}} = \chi_{\perp\,{\rm i}} = 1\,{\rm m^2\,s^{-1}}$, $m = 2$, $n = 1$, $\tau = 1$, $\eta_{\rm e} = \eta_{\rm i} = 1$, and $dp/dr = -p/a$. The plasma equilibrium is assumed to be of the Wesson type with $q(0) = 0.8$, $r_{\rm s}/a = 0.5$, and $r_{\rm w}/a \rightarrow \infty$ (see section 9.4).

13.4 Island phase evolution equations

It can easily be demonstrated that

$${\rm Im}(\Delta\hat{\Psi}_{\rm s}\,\hat{\Psi}_{\rm s}^{\,*}) = {\rm Im}\left(\frac{\Delta\hat{\Psi}_{\rm s}}{\hat{\Psi}_{\rm s}}\right)|\hat{\Psi}_{\rm s}|^2 = \left(\frac{W^4}{2^8\,L_{\rm s}^{\,2}\,R_0^{\,2}}\right)\left(\frac{W_{\rm c}}{W}\right)^2\sin\varphi. \tag{13.10}$$

Here, R_0 is the major radius of the plasma, and use has been made of equations (5.129) and (13.1). The previous equation can be combined with equations (3.189)–(3.191) and (13.2) to give the following set of equations that govern the time evolution of the magnetic island chain's helical phase:

$$\omega(t) = \omega_0 - \sum_{p=1,\infty}\left[\alpha_{\rm p}(t) + \beta_{\rm p}(t)\right], \tag{13.11}$$

$$\varphi(t) = \int_0^t \omega(t')\,dt', \tag{13.12}$$

$$\frac{d\alpha_{\rm p}}{dt} + \left(\frac{1}{\tau_\theta} + \frac{j_{1{\rm p}}^{\,2}}{\tau_{\rm M}}\right)\alpha_{\rm p} = \frac{1}{2^8\,\tau_{\rm H}^{\,2}}\left(\frac{a}{r_{\rm s}}\right)^2\hat{W}_{\rm c}^{\,2}\,\hat{W}^2\,\frac{[J_1(j_{1{\rm p}}\,r_{\rm s}/a)]^2}{[J_2(j_{1{\rm p}})]^2}\sin\varphi, \tag{13.13}$$

$$\frac{d\beta_{\rm p}}{dt} + \frac{j_{0{\rm p}}^{\,2}}{\tau_{\rm M}}\beta_{\rm p} = \frac{1}{2^8\,\tau_{\rm H}^{\,2}}\left(\frac{a}{r_{\rm s}}\right)^2\left(\frac{\epsilon_{\rm s}}{q_{\rm s}}\right)^2\hat{W}_{\rm c}^{\,2}\,\hat{W}^2\frac{[J_0(j_{0{\rm p}}\,r_{\rm s}/a)]^2}{[J_1(j_{0{\rm p}})]^2}\sin\varphi. \tag{13.14}$$

Here, ω_0 is the natural frequency of the magnetic island chain (i.e. the chain's rotation frequency in the absence of the resonant magnetic perturbation), $\tau_\theta = \tau_{\theta\,{\rm i}}(r_{\rm s})$ the ion neoclassical poloidal flow-damping time at the rational surface (see equation (2.332)), and $\tau_{\rm M} = (a/r_{\rm s})^2\,\tau_\varphi$. Moreover, $J_{\rm m}(z)$ is a Bessel function, and $j_{\rm mp}$ denotes its pth zero [1]. In deriving the previous equations, use has been made of equations (4.23), (5.43), and (5.129).

13.5 The analytic solution of the phase evolution equations

Table 13.2 gives estimates for the characteristic timescales that govern the time evolution of the width and phase of a magnetic island chain interacting with a static resonant magnetic perturbation in a low-field tokamak fusion reactor and a high-field tokamak fusion reactor. These estimates are made using the same assumptions used to produce the estimates shown in table 13.1. In addition, the natural frequency of the tearing mode is assumed to be equal to the diamagnetic frequency, ω_* (see equation (5.47)) (see section 5.15). The island phase evolves on the timescale ω_0^{-1}, while the island width evolves on the timescale $\tilde{\tau}_R$. It can be seen that the former timescale is very much shorter than the latter. Thus, to a first approximation, it is reasonable to suppose that the island width remains constant as the island phase evolves. This implies that the island phase evolution equations, (13.11)–(13.14), can be solved independently of the island width evolution equation, (13.7).

Table 13.1. Dimensionless parameters characterizing the generalized Rutherford equation in a low-field tokamak fusion reactor and a high-field tokamak fusion reactor.

	Low field	High field
B (T)	5.0	12.0
f_{sat}	7.47	7.47
f_{bs}	0.712	0.712
f_{curv}	3.99×10^{-2}	3.99×10^{-2}
f_{polz}	9.34×10^{-7}	9.34×10^{-7}
f_{polz1}	1.77×10^{-2}	7.38×10^{-3}
f_{polz2}	6.71×10^{4}	1.17×10^{4}

Table 13.2. Characteristic parameters governing the time evolution of the width and phase of a magnetic island chain in a low-field tokamak fusion reactor and a high-field tokamak fusion reactor.

	Low field	High field
B (T)	5.0	12.0
τ_H (s)	6.72×10^{-7}	2.80×10^{-7}
$\tilde{\tau}_R$ (s)	2.30×10^{3}	4.00×10^{2}
τ_φ (s)	1.60	0.278
τ_M (s)	6.41	1.11
τ_θ (s)	2.22×10^{-4}	3.85×10^{-5}
ω_0^{-1} (s)	5.72×10^{-4}	2.38×10^{-4}
ϵ_s/q_s	8.33×10^{-2}	8.33×10^{-2}
r_s/a	0.5	0.5

In this section, we shall solve the island phase evolution equations, (13.11)–(13.14), analytically via an expansion in a dimensionless small parameter, $\delta \propto \hat{W}_c^{\,2}\,\hat{W}^2$, that is defined in equation (13.29). To the lowest order, equations (13.13) and (13.14) yield $\alpha_p = \beta_p = 0$. Hence, equations (13.11) and (13.11) give

$$\omega = \omega_0, \tag{13.15}$$

$$\varphi = \omega_0\,t, \tag{13.16}$$

respectively. Of course, this lowest-order solution corresponds to steady rotation of the island chain at its natural frequency.

To the first order, equations (13.13), (13.14), and (13.16) yield

$$\frac{d\alpha_p}{dt} + \left(\frac{1}{\tau_\theta} + \frac{j_{1p}^{\,2}}{\tau_M}\right)\alpha_p = \frac{1}{2^8\,\tau_H^2}\left(\frac{a}{r_s}\right)^2 \hat{W}_c^{\,2}\,\hat{W}^2\,\frac{[J_1(j_{1p}\,r_s/a)]^2}{[J_2(j_{1p})]^2}\,\sin(\omega_0\,t), \tag{13.17}$$

$$\frac{d\beta_p}{dt} + \frac{j_{0p}^{\,2}}{\tau_M}\,\beta_p = \frac{1}{2^8\,\tau_H^2}\left(\frac{a}{r_s}\right)^2\left(\frac{\epsilon_s}{q_s}\right)^2 \hat{W}_c^{\,2}\,\hat{W}^2\,\frac{[J_0(j_{0p}\,r_s/a)]^2}{[J_1(j_{0p})]^2}\,\sin(\omega_0\,t). \tag{13.18}$$

Let us write

$$\alpha_p(t) = \mathrm{Re}\left(\hat{\alpha}_{p\,1}\,e^{-i\,\omega_0\,t}\right), \tag{13.19}$$

$$\beta_p(t) = \mathrm{Re}\left(\hat{\beta}_{p\,1}\,e^{-i\,\omega_0\,t}\right). \tag{13.20}$$

It follows that

$$\hat{\alpha}_{p\,1} = i\,\frac{\tau_\varphi}{2^8\,\tau_H^2}\left(\frac{a}{r_s}\right)^4 \hat{W}_c^{\,2}\,\hat{W}^2\,\frac{[J_1(j_{1p}\,r_s/a)]^2}{[J_2(j_{1p})]^2\,[(-i\,\omega_0 + 1/\tau_\theta)\,\tau_M + j_{1p}^{\,2}]}, \tag{13.21}$$

$$\hat{\beta}_{p\,1} = i\,\frac{\tau_\varphi}{2^8\,\tau_H^2}\left(\frac{a}{r_s}\right)^4\left(\frac{\epsilon_s}{q_s}\right)^2 \hat{W}_c^{\,2}\,\hat{W}^2\,\frac{[J_0(j_{0p}\,r_s/a)]^2}{[J_1(j_{0p})]^2\,(-i\,\omega_0\,\tau_M + j_{0p}^{\,2})}. \tag{13.22}$$

Equations (13.11), (13.19), and (13.20) give

$$\omega = \omega_0 - \mathrm{Re}\left[\sum_{p=1,\infty}\left(\hat{\alpha}_{p\,1} + \hat{\beta}_{p\,1}\right)e^{-i\,\omega_0\,t}\right]. \tag{13.23}$$

However, according to equations (13.21) and (13.22),

$$\sum_{p=1,\infty}\hat{\alpha}_{p\,1} = i\,\frac{\tau_\varphi}{2^8\,\tau_H^2}\left(\frac{a}{r_s}\right)^4 \hat{W}_c^{\,2}\,\hat{W}^2\sum_{p=1,\infty}\frac{[J_1(j_{1p}\,r_s/a)]^2}{[J_2(j_{1p})]^2\,[(-i\,\omega_0 + 1/\tau_\theta)\,\tau_M + j_{1p}^{\,2}]}, \tag{13.24}$$

$$\sum_{p=1,\infty} \hat{\beta}_{p\,1} = \mathrm{i}\,\frac{\tau_\varphi}{2^8\,\tau_H^2}\left(\frac{a}{r_s}\right)^4\left(\frac{\epsilon_s}{q_s}\right)^2 \hat{W}_c^{\,2}\,\hat{W}^2 \sum_{p=1,\infty}\frac{[J_0(j_{0p}\,r_s/a)]^2}{[J_1(j_{0p})]^2\,(-\mathrm{i}\,\omega_0\,\tau_M + j_{0p}^2)}. \tag{13.25}$$

It can be demonstrated that

$$\sum_{p=1,\infty}\frac{[J_1(j_{1p}\,r_s/a)]^2}{[J_2(j_{1p})]^2\,[(-\mathrm{i}\,\omega_0 + 1/\tau_\theta)\,\tau_M + j_{1p}^2]} \simeq \frac{1}{4}\,\frac{1}{(-\mathrm{i}\,\omega_0 + 1/\tau_\theta)^{1/2}\,\tau_\varphi^{1/2}} \tag{13.26}$$

in the limit that $|(-\mathrm{i}\,\omega_0 + 1/\tau_\theta)\,\tau_M| \gg 1$. As is clear from table 13.2, the previous inequality is easily satisfied in a tokamak fusion reactor. Likewise, it can be demonstrated that

$$\sum_{p=1,\infty}\frac{[J_0(j_{0p}\,r_s/a)]^2}{[J_1(j_{0p})]^2\,(-\mathrm{i}\,\omega_0\,\tau_M + j_{0p}^2)} \simeq \frac{1}{4}\,\frac{1}{(-\mathrm{i}\,\omega_0\,\tau_\varphi)^{1/2}} \tag{13.27}$$

in the limit that $\omega_0\,\tau_M \gg 1$. As is apparent from table 13.2, the previous inequality is also easily satisfied in a tokamak fusion reactor. Thus, we obtain

$$\omega(t) = \omega_0[1 + \delta\,C_1\cos(\omega_0\,t) - \delta\,S_1\sin(\omega_0\,t)], \tag{13.28}$$

where

$$\delta = \frac{1}{2^{10}}\left(\frac{a}{r_s}\right)^4\left(\frac{\sqrt{\tau_\theta\,\tau_\varphi}}{\omega_0\,\tau_H^2}\right)\hat{W}_c^{\,2}\,\hat{W}^2, \tag{13.29}$$

and

$$C_1 = \frac{1}{\sqrt{2}}\left(\frac{\sqrt{1+\omega_0^2\,\tau_\theta^2} - 1}{1+\omega_0^2\,\tau_\theta^2}\right)^{1/2} + \frac{1}{\sqrt{2}}\left(\frac{\epsilon_s}{q_s}\right)^2\frac{1}{\sqrt{\omega_0\,\tau_\theta}}, \tag{13.30}$$

$$S_1 = \frac{1}{\sqrt{2}}\left(\frac{\sqrt{1+\omega_0^2\,\tau_\theta^2} + 1}{1+\omega_0^2\,\tau_\theta^2}\right)^{1/2} + \frac{1}{\sqrt{2}}\left(\frac{\epsilon_s}{q_s}\right)^2\frac{1}{\sqrt{\omega_0\,\tau_\theta}}. \tag{13.31}$$

According to equation (13.28), the electromagnetic locking torque exerted by the static magnetic perturbation on the island chain generates a periodic modulation in the chain's rotation frequency that oscillates at the natural frequency. Equations (13.11) and (13.28) give

$$\varphi(t) = \omega_0\,t - \delta\,S_1 + \delta\,S_1\cos(\omega_0\,t) + \delta\,C_1\sin(\omega_0\,t). \tag{13.32}$$

Finally, we obtain

$$\begin{aligned}\sin\varphi \simeq{}& \sin(\omega_0\,t) + \frac{1}{2}\,\delta\,S_1 - \delta\,S_1\cos(\omega_0\,t) + \frac{1}{2}\,\delta\,C_1\sin(2\,\omega_0\,t)\\ &+ \frac{1}{2}\,\delta\,S_1\cos(2\,\omega_0\,t),\end{aligned} \tag{13.33}$$

$$\cos\varphi \simeq \cos(\omega_0\,t) - \frac{1}{2}\,\delta\,C_1 + \delta\,S_1\sin(\omega_0\,t) + \frac{1}{2}\,\delta\,C_1\cos(2\,\omega_0\,t) - \frac{1}{2}\,\delta\,S_1\sin(2\,\omega_0\,t). \tag{13.34}$$

To the second order, equations (13.13), (13.14), and (13.33) yield

$$\frac{d\alpha_{\rm p}}{dt} + \left(\frac{1}{\tau_\theta} + \frac{j_{1\rm p}^{\,2}}{\tau_{\rm M}}\right)\alpha_{\rm p} = \frac{1}{2^8\,\tau_{\rm H}^2}\left(\frac{a}{r_{\rm s}}\right)^2 \hat{W}_{\rm c}^{\,2}\,\hat{W}^2\,\frac{[J_1(j_{1\rm p}\,r_{\rm s}/a)]^2}{[J_2(j_{1\rm p})]^2}\left[\sin(\omega_0\,t) + \frac{1}{2}\,\delta\,S_1 - \delta\,S_1\cos(\omega_0\,t) + \frac{1}{2}\,\delta\,C_1\sin(2\,\omega_0\,t) + \frac{1}{2}\,\delta\,S_1\cos(2\,\omega_0\,t)\right], \tag{13.35}$$

$$\frac{d\beta_{\rm p}}{dt} + \frac{j_{0\rm p}^{\,2}}{\tau_{\rm M}}\,\beta_{\rm p} = \frac{1}{2^8\,\tau_{\rm H}^2}\left(\frac{a}{r_{\rm s}}\right)^2\left(\frac{\delta_{\rm s}}{q_{\rm s}}\right)^2 \hat{W}_{\rm c}^{\,2}\,\hat{W}^2\,\frac{[J_0(j_{0\rm p}\,r_{\rm s}/a)]^2}{[J_1(j_{0\rm p})]^2}\left[\sin(\omega_0\,t) + \frac{1}{2}\,\delta\,S_1 - \delta\,S_1\cos(\omega_0\,t) + \frac{1}{2}\,\delta\,C_1\sin(2\,\omega_0\,t) + \frac{1}{2}\,\delta\,S_1\cos(2\,\omega_0\,t)\right]. \tag{13.36}$$

Let us write

$$\alpha_{\rm p}(t) = \alpha_{\rm p\,0} + {\rm Re}\left(\hat{\alpha}_{\rm p\,1}\,{\rm e}^{-{\rm i}\,\omega_0\,t} + \hat{\alpha}_{\rm p\,2}\,{\rm e}^{-{\rm i}\,2\,\omega_0\,t}\right), \tag{13.37}$$

$$\beta_{\rm p}(t) = \beta_{\rm p\,0} + {\rm Re}\left(\hat{\beta}_{\rm p\,1}\,{\rm e}^{-{\rm i}\,\omega_0\,t} + \hat{\beta}_{\rm p\,2}\,{\rm e}^{-{\rm i}\,2\,\omega_0\,t}\right). \tag{13.38}$$

If follows that

$$\alpha_{\rm p\,0} = \frac{\tau_\varphi}{2^9\,\tau_{\rm H}^2}\left(\frac{a}{r_{\rm s}}\right)^4 \hat{W}_{\rm c}^{\,2}\,\hat{W}^2\,\delta\,S_1\,\frac{[J_1(j_{1\rm p}\,r_{\rm s}/a)]^2}{[J_2(j_{1\rm p})]^2\,(\tau_{\rm M}/\tau_\theta + j_{1\rm p}^{\,2})}, \tag{13.39}$$

$$\beta_{\rm p\,0} = \frac{\tau_\varphi}{2^9\,\tau_{\rm H}^2}\left(\frac{a}{r_{\rm s}}\right)^4\left(\frac{\epsilon_{\rm s}}{q_{\rm s}}\right)^2 \hat{W}_{\rm c}^{\,2}\,\hat{W}^2\,\delta\,S_1\,\frac{[J_0(j_{0\rm p}\,r_{\rm s}/a)]^2}{[J_1(j_{0\rm p})]^2\,j_{0\rm p}^{\,2}}, \tag{13.40}$$

$$\hat{\alpha}_{\rm p\,1} = \frac{\tau_\varphi}{2^8\,\tau_{\rm H}^2}\left(\frac{a}{r_{\rm s}}\right)^4 \hat{W}_{\rm c}^{\,2}\,\hat{W}^2({\rm i} - \delta\,S_1)\frac{[J_1(j_{1\rm p}\,r_{\rm s}/a)]^2}{[J_2(j_{1\rm p})]^2\,[(-{\rm i}\,\omega_0 + 1/\tau_\theta)\,\tau_{\rm M} + j_{1\rm p}^{\,2}]}, \tag{13.41}$$

$$\hat{\beta}_{\rm p\,1} = \frac{\tau_\varphi}{2^8\,\tau_{\rm H}^2}\left(\frac{a}{r_{\rm s}}\right)^4\left(\frac{\epsilon_{\rm s}}{q_{\rm s}}\right)^2 \hat{W}_{\rm c}^{\,2}\,\hat{W}^2({\rm i} - \delta\,S_1)\frac{[J_0(j_{0\rm p}\,r_{\rm s}/a)]^2}{[J_1(j_{0\rm p})]^2\,(-{\rm i}\,\omega_0\,\tau_{\rm M} + j_{0\rm p}^{\,2})}, \tag{13.42}$$

$$\hat{\alpha}_{\rm p\,2} = \frac{\tau_\varphi}{2^9\,\tau_{\rm H}^2}\left(\frac{a}{r_{\rm s}}\right)^4 \hat{W}_{\rm c}^{\,2}\,\hat{W}^2\,\delta(S_1 + {\rm i}\,C_1)\frac{[J_1(j_{1\rm p}\,r_{\rm s}/a)]^2}{[J_2(j_{1\rm p})]^2\,[(-{\rm i}\,2\,\omega_0 + 1/\tau_\theta)\,\tau_{\rm M} + j_{1\rm p}^{\,2}]}, \tag{13.43}$$

$$\hat{\beta}_{p\,2} = \frac{\tau_\varphi}{2^9\,\tau_H^2}\left(\frac{a}{r_s}\right)^4\left(\frac{\epsilon_s}{q_s}\right)^2 \hat{W}_c^{\,2}\,\hat{W}^2\,\delta(S_1 + {\rm i}\,C_1)\frac{[J_0(j_{0p}\,r_s/a)]^2}{[J_1(j_{0p})]^2\,(-{\rm i}\,2\,\omega_0\,\tau_M + j_{0p}^2)}. \tag{13.44}$$

Equations (13.37)–(13.44) can be combined with equations (7.34), (7.35), (13.11), (13.26), and (13.27) to give

$$\begin{aligned}\omega(t) = \omega_0\bigg[&1 - \frac{1}{2}\,\delta^2\,S_0\,S_1 + \delta\,(C_1 + \delta\,S_1\,S_1)\cos(\omega_0\,t) - \delta\,(S_1 - \delta\,S_1\,C_1)\sin(\omega_0\,t)\\ &+ \frac{1}{2}\,\delta^2\,(C_1\,C_2 - S_1\,S_2)\cos(2\,\omega_0\,t) - \frac{1}{2}\,\delta^2\,(S_1\,C_2 + C_1\,S_2)\sin(2\,\omega_0\,t)\bigg],\end{aligned} \tag{13.45}$$

where

$$S_0 = 1 + 2\left(\frac{\tau_\varphi}{\tau_\theta}\right)^{1/2}\left(\frac{\epsilon_s}{q_s}\right)^2 \ln\left(\frac{a}{r_s}\right), \tag{13.46}$$

$$C_2 = \frac{1}{\sqrt{2}}\left(\frac{\sqrt{1 + 4\,\omega_0^{\,2}\,\tau_\theta^{\,2}} - 1}{1 + 4\,\omega_0^{\,2}\,\tau_\theta^{\,2}}\right)^{1/2} + \frac{1}{\sqrt{2}}\left(\frac{\epsilon_s}{q_s}\right)^2\frac{1}{\sqrt{2\,\omega_0\,\tau_\theta}}, \tag{13.47}$$

$$S_2 = \frac{1}{\sqrt{2}}\left(\frac{\sqrt{1 + 4\,\omega_0^{\,2}\,\tau_\theta^{\,2}} + 1}{1 + 4\,\omega_0^{\,2}\,\tau_\theta^{\,2}}\right)^{1/2} + \frac{1}{\sqrt{2}}\left(\frac{\epsilon_s}{q_s}\right)^2\frac{1}{\sqrt{2\,\omega_0\,\tau_\theta}}. \tag{13.48}$$

Let us define the *period-average operator*:

$$\langle A\rangle \equiv \frac{1}{T}\int_0^T A(t)\,dt, \tag{13.49}$$

where $T = 2\pi/\omega_0$. It is clear from equation (13.45) that

$$\langle\omega\rangle = \omega_0\left[1 - \frac{1}{2}\,\delta^2\,S_0\,S_1 + \mathcal{O}(\delta^3)\right]. \tag{13.50}$$

In other words, the electromagnetic locking torque exerted by a static resonant magnetic perturbation on the rotating island chain causes an average reduction in the chain's rotation frequency that is proportional to $\hat{W}_c^{\,4}\,W^4$ (i.e. to the square of the current flowing in the external magnetic field coil that generates the perturbation) (see equations (13.3) and (13.29)).

Equation (13.33) yields

$$\langle\sin\varphi\rangle = \frac{1}{2}\,\delta\,S_1 + \mathcal{O}(\delta^2). \tag{13.51}$$

The electromagnetic locking torque exerted on the island chain is proportional to $-\sin\varphi$. Furthermore, as the island chain rotates past the static resonant magnetic perturbation, its rotation is alternately accelerated (when $\sin\varphi < 0$) and decelerated (when $\sin\varphi > 0$) by this torque. However, as is apparent from the previous equation, the uneven rotation of the island chain in the presence of the perturbation (see equation (13.45)) causes it to spend more time in the helical phase relative to the perturbation in which it is *decelerated* by the torque than in the phase in which it is accelerated [9, 14, 24, 28]. This explains the net slowing of the chain's rotation indicated in equation (13.50).

Equation (13.34) yields

$$\langle\cos\varphi\rangle = -\frac{1}{2}\,\delta\,C_1 + \mathcal{O}(\delta^2). \tag{13.52}$$

The direct interaction between the static magnetic perturbation and the island chain generates a term in the Rutherford island width evolution equation that is proportional to $\cos\varphi$ (see equation (13.7)). Furthermore, as the island chain rotates past the static resonant magnetic perturbation, this term alternately destabilizes (when $\cos\varphi > 0$) and stabilizes (when $\cos\varphi < 0$) the chain. However, as is apparent from the previous equation, the uneven rotation of the island chain in the presence of the perturbation (see equation (13.45)) causes it to spend more time in the helical phase relative to the perturbation in which it is *stabilized* by the interaction term than in the phase that it is destabilized [9, 14, 24, 28]. Thus, there is a net stabilizing effect when a rotating magnetic island chain interacts directly with a static resonant magnetic perturbation.

13.6 A numerical solution of the phase evolution equations

Unfortunately, in the case of a rotating island chain, no further meaningful progress regarding the solution of the island phase evolution equations can be achieved via analysis. Figure 13.1 shows a numerical solution of the island phase evolution equations, (13.11)–(13.14), obtained for a low-field tokamak fusion reactor (i.e. using the parameters given in the left-hand column of table 13.2), for a case in which the mode-locking parameter, δ, is ramped linearly in time. It can be seen that the numerical solution is consistent with the approximate analytic solution described in the previous section. As the mode-locking parameter increases, the electromagnetic locking torque exerted by the static resonant magnetic perturbation on the island chain causes the chain to rotate in an increasingly irregular manner (see equation (13.45)). Close inspection of the figure reveals that the irregular rotation leads the island chain to spend more time in the helical phase relative to the perturbation in which $\sin\varphi > 0$, rather than the phase in which $\sin\varphi < 0$, and more time in the phase in which $\cos\varphi < 0$, rather than the phase in which $\cos\varphi > 0$, as is consistent with equations (13.51) and (13.52), respectively. The distortions of the initially sinusoidal oscillations shown in the bottom two panels of figure 13.1 as δ increases are known as *mode-locking distortions*, and have been observed (by means of magnetic pickup coil data) in many tokamak experiments [6, 14, 24, 28].

Figure 13.1 indicates that as soon as the mode-locking parameter, δ, exceeds a critical value that is of order unity, the island chain *locks* to the static resonant magnetic perturbation. In other words, the island chain ceases to rotate and instead

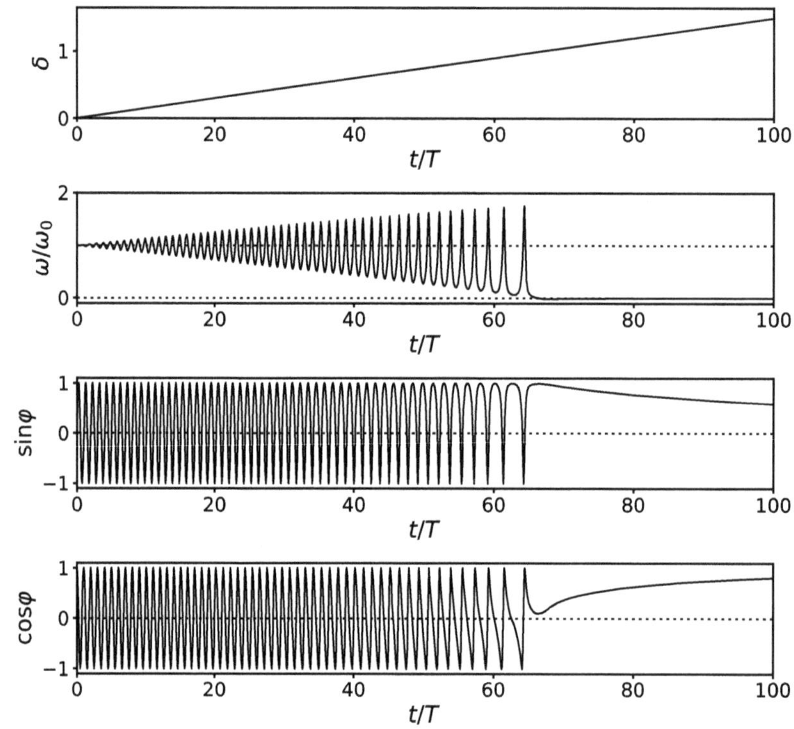

Figure 13.1. A numerical solution of the island phase evolution equations for a low-field tokamak fusion reactor.

maintains a fixed phase relation with respect to the perturbation. It can be seen that the island chain locks in a helical phase relative to the perturbation such that $\sin\varphi > 0$—that is, such that the electromagnetic locking torque opposes the viscous torque that acts to maintain the chain's rotation—and $\cos\varphi > 0$—that is, such that the resonant magnetic perturbation *destabilizes* the island chain [9, 14, 24, 28] (see section 13.7). Thus, we conclude that the direct interaction of a static resonant magnetic perturbation with a magnetic island chain is stabilizing when the chain is rotating, but becomes destabilizing as soon as the chain locks to the perturbation [9, 14, 24, 28].

Figure 13.2 shows the critical value of the locking parameter, δ, at which a rotating island chain locks to a static resonant magnetic perturbation, calculated as a function of the chain's natural frequency, for a low-field tokamak fusion reactor and a high-field tokamak fusion reactor. The calculations are performed using the parameters given in table 13.2, except that ω_0 is varied. It can be seen that the critical δ value exhibits some dependence on the exact phase at which the island chain locks. However, this dependence becomes less marked in the physically relevant limit in which the timescale on which the locking parameter is ramped is much longer than the rotation period of the chain (i.e. in the limit in which the ramp rate is decreased). It can also be seen that the critical δ value is approximately unity and only exhibits a fairly weak dependence on the natural frequency, ω_0. Adopting the rule of thumb

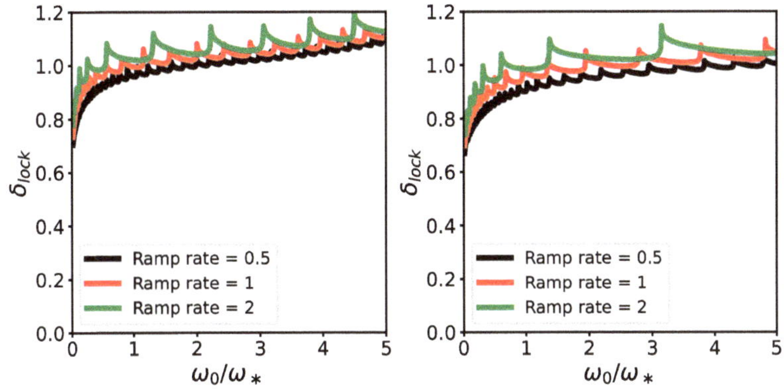

Figure 13.2. The critical value of the locking parameter, δ, at which a rotating island chain locks to a static resonant magnetic perturbation as a function of the chain's natural frequency for a low-field tokamak fusion reactor (left-hand panel) and a high-field tokamak fusion reactor (right-hand panel). The rate at which the locking parameter is ramped is either 0.5, 1, or 2 times that shown in figure 13.1.

that $\delta_{\rm lock} \simeq 1$, we deduce from equation (13.29) and table 13.2 that that the criterion for the locking of a rotating island chain to a static resonant magnetic perturbation is

$$\sqrt{\hat{W}_c\,\hat{W}} \gtrsim 0.040\left(\frac{\omega_0}{\omega_*}\right)^{1/4} \tag{13.53}$$

in a low-field tokamak fusion reactor, and

$$\sqrt{\hat{W}_c\,\hat{W}} \gtrsim 0.050\left(\frac{\omega_0}{\omega_*}\right)^{1/4} \tag{13.54}$$

in a high-field tokamak fusion reactor.

13.7 Locked magnetic island chains

In the case of a locked magnetic island chain, we can set $d/dt = 0$ in equations (13.13) and (13.14). It follows that

$$\alpha_{\rm p} = \frac{\tau_\varphi}{2^8\,\tau_{\rm H}^2}\left(\frac{a}{r_{\rm s}}\right)^4 \hat{W}_c^{\,2}\,\hat{W}^2\,\frac{[J_1(j_{1p}\,r_{\rm s}/a)]^2}{[J_2(j_{1p})]^2\,(\tau_{\rm M}/\tau_\varphi + j_{1p}^{\,2})}\sin\varphi, \tag{13.55}$$

$$\beta_{\rm p} = \frac{\tau_\varphi}{2^8\,\tau_{\rm H}^2}\left(\frac{a}{r_{\rm s}}\right)^4\left(\frac{\epsilon_{\rm s}}{q_{\rm s}}\right)^2 \hat{W}_c^{\,2}\,\hat{W}^2\,\frac{[J_0(j_{0p}\,r_{\rm s}/a)]^2}{[J_1(j_{0p})]^2\,j_{0p}^{\,2}}\sin\varphi. \tag{13.56}$$

Making use of equations (7.34) and (7.35), we get

$$\sum_{p=1,\infty}\alpha_{\rm p} = \frac{\sqrt{\tau_\varphi\,\tau_\theta}}{2^{10}\,\tau_{\rm H}^2}\left(\frac{a}{r_{\rm s}}\right)^4 \hat{W}_c^{\,2}\,\hat{W}^2\sin\varphi, \tag{13.57}$$

$$\sum_{p=1,\infty} \beta_p = \frac{\tau_\varphi}{2^9\, \tau_H^2}\left(\frac{a}{r_s}\right)^4\left(\frac{\epsilon_s}{q_s}\right)^2 \ln\left(\frac{a}{r_s}\right)\hat{W}_c^{\,2}\,\hat{W}^2 \sin\varphi. \tag{13.58}$$

Finally, setting $\omega = 0$ in equation (13.11), we obtain the torque balance criterion [9]

$$\delta\, S_0 \sin\varphi = 1. \tag{13.59}$$

Here, use has been made of equations (13.29) and (13.46).

The left-hand side of the previous equation is the normalized electromagnetic locking torque that acts to maintain a fixed helical phase relation between the island chain and the static resonant magnetic perturbation, whereas the right-hand side is the normalized viscous torque that attempts to force the island chain to rotate with respect to the perturbation [9]. In general, the previous equation possesses two solutions. However, only the solution in which $0 \leqslant \varphi \leqslant \pi/2$ is dynamically stable [9]. Hence, we conclude that the island chain locks in a helical phase relative to the perturbation such that $\sin\varphi > 0$ and $\cos\varphi > 0$, which is consistent with the numerical simulation shown in figure 13.1.

Actually, the previous equation only possesses solutions when $\delta > \delta_{\rm unlock}$, where

$$\delta_{\rm unlock} = \frac{1}{S_0}. \tag{13.60}$$

If the locking parameter, δ, falls below the critical value, $\delta_{\rm unlock}$, then torque balance breaks down, and the island chain is forced to rotate with respect to the static resonant magnetic perturbation.

Making use of the data given in table 13.2, we can now estimate that $S_0 = 1.81$ for both low-field tokamak fusion reactors and high-field tokamak fusion reactors. It follows that $\delta_{\rm unlock} \simeq 0.55$ in such reactors. According to figure 13.2, this value is about half the critical value of the locking parameter above which a rotating magnetic island chain locks to a static resonant magnetic perturbation. In other words, once the locking parameter has exceeded the critical value required to cause a rotating island chain to lock to a static resonant magnetic perturbation, it must be reduced by a factor of about two before the island chain will unlock from the perturbation and start to rotate again. This hysteresis between rotating and locked island states has been observed experimentally [14].

Finally, we deduce from equations (13.29) and (13.60) and table 13.2 that that the criterion for the unlocking of a rotating island chain from a static resonant magnetic perturbation is

$$\sqrt{\hat{W}_c\, \hat{W}} \lesssim 0.035\left(\frac{\omega_0}{\omega_*}\right)^{1/4} \tag{13.61}$$

in a low-field tokamak fusion reactor, and

$$\sqrt{\hat{W}_c\, \hat{W}} \lesssim 0.043\left(\frac{\omega_0}{\omega_*}\right)^{1/4} \tag{13.62}$$

in a high-field tokamak fusion reactor.

13.8 Island width evolution

Let us write the generalized Rutherford island width evolution equation, (13.7), in the form

$$\tilde{\tau}_R \frac{d\hat{W}}{dt} = \Delta - f_{sat}\, \hat{W} + \frac{f_{bs}}{\hat{W}} - \frac{f_{curv}}{\hat{W}} - \frac{f_{polz}}{\hat{W}^3} + F(\hat{W}_c, \hat{W}), \tag{13.63}$$

where

$$F(\hat{W}_c, \hat{W}) = f_{polz1}\, \hat{W}_c^{\,2} \sin\varphi + f_{polz2}\, \hat{W}_c^{\,4}\, \hat{W}^3 \sin^2\varphi + \frac{\hat{W}_c^{\,2}}{\hat{W}^2} \cos\varphi \tag{13.64}$$

represents the three additional terms that appear on the right-hand side of the equation when an island chain interacts with a static resonant magnetic perturbation.

First, let us consider a rotating magnetic island chain. Given that the island width evolves on a much longer timescale than the island phase (see table 13.2), it is a good approximation to replace $F(\hat{W}_c, \hat{W})$ by its period average in the generalized Rutherford equation [9, 10] (see equation (13.49)). We can now write

$$\langle F(\hat{W}_c, \hat{W})\rangle = f_{polz1}\, \hat{W}_c^{\,2} \langle\sin\varphi\rangle + f_{polz2}\, \hat{W}_c^{\,4}\, \hat{W}^3 \langle\sin^2\varphi\rangle + \frac{\hat{W}_c^{\,2}}{\hat{W}^2} \langle\cos\varphi\rangle. \tag{13.65}$$

However, to the lowest order, $\langle\sin^2\varphi\rangle = 1/2$, $\langle\sin\varphi\rangle = (1/2)\,\delta\, S_1$, and $\langle\cos\varphi\rangle = -(1/2)\,\delta\, C_1$ (see equations (13.51) and (13.52)). Thus, we obtain

$$\langle F(\hat{W}_c, \hat{W})\rangle = \frac{1}{2} f_{polz1}\, \hat{W}_c^{\,2}\, \delta\, S_1 + \frac{1}{2} f_{polz2}\, \hat{W}_c^{\,4}\, \hat{W}^3 - \frac{1}{2}\frac{\hat{W}_c^{\,2}}{\hat{W}^2}\, \delta\, C_1. \tag{13.66}$$

As is clear from table 13.1 and equations (13.29)–(13.31), the first two terms on the right-hand side of the previous equation, which are ion polarization terms that arise from the modified ion flow in the immediate vicinity of the island chain when the chain is subject to an electromagnetic locking torque, are destabilizing (i.e. positive). On the other hand, the third term, which is due to the direct interaction between the island chain and the static resonant magnetic perturbation, is stabilizing (i.e. negative).

If we use the data given in tables 13.1 and 13.2 to estimate the sizes of the three terms on the right-hand side of the previous equation in a tokamak fusion reactor, then we discover that the first two terms are completely negligible compared with the third. In other words, the polarization terms associated with the modified ion flow in the vicinity of the island chain are much smaller than the term due to the direct interaction between the island chain and the static resonant magnetic perturbation. In this limit, we can use equation (13.29) to write

$$\langle F(\hat{W}_c, \hat{W})\rangle \simeq -\frac{1}{2^{11}}\left(\frac{a}{r_s}\right)\left(\frac{\sqrt{\tau_\theta\, \tau_\varphi}}{\omega_0\, \tau_H^2}\right) C_1\, \hat{W}_c^{\,4}. \tag{13.67}$$

Thus, we conclude that the interaction between a rotating island chain and a static magnetic perturbation gives rise to a stabilizing term in the chain's Rutherford equation that is independent of the island width and scales as the square of the current flowing in the magnetic field coil that generates the perturbation (see equation (13.3)). Making use of the data given in table 13.2, we deduce that the stabilizing term takes the form

$$\langle F(\hat{W}_c, \hat{W})\rangle \simeq -\left(\frac{\hat{W}_c}{0.109}\right)^4 \tag{13.68}$$

in a low-field tokamak fusion reactor (with $\omega_0 = \omega_*$), and

$$\langle F(\hat{W}_c, \hat{W})\rangle \simeq -\left(\frac{\hat{W}_c}{7.33 \times 10^{-2}}\right)^4 \tag{13.69}$$

in a high-field tokamak fusion reactor. Of course, these estimates are only valid as long as the locking criteria (13.53) and (13.54) are not satisfied. Equations (13.53), (13.54), (13.68), and (13.69) suggest that the stabilizing effect of the interaction of a rotating magnetic island chain with a static resonant magnetic perturbation could completely stabilize a relatively narrow island chain but would not be able to completely stabilize a wide island chain, because the chain would lock to the perturbation before complete stabilization was achieved. The complete stabilization of a narrow rotating magnetic island chain by a static resonant magnetic perturbation has been observed experimentally [16].

Let us now consider a locked magnetic island chain. Equations (13.59), (13.60), and (13.64) yield

$$\begin{aligned} F(\hat{W}_c, \hat{W}) = {} & f_{\rm polz1}\left(\frac{\delta_{\rm unlock}}{\delta}\right)\hat{W}_c^{\,2} + f_{\rm polz2}\left(\frac{\delta_{\rm unlock}}{\delta}\right)^2 \hat{W}_c^{\,4}\,\hat{W}^3 \\ & + \left[1 - \left(\frac{\delta_{\rm unlock}}{\delta}\right)^2\right]^{1/2} \frac{\hat{W}_c^{\,2}}{\hat{W}^2}. \end{aligned} \tag{13.70}$$

Of course, the previous equation is only valid when $\delta > \delta_{\rm unlock}$ (i.e. provided that the island chain remains locked to the static resonant magnetic perturbation). Assuming that $\delta \sim \delta_{\rm unlock}$, $\hat{W}_c \lesssim 0.1$, and $\hat{W} \lesssim 0.1$, it is clear from the data given in table 13.1 that the first two terms on the right-hand side of the previous equation are completely negligible compared to the third. In other words, as in the case of a rotating island chain, the polarization terms in the Rutherford equation associated with the modified ion flow in the vicinity of the island chain are much smaller than the term due to the direct interaction between the island chain and the static resonant magnetic perturbation. In this case, we can write

$$F(\hat{W}_c, \hat{W}) \simeq \left[1 - \left(\frac{\delta_{\rm unlock}}{\delta}\right)^2\right]^{1/2} \left(\frac{\hat{W}_c}{\hat{W}}\right)^2. \tag{13.71}$$

Thus, it is clear that the destabilizing effect of a static resonant magnetic perturbation on a locked island chain can be significant, especially in the limit in which $\hat{W} \ll \hat{W}_c$. This destabilizing effect has been observed experimentally [14].

Finally, experiments have been performed in which the rotation frequency of a magnetic island chain in a tokamak plasma is controlled by locking the chain to a *rotating* resonant magnetic perturbation [23]. More ambitious experiments have been performed in which the helical phase of a rotating resonant magnetic perturbation is actively controlled in such a manner that a magnetic island chain in a tokamak plasma is held in a permanent stabilizing phase relation with respect to the perturbation [19, 21]. Unfortunately, neither of these control schemes is practical in a tokamak fusion reactor because the magnetic field coils that generate the rotating magnetic perturbation would have to be placed inside the vacuum vessel (otherwise the oscillating magnetic field generated by the coils would be shielded from the plasma by eddy currents excited in the vessel), where they would be unacceptably vulnerable to neutrons produced by fusion reactions within the plasma as well as large electromagnetic forces generated by plasma disruptions.

References

[1] Abramowitz M and Stegun I A (ed) 1964 *Handbook of Mathematical Functions with Formulas, Graphs, and Mathematical tables* (New York: Dover) https://store.doverpublications.com/0486612724.html

[2] Berge G, Sandal L K and Wesson J A 1989 Rotation and mode locking in tokamaks *Phys. Scr.* **40** 173

[3] Bondeson A and Persson M 1988 Stabilization by resistive walls and q-limit disruptions in tokamaks *Nucl. Fusion* **28** 1887

[4] Carrera R, Hazeltine R D and Kotschenreuther M 1986 Island bootstrap current modification of the nonlinear dynamics of the tearing mode *Phys. Fluids* **29** 899

[5] Chapman B E, Fitzpatrick R, Craig D, Martin P and Spizzo G 2004 Observation of tearing mode deceleration and locking due to eddy currents induced in a conducting shell *Phys. Plasmas* **11** 2156

[6] Chudnovskiy A N, Ivanov N V, Kakurin A M, Orlovskiy I I, Volkov V and the T-10 Team 2004 Tearing-mode natural frequency behaviour under mode-locking in T-10 tokamak *Nucl. Fusion* **44** 287

[7] Connor J W, Waelbroeck F L and Wilson H R 2001 The role of polarization current in magnetic island evolution *Phys. Plasmas* **8** 2835

[8] de Vries P C, Johnson M F, Alper B, Buratti P, Hender T C, Koslowski H R, Riccardo V and JET-EFDA Contributors 2011 Survey of disruption causes at JET *Nucl. Fusion* **51** 053018

[9] Fitzpatrick R 1993 Interaction of tearing modes with external structures in cylindrical geometry *Nucl. Fusion* **33** 1049

[10] Fitzpatrick R and Waelbroeck F L 2005 Two-fluid magnetic island dynamics in slab geometry. II. Islands interacting with resistive walls or resonant magnetic perturbations *Phys. Plasmas* **12** 022308

[11] Furth H P, Killeen J and Rosenbluth M N 1963 Finite-resistivity instabilities of a sheet pinch *Phys. Fluids* **6** 459

[12] Hastie R J, Militello F and Porcelli F 2005 Nonlinear saturation of tearing mode islands *Phys. Rev. Lett.* **95** 065001

[13] Hegna C C 1999 Nonlinear dynamics of pressure driven magnetic islands in low aspect ratio tokamaks *Phys. Plasmas* **6** 3980

[14] Hender T C *et al* 2091 Effect of resonant magnetic perturbations on COMPASS-C tokamak discharges *Nucl. Fusion* **32** 2091

[15] Hender T C, Gimblett C G and Robinson D C 1989 Effects of a resistive wall on magnetohydrodynamic instabilities *Nucl. Fusion* **29** 1279

[16] Hu Q, Rao B, Yu Q, Ding Y, Zhuang G, Jin W and Hu X 2013 Understanding the effect of resonant magnetic perturbations on tearing mode dynamics *Phys. Plasmas* **20** 092502

[17] Jensen T H and Chu M S 1983 A Linear model for the tearing mode of a tokamak plasma with flow and a resistive wall boundary condition *J. Plasma Phys.* **30** 57

[18] Kotschenreuther M, Hazeltine R D and Morrison P J 1985 Nonlinear dynamics of magnetic islands with curvature and pressure *Phys. Fluids* **28** 294

[19] Morris A W, Hender T C, Hugill J, Haynes P S, Johnson P C, Lloyd B, Robinson D C, Silvester C, Arshad S and Fishpool G M 1990 Feedback stabilization of disruption precursors in a tokamak *Phys. Rev. Lett.* **64** 1254

[20] Nave M F F and Wesson J A 1990 Mode locking in tokamaks *Nucl. Fusion* **30** 2575

[21] Navratil G A, Cates C, Mauel M E, Maurer D, Nadle D, Taylor E and Xiao Q 1998 Active control of 2/1 magnetic islands in a tokamak *Phys. Plasmas* **5** 1855

[22] Persson M and Bondeson A 1989 Wall locking and density limit disruptions *Nucl. Fusion* **29** 989

[23] Rao B *et al* 2013 First observation of rotation acceleration of magnetic island by using rotating resonant magnetic perturbation on the J-TEXT tokamak *Plasma Phys. Control. Fusion* **55** 122001

[24] Roberts D E, Sherwell D, Fletcher J D, Nothnagel G and de Villiers J A M 1991 Major disruptions induced by helical coils on the Tokoloshe tokamak *Nucl. Fusion* **31** 319

[25] Smolyakov A I 1993 Nonlinear evolution of tearing modes in inhomogeneous plasmas *Plasma Phys. Control. Fusion* **35** 657

[26] Smolyakov A I, Hirose A, Lazzaro E, Re G B and Callen J D 1995 Rotating nonlinear magnetic islands in a tokamak plasma *Phys. Plasmas* **2** 1581

[27] Waelbroeck F L and Fitzpatrick R 1997 Rotation and locking of magnetic islands *Phys. Rev. Lett.* **78** 1703

[28] Zohm H, Kallenbach A, Bruhns H, Fussmann G and Klüber O 1990 Plasma angular-momentum loss by MHD mode locking *Europhys. Lett.* **11** 745

Chapter 14

Toroidal tearing modes

14.1 Introduction

All of the analysis contained in chapters 3–13 is premised on the assumption that the geometry of a tokamak plasma can be adequately approximated by a periodic cylinder. In this final chapter, we shall move beyond this simplistic approach and take into account the fact that tokamak plasmas are actually *toroidal* in shape. Moreover, as illustrated in figure 14.1, the poloidal cross section of the equilibrium magnetic flux surfaces in a typical tokamak plasma are not circular, as assumed in the cylindrical model. Instead, they are highly elongated in the vertical direction and somewhat triangular. Finally, the outermost closed magnetic flux surface in a tokamak plasma possesses a magnetic X-point (see figure 14.1), which implies that the corresponding value of the safety factor is infinite (see figure 14.2). The aim of this chapter is to explore how the true geometry of the equilibrium magnetic flux surfaces in a tokamak plasma affects the dynamics of any tearing modes to which the plasma is subject.

14.2 Coordinate systems

Let X, Y, Z be a conventional right-handed Cartesian coordinate system (where Z measures vertical height).

Let R, φ, Z be the corresponding cylindrical coordinate system, where $R \equiv (X^2 + Y^2)^{1/2}$ and $\varphi \equiv \tan^{-1}(Y/X)$. It follows that

$$|\nabla\varphi| = \frac{1}{R}. \tag{14.1}$$

Finally, let us define a flux coordinate system: $r(R, Z)$, $\theta(R, Z)$, φ, where [5, 14]

$$(\nabla r \cdot \nabla\theta \times \nabla\varphi)^{-1} = \mathcal{J}, \tag{14.2}$$

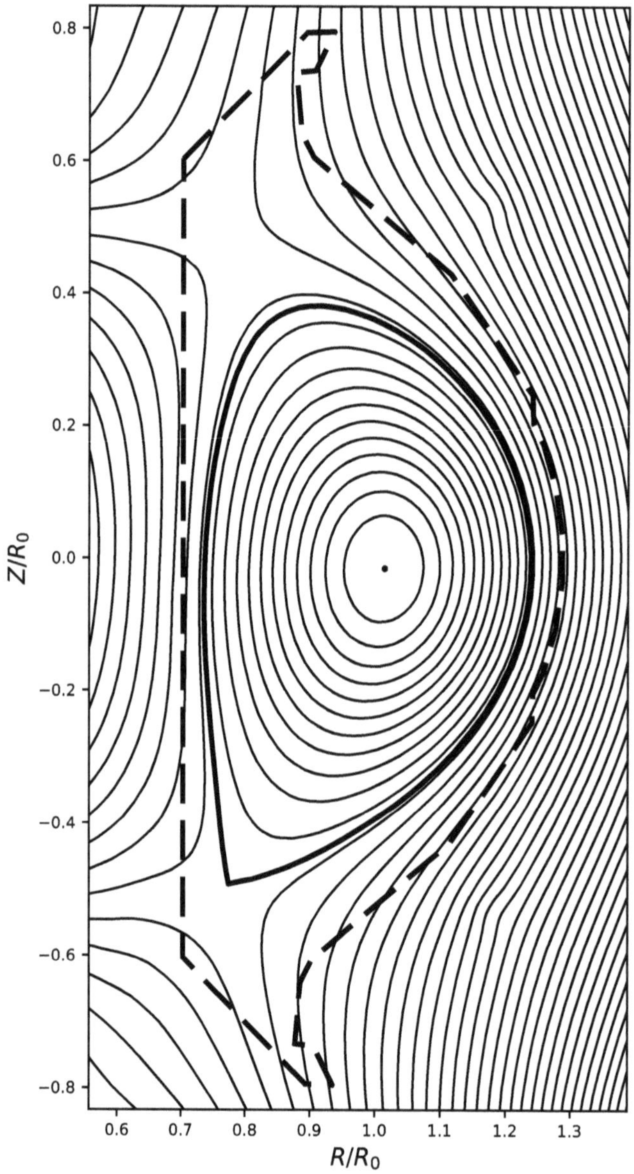

Figure 14.1. Poloidal magnetic flux contours in KSTAR discharge #18594 at time t = 6450 ms. R, φ, Z is a cylindrical coordinate system with a vertical axis, and R_0 = 1.8 m. The bold solid line shows the last closed magnetic flux surface. The bold dashed line shows the plasma-facing components.

$$\mathcal{J}(R,\,Z) = \frac{r\,R^2}{R_0}. \tag{14.3}$$

(Note that R, Z, θ, and φ are the same as the corresponding quantities defined in section 2.10). Here, R_0 is the major radius of the magnetic axis, the magnetic

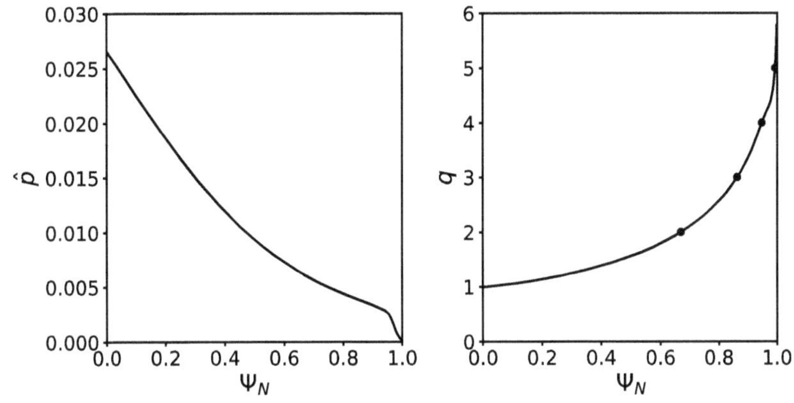

Figure 14.2. Experimental normalized pressure ($\hat{p} = \mu_0\, p/B_0^{\,2}$) and safety-factor profiles in KSTAR discharge #18594 at time $t = 6450$ ms. The points in the right-hand panel show the locations of the $n = 1$ rational surfaces.

flux-surface label r has units of length, and θ is is an angular coordinate that increases by 2π radians for every poloidal circuit of the magnetic axis. Let $r = 0$ correspond to the magnetic axis of the plasma, and let $r = r_{100}$ correspond to the last closed magnetic flux surface. Suppose that $\theta = 0$ at the outboard midplane of the plasma. It follows that $0 < \theta < \pi$ below the midplane.

14.3 Useful identities

The following identities are useful:

$$\mathbf{A} = A^r\, \mathcal{J}\nabla\theta\times\nabla\varphi + A^\theta\, \mathcal{J}\nabla\varphi\times\nabla r + A^\varphi\, \mathcal{J}\nabla r\times\nabla\theta, \tag{14.4}$$

$$\mathbf{A} = A_r\,\nabla r + A_\theta\,\nabla\theta + A_\varphi\,\nabla\varphi, \tag{14.5}$$

$$\mathbf{A}\cdot\mathbf{B} = A_r\, B^{\,r} + A_\theta\, B^{\,\theta} + A_\varphi\, B^{\,\varphi} = A^{\,r}\, B_r + A^{\,\theta}\, B_\theta + A^{\,\varphi}\, B_\varphi, \tag{14.6}$$

$$(\mathbf{A}\times\mathbf{B})_r = \mathcal{J}(A^{\,\theta}\, B^{\,\varphi} - A^{\,\varphi}\, B^{\,\theta}), \tag{14.7}$$

$$(\mathbf{A}\times\mathbf{B})_\theta = \mathcal{J}(A^{\,\varphi}\, B^{\,r} - A^{\,r}\, B^{\,\varphi}), \tag{14.8}$$

$$(\mathbf{A}\times\mathbf{B})_\varphi = \mathcal{J}(A^{\,r}\, B^{\,\theta} - A^{\,\theta}\, B^{\,r}), \tag{14.9}$$

$$\mathcal{J}(\mathbf{A}\times\mathbf{B})^{\,r} = A_\theta\, B_\varphi - A_\varphi\, B_\theta, \tag{14.10}$$

$$\mathcal{J}(\mathbf{A}\times\mathbf{B})^{\,\theta} = A_\varphi\, B_r - A_r\, B_\varphi, \tag{14.11}$$

$$\mathcal{J}(\mathbf{A}\times\mathbf{B})^{\,\varphi} = A_r\, B_\theta - A_\theta\, B_r, \tag{14.12}$$

$$\mathcal{J}\nabla\cdot\mathbf{A} = \frac{\partial\,(\mathcal{J}A^{\,r})}{\partial r} + \frac{\partial\,(\mathcal{J}A^{\,\theta})}{\partial\theta} + \frac{\partial\,(\mathcal{J}A^{\,\varphi})}{\partial\varphi}, \tag{14.13}$$

$$\mathcal{J}(\nabla\times\mathbf{A})^{\,r} = \frac{\partial A_\varphi}{\partial\theta} - \frac{\partial A_\theta}{\partial\varphi}, \tag{14.14}$$

$$\mathcal{J}(\nabla\times\mathbf{A})^{\,\theta} = \frac{\partial A_r}{\partial\varphi} - \frac{\partial A_\varphi}{\partial r}, \tag{14.15}$$

$$\mathcal{J}(\nabla\times\mathbf{A})^{\,\varphi} = \frac{\partial A_\theta}{\partial r} - \frac{\partial A_r}{\partial\theta}. \tag{14.16}$$

Here, **A** and **B** are arbitrary vector fields. Moreover, subscript/superscript r, θ, and φ refer to covariant/contravariant components of a vector in the r, θ, φ coordinate system.

14.4 The equilibrium magnetic field

The equilibrium magnetic field is written [5, 14] (see equation (2.130))

$$\mathbf{B} = B_0\,R_0[f(r)\,\nabla\varphi\times\nabla r + g(r)\,\nabla\varphi], \tag{14.17}$$

where B_0 is the vacuum toroidal magnetic field strength at the magnetic axis, and

$$q(r) = \frac{r\,g}{R_0\,f} \tag{14.18}$$

is the safety-factor profile (see equation (2.128)). Here, $g(r)$ and $f(r)$ are dimensionless functions.

It follows from equations (14.1)–(14.5), (14.17), and (14.18) that

$$B^{\,r} = 0, \tag{14.19}$$

$$B^{\,\theta} = B_0\,R_0^{\,2}\,\frac{f}{r\,R^2}, \tag{14.20}$$

$$B^{\,\varphi} = B_0\,R_0\,\frac{g}{R^2} = B_0\,R_0^{\,2}\,\frac{f\,q}{r\,R^2}, \tag{14.21}$$

$$B_r = -B_0\,r\,f\,\nabla r\cdot\nabla\theta, \tag{14.22}$$

$$B_\theta = B_0\,r\,f\,|\nabla r|^2, \tag{14.23}$$

$$B_\varphi = B_0\,R_0\,g. \tag{14.24}$$

The equilibrium poloidal magnetic flux (divided by 2π), $\Psi_{\rm p}(r)$, satisfies,

$$\frac{d\Psi_{\rm p}}{dr} = B_0\,R_0\,f(r), \tag{14.25}$$

where, by convention, $\Psi_{\rm p}(r_{100}) = 0$. (Note that $\Psi_{\rm p}$ is equivalent to the quantity ψ defined in equation (2.130).) The normalized poloidal magnetic flux, $\Psi_N(r)$, is defined such that

$$\Psi_N(r) = 1 - \frac{\Psi_p(r)}{\Psi_p(0)}. \tag{14.26}$$

Hence, $\Psi_N(0) = 0$ and $\Psi_N(r_{100}) = 1$. In general, $q(r_{100}) = \infty$. Finally, if $\Psi_N(r_{95}) = 0.95$, then $q_{95} = q(r_{95})$. Thus, q_{95} is the safety factor of the magnetic flux surface that encloses 95% of the poloidal magnetic flux enclosed by the last closed magnetic flux surface.

14.5 The equilibrium plasma current density

Let $\mathbf{j}$ be the equilibrium electric current density. The Maxwell equation $\mu_0\, \mathbf{j} = \nabla\times\mathbf{B}$ yields

$$\mu_0\, \mathcal{J} j^{\,r} = 0, \tag{14.27}$$

$$\mu_0\, \mathcal{J} j^{\,\theta} = -B_0\, R_0\, \frac{dg}{dr}, \tag{14.28}$$

$$\mu_0\, \mathcal{J} j^{\,\varphi} = B_0\, \frac{\partial}{\partial r}(r\, f\, |\nabla r|^2) + B_0\, \frac{\partial}{\partial\theta}(r\, f\, \nabla r\cdot\nabla\theta), \tag{14.29}$$

where use has been made of equations (14.14)–(14.16) and (14.22)–(14.24).

14.6 The Grad–Shafranov equation

Let $p(r)$ be the equilibrium plasma pressure profile. The equilibrium force balance relation (see section 2.25)

$$\mathbf{j}\times\mathbf{B} = \nabla p \tag{14.30}$$

gives

$$\mathcal{J}(j^{\,\theta}\, B^{\varphi} - j^{\,\varphi}\, B^{\theta}) = \frac{dp}{dr}, \tag{14.31}$$

where use has been made of equations (14.5) and (14.7). The previous equation reduces to the *Grad–Shafranov equation* [19, 26, 32],

$$\frac{f}{r}\frac{\partial}{\partial r}(r\, f\, |\nabla r|^2) + \frac{f}{r}\frac{\partial}{\partial\theta}(r\, f\, \nabla r\cdot\nabla\theta) + g\,\frac{dg}{dr} + \frac{\mu_0}{B_0^{\,2}}\left(\frac{R}{R_0}\right)^2\frac{dp}{dr} = 0, \tag{14.32}$$

where use has been made of equations (14.20), (14.21), (14.28), and (14.29). The Grad–Shafranov equation can be written in the alternative form

$$\frac{\partial^2\hat{\Psi}_p}{\partial\hat{R}^2} - \frac{1}{\hat{R}}\frac{\partial\hat{\Psi}_p}{\partial\hat{R}} + \frac{\partial^2\hat{\Psi}_p}{\partial\hat{Z}^2} + g\,\frac{dg}{d\hat{\Psi}_p} + \hat{R}^2\,\frac{d\hat{p}}{d\hat{\Psi}_p} = 0, \tag{14.33}$$

where $\hat{\Psi}_p = \Psi_p/(B_0\, R_0^{\,2})$, $\hat{R} = R/R_0$, $\hat{Z} = Z/R_0$, and $\hat{p} = \mu_0\, p/B_0^{\,2}$.

14.7 The perturbed magnetic field

Let $\delta\mathbf{B}$ be the perturbed magnetic field associated with a tearing mode to which the plasma is subject. Since $\nabla\cdot\delta\mathbf{B}=0$, we obtain

$$\frac{\partial(\mathcal{J}\,\delta B^{\,r})}{\partial r}+\frac{\partial(\mathcal{J}\,\delta B^{\,\theta})}{\partial\theta}+\frac{\partial(\mathcal{J}\,\delta B^{\,\varphi})}{\partial\varphi}=0, \tag{14.34}$$

where use has been made of equation (14.13).

It can easily be demonstrated from equations (14.1)–(14.5) that

$$\delta B_r=\left(\frac{1}{|\nabla r|^2}\right)\delta B^{\,r}-\left(\frac{\nabla r\cdot\nabla\theta}{|\nabla r|^2}\right)\delta B_\theta, \tag{14.35}$$

$$\delta B^{\,\theta}=\left(\frac{\nabla r\cdot\nabla\theta}{|\nabla r|^2}\right)\delta B^{\,r}+\left(\frac{R_0^{\,2}}{r^2\,R^2\,|\nabla r|^2}\right)\delta B_\theta. \tag{14.36}$$

Suppose, for the moment, that the tearing perturbation has m periods in the poloidal direction and n periods in the toroidal direction. Let us adopt the simplifying approximation that the perturbed current density, $\delta\mathbf{j}$, is negligible in the regions lying between the various rational surfaces in the plasma [16]. Given that $\mu_0\delta\mathbf{j}=\nabla\times\delta\mathbf{B}$, it follows from equations (14.13)–(14.15) that

$$\frac{\partial\,\delta B_\varphi}{\partial\theta}\simeq\frac{\partial\,\delta B_\theta}{\partial\varphi}, \tag{14.37}$$

$$\frac{\partial\,\delta B_r}{\partial\varphi}\simeq\frac{\partial\,\delta B_\varphi}{\partial r}, \tag{14.38}$$

$$\frac{\partial\,\delta B_\theta}{\partial r}\simeq\frac{\partial\,\delta B_r}{\partial\theta}. \tag{14.39}$$

Assuming that $\partial/\partial r\sim m/r$, the previous three equations imply that

$$r\,\delta B_r\sim\delta B_\theta\sim\frac{m}{n}\,\delta B_\varphi. \tag{14.40}$$

Hence, we deduce that

$$r\,\delta B^{\,r}\sim r^2\,\delta B^{\,\theta}\sim\frac{m}{n}\,R^2\,\delta B^{\,\varphi}. \tag{14.41}$$

Consequently, the final term on the left-hand side of equation (14.34) is of the order of $(n/m)^2\,(r/R)^2$ smaller than the other two terms. Let us assume that this final term is negligible, as would be the case in a large-aspect-ratio (i.e. $R_0\gg r_{100}$) torus. It follows that

$$r\,\frac{\partial}{\partial r}\!\left(\frac{r\,R^2\,\delta B^{\,r}}{R_0^{\,2}}\right)\simeq-\frac{\partial}{\partial\theta}\!\left[\left(\frac{r\,\nabla r\cdot\nabla\theta}{|\nabla r|^2}\right)\!\left(\frac{r\,R^2\,\delta B^{\,r}}{R_0^{\,2}}\right)+\left(\frac{1}{|\nabla r|^2}\right)\delta B_\theta\right], \tag{14.42}$$

where use has been made of equation (14.3), (14.35), and (14.36). Finally, equations (14.35) and (14.39) yield

$$r\,\frac{\partial\,\delta B_\theta}{\partial r} \simeq \frac{\partial}{\partial\theta}\left[\left(\frac{{R_0}^2}{R^2\,|\nabla r|^2}\right)\left(\frac{r\,R^2\,\delta B^{\,r}}{{R_0}^2}\right) - \left(\frac{r\,\nabla r\cdot\nabla\theta}{|\nabla r|^2}\right)\delta B_\theta\right]. \tag{14.43}$$

Let

$$\frac{r\,R^2\,\delta B^{\,r}(r,\,\theta,\,\varphi,\,t)}{{R_0}^2} = \mathrm{i}\sum_j \psi_j(r,\,t)\,\mathrm{e}^{\,\mathrm{i}\,(m_j\,\theta - n\,\varphi)}, \tag{14.44}$$

$$\delta B_\theta(r,\,\theta,\,\varphi,\,t) = -\sum_j \chi_j(r,\,t)\,\mathrm{e}^{\,\mathrm{i}\,(m_j\,\theta - n\,\varphi)}, \tag{14.45}$$

where the sum is over all relevant poloidal harmonics of the perturbed magnetic field. Here, we are now taking account of the fact that a tearing mode in a toroidal tokamak plasma possesses a unique toroidal mode number, but consists of many coupled poloidal harmonics with different poloidal mode numbers [5, 6, 11, 14, 22, 31]. By operating on equations (14.42) and (14.43) with $\oint(\cdots)\,\mathrm{e}^{-\mathrm{i}\,m_j\,\theta}\,d\theta/(2\pi)$, we obtain

$$r\,\frac{\partial\psi_j}{\partial r} \simeq m_j\sum_{j'}\left(-c_{jj'}\,\psi_{j'} + a_{jj'}\,\chi_{j'}\right), \tag{14.46}$$

$$r\,\frac{\partial\chi_j}{\partial r} \simeq m_j\sum_{j'}\left(-c_{jj'}\,\chi_{j'} + b_{jj'}\,\psi_{j'}\right), \tag{14.47}$$

where

$$a_{jj'}(r) = \oint\frac{1}{|\nabla r|^2}\,\mathrm{e}^{-\mathrm{i}\,(m_j - m_{j'})\,\theta}\,\frac{d\theta}{2\pi}, \tag{14.48}$$

$$b_{jj'}(r) = \oint\frac{{R_0}^2}{R^2\,|\nabla r|^2}\,\mathrm{e}^{-\mathrm{i}\,(m_j - m_{j'})\,\theta}\,\frac{d\theta}{2\pi}, \tag{14.49}$$

$$c_{jj'}(r) = \oint\frac{\mathrm{i}\,r\,\nabla r\cdot\nabla\theta}{|\nabla r|^2}\,\mathrm{e}^{-\mathrm{i}\,(m_j - m_{j'})\,\theta}\,\frac{d\theta}{2\pi}. \tag{14.50}$$

Incidentally, we can recover the approximate relations (14.46) and (14.47) from the completely general analysis of reference [11] by neglecting the equilibrium plasma current as well as by taking the limit that $(n/m)^2\,(r/R_0)^2 \ll 1$. This procedure is roughly equivalent to neglecting the term involving J_z' in the cylindrical tearing-mode equation, (3.60). Hence, by analogy with this equation, we would expect our toroidal tearing mode to be classically stable (given that the drive for the classical tearing instability in the cylindrical tearing-mode equation derives from the term involving J_z'). However, this does not preclude the possibility that our toroidal tearing mode could be unstable as a neoclassical tearing mode (see chapter 12).

Finally, can readily be demonstrated that

$$\frac{r\,R^2\,\delta B^{\,\theta}(r,\,\theta,\,\varphi,\,t)}{R_0^{\,2}} = -\sum_j \frac{1}{m_j}\,\frac{\partial\psi_j}{\partial r}\,\mathrm{e}^{\,\mathrm{i}\,(m_j\,\theta - n\,\varphi)}, \tag{14.51}$$

$$R^2\,\delta B^{\,\varphi}(r,\,\theta,\,\varphi,\,t) = n\sum_j \frac{\chi_j}{m_j}\,\mathrm{e}^{\,\mathrm{i}\,(m_j\,\theta - n\,\varphi)}, \tag{14.52}$$

$$\delta B_r(r,\,\theta,\,\varphi,\,t) = \mathrm{i}\sum_j \frac{1}{m_j}\,\frac{\partial\chi_j}{\partial r}\,\mathrm{e}^{\,\mathrm{i}\,(m_j\,\theta - n\,\varphi)}, \tag{14.53}$$

$$\delta B_\varphi(r,\,\theta,\,\varphi,\,t) = n\sum_j \frac{\chi_j}{m_j}\,\mathrm{e}^{\,\mathrm{i}\,(m_j\,\theta - n\,\varphi)}. \tag{14.54}$$

14.8 The perturbed current density

Let $\delta\mathbf{j}$ be the perturbed current density associated with the tearing mode. We can write

$$\mathcal{J}\,\mu_0\,\delta j^r = \frac{\partial\,\delta B_\varphi}{\partial\theta} - \frac{\partial\,\delta B_\theta}{\partial\varphi}, \tag{14.55}$$

$$\mathcal{J}\,\mu_0\,\delta j^\theta = \frac{\partial\,\delta B_r}{\partial\varphi} - \frac{\partial\,\delta B_\varphi}{\partial r}, \tag{14.56}$$

$$\mathcal{J}\,\mu_0\,\delta j^\varphi = \frac{\partial\,\delta B_\theta}{\partial r} - \frac{\partial\,\delta B_r}{\partial\theta}. \tag{14.57}$$

Normally, according to our previous assumptions, all three contravariant components of $\delta\mathbf{j}$ are zero. Consider, however, the behavior in the vicinity of the kth rational surface, $r = r_k$, at which $n\,q(r_k) = m_k$ (see section 3.7). In general, ψ_k, $\psi_{j\neq k}$, and $\chi_{j\neq k}$ are continuous across the surface, while χ_k is discontinuous [11, 14]. Hence, we deduce that

$$\mathcal{J}\,\mu_0\,\delta j^r(r,\,\theta,\,\varphi,\,t) = 0, \tag{14.58}$$

$$\mathcal{J}\,\mu_0\,\delta j^\theta(r,\,\theta,\,\varphi,\,t) = -\sum_{k=1,K}\frac{n}{m_k}\,[\chi_k]_{r_{k-}}^{r_{k+}}\,\delta(r - r_k)\,\mathrm{e}^{\mathrm{i}\,(m_k\,\theta - n\,\varphi)}, \tag{14.59}$$

$$\mathcal{J}\,\mu_0\,\delta j^\varphi(r,\,\theta,\,\varphi,\,t) = -\sum_{k=1,K}[\chi_k]_{r_{k-}}^{r_{k+}}\,\delta(r - r_k)\,\mathrm{e}^{\mathrm{i}\,(m_k\,\theta - n\,\varphi)}, \tag{14.60}$$

where use has been made of equations (14.45) and (14.54). Here, it is assumed that there are K rational surfaces in the plasma, numbered sequentially from one to K from the innermost to the outermost. It can easily be demonstrated from equations (14.7)–(14.9), (14.19)–(14.22), and (14.58)–(14.60) that $\delta\mathbf{j} \times \mathbf{B} = 0$ at a given rational surface. Thus, we conclude that a current sheet forms at each rational surface in the plasma. Moreover, each sheet is made up of current filaments that flow *parallel* to the local equilibrium magnetic field.

14.9 Electromagnetic torques

The flux-surface integrated poloidal and toroidal electromagnetic torque densities acting on the plasma can be written as

$$T_\theta(r,\, t) = \oint\oint (\delta\mathbf{j} \times \delta\mathbf{B})_\theta\, \mathcal{J}\, d\theta\, d\varphi, \tag{14.61}$$

$$T_\varphi(r,\, t) = \oint\oint (\delta\mathbf{j} \times \delta\mathbf{B})_\varphi\, \mathcal{J}\, d\theta\, d\varphi, \tag{14.62}$$

respectively. Equations (14.3), (14.8), (14.9), (14.44), and (14.58)–(14.60) yield

$$T_\theta = -\frac{2\pi^2 R_0}{\mu_0} \sum_{k=1,K} \mathrm{Im}\left([\chi_k]_{r_{k-}}^{r_{k+}}\, \psi_k^{\,*}\right) \delta(r - r_k), \tag{14.63}$$

$$T_\varphi = \frac{2\pi^2 R_0}{\mu_0} \sum_{k=1,K} \mathrm{Im}\left(\frac{n}{m_k}\, [\chi_k]_{r_{k-}}^{r_{k+}}\, \psi_k^{\,*}\right) \delta(r - r_k). \tag{14.64}$$

Let

$$\Psi_k(t) = \frac{\psi_k(r_k,\, t)}{m_k}, \tag{14.65}$$

$$\Delta\Psi_k(t) = [\chi_k]_{r_{k-}}^{r_{k+}}. \tag{14.66}$$

Note that these quantities are, in general, complex. It follows that [11, 14, 16]

$$T_\theta(r,t) = \sum_{k=1,K} T_{\theta\, k}(t)\, \delta(r - r_k), \tag{14.67}$$

$$T_\varphi(r,t) = \sum_{k=1,K} T_{\varphi\, k}(t)\, \delta(r - r_k), \tag{14.68}$$

where

$$T_{\theta\, k} = -\frac{2\pi^2 R_0\, m_k}{\mu_0}\, \mathrm{Im}(\Delta\Psi_k\, \Psi_k^{\,*}), \tag{14.69}$$

$$T_{\varphi\,k} = \frac{2\pi^2\,R_0\,n}{\mu_0}\,\mathrm{Im}(\Delta\Psi_k\,\Psi_k^{\,*}). \tag{14.70}$$

It can be seen, by analogy with the analysis of sections 3.3, 3.8, and 3.13, that Ψ_k is the reconnected magnetic flux at the kth rational surface, while $\Delta\Psi_k$ parameterizes the amplitude and phase of the current sheet flowing (parallel to the equilibrium magnetic field) at the same surface.

14.10 Magnetic island chains

The analysis of section 5.16 suggests that a nonzero value of the reconnected magnetic flux at the kth rational surface, Ψ_k, causes a helical magnetic island chain, which has m_k periods in the poloidal direction and n periods in the toroidal direction, to open in the immediate vicinity of the surface. Let us investigate the properties of such a chain.

We can write

$$\delta\mathbf{B} = \nabla\times\delta\mathbf{A}, \tag{14.71}$$

where

$$\nabla\cdot\delta\mathbf{A} = 0. \tag{14.72}$$

Suppose that all terms in the previous equation are of equal importance. It follows from equation (14.13) that

$$r\,\delta A^r \sim r^2\,\delta A^\theta \sim \frac{n}{m}\,r^2\,\delta A^\varphi, \tag{14.73}$$

and hence that

$$r\,\delta A_r \sim \delta A_\theta \sim \frac{n}{m}\,\frac{r^2}{R^2}\,\delta A_\varphi. \tag{14.74}$$

Thus, equations (14.14)–(14.16) and (14.71) yield

$$\mathcal{J}\,\delta B^{\,r} \simeq \frac{\partial\,\delta A_\varphi}{\partial\theta}, \tag{14.75}$$

$$\mathcal{J}\,\delta B^{\,\theta} \simeq -\frac{\partial\,\delta A_\varphi}{\partial r}, \tag{14.76}$$

$$\mathcal{J}\,\delta B^{\,\varphi} \simeq 0, \tag{14.77}$$

where the neglected terms are at least of order $(n/m)\,(r/R)$ smaller than the retained terms. The previous three equations are consistent with equations (14.44), (14.51), and (14.52) provided that

$$\delta A_\varphi(r,\,\theta,\,\varphi,\,t) \simeq R_0\sum_j\frac{\psi_j(r,\,t)}{m_j}\,\mathrm{e}^{\,\mathrm{i}\,(m_j\,\theta - n\,\varphi)}. \tag{14.78}$$

Let us search for a function, $F(r, \theta, \varphi, t)$, such that

$$(\mathbf{B} + \delta\mathbf{B}) \cdot \nabla F = 0. \tag{14.79}$$

It follows from equations (14.2), (14.4), and (14.6) that

$$(B^r + \delta B^r)\,\frac{\partial F}{\partial r} + (B^\theta + \delta B^\theta)\,\frac{\partial F}{\partial \theta} + (B^\varphi + \delta B^\varphi)\,\frac{\partial F}{\partial \varphi} = 0. \tag{14.80}$$

Equations (14.3), (14.19)–(14.21), and (14.75)–(14.77) yield

$$\frac{\partial F}{\partial r}\,\frac{\partial \delta A_\varphi}{\partial \theta} - \frac{\partial \delta A_\varphi}{\partial r}\,\frac{\partial F}{\partial \theta} + B_0\,R_0\,f\left(\frac{\partial F}{\partial \theta} + q\,\frac{\partial F}{\partial \varphi}\right) = 0. \tag{14.81}$$

Suppose that

$$F(r, \theta, \varphi, t) = F_0(r) + \delta A_\varphi(r, \theta, \varphi, t). \tag{14.82}$$

The previous two equations give

$$\frac{dF_0}{dr}\,\frac{\partial \delta A_\varphi}{\partial \theta} + B_0\,R_0\,f\left(\frac{\partial \delta A_\varphi}{\partial \theta} + q\,\frac{\partial \delta A_\varphi}{\partial \varphi}\right) = 0. \tag{14.83}$$

According to equation (14.78), we can write

$$\delta A_\varphi(r, \theta, \varphi, t) \simeq R_0\,\frac{\psi_k(r, t)}{m_k}\,\mathrm{e}^{\,\mathrm{i}\,(m_k\,\theta - n\,\varphi)} \tag{14.84}$$

in the vicinity of the kth rational surface, where we have neglected the nonresonant components of δA_φ (because we do not expect them to open up an island chain at this surface) [26]. The previous two equations yield

$$\frac{dF_0}{dr} = -\frac{B_0\,R_0\,f}{m_k}\,(m_k - n\,q) \simeq B_0\left(\frac{s\,g}{q}\right)_{r_k}(r - r_k), \tag{14.85}$$

where $s = d\ln q/d\ln r$ is the magnetic shear, and use has been made of equation (14.18) as well as the fact that $q(r_k) = m_k/n$. Finally, equations (14.65), (14.82), (14.84), and (14.85) can be combined to give

$$F(r, \theta, \varphi, t) = \frac{B_0}{2}\left(\frac{s\,g}{q}\right)_{r_k}(r - r_k)^2 + R_0\,|\Psi_k|\cos\xi, \tag{14.86}$$

where $\xi = m_k\,\theta - n\,\varphi + \arg(\Psi_k)$. Here, we have taken the (physical) real part of F, and use has been made of the constant-ψ approximation that $\psi_k(r, t) \simeq m_k\,\Psi_k(t)$ in the immediate vicinity of the kth rational surface (see chapter 5).

The previous equation can be written

$$\frac{F(r, \xi, t)}{R_0\,|\Psi_k|} = 8\left(\frac{r - r_k}{W_k}\right)^2 + \cos\xi, \tag{14.87}$$

where

$$W_k = 4\,R_0\left[\left(\frac{q}{g\,s}\right)_{r_k}\frac{|\Psi_k|}{B_0\,R_0}\right]^{1/2}. \tag{14.88}$$

According to equation (14.79), the contours of the function $F(r,\,\xi,\,t)$ map out the perturbed magnetic flux surfaces in the immediate vicinity of the kth rational surface. These contours are shown in figure 5.7 (with $(r - r_k)/W_k$ playing the role of $\hat{x}/\hat{W}$). It can be seen that the reconnected magnetic flux at the kth rational surface has indeed opened up a helical magnetic island chain, which has m_k periods in the poloidal direction and n periods in the toroidal direction, at the surface. Moreover, the full radial width of the island chain (in r) is W_k. Incidentally, the previous equation is the toroidal generalization of the cylindrical result, (5.129).

14.11 The inductance matrix

In chapter 3, we investigated a tearing mode in a cylindrical tokamak plasma. We found that such a mode resonates with the equilibrium magnetic field at a single magnetic flux surface, known as the rational surface. The minor radius of the rational surface, r_s, satisfies $n\,q(r_s) = m$, where m and n are the unique poloidal and toroidal mode numbers, respectively, of the mode. We also found that the mode reconnects magnetic flux at the rational surface and simultaneously generates a current sheet at the surface that flows parallel to the local magnetic field.

In this chapter, we started to investigate a tearing mode in a toroidal tokamak plasma. We found that such a mode possesses a unique toroidal mode number but consists of many coupled poloidal harmonics (see equations (14.46) and (14.47)). Moreover, the mode resonates at multiple rational surfaces lying within the plasma. The kth rational surface, whose minor radius is r_k, satisfies $m_k = n\,q(r_k)$, where m_k is a positive integer. In principle, the mode can reconnect magnetic flux at each rational surface in the plasma and can also generate a current sheet flowing parallel to the equilibrium magnetic field at each surface. We wish to find the relationship between the reconnected magnetic fluxes and the current sheet strengths at the various rational surfaces in the plasma.

Let

$$\Phi_k(t) = \frac{1}{2\pi\,m_k}\oint\oint \mathrm{i}\,\delta\mathbf{B}\cdot\nabla r\;\mathrm{e}^{-\mathrm{i}\,(m_k\,\theta - n\,\varphi)}\,\mathcal{J}\,d\theta\,d\varphi\,\bigg|_{r=r_k} \tag{14.89}$$

be the (complex) m_k/n helical component of the radial magnetic flux at the kth rational surface. It follows from equations (14.2)–(14.4), (14.44), and (14.65) that

$$\Phi_k = -2\pi\,R_0\,\Psi_k. \tag{14.90}$$

Thus, Φ_k is related to the reconnected magnetic flux, Ψ_k (which is actually a flux per unit length), at the kth rational surface. Let

$$I_k(t) = \frac{1}{2\pi} \int_{r_{k-}}^{r_{k+}} \oint \oint \delta \mathbf{j} \cdot \nabla\varphi \, \mathcal{J} \, dr \, d\theta \, d\varphi \tag{14.91}$$

be the (complex) m_k/n helical component of the localized toroidal plasma current flowing at the kth rational surface. It follows from equations (14.2)–(14.4), (14.60), and (14.66) that

$$I_k = -\frac{2\pi}{\mu_0} \, \Delta\Psi_k. \tag{14.92}$$

Thus, I_k is related to the parallel current sheet strength, $\Delta\Psi_k$, at the kth rational surface.

In general, we expect Φ_k and I_k to be related to one another via a complex inductance matrix:

$$\Phi_k = \sum_{k'=1,K} L_{kk'} \, I_{k'}, \tag{14.93}$$

for $k = 1, K$. Equations (14.90), (14.92), and (14.93) yield [6, 14]

$$\Psi_k = \sum_{k'=1,K} F_{kk'} \, \Delta\Psi_{k'}, \tag{14.94}$$

where

$$L_{kk'} = \mu_0 \, R_0 \, F_{kk'}. \tag{14.95}$$

Hence, we deduce that the complex matrix, $F_{kk'}$, which relates the reconnected fluxes to the current sheet strengths at the various rational surfaces within the plasma, can be regarded as a dimensionless inductance matrix.

The magnetic energy associated with the tearing perturbation is written as

$$\mathcal{E}(t) = \frac{1}{4\,\mu_0} \int_0^{\infty} \oint \oint \delta \mathbf{B} \cdot \delta \mathbf{B}^* \, \mathcal{J} \, dr \, d\theta \, d\varphi. \tag{14.96}$$

Making use of equations (14.6), (14.44), (14.45), and (14.51)–(14.54), we obtain

$$\mathcal{E} = \frac{\pi^2 \, R_0}{\mu_0} \sum_j \frac{1}{m_j} \int_0^{\infty} \frac{\partial}{\partial r}(\chi_j \, \psi_j^*) \, dr + \frac{\pi^2 \, n^2}{\mu_0 \, R_0} \sum_j \frac{1}{m_j^{\,2}} \int_0^{\infty} r \, |\chi_j|^2 \, dr. \tag{14.97}$$

The second term on the right-hand side of the previous equation is negligible compared to the first, because we are assuming that $(n/m_j)^2 \, (r/R_0)^2 \ll 1$. We expect that $\chi_j(r, t) \, \psi_j^*(r, t) \to 0$ as $r \to 0$. Furthermore, assuming that the plasma is isolated (i.e. it is not interacting with a resistive wall or an externally generated resonant magnetic perturbation), we also expect that $\chi_j(r, t) \, \psi_j^*(r, t) \to 0$ as $r \to \infty$. Recalling, however, that $\chi_k(r, t)$ is discontinuous across the kth rational surface, while $\chi_{j\neq k}(r, t)$ and $\psi_j(r, t)$ are continuous, and making use of equations (14.65), (14.66), (14.90), and (14.92), we arrive at

$$\mathcal{E} \simeq \frac{\pi^2 R_0}{\mu_0} \sum_{k=1,K} \Delta\Psi_k \Psi_k^* = \frac{1}{4} \sum_{k=1,K} I_k \Phi_k^*. \tag{14.98}$$

Finally, making use of equation (14.93) and (14.94), we get

$$\mathcal{E} \simeq \frac{\pi^2 R_0}{\mu_0} \sum_{k,k'=1,K} \Delta\Psi_k F_{kk'}^* \Delta\Psi_{k'}^* = \frac{1}{4} \sum_{k,k'=1,K} I_k L_{kk'}^* I_{k'}^*. \tag{14.99}$$

Now, $\mathcal{E}$ is a manifestly real quantity. Hence, it follows from the previous equation that the inductance matrix, $L_{kk'}$, must be Hermitian, which implies that the normalized inductance matrix, $F_{kk'}$, is also Hermitian: that is,

$$F_{k'k} = F_{kk'}^*. \tag{14.100}$$

There is another reason why the F-matrix, $F_{kk'}$, must be Hermitian. According to equations (14.70) and (14.94), the net toroidal electromagnetic torque acting on the plasma is

$$\sum_{k=1,K} T_{\varphi\,k} = \frac{2\pi^2 R_0\, n}{\mu_0} \sum_{k=1,K} \mathrm{Im}(\Delta\Psi_k \Psi_k^*) = \frac{2\pi^2 R_0\, n}{\mu_0} \sum_{k,k'=1,K} \mathrm{Im}(\Delta\Psi_k F_{kk'}^* \Delta\Psi_{k'}^*). \tag{14.101}$$

However, given that we are assuming that the plasma is isolated, and given that the plasma equilibrium is axisymmetric, this torque must be zero, otherwise toroidal angular momentum would not be conserved [14]. The fact that the F-matrix is Hermitian ensures that the net toroidal torque is indeed zero.

14.12 The calculation of the inductance matrix

In principle, we could determine the relationship between $\Delta\Psi_k$ and Ψ_k (which is equivalent to determining the relationship between I_k and Φ_k) by solving equations (14.46) and (14.47) subject to suitable spatial boundary conditions at $r = 0$ and $r = r_{100}$ [11, 14, 22]. However, in this chapter, we shall adopt a more direct approach [12, 13, 15, 16].

According to the Biot–Savart law [27]:

$$\delta A_\varphi(\mathbf{x}, t) = \frac{\mu_0}{4\pi} \int \frac{R\, R'\, \delta\mathbf{j}(\mathbf{x}', t) \cdot \nabla\varphi}{|\mathbf{x} - \mathbf{x}'|}\, d^3\mathbf{x}'. \tag{14.102}$$

Let us assume that

$$\delta A_\varphi(R, \varphi, Z, t) = \delta A_\varphi(R, 0, Z, t)\, \mathrm{e}^{-\mathrm{i}\, n\, \varphi}. \tag{14.103}$$

It follows that we can evaluate the integral on the right-hand side of equation (14.102) at $\varphi = 0$ without loss of generality. We can now write

$$\delta\mathbf{j}(R', \varphi', Z', t) \cdot \nabla\varphi = \delta j^{\,\varphi}(R', 0, Z', t)\, \mathrm{e}^{-\mathrm{i}\, n\, \varphi'} \cos\varphi', \tag{14.104}$$

so that we get

$$\delta A_\varphi(R,0,Z,t) = \frac{\mu_0}{4\pi}\int_0^\infty \oint R\,R'\,\delta j^{\,\varphi}(R',0,Z',t)\,G(R,Z;R',Z')\,\mathcal{J}\,dr'\,d\theta', \tag{14.105}$$

where

$$G(R,Z;R',Z') = \frac{1}{2}\oint \frac{(\cos[(n-1)\,\varphi'] + \cos[(n+1)\,\varphi'])\,d\varphi'}{[R^2 + R'^{\,2} + (Z-Z')^2 - 2\,R\,R'\cos\varphi']^{1/2}}. \tag{14.106}$$

Finally, making use of the standard definition of a toroidal function [23],

$$P^{\,n}_{-1/2}(\cosh\eta) = \frac{(-1)^n\,\Gamma(1/2)\,\Gamma(1/2+n)}{2\pi^{\,2}}\oint \frac{\cos(n\,\varphi)\,d\varphi}{(\cosh\eta - \sinh\eta\cos\varphi)^{1/2}}, \tag{14.107}$$

where $\Gamma(x)$ denotes a gamma function [1], we arrive at

$$G(R,Z;R',Z') = \frac{(-1)^{n+1}\,\pi^2}{\Gamma(1/2)\,\Gamma(n+1/2)}\left[\frac{\cosh\eta}{R^2 + R'^{\,2} + (Z-Z')^2}\right]^{1/2} \times\left[(n-1/2)\,P^{\,n-1}_{-1/2}(\cosh\eta) + \frac{P^{\,n+1}_{-1/2}(\cosh\eta)}{n+1/2}\right], \tag{14.108}$$

where

$$\eta = \tanh^{-1}\left[\frac{2\,R\,R'}{R^2 + R'^{\,2} + (Z-Z')^2}\right]. \tag{14.109}$$

According to equations (14.60) and (14.66),

$$\mathcal{J}\,\mu_0\,\delta j^{\,\varphi}(r,\theta,0,t) = -\sum_{k=1,K}\Delta\Psi_k(t)\,\delta(r-r_k)\,{\rm e}^{\,{\rm i}\,m_k\,\theta}. \tag{14.110}$$

Furthermore, equations (14.65) and (14.78) yield

$$\Psi_k(t) = \frac{1}{R_0}\oint \delta A_\varphi(r_k,\theta,0,t)\,{\rm e}^{-{\rm i}\,m_k\,\theta}\,\frac{d\theta}{2\pi}. \tag{14.111}$$

Hence, combining the previous two expressions with equation (14.105), we obtain the expected normalized inductance relation (see equation (14.94)),

$$\Psi_k = \sum_{k'=1,K} F_{kk'}\,\Delta\Psi_{k'}, \tag{14.112}$$

for $k = 1, K$, where [12, 13, 15, 16]

$$F_{kk'} = \oint\oint \mathcal{G}(R_k, Z_k; R_{k'}, Z_{k'})\,{\rm e}^{-{\rm i}\,(m_k\,\theta_k - m_{k'}\,\theta_{k'})}\,\frac{d\theta_k}{2\pi}\,\frac{d\theta_{k'}}{2\pi}, \tag{14.113}$$

and

$$\mathcal{G}(R_k, Z_k; R_{k'}, Z_{k'}) = \frac{(-1)^n \pi^2 R_k R_{k'}/R_0}{2\,\Gamma(1/2)\,\Gamma(n+1/2)} \left[\frac{\cosh\eta_{kk'}}{R_k^{\,2} + R_{k'}^{\,2} + (Z_k - Z_{k'})^2}\right]^{1/2} \times \left[(n-1/2)\,P_{-1/2}^{\,n-1}(\cosh\eta_{kk'}) + \frac{P_{-1/2}^{\,n+1}(\cosh\eta_{kk'})}{n+1/2}\right], \tag{14.114}$$

with

$$\eta_{kk'} = \tanh^{-1}\left[\frac{2\,R_k\,R_{k'}}{R_k^{\,2} + R_{k'}^{\,2} + (Z_k - Z_{k'})^2}\right]. \tag{14.115}$$

Here, k and k' index the various rational surfaces in the plasma. Moreover, the double integral in equation (14.113) is taken around the kth rational surface (cylindrical coordinates R_k, 0, Z_k; flux coordinates r_k, θ_k, 0, with r_k constant; resonant poloidal mode number m_k) and the k'th rational surface (cylindrical coordinates $R_{k'}$, 0, $Z_{k'}$; flux coordinates $r_{k'}$, $\theta_{k'}$, 0, with $r_{k'}$ constant; resonant poloidal mode number $m_{k'}$).

Note that

$$\mathcal{G}(R_{k'}, Z_{k'}; R_k, Z_k) = \mathcal{G}(R_k, Z_k; R_{k'}, Z_{k'}), \tag{14.116}$$

which, from equation (14.113), implies that the F-matrix is Hermitian (see equation (14.100)), as must be the case.

Finally, according to equations (14.95) and (14.113), the unnormalized inductance matrix takes the form

$$L_{kk'} = \mu_0\,R_0 \oint\oint \mathcal{G}(R_k, Z_k; R_{k'}, Z_{k'})\,\mathrm{e}^{-\mathrm{i}\,(m_k\,\theta_k - m_{k'}\,\theta_{k'})}\,\frac{d\theta_k}{2\pi}\,\frac{d\theta_{k'}}{2\pi}. \tag{14.117}$$

The Hermitian L-matrix, $L_{kk'}$, specifies the self and mutual inductances of the helical current sheets that flow at the various rational surfaces within the plasma.

Note that the calculation of the F-matrix outlined in this section is only approximate. The exact calculation is specified in references [11] and [22].

14.13 The toroidal tearing-mode dispersion relation

The toroidal tearing-mode dispersion relation

$$\Psi_k = \sum_{k'=1,K} F_{kk'}\,\Delta\Psi_{k'}, \tag{14.118}$$

for $k = 1, K$ (see equation (14.112)), can also be written in the form [6, 14]

$$\Delta\Psi_k = \sum_{k'=1,K} E_{kk'}\,\Psi_{k'}, \tag{14.119}$$

for $k = 1, K$, where the E-matrix, $E_{kk'}$, is the inverse of the F-matrix. Because the F-matrix is Hermitian, so is the E-matrix. Given that the F-matrix is a dimensionless *inductance* matrix, it is clear that the E-matrix is a dimensionless *reluctance* matrix.

In cylindrical geometry, the tearing-mode dispersion relation takes the form

$$\Delta\Psi_k = E_{kk}\,\Psi_k, \tag{14.120}$$

for $k = 1, K$ (see equation (3.74)). Here, E_{kk} is the tearing stability index at the kth rational surface [20]. A comparison between the previous two equations reveals that the main difference between the tearing-mode dispersion relation in toroidal geometry and that in cylindrical geometry is the presence of nonzero off-diagonal elements of the E-matrix in the former case. These off-diagonal elements couple different poloidal harmonics.

The nonzero off-diagonal elements of the E-matrix are a consequence of nonzero off-diagonal elements of the F-matrix. (If the F-matrix were diagonal, then the E-matrix would also be diagonal.) The off-diagonal elements of the F-matrix are determined by double integrals of the form (14.113) which is used to calculate the normalized mutual inductances of current sheets flowing parallel to the local equilibrium magnetic field at different rational magnetic flux surfaces. In the case of a cylindrical plasma, the rational surfaces are concentric cylindrical surfaces of circular cross section, and the current sheets have different poloidal periods. Consequently, the mutual inductance integrals all average to zero. The same is not generally true in a toroidal plasma, for two reasons. First, in a toroidal plasma, the different rational magnetic flux surfaces are not concentric, as a consequence of toroidicity and pressure gradients [19, 34] (see figure 14.1). This effect, which is known as the *Shafranov shift* [34], gives rise to a coupling between poloidal harmonics whose poloidal mode numbers differ by unity [5, 14]. Second, in a realistic tokamak plasma, the rational magnetic flux surfaces do not have circular cross sections. Instead, they are highly elongated in the vertical direction and slightly triangular (see figure 14.1). These shaping effects give rise to couplings between poloidal harmonics whose poloidal mode numbers differ by two and three, respectively [14]. Of course, there is no coupling between toroidal harmonics with different toroidal mode numbers because the rational magnetic flux surfaces are all axisymmetric. This accounts for the fact that a general toroidal tearing mode possesses a unique toroidal mode number but does not possess a unique poloidal mode number.

Suppose that a toroidal tearing mode reconnects an amount of magnetic flux, Ψ_k, at the kth rational surface and generates a current sheet of normalized strength, $\Delta\Psi_k$ at the same surface. What is the response of the other rational surfaces in the plasma? In fact, the response at the k'th rational surface (where $k' \neq k$) can fall between two extremes. Either the response can be such as to induce a current sheet at the k'th rational surface that completely suppresses driven magnetic reconnection at the surface (i.e. $\Psi_{k'} = 0$), or the response can be such that no current sheet is induced at the k'th rational surface (i.e. $\Delta\Psi_{k'} = 0$), with the result that the maximum amount of reconnection is driven at the surface. We shall refer to the first type of response as *full shielding* and to the second type as *no shielding*.

Suppose that the responses at all of the other rational surfaces in the plasma are of the full shielding type: that is, $\Psi_{k'\neq k} = 0$. It follows from equation (14.119) that [14]

$$\Delta\Psi_k = E_{kk}\,\Psi_k. \tag{14.121}$$

In other words, in this limit, we essentially reproduce the cylindrical result, (14.120). We conclude that if all of the rational surfaces in the plasma are fully shielded from one another, then a toroidal tearing mode does, in fact, possess a unique poloidal mode number, which is the mode number of the single rational surface in the plasma at which it reconnects magnetic flux. (Note that we are not saying that the magnetic perturbation associated with the mode consists of a single poloidal harmonic.) Moreover, the diagonal element of the E-matrix at that surface is the effective tearing stability index for the mode.

Suppose that the responses at all of the other rational surfaces in the plasma are of the no-shielding type: that is, $\Delta\Psi_{k'\neq k} = 0$. It follows from equation (14.118) that [14]

$$\Delta\Psi_k = \frac{1}{F_{kk}}\,\Psi_k. \tag{14.122}$$

In other words, in this limit, we again reproduce the cylindrical result, (14.120). We conclude that if there is no shielding in the plasma, then a toroidal tearing mode does, in fact, possess a unique poloidal mode number, which is the mode number of the single rational surface in the plasma at which it generates a current sheet. Moreover, the inverse of the diagonal element of the F-matrix at that surface is the effective tearing stability index for the mode. In general, we expect $1/F_{kk} > E_{kk}$ [14]. In other words, in the absence of shielding, we expect the mutual coupling of the different resonant surfaces in the plasma to destabilize the tearing mode relative to the case in which there is full shielding.

The question that we now need to address is which of the two extreme response regimes just outlined most accurately describes a tearing mode in a realistic tokamak plasma. In order to determine the answer to this question, let us examine an example tokamak discharge.

14.14 An example tokamak discharge

KSTAR discharge #18594 [29] is a typical H-mode [34] discharge in a midsized tokamak. Figure 14.1 shows the equilibrium magnetic flux surfaces in this discharge at time $t = 6450$ ms, at which time $B_0 = 1.79$ T, $R_0 = 1.80$ m, $r_{100} = 0.596$ m, and $q_{95} = 4.04$. Figures 14.2 and 14.3 show the corresponding pressure, safety-factor, and g profiles [29]. Of course, the equilibrium flux surfaces shown in figure 14.1, combined with the p and g profiles specified in figures 14.2 and 14.3, constitute a solution of the Grad–Shafranov equation, (14.33). Note that the safety factor becomes infinite at the edge of the plasma due to the presence of a magnetic X-point on the bounding magnetic flux surface (see figure 14.1). In principle, there are an infinite number of $n = 1$ rational surfaces lying within the plasma. However, if we truncate the plasma at $\Psi_N = 0.995$ (i.e. at the magnetic flux surface that contains 99.5% of the poloidal magnetic flux contained by the last closed flux surface), which

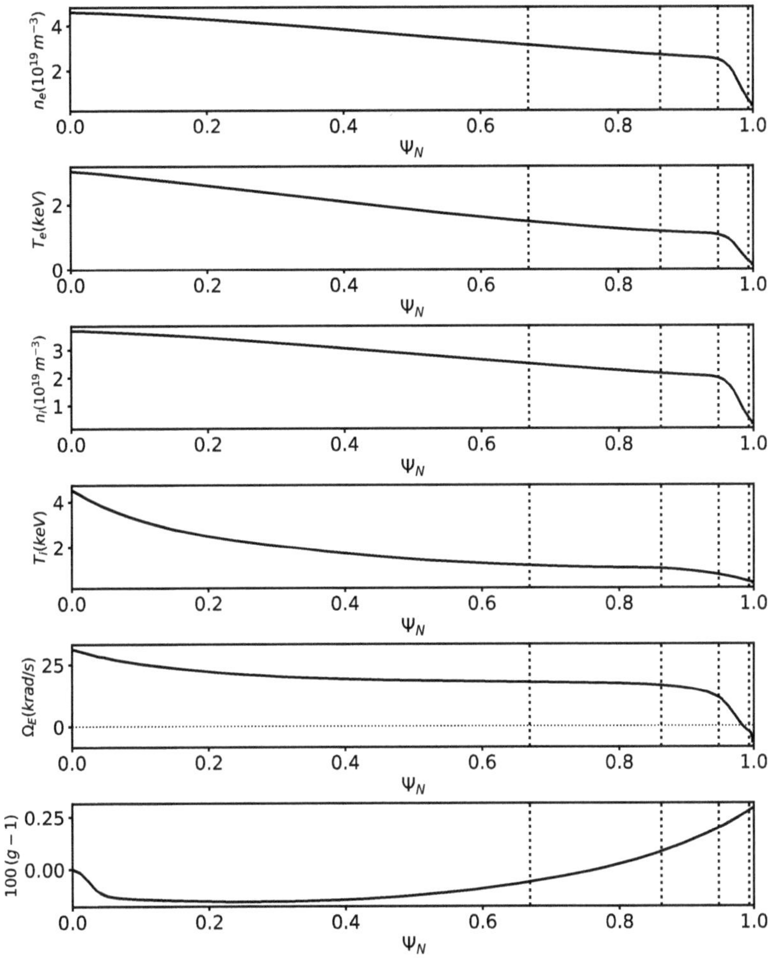

Figure 14.3. Experimental electron number density, electron temperature, majority ion number density, majority ion temperature, E-cross-B frequency, and g profiles in KSTAR discharge #18594 at time t = 6450 ms. The vertical dotted lines show the locations of the n = 1 rational surfaces.

is the standard approach, then there are only four such surfaces. The properties of these surfaces are listed in table 14.1.

Table 14.2 specifies the elements of the n = 1 E-matrix in KSTAR discharge #18594 at time t = 6450 ms, calculated according to the procedure set out in section 14.12. As expected, it can be seen that the matrix is Hermitian. Moreover, all of the diagonal elements of the matrix are negative, which indicates that the n = 1 tearing modes are all classically stable. (This is not surprising, because the classical drive is absent from our calculation.) Finally, the off-diagonal elements of the matrix are all substantial, indicating that there is significant coupling between different poloidal harmonics.

In order to determine the effective tearing stability index for an n = 1 toroidal tearing mode that reconnects magnetic flux at a particular rational surface in the

Table 14.1. Properties of the $n = 1$ rational surfaces in KSTAR discharge #18594 at time $t = 6450$ ms. Here, $s(r)$ is defined in equation (14.129). Only those surfaces with $\Psi_N(r_k) < 0.995$ are listed.

k	m_k	r_k/r_{100}	$\Psi_N(r_k)$	$s(r_k)$	$g(r_k)$
1	2	0.692	0.669	1.51	0.998
2	3	0.852	0.862	2.53	0.999
3	4	0.938	0.947	3.06	0.999
4	5	0.993	0.993	1.27	0.999

Table 14.2. Elements of the $n = 1$ E-matrix in KSTAR discharge #18594 at time $t = 6450$ ms.

k	k'	$\mathrm{Re}(E_{kk'})$	$\mathrm{Im}(E_{kk'})$	k	k'	$\mathrm{Re}(E_{kk'})$	$\mathrm{Im}(E_{kk'})$
1	1	−4.81	+0.00	3	1	+0.589	+1.22
1	2	−0.133	+1.41	3	2	+2.04	−3.19
1	3	+0.589	−1.22	3	3	−15.3	+0.00
1	4	+1.02	−0.646	3	4	+8.13	+2.84
2	1	−0.133	−1.41	4	1	+1.02	+0.646
2	2	−7.92	+0.00	4	2	−0.726	+0.399
2	3	+2.04	+3.19	4	3	+8.13	−2.84
2	4	−0.726	−0.399	4	4	−17.4	+0.00

plasma, let us assume that the responses of the other rational surfaces exhibit strong shielding. This assumption can be justified *a posteriori*. Our assumption implies that very little magnetic reconnection is driven at the other rational surfaces. In this case, it is reasonable to calculate the responses of these surfaces using *linear* theory. (The linear approximation is valid as long as the widths of the magnetic island chains driven at the other rational surfaces are less than the corresponding linear layer widths. See section 5.16.)

According to the analysis given in chapter 5, the linear response of the kth rational surface is characterized by

$$\Delta\Psi_k = S_k^{1/3}\,\hat{\Delta}(Q_k,\,Q_{E\,k},\,Q_{e\,k},\,Q_{i\,k},\,\tau_k,\,D_k,\,P_{\varphi\,k},\,P_{\perp\,k})\,\Psi_k, \tag{14.123}$$

where

$$S_k = \frac{\tau_{R\,k}}{\tau_{H\,k}}, \tag{14.124}$$

$$\tau_{R\,k} = \mu_0\,r_k^2\,\sigma_{ee}(r_k)\,Q_{ee}(r_k), \tag{14.125}$$

$$\sigma_{ee}(r) = \frac{n_e(r)\,e^2\,\tau_{ee}(r)}{m_e}, \tag{14.126}$$

$$\tau_{\mathrm{H}\,k} = \frac{R_0}{B_0\, g(r_k)}\,\frac{\sqrt{\mu_0\,\rho(r_k)}}{n\, s(r_k)}, \tag{14.127}$$

$$\rho(r) = m_{\mathrm{i}}\, n_{\mathrm{i}} + m_{\mathrm{I}}\, n_{\mathrm{I}}, \tag{14.128}$$

$$s(r) = \frac{d\ln q}{d\ln r}, \tag{14.129}$$

$$Q_k = S_k^{\,1/3}\,\omega\,\tau_{\mathrm{H}\,k}, \tag{14.130}$$

$$Q_{\mathrm{E}\,k} = -S_k^{\,1/3}\,n\,\Omega_{\mathrm{E}}(r_k)\,\tau_{\mathrm{H}\,k}, \tag{14.131}$$

$$Q_{\mathrm{e}\,k} = -S_k^{\,1/3}\,n\,\Omega_{*\,\mathrm{e}}(r_k)\,\tau_{\mathrm{H}\,k}, \tag{14.132}$$

$$Q_{\mathrm{i}\,k} = -S_k^{\,1/3}\,n\,\Omega_{*\,\mathrm{i}}(r_k)\,\tau_{\mathrm{H}\,k}, \tag{14.133}$$

$$\tau_k = -\frac{\Omega_{*\,\mathrm{e}}(r_k)}{\Omega_{*\,\mathrm{i}}(r_k)}, \tag{14.134}$$

$$D_k = S_k^{\,1/3}\left(\frac{\tau_k}{1+\tau_k}\right)^{1/2}\frac{d_\beta(r_k)}{r_k}, \tag{14.135}$$

$$d_\beta(r) = \frac{\sqrt{(5/3)\,m_{\mathrm{i}}\,[T_{\mathrm{e}} + (n_{\mathrm{i}}/n_{\mathrm{e}})\,T_{\mathrm{i}} + (n_{\mathrm{I}}/n_{\mathrm{e}})\,T_{\mathrm{I}}]}}{e\,B_0\,g}, \tag{14.136}$$

$$P_{\varphi\,k} = \frac{\tau_{\mathrm{R}\,k}}{\tau_{\varphi\,k}}, \tag{14.137}$$

$$P_{\perp\,k} = \frac{\tau_{\mathrm{R}\,k}}{\tau_{\perp\,k}}, \tag{14.138}$$

$$\tau_{\varphi\,k} = \frac{r_k^{\,2}}{\Xi_{\perp\,\mathrm{i}}(r_k)}, \tag{14.139}$$

$$\tau_{\perp\,k} = \frac{r_k^{\,2}}{D_\perp(r_k)}. \tag{14.140}$$

Here, S_k is the Lundquist number, $\tau_{\mathrm{R}\,k}$ the resistive diffusion time (see equation (5.49)), and $\tau_{\mathrm{H}\,k}$ the hydrodynamic time (see equation (5.43)) at the kth rational surface. Moreover, $n_{\mathrm{e}}(r)$, $n_{\mathrm{i}}(r)$, and $n_{\mathrm{I}}(r)$ are the equilibrium number density profiles of the electrons, majority ions, and impurity ions, respectively, while $T_{\mathrm{e}}(r)$, $T_{\mathrm{i}}(r)$, and $T_{\mathrm{I}}(r)$ are the corresponding temperature profiles, and m_{e}, m_{i}, and m_{I} the corresponding species masses. Furthermore, ω is the real frequency of the tearing mode in the laboratory frame, while the E-cross-B, electron diamagnetic, and majority ion diamagnetic frequency profiles, $\Omega_E(r)$, $\Omega_e(r)$, and $\Omega_i(r)$, respectively, are defined in equations (A.77) and (A.78). The electron–electron collision time, $\tau_{ee}(r)$, is specified in equation (A.23), and e is the magnitude of the electron charge. The dimensionless function $\mathcal{Q}_{ee}(r)$, defined in equation (A.81), specifies the reduction in the plasma's electrical conductivity due to the presence of impurity ions and trapped particles (see section 2.20). In addition, $\tau_{\varphi\,k}$ and $\tau_{\perp\,k}$ are the toroidal momentum confinement time (see equation (5.50)) and the particle confinement time, respectively, at the kth rational surface, while $\Xi_{\perp\,i}(r)$ and $D_\perp(r)$ are the ion perpendicular momentum diffusivity and perpendicular particle diffusivity profiles, respectively.

Equation (14.123) states that the linear response of the kth rational surface to a magnetic perturbation generated at another rational surface is governed by *nine* dimensionless parameters. These parameters are the Lundquist number, S_k, the normalized mode frequency, Q_k, the normalized E-cross-B frequency, $Q_{E\,k}$, the normalized electron diamagnetic frequency, $Q_{e\,k}$, the normalized ion diamagnetic frequency, $Q_{i\,k}$, the pressure gradient ratio parameter, τ_k, the semi-collisional parameter, D_k, and the two magnetic Prandtl numbers, $P_{\varphi\,k}$ and $P_{\perp\,k}$. (Note that these parameters are called Q, Q_E, Q_e, Q_i, τ, D, P_φ, and $P_\perp$, respectively, in chapter 5). The dimensionless layer response index, $\hat{\Delta}_k$, can be calculated numerically as a function of these nine parameters by solving the Riccati differential equation, (5.121), subject to the boundary conditions (5.122) and (5.123).

Figure 14.3 shows the experimental number density, temperature, and E-cross-B frequency profiles in KSTAR discharge #18594 at time t = 6450 ms [29]. The majority ions are deuterium, while the impurity ions are carbon-VI (i.e. $Z_I = 6$). The majority ion and impurity ion number density profiles are calculated from the measured electron number density profile (see equations (A.4) and (A.5)) on the assumption that the effective ion charge number, $Z_{\rm eff}$ (see equation (A.3)), takes the value 2.0 throughout the plasma. (This value is a best guess based on the measured stored energy.) The impurity ions are assumed to have the same temperature as the measured temperature of the majority ions. The E-cross-B frequency profile is deduced from the measured impurity ion toroidal angular velocity profile using the neoclassical theory outlined in appendix A [15]. In particular, the impurity ion poloidal angular velocity profile is assumed to take its neoclassical value (see section A.7). Furthermore, the ion perpendicular momentum and perpendicular particle diffusivities are given the plausible values of 1.0 $\text{m}^2\,\text{s}^{-1}$ and 0.2 $\text{m}^2\,\text{s}^{-1}$, respectively, throughout the plasma [34]. Finally, the values of the various n = 1 resistive-layer parameters, determined from the data shown in figure 14.3, as well as the aforementioned assumptions, are given in table 14.3.

14.15 Linear calculation

Consider an n = 1 tearing mode that principally reconnects magnetic flux at the kth rational surface in our example tokamak plasma. Let us suppose that the amount of reconnected magnetic flux is sufficiently small that the plasma response at this surface lies in the *linear* regime, which is equivalent to supposing that the magnetic

Table 14.3. The $n = 1$ resistive-layer parameters in KSTAR discharge #18594 at time $t = 6450$ ms.

k	m_k	$\tau_{\rm H\,k}$	S_k	τ_k	$P_{\perp\,k}$	$P_{\varphi\,k}$	D_k	$Q_{\rm E\,k}$	$Q_{\rm e\,k}$	$Q_{\rm i\,k}$
1	2	2.41×10^{-7}	1.52×10^{7}	1.51	8.05	40.24	2.41	−3.85	1.11	−0.735
2	3	1.34×10^{-7}	2.87×10^{7}	0.917	5.67	28.4	1.96	−1.96	0.475	−0.518
3	4	1.06×10^{-7}	3.98×10^{7}	1.00	4.91	24.5	1.86	−0.78	1.78	−1.78
4	5	1.30×10^{-8}	5.34×10^{7}	0.877	0.513	2.57	1.13	+0.33	0.959	−1.09

island width at the kth rational surface is much less than the linear layer width (see section 5.16). Given that the mode is assumed not to interact strongly with the other rational surfaces in the plasma, due to the assumed strong shielding present at these surfaces, the analysis given in chapter 6 implies that the frequency of the mode can be written as

$$\omega = {\rm i}\,\gamma_k + \omega_{{\rm linear}\,k}, \tag{14.141}$$

where $|\gamma_k| \ll |\omega_{{\rm linear}\,k}|$, and

$$\omega_{{\rm linear}\,k} = -n(\Omega_{\rm E} + \Omega_{*\,{\rm e}})_{r=r_k}. \tag{14.142}$$

The previous two equations, which are equivalent to equation (6.1), imply that a linear tearing mode corotates with the *electron* fluid at the kth rational surface. It follows from equation (14.130) that the normalized mode rotation frequency at the k' th rational surface, where $k' \neq k$, is

$$Q_{k'} = S_{k'}\,\omega_{{\rm linear}\,k}\,\tau_{{\rm H}\,k'} = \frac{S_{k'}^{\,1/3}}{S_k^{\,1/3}}\,(Q_{{\rm E}\,k} + Q_{{\rm e}\,k})\,\frac{\tau_{{\rm H}\,k'}}{\tau_{{\rm H}\,k}}. \tag{14.143}$$

Here, we have neglected the comparatively small growth rate of the mode with respect to its comparatively large real frequency in the laboratory frame.

According to equations (14.120) and (14.123), the linear dispersion relation of our tearing mode can be written as

$$\Delta\Psi_k = E_{kk}\,\Psi_k + \sum_{k'=1,K}^{k'\neq k} E_{kk'}\,\Psi_{k'}, \tag{14.144}$$

with

$$(\Delta_{k'} - E_{k'k'})\,\Psi_{k'} = E_{k'k}\,\Psi_k + \sum_{k''=1,K}^{k''\neq k,k'} E_{k'k''}\,\Psi_{k''}, \tag{14.145}$$

for $k' \neq k$, where

$$\Delta_{k'} = S_{k'}^{\,1/3}\,\hat{\Delta}_{k'}(Q_{k'}, Q_{{\rm E}\,k'}, Q_{{\rm e}\,k'}, Q_{{\rm i}\,k'}, \tau_{k'}, D_{k'}, P_{\varphi\,k'}, P_{\perp\,k'}). \tag{14.146}$$

Table 14.4 shows the dimensionless layer-matching parameters, $\Delta_{k'}$, where $k' \neq k$, calculated from the layer parameters given in table 14.3, with $Q_{k'}$ as specified by equation (14.143). It can be seen that all $\Delta_{k'}$ have magnitudes that are much

Table 14.4. The $n = 1$ layer-matching parameters in KSTAR discharge #18594 at time $t = 6450$ ms.

k	k'	$\text{Re}(\Delta_{k'})$	$\text{Im}(\Delta_{k'})$	k	k'	$\text{Re}(\Delta_{k'})$	$\text{Im}(\Delta_{k'})$
1	2	−47.5	+234	3	1	−21.2	−128
1	3	−713	+249	3	2	+18.2	−257
1	4	−419	+826	3	4	−229	+830
2	1	−57.1	−206	4	1	+0.06	−39.0
2	3	−728	+243	4	2	+0.41	−73.7
2	4	−403	+830	4	3	−0.07	−95.9

Table 14.5. Natural frequencies of the $n = 1$ tearing modes in KSTAR discharge #18594 at time $t = 6450$ ms.

k	m_k	$w_{\text{linear}\,k}$ (krad s^{-1})	$w_{\text{nonlinear}\,k}$ (krad s^{-1})
1	2	−45.9	−72.4
2	3	−36.2	−55.9
3	4	+27.6	−56.6
4	5	+263	−114

greater than unity, indicating a strong shielding response at the rational surfaces other than the kth surface. This strong shielding is a consequence of *sheared rotation* in the plasma [14]. Roughly speaking, a given rational surface can only reconnect magnetic flux that corotates with the local electron fluid. However, as a consequence of sheared rotation, if a tearing perturbation corotates with the local electron fluid at a given rational surface in the plasma, then it does not corotate with the local electron fluids at any of the other surfaces. This is illustrated in table 14.5, which specifies the linear natural frequencies, $\omega_{\text{linear}\,k}$, associated with the various $n = 1$ rational surfaces present in the plasma. It is clear that the natural frequencies all differ substantially from one another. Consequently, a linear $n = 1$ tearing mode can only reconnect magnetic flux at one rational surface at a time in the plasma.

Treating $|\Delta_{k'} - E_{k'k'}|^{-1}$ as a small parameter, equation (14.145), yields

$$\frac{\Psi_{k'}}{\Psi_k} \simeq \frac{E_{k'k}}{\Delta_{k'} - E_{k'k'}}, \tag{14.147}$$

for $k' \neq k$, which implies that $|\Psi_{k'}|/|\Psi_k| \ll 1$. In other words, the strong rotational shielding present at the k'th rational surface does indeed ensure that very little magnetic flux is reconnected at that surface compared to that reconnected at the kth surface (i.e. the rational surface at which the mode corotates with the local electron fluid). The previous equation can be substituted back into equation (14.144) to give

$$\frac{\Delta\Psi_k}{\Psi_k} \simeq E_{kk} + \delta E_{kk}, \tag{14.148}$$

Table 14.6. Linear $n = 1$ tearing stability indices in KSTAR discharge #18594 at time $t = 6450$ ms.

k	E_{kk}	$\mathrm{Re}(\delta E_{kk})$	$\mathrm{Im}(\delta E_{kk})$
1	−4.81	-4.43×10^{-3}	-1.05×10^{-2}
2	−7.92	-2.07×10^{-2}	$+2.26 \times 10^{-3}$
3	−15.2	-1.75×10^{-2}	-1.45×10^{-2}
4	−17.4	$+1.24 \times 10^{-1}$	$+8.01 \times 10^{-1}$

where

$$\delta E_{kk} = \sum_{k'=1,K}^{k'\neq k} \frac{|E_{kk'}|^2}{\Delta_{k'} - E_{k'k'}}. \tag{14.149}$$

Here, E_{kk} is the tearing stability index of the mode that reconnects magnetic flux at the kth rational surface in the limit of full shielding (i.e. $\Delta\Psi_{k'} = 0$) at the other rational surfaces, while δE_{kk} is the correction to this index due to the fact that the shielding is not actually perfect. Table 14.6 gives the E_{kk} and δE_{kk} values calculated from the data in tables 14.2 and 14.4 for the four possible linear $n = 1$ tearing modes in our example tokamak discharge. It can be seen that $|\delta E_{kk}| \ll |E_{kk}|$ for all modes. In other words, in all cases, the correction due to residual reconnection at the other rational surfaces is essentially negligible.

14.16 Nonlinear calculation

Let us again consider an $n = 1$ tearing mode that principally reconnects magnetic flux at the kth rational surface in our example tokamak plasma. Let us now suppose that the amount of reconnected magnetic flux is sufficiently large that the plasma response at this surface lies in the *nonlinear* regime, which is equivalent to supposing that the magnetic island width at the kth rational surface is much greater than the linear layer width (see section 5.16). Given that the mode is assumed not to interact strongly with the other rational surfaces in the plasma, due to the assumed strong shielding present at these surfaces, the analysis given in chapter 12 implies that the frequency of the mode can be written as [10, 18]

$$\omega = \mathrm{i}\,\gamma_k + \omega_{\mathrm{nonlinear}\,k}, \tag{14.150}$$

where $|\gamma_k| \ll |\omega_{\mathrm{nonlinear}\,k}|$, and

$$\omega_{\mathrm{nonlinear}\,k} = -n(\Omega_{\mathrm{E}} + \Omega_{*\,\mathrm{i}} - \mathcal{L}_{\mathrm{ii}}\,\Omega_{*\,\mathrm{i}} - \mathcal{L}_{\mathrm{iI}}\,\Omega_{*\,\mathrm{I}})_{r=r_k}. \tag{14.151}$$

Here, the impurity ion diamagnetic frequency profile, $\Omega_{*\,\mathrm{I}}(r)$, is defined in equation (A.78). Moreover, the dimensionless neoclassical quantities, $\mathcal{L}_{\mathrm{ii}}(r)$ and $\mathcal{L}_{\mathrm{iI}}(r)$, are defined in equation (A.80). The previous two equations are a generalization of equation (12.11) that takes the presence of impurity ions in the plasma into account, does not assume a small fraction of trapped particles, and does not assume that the plasma at the kth rational surface is in the banana collisionality regime [11, 14, 16].

Table 14.7. Nonlinear $n = 1$ tearing stability indices in KSTAR discharge #18594 at time $t = 6450$ ms.

k	E_{kk}	$\mathrm{Re}(\delta E_{kk})$	$\mathrm{Im}(\delta E_{kk})$
1	−4.81	-4.60×10^{-3}	-5.57×10^{-3}
2	−7.92	-2.00×10^{-2}	-1.40×10^{-2}
3	−15.2	-5.30×10^{-2}	-1.07×10^{-1}
4	−17.4	-2.66×10^{-3}	-1.38×10^{-1}

The previous two equations imply that a nonlinear tearing mode is entrained by the *ion* fluid at the kth rational surface. It follows from equation (14.130) that the normalized mode rotation frequency at the k'th rational surface, where $k' \neq k$, is

$$Q_{k'} = S_{k'}\, \omega_{\mathrm{nonlinear}\,k}\, \tau_{\mathrm{H}\,k'}, \tag{14.152}$$

where the nonlinear natural frequencies, $\omega_{\mathrm{nonlinear}\,k}$, are specified in table 14.5. Here, we have again neglected the comparatively small growth rate of the mode with respect to its comparatively large real frequency in the laboratory frame. Note that the nonlinear natural frequencies are all significantly different from one another and are also significantly different from the linear natural frequencies.

If we repeat the calculation performed in the previous section, replacing the $Q_{k'}$ values specified in equation (14.143) with those specified in equation (14.152), then we obtain the results shown in table 14.7. As before, it can be seen that $|\delta E_{kk}| \ll |E_{kk}|$ for all possible nonlinear $n = 1$ tearing modes. Thus, the fact that nonlinear tearing modes effectively corotate with the ion fluid, rather than the electron fluid, does not change the conclusion of the previous section that there is sufficient rotational shear present in the plasma to ensure that a nonlinear $n = 1$ tearing mode can only reconnect magnetic flux at one rational surface at a time.

Thus, we have effectively answered the question posed at the end of section 14.13. The *full shielding* response model most accurately describes a tearing mode in a realistic tokamak plasma. Note that plasmas in tokamak fusion reactor reactors are likely to undergo significantly less rotational shear than KSTAR plasmas (because they do not rotate as fast as KSTAR plasmas—see section 5.15) but are likely to possess much larger Lundquist numbers. The two effects largely offset one another. Hence, we conclude that the full shielding response model is also likely to apply to tokamak fusion reactors.

14.17 The effect of electromagnetic torques

We have seen that a nonlinear $n = 1$ tearing mode that principally reconnects magnetic flux at a given rational surface in our example tokamak discharge generates comparatively small reconnected fluxes at the other rational surfaces, as a consequence of sheared plasma rotation. However, the small, but nonzero, reconnected fluxes driven at the other surfaces give rise to localized electromagnetic torques (see equations (14.69) and (14.70)). In principle, such torques can modify the plasma rotation and may even lead to the collapse of the rotation shear that is responsible for the strong shielding of different rational surfaces from one another [14]. Let us investigate this effect.

We shall adopt model plasma equations of poloidal and toroidal angular motion that are analogous to those introduced in section 3.14. Let us write

$$\Omega_\theta(r,\,t) = \Omega_{\theta\,0}(r) + \Delta\Omega_\theta(r,\,t), \tag{14.153}$$

$$\Omega_\varphi(r,\,t) = \Omega_{\varphi\,0}(r) + \Delta\Omega_\varphi(r,\,t). \tag{14.154}$$

Here, r is the flux-surface label introduced in section 14.2. Moreover, $\Omega_\theta(r,\,t)$ and $\Omega_\varphi(r,\,t)$ are the majority ion poloidal and toroidal angular velocity profiles, respectively. Furthermore, $\Omega_{\theta\,0}(r)$ and $\Omega_{\varphi\,0}(r)$ are the majority ion poloidal and toroidal angular velocity profiles, respectively, in the absence of electromagnetic torques at the rational surfaces, while $\Delta\Omega_\theta(r,\,t)$ and $\Delta\Omega_\varphi(r,\,t)$ are the respective changes in these profiles induced by the electromagnetic torques. The modifications to the angular velocity profiles are governed by poloidal and toroidal angular equations of motion that take the respective forms [8, 16]:

$$4\pi^2\,R_0\left[\rho\,r^3\,\frac{\partial\Delta\Omega_\theta}{\partial t} + \frac{\rho}{\tau_\theta}\,r^3\,\Delta\Omega_\theta - \frac{\partial}{\partial r}\left(\rho\,\Xi_{\perp\,i}\,r^3\,\frac{\partial\Delta\Omega_\theta}{\partial r}\right)\right] = \sum_{k=1,K} T_{\theta\,k}\,\delta(r-r_k), \tag{14.155}$$

$$4\pi^2\,R_0^{\,3}\left[\rho\,r\,\frac{\partial\Delta\Omega_\varphi}{\partial t} - \frac{\partial}{\partial r}\left(\rho\,\Xi_{\perp\,\mathrm{i}}\,r\,\frac{\partial\Delta\Omega_\varphi}{\partial r}\right)\right] = \sum_{k=1,K} T_{\varphi\,k}\,\delta(r-r_k), \tag{14.156}$$

and are subject to the spatial boundary conditions

$$\frac{\partial\Delta\Omega_\theta(0,\,t)}{\partial r} = \frac{\partial\Delta\Omega_\varphi(0,\,t)}{\partial r} = 0, \tag{14.157}$$

$$\Delta\Omega_\theta(r_{100},\,t) = \Delta\Omega_\varphi(r_{100},\,t) = 0. \tag{14.158}$$

Here,

$$\tau_\theta(r) = \frac{r^2}{q^2\,R_0^{\,2}}\,\frac{\tau_{\mathrm{ii}}}{\mu_{\mathrm{i}\,11}}, \tag{14.159}$$

where τ_{ii} and $\mu_{\mathrm{i}\,11}$ are defined in equations (A.23) and (A.53), respectively, and $\tau_\theta(r)$ is the poloidal flow-damping time profile. Equation (14.159) is a generalization of equation (2.332) that does not assume that the faction of trapped particles is small or that the plasma is in the banana collisionality regime. Finally, the electromagnetic torques, $T_{\theta\,k}(t)$ and $T_{\varphi\,k}(t)$, are specified in equations (14.69) and (14.70), respectively.

Following the analysis of section 3.15, it is convenient to write

$$\Delta\Omega_\theta(r,\,t) = \sum_{k=1,K}\Delta\Omega_{\theta\,k}(r,\,t), \tag{14.160}$$

$$\Delta\Omega_\varphi(r,\,t) = \sum_{k=1,K}\Delta\Omega_{\varphi\,k}(r,\,t), \tag{14.161}$$

where

$$4\pi^2\,R_0\left[\rho r^3\,\frac{\partial\Delta\Omega_{\theta\,k}}{\partial t} + \frac{\rho}{\tau_\theta}\,r^3\,\Delta\Omega_{\theta\,k} - \frac{\partial}{\partial r}\!\left(\rho\,\Xi_{\perp\,\mathrm{i}}\,r^3\,\frac{\partial\Delta\Omega_{\theta\,k}}{\partial r}\right)\right] = T_{\theta\,k}\,\delta(r-r_k), \tag{14.162}$$

$$4\pi^2\,R_0^3\left[\rho\,r\,\frac{\partial\Delta\Omega_{\varphi\,k}}{\partial t} - \frac{\partial}{\partial r}\!\left(\rho\,\Xi_{\perp\,\mathrm{i}}\,r\,\frac{\partial\Delta\Omega_{\varphi\,k}}{\partial r}\right)\right] = T_{\varphi\,k}\,\delta(r-r_k), \tag{14.163}$$

and

$$\frac{\partial\Delta\Omega_{\theta\,k}(0,\,t)}{\partial r} = \frac{\partial\Delta\Omega_{\varphi\,k}(0,\,t)}{\partial r} = 0, \tag{14.164}$$

$$\Delta\Omega_{\theta\,k}(r_{100},\,t) = \Delta\Omega_{\varphi\,k}(r_{100},\,t) = 0. \tag{14.165}$$

The modified angular velocity profiles, $\Delta\Omega_{\theta\,k}$ and $\Delta\Omega_{\varphi\,k}$, are mostly localized in the vicinity of the kth rational surface. Hence, it is a reasonable approximation to express equations (14.162) and (14.163) in the simplified forms [16]

$$\begin{aligned}4\pi^2\,R_0&\left[\rho_k\,r^3\,\frac{\partial\Delta\Omega_{\theta\,k}}{\partial t} + \frac{\rho_k}{\tau_{\theta\,k}}\,r^3\,\Delta\Omega_{\theta\,k} - \rho_k\,\Xi_k\,\frac{\partial}{\partial r}\!\left(r^3\,\frac{\partial\Delta\Omega_{\theta\,k}}{\partial r}\right)\right]\\ &= T_{\theta\,k}\,\delta(r-r_k),\end{aligned} \tag{14.166}$$

$$4\pi^2\,R_0^3\left[\rho_k r\,\frac{\partial\Delta\Omega_{\varphi\,k}}{\partial t} - \rho_k\,\Xi_k\,\frac{\partial}{\partial r}\!\left(r\,\frac{\partial\Delta\Omega_{\varphi\,k}}{\partial r}\right)\right] = T_{\varphi\,k}\,\delta(r-r_k), \tag{14.167}$$

where $\rho_k = \rho(r_k)$, $\tau_{\theta\,k} = \tau_\theta(r_k)$, and $\Xi_k = \Xi_{\perp\,\mathrm{i}}(r_k)$.

Let us write [4, 16, 17]

$$\Delta\Omega_{\theta\,k}(r,\,t) = -\frac{1}{m_k}\sum_{p=1,\infty}\alpha_{k,p}(t)\,\frac{y_\mathrm{p}(r)}{y_\mathrm{p}(r_k)}, \tag{14.168}$$

$$\Delta\Omega_{\varphi\,k}(r,\,t) = \frac{1}{n}\sum_{p=1,\infty}\beta_{k,p}(t)\,\frac{z_\mathrm{p}(r)}{z_\mathrm{p}(r_k)}, \tag{14.169}$$

where

$$y_\mathrm{p}(r) = \frac{J_1(j_{1\mathrm{p}}\,r/r_{100})}{r/r_{100}}, \tag{14.170}$$

$$z_\mathrm{p}(r) = J_0(j_{0\mathrm{p}}\,r/r_{100}). \tag{14.171}$$

Here, $J_m(z)$ is a Bessel function, and $j_{m\mathrm{p}}$ denotes its pth zero [1]. Note that equations (14.168)–(14.171) automatically satisfy the boundary conditions (14.164) and (14.165).

It can easily be demonstrated that [33]

$$\frac{d}{dr}\left(r^3\,\frac{dy_\mathrm{p}}{dr}\right) = -\frac{j_{1\mathrm{p}}^{\,2}\,r^3\,y_\mathrm{p}}{r_{100}^2}, \tag{14.172}$$

$$\frac{d}{dr}\left(r\,\frac{dz_\mathrm{p}}{dr}\right) = -\frac{j_{0\mathrm{p}}^{\,2}\,r\,z_\mathrm{p}}{r_{100}^2}, \tag{14.173}$$

and

$$\int_0^{r_{100}} r^3\,y_\mathrm{p}(r)\,y_\mathrm{q}(r)\,dr = \frac{r_{100}^4}{2}\,[J_2(j_{1\mathrm{p}})]^2\,\delta_{pq}, \tag{14.174}$$

$$\int_0^{r_{100}} r\,z_\mathrm{p}(r)\,z_\mathrm{q}(r)\,dr = \frac{r_{100}^2}{2}\,[J_1(j_{0\mathrm{p}})]^2\,\delta_{pq}. \tag{14.175}$$

Equations (14.69), (14.70), and (14.166)–(14.175) yield

$$\frac{d\alpha_{k,\mathrm{p}}}{dt} + \left(\frac{1}{\tau_{\theta\,k}} + \frac{j_{1\mathrm{p}}^{\,2}}{\tau_{\mathrm{M}\,k}}\right)\alpha_{k,\mathrm{p}} = \frac{m_k^2\,[J_1(j_{1\mathrm{p}}\,r_k/r_{100})]^2}{\tau_{\mathrm{A}\,k}^2\,\epsilon_k^{\,2}\,[J_2(j_{1\mathrm{p}})]^2}\,\mathrm{Im}(\Delta\hat{\Psi}_k\,\hat{\Psi}_k^{\,*}), \tag{14.176}$$

$$\frac{d\beta_{k,\mathrm{p}}}{dt} + \frac{j_{0\mathrm{p}}^{\,2}}{\tau_{\mathrm{M}\,k}}\,\beta_{k,\mathrm{p}} = \frac{n^2\,[J_0(j_{0\mathrm{p}}\,r_k/r_{100})]^2}{\tau_{\mathrm{A}\,k}^2\,[J_1(j_{0\mathrm{p}})]^2}\,\mathrm{Im}(\Delta\hat{\Psi}_k\,\hat{\Psi}_k^{\,*}). \tag{14.177}$$

Here,

$$\tau_{\mathrm{M}\,k} = \frac{r_{100}^2}{\Xi_k}, \tag{14.178}$$

$$\tau_{\mathrm{A}\,k} = \left(\frac{\mu_0\,\rho_k\,r_{100}^2}{B_0^{\,2}}\right)^{1/2}, \tag{14.179}$$

$$\epsilon_k = \frac{r_k}{R_0}, \tag{14.180}$$

$$\Delta\hat{\Psi}_k = \frac{\Delta\Psi_k}{R_0\,B_0}, \tag{14.181}$$

$$\hat{\Psi}_k = \frac{\Psi_k}{R_0\,B_0}. \tag{14.182}$$

The values of $\tau_{\mathrm{A}\,k}$, $\tau_{\mathrm{M}\,k}$, $\tau_{\theta\,k}$, and ϵ_k at the $n = 1$ rational surfaces in our example tokamak discharge are specified in table 14.8. Incidentally, equations (14.176) and (14.177) are the toroidal generalizations of the cylindrical equations (3.190) and (3.191), respectively.

Let us define the frequency shifts that develop at the various rational surfaces in the plasma in response to the electromagnetic torques:

$$\Delta\Omega_k(t) = m_k\,\Delta\Omega_\theta(r_k,\,t) - n\,\Delta\Omega_\varphi(r_k,\,t). \tag{14.183}$$

It follows from equations (14.160), (14.161), and (14.168)–(14.171) that

$$\begin{aligned}\Delta\Omega_k = &- \sum_{k'=1,K}\sum_{p=1,\infty} \alpha_{k',p}\,\frac{m_k}{m_{k'}}\,\frac{r_{k'}}{r_k}\,\frac{J_1(j_{1\mathrm{p}}\,r_k/r_{100})}{J_1(j_{1\mathrm{p}}\,r_{k'}/r_{100})}\\ &- \sum_{k'=1,K}\sum_{p=1,\infty} \beta_{k',\mathrm{p}}\,\frac{J_0(j_{0\mathrm{p}}\,r_k/r_{100})}{J_0(j_{0\mathrm{p}}\,r_{k'}/r_{100})}.\end{aligned} \tag{14.184}$$

The frequency shifts are ultimately due to changes in the E-cross-B rotational frequency at the various rational surfaces (because the diamagnetic components of the majority ion poloidal and toroidal rotation frequencies are not directly affected by the electromagnetic torques) (see section A.7). It follows that

$$-n\,\Delta\Omega_{\mathrm{E}}(r_k,\,t) = \Delta\Omega_k(t) \tag{14.185}$$

at each rational surface in the plasma. This is the essence of the *no-slip constraint* introduced in section 3.16 [8].

Consider an $n = 1$ tearing mode that principally reconnects magnetic flux at the kth rational surface in our example tokamak discharge. Let $\hat{\Psi}_k$ be the normalized magnetic flux reconnected at the kth surface, and let us suppose that $|\hat{\Psi}_k|$ is sufficiently large that the plasma response at the kth surface lies in the nonlinear regime. Because of the assumed strong shielding at the other rational surfaces in the plasma, we expect the plasma responses at these surfaces to lie in the linear regime.

We can determine the normalized reconnected magnetic fluxes, $\hat{\Psi}_{k'}$, driven at the other rational surfaces from equations (14.119), (14.123), (14.146), and (14.182). We find that

Table 14.8. Alfvén times, momentum confinement times, poloidal flow-damping times, and inverse aspect ratios at the $n = 1$ rational surfaces in KSTAR discharge #18594 at time t = 6450 ms.

k	$\tau_{\mathrm{A}\,k}$ (s)	$\tau_{\mathrm{M}\,k}$ (s)	$\tau_{\theta\,k}$ (s)	ϵ_k
1	1.21×10^{-7}	3.55×10^{-1}	2.30×10^{-5}	0.229
2	1.11×10^{-7}	3.55×10^{-1}	1.81×10^{-5}	0.282
3	1.07×10^{-7}	3.55×10^{-1}	1.74×10^{-5}	0.311
4	5.47×10^{-8}	3.55×10^{-1}	5.22×10^{-5}	0.329

$$\sum_{k''=1,K}^{k''\neq k} (\Delta_{k'}\,\delta_{k'k''} - E_{k'k''})\,\hat{\Psi}_{k''} = E_{k'k}\,\hat{\Psi}_k \tag{14.186}$$

for $k' \neq k$. Hence,

$$\hat{\Psi}_{k'}(t) = p_{k'}\,\hat{\Psi}_k \tag{14.187}$$

for $k' \neq k$, where

$$p_{k'}(t) = \sum_{k''=1,K}^{k''\neq k} (\Delta_{k'}\,\delta_{k'k''} - E_{k'k''})^{-1}\,E_{k''k}. \tag{14.188}$$

The normalized electromagnetic torque acting at the k'th (where $k' \neq k$) rational surface is

$$\mathrm{Im}(\Delta\hat{\Psi}_{k'}\,\hat{\Psi}_{k'}^{*}) = \mathrm{Im}(\Delta_{k'})\,|\hat{\Psi}_{k'}|^2 = \mathrm{Im}(\Delta_{k'})\,|p_{k'}|^2\,|\hat{\Psi}_k|^2, \tag{14.189}$$

where use has been made of equations (14.123), (14.146), (14.181), (14.182), and (14.187). The normalized electromagnetic torque acting at the kth rational surface is obtained from angular momentum conservation (see section 14.11):

$$\mathrm{Im}(\Delta\hat{\Psi}_k\,\hat{\Psi}_k^{*}) = -\sum_{k'=1,K}^{k'\neq k} \mathrm{Im}(\Delta\hat{\Psi}_{k'}\,\hat{\Psi}_{k'}^{*}) = -\sum_{k'=1,K}^{k'\neq k} \mathrm{Im}(\Delta_{k'})\,|p_{k'}|^2\,|\hat{\Psi}_k|^2. \tag{14.190}$$

According to equation (14.151), the frequency shift at the kth rational surface modifies the real frequency of the tearing mode. In fact,

$$\omega = \omega_{\mathrm{nonlinear}\,k} + \Delta\Omega_k. \tag{14.191}$$

The normalized linear layer response indices, $\Delta_{k'}$ (where $k' \neq k$), that appear in equation (14.188) are functions of nine normalized layer parameters (see table 14.3). Two of these parameters, $Q_{k'}$ and $Q_{\mathrm{E}\,k'}$, are modified by the frequency shifts induced by the electromagnetic torques. In fact, it is clear from equations (14.130), (14.131), (14.185), and (14.191) that

$$Q_{k'} = Q_{k'}^{(0)} + S_{k'}^{1/3}\,\Delta\Omega_k\,\tau_{\mathrm{H}\,k'}, \tag{14.192}$$

$$Q_{\mathrm{E}\,k'} = Q_{\mathrm{E}\,k'}^{(0)} + S_{k'}^{1/3}\,\Delta\Omega_{k'}\,\tau_{\mathrm{H}\,k'}, \tag{14.193}$$

where the superscript (0) indicates a quantity that is unaffected by the electromagnetic torques.

Equations (14.176), (14.177), (14.184), (14.188)–(14.190), (14.192), and (14.193) —together with the numerical solution of the Riccati differential equation, (5.121), subject to the boundary conditions (5.122) and (5.123) which determine $\Delta_{k'}(Q_{k'}, Q_{\mathrm{E}\,k'})$—form a closed set of equations that allow us to determine the reconnected magnetic fluxes driven at the various rational surfaces in our example tokamak discharge by a tearing mode that primarily reconnects magnetic flux at the

kth rational surface. The only free parameter in the model is the normalized reconnected magnetic flux at the kth rational surface, $|\hat{\Psi}_k|(t)$.

Consider an $n = 1$ tearing mode in our example tokamak discharge that principally reconnects magnetic flux at the $m = 2$ (i.e. $k = 1$) rational surface. Such a mode could represent an $m = 2/n = 1$ neoclassical tearing mode. It is helpful to define the natural frequencies at the various $n = 1$ rational surfaces in the plasma:

$$\omega_{0\,1} = \omega_{\text{nonlinear}\,1} + \Delta\Omega_1, \tag{14.194}$$

$$\omega_{0\,k'} = \omega_{\text{linear}\,k'} + \Delta\Omega_{k'}, \tag{14.195}$$

for $k' = 2, 3, 4$, where $\omega_{\text{nonlinear}\,k}$ and $\omega_{\text{linear}\,k}$ are specified in table 14.5. Here, we are taking into account the fact that the plasma response at the $m = 2$ rational surface lies in the nonlinear regime (because the $m = 2/n = 1$ island width is assumed to be greater than the corresponding linear layer width), while the plasma responses at the other rational surfaces lie in the linear regime (because the driven island widths are assumed to be less than the corresponding linear layer widths). Moreover, we are also taking into account the modifications to the natural frequencies generated by the electromagnetic torques that develop at the rational surfaces. Recall that the natural frequency at a given rational surface is the preferred rotational frequency of reconnected magnetic flux at that surface.

Figure 14.4 shows the $n = 1$ natural frequencies in our example tokamak discharge calculated as functions of the normalized $m = 2/n = 1$ magnetic island width, W_1/r_{100}, in a simulation in which the island width is slowly ramped up. This

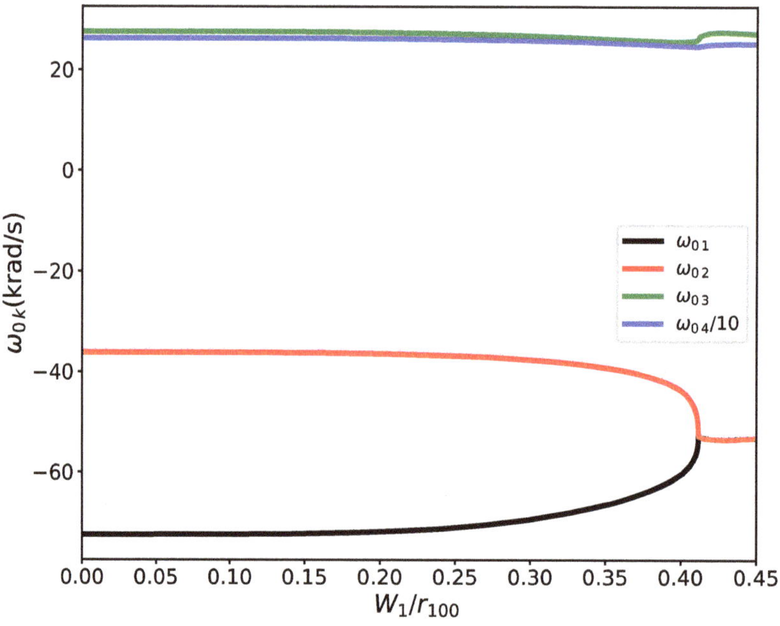

Figure 14.4. Natural $n = 1$ frequencies in KSTAR discharge #18594 at time $t = 6450$ ms calculated as functions of the normalized $m = 2/n = 1$ magnetic island width.

calculation is made using the previously mentioned closed set of equations, as well as the data given in tables 14.1, 14.2, 14.3, 14.5, and 14.8. It can be seen that if W_1/r_{100} lies close to zero, then the natural frequencies all take the unperturbed values specified in table 14.5. On the other hand, as W_1/r_{100} increases, the electromagnetic torques that develop at the rational surfaces modify the natural frequencies. In particular, the torques cause the $m = 2$ natural frequency, $\omega_{0\,1}$, and the $m = 3$ natural frequency, $\omega_{0\,2}$, to approach one another. At a critical value of W_1/r_{100}, which is approximately 0.41, the two natural frequencies suddenly snap together, indicating a sudden loss of shielding at the $m = 3$ rational surface [14].

The aforementioned sudden loss of shielding is illustrated in figure 14.5, which shows the driven normalized magnetic island widths at the $m = 3$, 4, and 5 rational surfaces in our example discharge as functions of the normalized $m = 2$ island width. It can be seen that, prior to the loss of shielding, comparatively narrow magnetic island chains are driven at the $m = 3$, 4, and 5 rational surfaces. However, as soon as the shielding at the $m = 3$ rational surface is lost, there is a very significant increase in the width of the magnetic island chain driven at the $m = 3$ surface.

Finally, figure 14.6 shows the normalized electromagnetic torques that develop at the $n = 1$ rational surfaces in our example discharge close to the time at which shielding is lost at the $m = 3$ rational surface. It can be seen that the sudden loss of shielding is due to comparatively large transient electromagnetic torques that develop at the $m = 2$ and $m = 3$ rational surfaces and drive the corresponding natural frequencies together.

The calculation that we have just performed indicates that the shielding of $n = 1$ rational surfaces from one another, as a consequence of sheared rotation, is a very

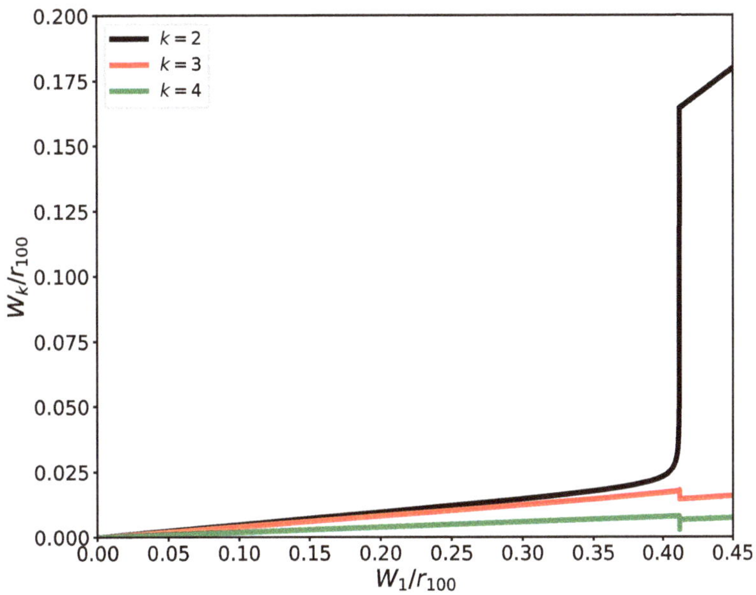

Figure 14.5. Normalized $n = 1$ magnetic island widths in KSTAR discharge #18594 at time $t = 6450$ ms calculated as functions of the normalized $m = 2/n = 1$ magnetic island width.

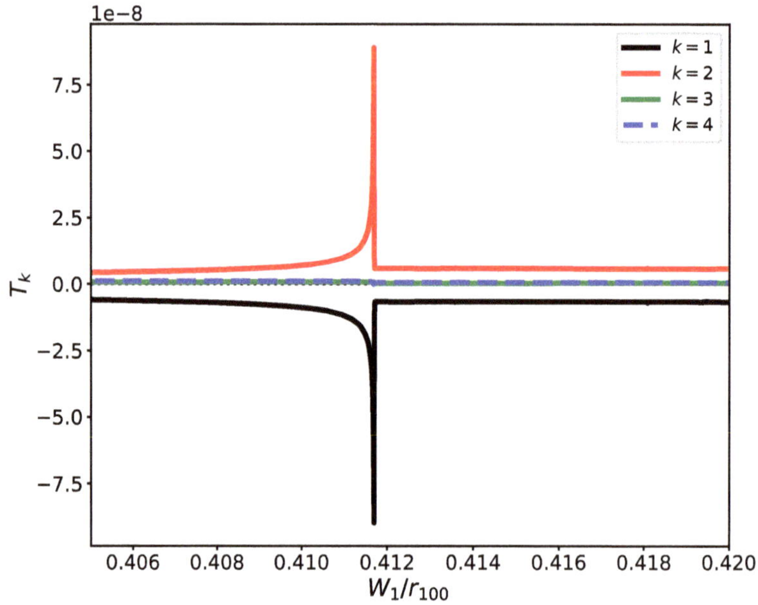

Figure 14.6. Normalized electromagnetic torques, $T_k \equiv \mathrm{Im}(\Delta\hat{\Psi}_k\,\hat{\Psi}_k^{\,*})$, at the $n = 1$ rational surfaces in KSTAR discharge #18594 at time $t = 6450$ ms calculated as functions of the normalized $m = 2/n = 1$ magnetic island width.

robust effect in our example tokamak discharge. In fact, the shielding only breaks down when a tearing mode grows to a sufficient amplitude that the width of its magnetic island chain becomes a substantial fraction of the plasma's minor radius. However, when shielding breaks down, it does so in a sudden and catastrophic manner [14]. In fact, in our example calculation, the loss of shielding at the $m = 3$ rational surface drives a magnetic island chain at that surface whose width is sufficient that the chain would overlap with the chain at the $m = 2$ rational surface, leading to the sudden destruction of magnetic flux surfaces [26], which could quite conceivably trigger a disruption [7].

14.18 Neoclassical tearing modes

The aim of this section is to provide accurate expressions for the bootstrap and curvature terms appearing in the generalized Rutherford equation that controls the growth of a magnetic island chain in a toroidal tokamak plasma (see section 12.4).

Consider a magnetic island chain of width W_k that reconnects magnetic flux at the kth rational surface in the plasma. The toroidal equivalent of the cylindrical island width evolution equation, (12.15), is

$$I_1\,\tau_{\mathrm{R}\,k}\,\frac{d}{dt}\!\left(\frac{W_k}{r_k}\right) = E_{kk} + 4\,I_2\,D_{\mathrm{B}\,k}\,\frac{r_k}{W_k} + 4\,I_2\,D_{\mathrm{R}\,k}\,\frac{r_k}{W_k}, \qquad (14.196)$$

where $I_1 = 0.8227$, $I_2 = 1.5835$, $\tau_{\mathrm{R}\,k}$ is specified in equation (14.125), $D_{\mathrm{R}\,k} = D_{\mathrm{R}}(r_k)$, where $D_{\mathrm{R}}(r)$ is defined in equation (A.98), and

$$D_{\mathrm{B}\,k} = \frac{q(r_k)}{s(r_k)}\,\frac{\mu_0\,R_0^{\,3}\,\langle\mathbf{j}_{\mathrm{bs}}\cdot\mathbf{B}\rangle(r_k)}{[I(r_k)]^2}. \tag{14.197}$$

Here, $\mathbf{j}_{\mathrm{bs}}$ denotes the noninductive 'bootstrap' component (i.e. the component that is driven by pressure gradients, rather than the parallel electric field) of the equilibrium plasma current (see section 2.20), $\langle\cdots\rangle$ is a flux-surface average operator (see equation (A.82)), and $I(r) = B_0\,R_0\,g(r)$. In writing equation (14.196), we have neglected the relatively unimportant (for neoclassical tearing modes) island saturation term in the Rutherford equation (see section 12.2). We have also neglected the ion polarization term (because we have previously shown that this term is negligible) (see section 12.4). Finally, in accordance with the previous analysis in this chapter, we have assumed that there is perfect shielding at all of the other rational surfaces in the plasma (i.e. $\Psi_{k'} = 0$ for $k' \neq k$).

Equation (14.196) states that the width of the magnetic island chain evolves on the local (to the kth rational surface) resistive timescale, $\tau_{\mathrm{R}\,k}$, in response to the effective tearing stability index, E_{kk} [20], the destabilizing (because $D_{\mathrm{B}k}$ is usually positive) effect of the loss of the bootstrap current inside the chain's magnetic separatrix [3], and the stabilizing (because $D_{\mathrm{R}k}$ is usually negative) effect of magnetic field-line curvature [28].

Making use of equation (A.99), it can be seen that the curvature term (i.e. the third term on the right-hand side) in equation (14.196) is entirely consistent with the corresponding term in equation (12.15). However, our new curvature term is more general than our previous cylindrical version because the calculation of the dimensionless curvature parameter $D_{\mathrm{R}}(r)$ outlined in section A.8 makes no assumptions about the geometry of the equilibrium magnetic flux surfaces (other than that they are axisymmetric) [21].

Making use of equation (14.197), the bootstrap term (i.e. the second term on the right-hand side) in equation (14.196) is equivalent to the corresponding term in equation (12.15) provided that the parallel bootstrap current takes the form (see equation (2.265))

$$j_{\mathrm{bs}}(r) = -\frac{f_{\mathrm{t}}}{B_\theta}\left(\beta_{11}\,\frac{dp_{\mathrm{e}}}{dr} - \beta_{12}\,n_{\mathrm{e}}\,\frac{dT_{\mathrm{e}}}{dr} + \beta_{11}\,\frac{dp_{\mathrm{i}}}{dr} - \alpha_1\,\beta_{11}\,n_{\mathrm{e}}\frac{dT_{\mathrm{i}}}{dr}\right). \tag{14.198}$$

Here, $f_{\mathrm{t}}(r)$ is the fraction of trapped particles (see equation (2.202)), $B_\theta(r)$ the equilibrium poloidal magnetic field, $p_{\mathrm{s}}(r) = n_{\mathrm{s}}\,T_{\mathrm{s}}$ the species-s equilibrium pressure profile, $\alpha_1 = 1.17$ (see equation (2.217)), $\beta_{11} = 1.64$ (see equation (2.243)), and $\beta_{12} = 1.23$ (see equation (2.244)). The previous expression is only valid when the fraction of trapped particles is small, the magnetic flux surfaces have circular cross sections, there are no impurity ions, and the plasma lies in the banana collisionality regime. However, we can formulate a more general expression for the bootstrap current using the analysis given in section A.7. If we define

$$\omega_{\mathrm{B}}(r) = \frac{\langle\mathbf{j}_{\mathrm{bs}}\cdot\mathbf{B}\rangle}{n_{\mathrm{e}}\,e\,I}, \tag{14.199}$$

which has the dimensions of a frequency, then we find that

$$\begin{aligned}\omega_B(r) = &- (\Omega_{*\,\mathrm{e}} - \mathcal{L}_{\mathrm{ee}}\,\Omega_{*\,\mathrm{e}} - \mathcal{L}_{\mathrm{ei}}\,\Omega_{*\,\mathrm{i}} - \mathcal{L}_{\mathrm{eI}}\,\Omega_{*\,\mathrm{I}})\\ &+ \frac{n_{\mathrm{i}}}{n_{\mathrm{e}}}(\Omega_{*\,\mathrm{i}} - \mathcal{L}_{\mathrm{ii}}\,\Omega_{*\,\mathrm{i}} - \mathcal{L}_{\mathrm{iI}}\,\Omega_{*\,\mathrm{I}})\\ &+ \frac{Z_{\mathrm{I}}\,n_{\mathrm{I}}}{n_{\mathrm{e}}}(\Omega_{*\,\mathrm{I}} - \mathcal{L}_{\mathrm{II}}\,\Omega_{*\,\mathrm{I}} - \mathcal{L}_{\mathrm{Ii}}\,\Omega_{*\,\mathrm{i}}).\end{aligned} \tag{14.200}$$

Here, Z_{I} is the impurity ion charge number, the species-s number densities, $n_{\mathrm{s}}(r)$, are defined in equations (A.4) and (A.5), the diamagnetic frequencies, $\Omega_{*\,\mathrm{s}}(r)$, are specified in equation (A.78), and the dimensionless neoclassical parameters, $\mathcal{L}_{\mathrm{ss}'}(r)$, are defined in equation (A.80). Moreover, use has been made of the quasi-neutrality constraint (A.1). The previous expression makes no assumption about the geometry of the equilibrium magnetic flux surfaces, does not assume that the fraction of trapped particles is small or that the plasma is in the banana collisionality regime, and allows for the presence of impurity ions. Equations (14.197) and (14.199) yield

$$D_{\mathrm{B}}(r) = \frac{\omega_{\mathrm{B}}}{\omega_\beta}, \tag{14.201}$$

where

$$\omega_\beta(r) = \frac{s\,g\,B_0}{\mu_0\,n_{\mathrm{e}}\,e\,R_0^{\,2}\,q} \tag{14.202}$$

has the dimensions of frequency. Finally, $D_{\mathrm{B}\,k} = D_{\mathrm{B}}(r_k)$. As mentioned in section 12.4, the bootstrap and curvature terms in the generalized Rutherford equation, (14.196), depend crucially on an assumed flattening of the plasma pressure profile inside the magnetic separatrix of the island chain. However, if the island width falls below a certain threshold value, then the transport of heat and particles along magnetic field lines cannot compete with the anomalous transport of heat and particles across magnetic flux surfaces, and the pressure flattening within the magnetic separatrix is lost [9]. Under these circumstances, we would expect a modification of the bootstrap and curvature terms in the generalized Rutherford equation [9, 30]. According to the analysis of reference [9], the critical value of W_k below which the *electron temperature* fails to flatten inside the magnetic separatrix of our magnetic island chain can be written as

$$\frac{W_{T_{\mathrm{e}}\,k}}{r_k} = \sqrt{8}\left[\frac{\chi_{\perp\,\mathrm{e}}(r_k)}{\chi_{\parallel\,\mathrm{e}}(r_k)}\right]^{1/4}\left[\frac{1}{\epsilon(r_k)\,s(r_k)\,n}\right]^{1/2}, \tag{14.203}$$

where $\epsilon(r) = r/R_0$, and

$$\chi_{\parallel\,\mathrm{e}}(r) = \frac{\chi_{\parallel\,\mathrm{e}}^{\mathrm{smfp}}\,\chi_{\parallel\,\mathrm{e}}^{\mathrm{lmfp}}}{\chi_{\parallel\,\mathrm{e}}^{\mathrm{smfp}} + \chi_{\parallel\,\mathrm{e}}^{\mathrm{lmfp}}}, \tag{14.204}$$

$$\chi_{\parallel\,\mathrm{e}}^{\mathrm{smfp}}(r) = \frac{1.581\,\tau_{\mathrm{e}}\,v_{\mathrm{t\,e}}^{2}}{1 + 0.2535\,Z_{\mathrm{eff}}}, \tag{14.205}$$

$$\chi_{\parallel\,\mathrm{e}}^{\mathrm{lmfp}}(r) = \frac{2\,R_0\,v_{\mathrm{t\,e}}}{\pi^{1/2}\,n\,s}\,\frac{r_k}{W_{T_{\mathrm{e}}\,k}}. \tag{14.206}$$

Here, $\tau_{\mathrm{e}}(r)$ is the electron–ion collision time (see equation (2.20)), $v_{\mathrm{t\,e}}(r)$ the electron thermal speed (see equation (2.17)), $\chi_{\perp\,\mathrm{e}}(r)$ the (anomalous) electron perpendicular energy diffusivity, and $\chi_{\parallel\,\mathrm{e}}(r)$ the electron parallel energy diffusivity. Equation (14.205) specifies the parallel diffusivity predicted by the short mean-free-path theory of Braginskii [2] (see equation (2.54)). Note that this expression has been corrected into order to account for the presence of impurity ions in the plasma. Equation (14.206) specifies the parallel diffusivity predicted by the long mean-free-path theory of section 2.23 [24, 25] on the assumption that

$$k_{\parallel} \simeq \frac{n\,s}{R_0}\,\frac{W_{T_{\mathrm{e}}\,k}}{2\,r_k}, \tag{14.207}$$

which is the typical parallel wavenumber of the resonant harmonic of the tearing mode at the edge of a magnetic island chain, centered on the kth rational surface, of width $W_{T_{\mathrm{e}}\,k}$. Finally, equation (14.204) ensures that the parallel electron energy diffusivity never exceeds the limiting value set by long mean-free-path theory. Equations (14.203)–(14.206) can be solved for $W_{T_{\mathrm{e}}\,k}$ via iteration.

According to the analysis given in [9], the critical value of W_k below which the *ion temperature* fails to flatten inside the magnetic separatrix of our magnetic island chain can be written as

$$\frac{W_{T_{\mathrm{i}}\,k}}{r_k} = \sqrt{8}\left[\frac{\chi_{\perp\,\mathrm{i}}(r_k)}{\chi_{\parallel\,\mathrm{i}}(r_k)}\right]^{1/4}\left[\frac{1}{\epsilon(r_k)\,s(r_k)\,n}\right]^{1/2}, \tag{14.208}$$

where

$$\chi_{\parallel\,\mathrm{i}}(r) = \frac{\chi_{\parallel\,\mathrm{i}}^{\mathrm{smfp}}\,\chi_{\parallel\,\mathrm{i}}^{\mathrm{lmfp}}}{\chi_{\parallel\,\mathrm{i}}^{\mathrm{smfp}} + \chi_{\parallel\,\mathrm{i}}^{\mathrm{lmfp}}}, \tag{14.209}$$

$$\chi_{\parallel\,\mathrm{i}}^{\mathrm{smfp}}(r) = \frac{1.953\,\tau_{\mathrm{i}}\,v_{\mathrm{t\,i}}^{2}}{Z_{\mathrm{eff}}}, \tag{14.210}$$

$$\chi_{\parallel\,\mathrm{i}}^{\mathrm{lmfp}}(r) = \frac{2\,R_0\,v_{\mathrm{t\,i}}}{\pi^{1/2}\,n\,s}\,\frac{r_k}{W_{T_{\mathrm{i}}\,k}}. \tag{14.211}$$

Here, $\tau_{\mathrm{i}}(r)$ is the ion–ion collision time (see equation (2.21)), $v_{\mathrm{t\,i}}(r)$ the ion thermal speed (see equation (2.17)), $\chi_{\perp\,\mathrm{i}}(r)$ the (anomalous) ion perpendicular energy diffusivity, and $\chi_{\parallel\,\mathrm{i}}(r)$ the ion parallel energy diffusivity. The previous four equations must be solved iteratively to determine $W_{T_{\mathrm{i}}\,k}$.

Finally, according to the analysis given in [9], the critical value of W_k below which the *electron number density* fails to flatten inside the magnetic separatrix of our magnetic island chain can be written as

$$\frac{W_{n_e\,k}}{r_k} = \sqrt{8}\left[\frac{D_\perp(r_k)}{\chi_{\parallel\,i}(r_k)}\right]^{1/4}\left[\frac{1}{\epsilon(r_k)\,s(r_k)\,n}\right]^{1/2}, \tag{14.212}$$

where

$$\chi_{\parallel\,i}(r) = \frac{\chi_{\parallel\,i}^{\rm smfp}\,\chi_{\parallel\,i}^{\rm lmfp}}{\chi_{\parallel\,i}^{\rm smfp} + \chi_{\parallel\,i}^{\rm lmfp}}, \tag{14.213}$$

$$\chi_{\parallel\,i}^{\rm smfp}(r) = \frac{1.953\,\tau_i\,v_{t\,i}^2}{Z_{\rm eff}}, \tag{14.214}$$

$$\chi_{\parallel\,i}^{\rm lmfp}(r) = \frac{2\,R_0\,v_{t\,i}}{\pi^{1/2}\,n\,s}\,\frac{r_k}{W_{n_e\,k}}. \tag{14.215}$$

Here, $D_\perp(r)$ is the (anomalous) perpendicular particle diffusivity, and we have taken account of the fact that parallel particle transport is constrained by the need to maintain amipolarity (which implies that the parallel particle diffusivity cannot exceed the parallel ion diffusivity).

Table 14.9 specifies the critical $n = 1$ island widths below which the local electron temperature, the ion temperature, and the electron number density profiles fail to flatten in our example tokamak discharge. These critical widths are calculated from the equilibrium and profile data shown in figures 14.1–14.3. It can be seen that the critical island width below which the electron temperature profile fails to flatten, $W_{T_e\,k}$, is of the order of 2% of the plasma's minor radius. On the other hand, the critical island width below which the ion temperature profile fails to flatten, $W_{T_i\,k}$, is significantly larger than $W_{T_e\,k}$ (because ions stream along magnetic field lines at a considerably slower rate than electrons) [9]. Finally, the critical island width below which the electron number density fails to flatten, $W_{n_e\,k}$, lies between $W_{T_e\,k}$ and $W_{T_i\,k}$.

Let us generalize the right-hand side of the generalized Rutherford equation, (14.196), to take account of the incomplete flattening of the electron temperature,

Table 14.9. Critical $n = 1$ island widths below which the local electron temperature, the ion temperature, and the electron number density profiles fail to flatten in KSTAR discharge #18594 at time $t = 6450$ ms.

k	$W_{T_e\,k}/r_{100}$	$W_{T_i\,k}/r_{100}$	$W_{n_e\,k}/r_{100}$
1	1.83×10^{-2}	7.34×10^{-2}	4.31×10^{-2}
2	1.72×10^{-2}	6.77×10^{-2}	3.97×10^{-2}
3	1.69×10^{-2}	6.91×10^{-2}	4.06×10^{-2}
4	1.40×10^{-2}	4.74×10^{-2}	2.78×10^{-2}

ion temperature, and electron number density profiles within the magnetic separatrix of the island chain. In order to achieve this goal, we need to identify the components of the bootstrap and curvature terms that are associated with the flattening of the electron temperature, ion temperature, and electron number density profiles and then modify these components in the appropriate manner. We shall assume that the majority ion and impurity ion temperature profiles both fail to flatten below the same critical island width. Likewise, we shall assume that the electron, majority ion, and impurity ion number density profiles all fail to flatten below the same critical island width. Taking these considerations into account, and making use of the analysis of references [9] and [30], we arrive at

$$\begin{aligned} I_1\,\tau_{R\,k}\,\frac{d}{dt}\left(\frac{W_k}{r_k}\right) &= E_{kk} + 4\,I_2\left(\frac{D_{B\,T_e\,k}\,W_k\,r_k}{W_k^{\,2}+W_{T_e\,k}^{\,2}} + \frac{D_{B\,T_i\,k}\,W_k\,r_k}{W_k^{\,2}+W_{T_i\,k}^{\,2}} + \frac{D_{B\,n_e\,k}\,W_k\,r_k}{W_k^{\,2}+W_{n_e\,k}^{\,2}}\right)\\ &\quad + 4\,I_2\left(\frac{D_{R\,T_e\,k}\,r_k}{\sqrt{W_k^{\,2}+W_{T_e\,k}^{\,2}}} + \frac{D_{R\,T_i\,k}\,r_k}{\sqrt{W_k^{\,2}+W_{T_i\,k}^{\,2}}} + \frac{D_{R\,n_e\,k}\,r_k}{\sqrt{W_k^{\,2}+W_{n_e\,k}^{\,2}}}\right), \end{aligned} \tag{14.216}$$

where

$$D_{B\,T_e}(r) = \frac{\omega_{B\,e}^{(2)}}{\omega_\beta}, \tag{14.217}$$

$$D_{B\,T_i}(r) = \frac{\omega_{B\,i}^{(2)} + \omega_{B\,I}^{(2)}}{\omega_\beta}, \tag{14.218}$$

$$D_{B\,n_e}(r) = \frac{\omega_{B\,e}^{(1)} + \omega_{B\,i}^{(1)} + \omega_{B\,I}^{(1)}}{\omega_\beta}, \tag{14.219}$$

$$\omega_{B\,e}^{(1)}(r) = -(1 - L_{ee}^{(1)})\Omega_{*\,e}^{(1)}, \tag{14.220}$$

$$\omega_{B\,e}^{(2)}(r) = -(1 - L_{ee}^{(2)})\Omega_{*\,e}^{(2)}, \tag{14.221}$$

$$\omega_{B\,i}^{(1)}(r) = L_{ei}^{(1)}\,\Omega_{*\,i}^{(1)} + \frac{n_i}{n_e}(1 - L_{ii}^{(1)})\Omega_{*\,i}^{(1)} - \frac{Z_I\,n_I}{n_e}\,L_{Ii}^{(1)}\,\Omega_{*\,i}^{(1)}, \tag{14.222}$$

$$\omega_{B\,i}^{(2)}(r) = L_{ei}^{(2)}\,\Omega_{*\,i}^{(2)} + \frac{n_i}{n_e}(1 - L_{ii}^{(2)})\Omega_{*\,i}^{(2)} - \frac{Z_I\,n_I}{n_e}\,L_{Ii}^{(2)}\,\Omega_{*\,i}^{(2)}, \tag{14.223}$$

$$\omega_{B\,I}^{(1)}(r) = L_{eI}^{(1)}\,\Omega_{*\,I}^{(1)} - \frac{n_i}{n_e}\,L_{iI}^{(1)}\,\Omega_{*\,I}^{(1)} + \frac{Z_I\,n_I}{n_e}(1 - L_{II}^{(1)})\Omega_{*\,I}^{(1)}, \tag{14.224}$$

$$\omega_{B\,I}^{(2)}(r) = L_{eI}^{(2)}\,\Omega_{*\,I}^{(2)} - \frac{n_i}{n_e}\,L_{iI}^{(2)}\,\Omega_{*\,I}^{(2)} + \frac{Z_I\,n_I}{n_e}(1 - L_{II}^{(2)})\Omega_{*\,I}^{(2)}, \tag{14.225}$$

$$\Omega_{*\,s}^{(1)} = \left(\frac{1}{1+\eta_s}\right)\Omega_{*\,s}, \tag{14.226}$$

$$\Omega_{*\,s}^{(2)} = \left(\frac{\eta_s}{1+\eta_s}\right)\Omega_{*\,s}, \tag{14.227}$$

$$L_{ss'}^{(1)} = L_{ss'\,11}, \tag{14.228}$$

$$L_{ss'}^{(2)} = L_{ss'\,11} - L_{ss'\,12}, \tag{14.229}$$

$$D_{R\,T_e}(r) = \left(\frac{\tilde{\tau}}{1+\tilde{\tau}}\right)\left(\frac{\eta_e}{1+\eta_e}\right)D_R, \tag{14.230}$$

$$D_{R\,T_i}(r) = \left(\frac{1}{1+\tilde{\tau}}\right)\left(\frac{\eta_i}{1+\eta_i}\right)D_R, \tag{14.231}$$

$$D_{R\,n_e}(r) = \left[\left(\frac{\tilde{\tau}}{1+\tilde{\tau}}\right)\left(\frac{1}{1+\eta_e}\right) + \left(\frac{1}{1+\tilde{\tau}}\right)\left(\frac{1}{1+\eta_i}\right)\right]D_R, \tag{14.232}$$

$$\tilde{\tau}(r) = \frac{n_e\,T_e}{n_i\,T_i}\left(\frac{1+\eta_e}{1+\eta_i}\right) = -\frac{n_e}{n_i}\frac{\Omega_{*\,e}}{\Omega_{*\,i}}. \tag{14.233}$$

Here, $\eta_s(r)$ is defined in equation (A.76), and the dimensionless neoclassical matrix elements $L_{ss'\,ij}(r)$ are specified in section A.6. Of course, $D_{B\,T_e\,k} = D_{B\,T_e}(r_k)$, etc.

Table 14.10 specifies the bootstrap and curvature parameters that appear on the right-hand sides of the generalized Rutherford equations, (14.216), of the $n = 1$ tearing modes in our example tokamak discharge. These parameters are calculated from the equilibrium and profile data shown in figures 14.1–14.3 as well as the neoclassical theory set out in appendix A. It can be seen that the $D_{B\,T_e\,k}$, $D_{B\,T_i\,k}$, and $D_{B\,n_e\,k}$ parameters are all positive, indicating that the bootstrap terms in the Rutherford equations are destabilizing. On the other hand, the $D_{R\,T_e\,k}$, $D_{R\,T_i\,k}$, and $D_{R\,n_e\,k}$ parameters are all negative, indicating that the curvature terms in the

Table 14.10. Parameters that characterize the right-hand sides of the generalized Rutherford equations of $n = 1$ tearing modes in KSTAR discharge #18594 at time $t = 6450$ ms.

k	$D_{B\,T_e}$	$D_{B\,T_i}$	$D_{B\,n_e}$	$D_{R\,T_e}$	$D_{R\,T_i}$	$D_{R\,n_e}$
1	8.75×10^{-2}	2.94×10^{-2}	9.96×10^{-2}	-3.18×10^{-2}	-1.46×10^{-2}	-3.13×10^{-2}
2	4.15×10^{-2}	3.01×10^{-2}	5.70×10^{-2}	-2.20×10^{-2}	-2.17×10^{-2}	-2.80×10^{-2}
3	1.59×10^{-1}	1.26×10^{-1}	2.49×10^{-1}	-9.09×10^{-2}	-9.09×10^{-2}	-1.40×10^{-1}
4	4.49×10^{-2}	1.07×10^{-2}	6.47×10^{-2}	-5.32×10^{-2}	-1.92×10^{-2}	-1.12×10^{-1}

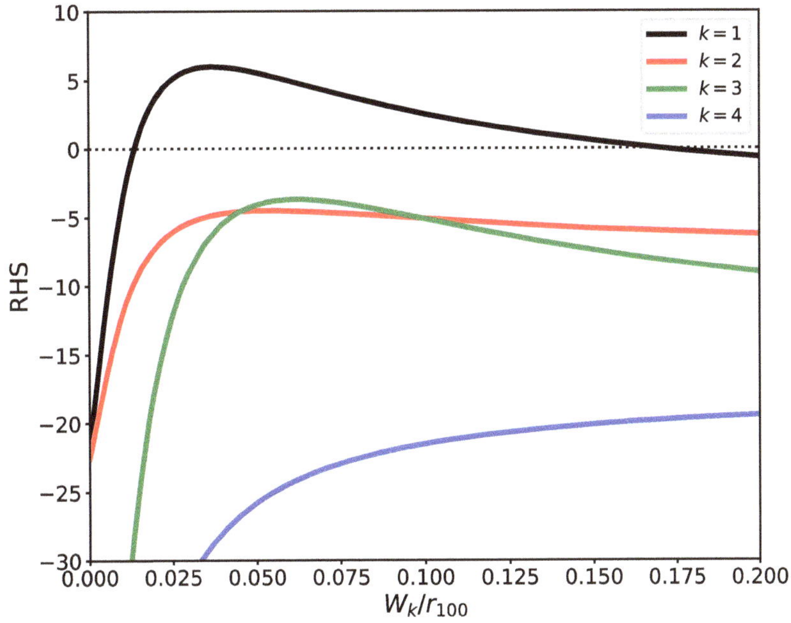

Figure 14.7. Right-hand sides of the generalized Rutherford equations for the $n = 1$ tearing modes in KSTAR discharge #18594 at time $t = 6450$ ms calculated as functions of the normalized magnetic island widths.

Rutherford equations are stabilizing. It is also clear that the magnitudes of the $D_{\mathrm{B}\,T_e\,k}$, $D_{\mathrm{B}\,T_i\,k}$, and $D_{\mathrm{B}\,n_e\,k}$ parameters generally exceed those of the $D_{\mathrm{R}\,T_e\,k}$, $D_{\mathrm{R}\,T_i\,k}$, and $D_{\mathrm{R}\,n_e\,k}$ parameters, which implies that the destabilizing effect of the perturbed bootstrap current is larger than the stabilizing effect of magnetic field-line curvature.

Finally, the right-hand sides of the generalized Rutherford equations of the $n = 1$ tearing modes in our example tokamak discharge are plotted as functions of the normalized island widths in figure 14.7. These right-hand sides are calculated from equation (14.216) using the data set out in tables 14.2, 14.9, and 14.10. It can be seen that only the $m = 2$ (i.e. $k = 1$) mode has a right-hand side that rises above zero. It follows that only the $m = 2/n = 1$ neoclassical tearing mode is potentially unstable (see section 12.4). The critical island width above which the mode is triggered (i.e. the smaller zero crossing of the right-hand side) is about 1% of the plasma's minor radius. The saturated island width (i.e. the larger zero crossing of the right-hand side) is about 17% of the plasma's minor radius.

References

[1] Abramowitz M and Stegun I A (ed) 1964 *Handbook of Mathematical Functions with Formulas, Graphs, and Mathematical tables* (New York: Dover)

[2] Braginskii S I 1965 Transport processes in a plasma *Reviews of Plasma Physics* vol 1 (New York: Consultants Bureau) p 205

[3] Carrera R, Hazeltine R D and Kotschenreuther M 1986 Island bootstrap current modification of the nonlinear dynamics of the tearing mode *Phys. Fluids* **29** 899

[4] Chapman B E, Fitzpatrick R, Craig D, Martin P and Spizzo G 2004 Observation of tearing mode deceleration and locking due to eddy currents induced in a conducting shell *Phys. Plasmas* **11** 2156
[5] Connor J W, Cowley S C, Hastie R J, Hender T C, Hood A and Martin T J 1988 Tearing modes in toroidal geometry *Phys. Fluids* **31** 577
[6] Connor J W, Hastie R J and Taylor J B 1991 Resonant magnetohydrodynamic modes with toroidal coupling. Part I: tearing modes *Phys. Fluids* B **3** 1532
[7] de Vries P C, Johnson M F, Alper B, Buratti P, Hender T C, Koslowski H R and Riccardo V and JET-EFDA Contributors 2011 Survey of disruption causes at JET *Nucl. Fusion* **51** 053018
[8] Fitzpatrick R 1993 Interaction of tearing modes with external structures in cylindrical geometry *Nucl. Fusion* **33** 1049
[9] Fitzpatrick R 1995 Helical temperature perturbations associated with tearing modes in tokamak plasmas *Phys. Plasmas* **2** 825
[10] Fitzpatrick R 2016 An improved neoclassical drift-magnetohydrodynamical fluid model of helical magnetic island equilibria in tokamak plasmas *Phys. Plasmas* **23** 052506
[11] Fitzpatrick R 2017 Determination of the non-ideal response of a high temperature tokamak plasma to a static external resonant magnetic perturbation via asymptotic matching *Phys. Plasmas* **24** 072506
[12] Fitzpatrick R 2020 Modeling q_{95} windows for the suppression of edge localized modes by resonant magnetic perturbations in the DIII-D tokamak *Phys. Plasmas* **27** 102511
[13] Fitzpatrick R 2021 Further modeling of q_{95} windows for the suppression of edge localized modes by resonant magnetic perturbations in the DIII-D tokamak *Phys. Plasmas* **28** 022503
[14] Fitzpatrick R, Hastie R J, Martin T J and Roach C M 1993 Stability of coupled tearing modes in tokamaks *Nucl. Fusion* **33** 1533
[15] Fitzpatrick R, Kim S and Lee J 2021 Modeling of q_{95} windows for the suppression of edge localized modes by resonant magnetic perturbations in the KSTAR tokamak *Phys. Plasmas* **28** 082511
[16] Fitzpatrick R and Nelson A O 2020 An improved theory of the response of DIII-D H-mode discharges to static resonant magnetic perturbations and its implications for the suppression of edge localized modes *Phys. Plasmas* **27** 072501
[17] Fitzpatrick R, Rossi E and Yu E P 2001 Improved evolution equations for magnetic island chains in toroidal pinch plasmas subject to externally applied resonant magnetic perturbations *Phys. Plasmas* **8** 4489
[18] Fitzpatrick R and Waelbroeck F L 2009 Effect of flow damping on drift-tearing magnetic islands in tokamak plasmas *Phys. Plasmas* **16** 072507
[19] Freidberg J P 1987 *Ideal Magnetohydrodynamics* (New York: Plenum)
[20] Furth H P, Killeen J and Rosenbluth M N 1963 Finite-resistivity instabilities of a sheet pinch *Phys. Fluids* **6** 459
[21] Glasser A H, Greene J M and Johnson J L 1975 Resistive instabilities in general toroidal plasma configurations *Phys. Fluids* **18** 875
[22] Glasser A H, Wang Z R and Park J-K 2016 Computation of resistive instabilities by matched asymptotic expansions *Phys. Plasmas* **23** 112056
[23] Gradshteyn A S and Ryzhik I M 1980 *Table of Integrals, Series, and Products* 4th edn (New York: Academic) https://www.elsevier.com/books/handbook-of-mathematical-formulas-and-integrals/jeffrey/978-0-12-374288-9

[24] Hammett B W and Perkins F W 1990 Fluid moment models for Landau damping with application to the ion-temperature-gradient instability *Phys. Rev. Lett.* **64** 3019

[25] Hazeltine R D 1998 Transport in the collisionless limit *Phys. Plasmas* **5** 3282

[26] Hazeltine R D and Meiss J D 2003 *Plasma Confinement* (New York: Dover) https://store.doverpublications.com/0486151034.html

[27] Jackson J D 1998 *Classical Electrodynamics* 3rd edn (New York: Wiley) https://www.wiley.com/en-us/Classical+Electrodynamics%2C+3rd+Edition-p-9780471309321

[28] Kotschenreuther M, Hazeltine R D and Morrison P J 1985 Nonlinear dynamics of magnetic islands with curvature and pressure *Phys. Fluids* **28** 294

[29] Lee J *et al* 2019 Direct evidence of $E \times B$ flow changes at the onset of resonant magnetic perturbation-driven edge-localized mode crash suppression *Nucl. Fusion* **59** 066033

[30] Lütjens H, Luciani J-F and Garbet X 2001 Curvature effects on the dynamics of tearing modes in tokamaks *Phys. Plasmas* **8** 4267

[31] Pletzer A and Dewar R L 1991 Non-ideal stability: variational method for the determination of the outer-region matching data *J. Plasma Phys.* **45** 427

[32] Shafranov V D 1966 Plasma equilibrium in a magnetic field *Reviews of Plasma Physics* vol 2 (New York: Consultants Bureau) p 103

[33] Watson G N 1922 *A Treatise on the Theory of Bessel Functions* (Cambridge: Cambridge University Press)

[34] Wesson J A 2011 *Tokamaks* 4th edn (Oxford: Oxford University Press) https://global.oup.com/academic/product/tokamaks-9780199592234?cc=gb&lang=en&

Appendix A

Neoclassical theory

A.1 Introduction

The main aim of this appendix is to generalize the neoclassical theory introduced in chapter 2 in order to take account of the presence of impurity ions in the plasma. We shall make no assumptions about the geometry of the equilibrium magnetic flux surfaces in this study, neither shall we assume that the plasma lies in the banana collisionality regime.

A.2 Plasma species

The plasma is assumed to consist of three (charged) species; namely, electrons (e), majority ions (i), and impurity ions (I). The charges of the three species are $e_e = -e$, $e_i = e$, and $e_I = Z_I\, e$, respectively, where e is the magnitude of the electron charge. Quasi-neutrality [1] demands that

$$n_e = n_i + Z_I\, n_I, \tag{A.1}$$

where $n_s(r)$ is the species-s number density. Here, r is the flux-surface label introduced in section 14.2. Let

$$\alpha_I(r) = \frac{Z_I\,(Z_{\rm eff} - 1)}{Z_I - Z_{\rm eff}}, \tag{A.2}$$

where

$$Z_{\rm eff}(r) = \frac{n_i + {Z_I}^2\, n_I}{n_e} \tag{A.3}$$

is the *effective ion charge number* (see section 1.6). It follows that

$$\frac{n_i}{n_e} = \frac{Z_I - Z_{\rm eff}}{Z_I - 1}, \tag{A.4}$$

doi:10.1088/978-0-7503-5367-0ch15

$$\frac{n_{\rm I}}{n_{\rm e}} = \frac{Z_{\rm eff} - 1}{Z_{\rm I}\,(Z_{\rm I} - 1)}. \tag{A.5}$$

Finally, let

$$Z_{\rm eff\,i}(r) = \frac{Z_{\rm I} - Z_{\rm eff}}{Z_{\rm I} - 1}, \tag{A.6}$$

$$Z_{\rm eff\,I}(r) = \frac{Z_{\rm I}\,(Z_{\rm eff} - 1)}{Z_{\rm I} - 1}. \tag{A.7}$$

Note that $Z_{\rm eff} = Z_{\rm eff\,i} + Z_{\rm eff\,I}$.

A.3 Collisionality parameters

Consider an equilibrium magnetic flux surface whose label is r. Let

$$\frac{1}{\gamma(r)} = \frac{q}{g}\oint \frac{B\,R^2}{B_0\,R_0^{\,2}}\,\frac{d\theta}{2\pi}, \tag{A.8}$$

where $B = |\mathbf{B}|$, and $\mathbf{B}$ is the equilibrium magnetic field. Here, $q(r)$, $g(r)$, R, θ, B_0, and R_0 are specified in sections 14.2 and 14.4. It is helpful to define a new poloidal angle Θ such that

$$\frac{d\Theta}{d\theta} = \frac{\gamma\,q}{g}\,\frac{B\,R^2}{B_0\,R_0^{\,2}}. \tag{A.9}$$

Let

$$I_1(r) = \oint \frac{B_0}{B}\,\frac{d\Theta}{2\pi}, \tag{A.10}$$

$$I_2(r) = \oint \frac{B}{B_0}\,\frac{d\Theta}{2\pi}, \tag{A.11}$$

$$I_3(r) = \oint \left(\frac{\partial B}{\partial\Theta}\right)^2 \frac{1}{B_0\,B}\,\frac{d\Theta}{2\pi}, \tag{A.12}$$

$$I_{4,j}(r) = \sqrt{2\,j}\oint \frac{\cos(j\,\Theta)}{B/B_0}\,\frac{d\Theta}{2\pi}, \tag{A.13}$$

$$I_{5,j}(r) = \sqrt{2\,j}\oint \frac{\cos(j\,\Theta)}{2\,(B/B_0)^{\,2}}\,\frac{d\Theta}{2\pi}, \tag{A.14}$$

$$I_6(r,\,\lambda) = \oint \frac{\sqrt{1 - \lambda\,B/B_{\rm max}}}{B/B_0}\,\frac{d\Theta}{2\pi}, \tag{A.15}$$

where the integrals are taken at constant r, $B_{\rm max}(r)$ is the maximum value of B on the magnetic flux surface, and j is a positive integer. The species-s *transit frequency* is written [7]

$$\omega_{\rm t\,s}(r) = K_{\rm t}\,\gamma\,v_{\rm t\,s}, \tag{A.16}$$

where

$$K_{\rm t}(r) = \frac{I_1^{\,2}\,I_3}{I_2^{\,2}\,\sum_{j=1,\infty} I_{4,j}\,I_{5,j}}, \tag{A.17}$$

and

$$v_{\rm t\,s}(r) = \sqrt{\frac{2\,T_{\rm s}}{m_{\rm s}}}. \tag{A.18}$$

Here, $m_{\rm s}$ is the species-s mass, and $T_{\rm s}(r)$ is the species-s temperature (in energy units). The fraction of passing particles is [7]

$$f_{\rm p}(r) = \frac{3\,I_2}{4}\,\frac{B_0^{\,2}}{B_{\rm max}^{\,2}}\int_0^1 \frac{\lambda\,d\lambda}{I_6(r,\,\lambda)} \tag{A.19}$$

(see equation (2.200)). Finally, the dimensionless species-s *collisionality parameter* (see equation (2.95)) is written [7]

$$\nu_{*\,\rm s}(r) = \frac{K_*\,g_{\rm t}}{\omega_{\rm t\,s}\,\tau_{\rm ss}}, \tag{A.20}$$

where

$$g_{\rm t}(r) = \frac{f_{\rm p}}{1 - f_{\rm p}}, \tag{A.21}$$

$$K_*(r) = \frac{3}{8\pi}\,\frac{I_2}{I_3}\,K_{\rm t}^{\,2}, \tag{A.22}$$

$$\frac{1}{\tau_{\rm ss}(r)} = \frac{4}{3\sqrt{\pi}}\,\frac{4\pi\,n_{\rm s}\,e_{\rm s}^{\,4}\,\ln\Lambda}{(4\pi\,\epsilon_0)^2\,m_{\rm s}^{\,2}\,v_{\rm t\,s}^{\,3}}. \tag{A.23}$$

(See equation (2.190).) Here, the Coulomb logarithm, $\ln\Lambda$ [1], is assumed to take the same large constant value (i.e. $\ln\Lambda \simeq 16$), independent of species.

A.4 Friction force matrices

Let

$$x_{\rm ss'} = \frac{v_{\rm t\,s'}}{v_{\rm t\,s}}. \tag{A.24}$$

In the following, all quantities that are of order $(m_e/m_i)^{1/2}$, $(m_e/m_I)^{1/2}$, or smaller are neglected with respect to unity. The 2×2 dimensionless ion collisional friction force matrices, $[F_{ii}](r)$, $[F_{iI}](r)$, $[F_{Ii}](r)$, and $[F_{II}](r)$ are defined to have the following elements (see section 2.16) [7, 9]:

$$F_{ii\,11} = \frac{\alpha_I\,(1 + m_i/m_I)}{(1 + x_{iI}^2)^{3/2}}, \tag{A.25}$$

$$F_{ii\,12} = \frac{3}{2}\,\frac{\alpha_I\,(1 + m_i/m_I)}{(1 + x_{iI}^2)^{5/2}}, \tag{A.26}$$

$$F_{ii\,21} = \frac{3}{2}\,\frac{\alpha_I\,(1 + m_i/m_I)}{(1 + x_{iI}^2)^{5/2}}, \tag{A.27}$$

$$F_{ii\,22} = \sqrt{2} + \frac{\alpha_I\,[13/4 + 4\,x_{iI}^2 + (15/2)\,x_{iI}^4]}{(1 + x_{iI}^2)^{5/2}}, \tag{A.28}$$

$$F_{iI\,11} = \frac{\alpha_I\,(1 + m_i/m_I)}{(1 + x_{iI}^2)^{3/2}}, \tag{A.29}$$

$$F_{iI\,12} = \frac{3}{2}\,\frac{T_i}{T_I}\,\frac{\alpha_I\,(1 + m_I/m_i)}{x_{iI}\,(1 + x_{Ii}^2)^{5/2}}, \tag{A.30}$$

$$F_{iI\,21} = \frac{3}{2}\,\frac{\alpha_I\,(1 + m_i/m_I)}{(1 + x_{iI}^2)^{5/2}}, \tag{A.31}$$

$$F_{iI\,22} = \frac{27}{4}\,\frac{T_i}{T_I}\,\frac{\alpha_I\,x_{iI}^2}{(1 + x_{iI}^2)^{5/2}}, \tag{A.32}$$

$$F_{Ii\,11} = \frac{\alpha_I\,(1 + m_i/m_I)}{(1 + x_{iI}^2)^{3/2}}, \tag{A.33}$$

$$F_{Ii\,12} = \frac{3}{2}\,\frac{\alpha_I\,(1 + m_i/m_I)}{(1 + x_{iI}^2)^{5/2}}, \tag{A.34}$$

$$F_{iI\,21} = \frac{3}{2}\,\frac{T_i}{T_I}\,\frac{\alpha_I\,(1 + m_I/m_i)}{x_{iI}\,(1 + x_{Ii}^2)^{5/2}}, \tag{A.35}$$

$$F_{Ii\,22} = \frac{27}{4}\,\frac{\alpha_I\,x_{iI}^2}{(1 + x_{iI}^2)^{5/2}}, \tag{A.36}$$

$$F_{II\,11} = \frac{\alpha_I\,(1 + m_i/m_I)}{(1 + x_{iI}^2)^{3/2}}, \tag{A.37}$$

$$F_{\mathrm{II}\,12} = \frac{3}{2}\,\frac{T_{\mathrm{i}}}{T_{\mathrm{I}}}\,\frac{\alpha_{\mathrm{I}}\,(1 + m_{\mathrm{I}}/m_{\mathrm{i}})}{x_{\mathrm{iI}}\,(1 + x_{\mathrm{Ii}}{}^{2})^{5/2}}, \tag{A.38}$$

$$F_{\mathrm{II}\,21} = \frac{3}{2}\,\frac{T_{\mathrm{i}}}{T_{\mathrm{I}}}\,\frac{\alpha_{\mathrm{I}}\,(1 + m_{\mathrm{I}}/m_{\mathrm{i}})}{x_{\mathrm{iI}}\,(1 + x_{\mathrm{Ii}}{}^{2})^{5/2}}, \tag{A.39}$$

$$F_{\mathrm{II}\,22} = \frac{T_{\mathrm{i}}}{T_{\mathrm{I}}}\left\{\sqrt{2}\,\alpha_{\mathrm{I}}{}^{2}\,x_{\mathrm{Ii}} + \frac{\alpha_{\mathrm{I}}\,[15/2 + 4\,x_{\mathrm{iI}}{}^{2} + (13/4)\,x_{\mathrm{iI}}{}^{4}]}{(1 + x_{\mathrm{iI}}{}^{2})^{5/2}}\right\}. \tag{A.40}$$

The 2×2 dimensionless electron collisional friction force matrices, $[F_{\mathrm{ee}}](r)$, $[F_{\mathrm{ei}}](r)$, and $F_{\mathrm{eI}}(r)]$ are defined to have the following elements (see section 2.16) [7, 9]:

$$F_{\mathrm{ee}\,11} = Z_{\mathrm{eff}}, \tag{A.41}$$

$$F_{\mathrm{ee}\,12} = \frac{3}{2}\,Z_{\mathrm{eff}}, \tag{A.42}$$

$$F_{\mathrm{ee}\,21} = \frac{3}{2}\,Z_{\mathrm{eff}}, \tag{A.43}$$

$$F_{\mathrm{ee}\,22} = \sqrt{2} + \frac{13}{4}\,Z_{\mathrm{eff}}, \tag{A.44}$$

$$F_{\mathrm{ei}\,11} = Z_{\mathrm{eff\,i}}, \tag{A.45}$$

$$F_{\mathrm{ei}\,12} = 0, \tag{A.46}$$

$$F_{\mathrm{ei}\,21} = \frac{3}{2}\,Z_{\mathrm{eff\,i}}, \tag{A.47}$$

$$F_{\mathrm{ei}\,22} = 0, \tag{A.48}$$

$$F_{\mathrm{eI}\,11} = Z_{\mathrm{eff\,I}}, \tag{A.49}$$

$$F_{\mathrm{eI}\,12} = 0, \tag{A.50}$$

$$F_{\mathrm{eI}\,21} = \frac{3}{2}\,Z_{\mathrm{eff\,I}}, \tag{A.51}$$

$$F_{\mathrm{eI}\,22} = 0. \tag{A.52}$$

A.5 Neoclassical viscosity matrices

The 2×2 dimensionless species-s neoclassical viscosity matrix, $[\mu_{\mathrm{s}}](r)$ (see section 2.17), is defined to have the following elements [7]:

$$\mu_{s\,11} = K_{s\,11}, \tag{A.53}$$

$$\mu_{s\,12} = \frac{5}{2}\,K_{s\,11} - K_{s\,12}, \tag{A.54}$$

$$\mu_{s\,21} = \frac{5}{2}\,K_{s\,11} - K_{s\,12}, \tag{A.55}$$

$$\mu_{s\,22} = K_{s\,22} - 5\,K_{s\,12} + \frac{25}{4}\,K_{s\,11}. \tag{A.56}$$

(Note that the viscosity matrix elements defined here differ from those defined in section 2.17 by a factor of g_t.) Here,

$$K_{e\,jk}(r) = g_t\,\frac{8}{3\sqrt{\pi}}\int_0^\infty \frac{e^{-x^2}\,x^{5+2\,(j+k)}\,\nu_{D\,e}(x)\,dx}{[x^4 + \nu_{*\,e}\,\nu_{D\,e}(x)]\,[x^4 + (5\pi/8)\,(\omega_{t\,e}\,\tau_{ee})^{-1}\,\nu_{T\,e}(x)]}, \tag{A.57}$$

$$\nu_{D\,e}(x) = \frac{3\sqrt{\pi}}{4}\left[\left(1 - \frac{1}{2\,x^2}\right)Y(x) + Y'(x)\right] + \frac{3\sqrt{\pi}}{4}\,Z_{\rm eff}, \tag{A.58}$$

$$\nu_{\epsilon\,e}(x) = \frac{3\sqrt{\pi}}{2}[Y(x) - Y'(x)], \tag{A.59}$$

$$\nu_{T\,s}(x) = 3\,\nu_{D\,s}(x) + \nu_{\epsilon\,s}(x), \tag{A.60}$$

and

$$Y(x) = \frac{2}{\sqrt{\pi}}\int_0^{x^2} \sqrt{t}\,e^{-t}\,dt, \tag{A.61}$$

$$Y'(x) = \frac{2}{\sqrt{\pi}}\,x\,e^{-x^2}. \tag{A.62}$$

Furthermore,

$$K_{i\,jk}(r) = g_t\,\frac{8}{3\sqrt{\pi}}\int_0^\infty \frac{e^{-x^2}\,x^{4+2\,(j+k)}\,\nu_{D\,i}(x)\,dx}{[x^3 + \nu_{*\,i}\,\nu_{D\,i}(x)]\,[x^3 + (5\pi/8)\,(\omega_{t\,i}\,\tau_{ii})^{-1}\,\nu_{T\,i}(x)]}, \tag{A.63}$$

$$\begin{aligned}\nu_{D\,i}(x) = {} & \frac{3\sqrt{\pi}}{4}\left[\left(1 - \frac{1}{2\,x^2}\right)Y(x) + Y'(x)\right]\frac{1}{x} \\ & + \frac{3\sqrt{\pi}}{4}\,\alpha_I\left[\left(1 - \frac{x_{iI}^{\,2}}{2\,x^2}\right)Y\!\left(\frac{x}{x_{iI}}\right) + Y\!\left(\frac{x}{x_{iI}}\right)\right]\frac{1}{x},\end{aligned} \tag{A.64}$$

$$\nu_{\epsilon\,i}(x) = \frac{3\sqrt{\pi}}{2}[Y(x) - Y'(x)]\frac{1}{x} + \frac{3\sqrt{\pi}}{2}\,\alpha_I\left[\frac{m_i}{m_I}\,Y\!\left(\frac{x}{x_{iI}}\right) - Y'\!\left(\frac{x}{x_{iI}}\right)\right]\frac{1}{x}, \tag{A.65}$$

and finally,

$$K_{\mathrm{I}\,jk}(r) = g_{\mathrm{t}}\,\frac{8}{3\sqrt{\pi}}\int_0^\infty \frac{\mathrm{e}^{-x^2}\,x^{4+2\,(j+k)}\,\nu_{\mathrm{D\,I}}(x)\,dx}{[x^3+\nu_{*\,\mathrm{I}}\,\nu_{\mathrm{D\,I}}(x)]\,[x^3+(5\pi/8)\,(\omega_{\mathrm{t\,I}}\,\tau_{\mathrm{II}})^{-1}\,\nu_{\mathrm{T\,I}}(x)]}, \tag{A.66}$$

$$\begin{aligned}\nu_{\mathrm{D\,I}}(x) &= \frac{3\sqrt{\pi}}{4}\left[\left(1-\frac{1}{2\,x^2}\right)Y(x)+Y'(x)\right]\frac{1}{x}\\ &\quad+\frac{3\sqrt{\pi}}{4}\,\frac{1}{\alpha_{\mathrm{I}}}\left[\left(1-\frac{x_{\mathrm{Ii}}^{\,2}}{2\,x^2}\right)Y\!\left(\frac{x}{x_{\mathrm{Ii}}}\right)+Y'\!\left(\frac{x}{x_{\mathrm{Ii}}}\right)\right]\frac{1}{x},\end{aligned} \tag{A.67}$$

$$\nu_{\epsilon\,\mathrm{I}}(x) = \frac{3\sqrt{\pi}}{2}[Y(x)-Y'(x)]\frac{1}{x}+\frac{3\sqrt{\pi}}{2}\,\frac{1}{\alpha_{\mathrm{I}}}\left[\frac{m_{\mathrm{I}}}{m_{\mathrm{i}}}\,Y\!\left(\frac{x}{x_{\mathrm{Ii}}}\right)-Y'\!\left(\frac{x}{x_{\mathrm{Ii}}}\right)\right]\frac{1}{x}. \tag{A.68}$$

Note that our expressions for the neoclassical viscosity matrices interpolate between the three standard neoclassical collisionality regimes (i.e. the banana, plateau, and Pfirsch–Schlüter regimes [8]) in the most accurate manner possible [7].

A.6 Parallel force and heat balance

Let

$$[\tilde{\mu}_{\mathrm{I}}] = \alpha_{\mathrm{I}}^{\,2}\,\frac{T_{\mathrm{i}}}{T_{\mathrm{I}}}\,x_{\mathrm{Ii}}\,[\mu_{\mathrm{I}}]. \tag{A.69}$$

The requirement of equilibrium force and heat balance parallel to the magnetic field (see sections 2.18 and 2.19) leads us to define four 2 × 2 dimensionless ion matrices, $[L_{\mathrm{ii}}](r)$, $[L_{\mathrm{iI}}](r)$, $[L_{\mathrm{Ii}}](r)$, and $[L_{\mathrm{II}}](r)$, where [7, 9]

$$\begin{pmatrix}[L_{\mathrm{ii}}], & [L_{\mathrm{iI}}]\\ [L_{\mathrm{Ii}}], & [L_{\mathrm{II}}]\end{pmatrix} = \begin{pmatrix}[F_{\mathrm{ii}}+\mu_{\mathrm{i}}], & -[F_{\mathrm{iI}}]\\ -[F_{\mathrm{Ii}}], & [F_{\mathrm{II}}+\tilde{\mu}_{\mathrm{I}}]\end{pmatrix}^{-1}\begin{pmatrix}[F_{\mathrm{ii}}], & -[F_{\mathrm{iI}}]\\ -[F_{\mathrm{Ii}}], & [F_{\mathrm{II}}]\end{pmatrix}, \tag{A.70}$$

and the 2 × 2 dimensionless electron matrices, $[Q_{\mathrm{ee}}](r)$, $[L_{\mathrm{ee}}](r)$, $[L_{\mathrm{ei}}](r)$, and $[L_{\mathrm{eI}}](r)$, where

$$[Q_{\mathrm{ee}}] = [F_{\mathrm{ee}}+\mu_{\mathrm{e}}]^{-1}, \tag{A.71}$$

$$[L_{\mathrm{ee}}] = [Q_{\mathrm{ee}}]\,[F_{\mathrm{ee}}], \tag{A.72}$$

$$[L_{\mathrm{ei}}] = [Q_{\mathrm{ee}}]\{[F_{\mathrm{ei}}]\,[L_{\mathrm{ii}}]-[F_{\mathrm{ei}}]+[F_{\mathrm{eI}}]\,[L_{\mathrm{Ii}}]\}, \tag{A.73}$$

$$[L_{\mathrm{eI}}] = [Q_{\mathrm{ee}}]\{[F_{\mathrm{eI}}]\,[L_{\mathrm{II}}]-[F_{\mathrm{eI}}]+[F_{\mathrm{ei}}]\,[L_{\mathrm{iI}}]\}. \tag{A.74}$$

A.7 Neoclassical fluid velocities

Let

$$I(r) = B_0\,R_0\,g, \tag{A.75}$$

$$\eta_s(r) = \frac{d\ln T_s}{d\ln n_s}, \tag{A.76}$$

$$\Omega_E(r) = -\frac{d\Phi}{d\Psi_p}, \tag{A.77}$$

$$\Omega_{*\,s}(r) = -\frac{1}{e_s\, n_s}\,\frac{dp_s}{d\Psi_p}, \tag{A.78}$$

$$\Omega_s(r) = \frac{e_s\,\langle B\rangle}{m_s}, \tag{A.79}$$

where $\Phi(r)$ is the equilibrium electric scalar potential, $\Psi_p(r)$ the equilibrium poloidal magnetic flux (see equation (14.25)), and $p_s(r) = n_s\,T_s$ the species-s pressure. Note that Ω_E, $\Omega_{*\,s}$, and Ω_s all have the dimensions of frequency. Furthermore, let

$$\mathcal{L}_{ss'}(r) = L_{ss'\,11} - \frac{\eta_{s'}}{1+\eta_{s'}}\,L_{ss'\,12}, \tag{A.80}$$

$$\mathcal{Q}_{ee}(r) = Q_{ee\,11}. \tag{A.81}$$

Finally,

$$\langle A\rangle(r) = \oint R^2\,A(r,\,\theta)\,\frac{d\theta}{2\pi}\Big/\oint R^2\,\frac{d\theta}{2\pi} = \oint\frac{A(r,\,\Theta)}{B}\,\frac{d\Theta}{2\pi}\Big/\oint\frac{1}{B}\,\frac{d\Theta}{2\pi}. \tag{A.82}$$

(See equation (2.148).)

As described in reference [7] and illustrated in a simplified form in sections 2.14–2.20, we can determine the equilibrium fluid velocities of the various species that make up the plasma by balancing the parallel viscous force density against the parallel friction force density and the force density due to the parallel component of the electric field. We obtain [2–5]:

$$\frac{\mathbf{V}_i\cdot\nabla\theta}{\mathbf{B}\cdot\nabla\theta}\,\frac{\langle B^2\rangle}{I} = -\mathcal{L}_{ii}\,\Omega_{*\,i} - \mathcal{L}_{iI}\,\Omega_{*\,I}, \tag{A.83}$$

$$\frac{\langle\mathbf{V}_i\cdot\mathbf{B}\rangle}{I} = \Omega_E + \Omega_{*\,i} - \mathcal{L}_{ii}\,\Omega_{*\,i} - \mathcal{L}_{iI}\,\Omega_{*\,I}, \tag{A.84}$$

$$\frac{\mathbf{V}_I\cdot\nabla\theta}{\mathbf{B}\cdot\nabla\theta}\,\frac{\langle B^2\rangle}{I} = -\mathcal{L}_{II}\,\Omega_{*\,I} - \mathcal{L}_{Ii}\,\Omega_{*\,i}, \tag{A.85}$$

$$\frac{\langle\mathbf{V}_I\cdot\mathbf{B}\rangle}{I} = \Omega_E + \Omega_{*\,I} - \mathcal{L}_{II}\,\Omega_{*\,I} - \mathcal{L}_{Ii}\,\Omega_{*\,i}, \tag{A.86}$$

$$\frac{\mathbf{V}_e\cdot\nabla\theta}{\mathbf{B}\cdot\nabla\theta}\,\frac{\langle B^2\rangle}{I} = -\mathcal{L}_{ee}\,\Omega_{*\,e} - \mathcal{L}_{ei}\,\Omega_{*\,i} - \mathcal{L}_{eI}\,\Omega_{*\,I} + \mathcal{Q}_{ee}\,\Omega_e\,\tau_{ee}\,\frac{\langle \mathbf{E}\cdot\mathbf{B}\rangle}{I\,\langle B\rangle}, \tag{A.87}$$

$$\begin{aligned}\frac{\langle \mathbf{V}_e\cdot\mathbf{B}\rangle}{I} = \Omega_E + \Omega_{*\,e} - \mathcal{L}_{ee}\,\Omega_{*\,e} - \mathcal{L}_{ei}\,\Omega_{*\,i} - \mathcal{L}_{eI}\,\Omega_{*\,I}\\ + \mathcal{Q}_{ee}\,\Omega_e\,\tau_{ee}\,\frac{\langle \mathbf{E}\cdot\mathbf{B}\rangle}{I\,\langle B\rangle}.\end{aligned} \tag{A.88}$$

Here, $\mathbf{V}_s$ is the species-s equilibrium fluid velocity. The previous equations are generalizations of equations (2.219), (2.221), (2.255), and (2.257). The additional effects included in the generalized equations are a frictional drag due to impurity ions, a fraction of trapped particles that is not necessarily small, and a plasma collisionality that is not necessarily in the banana regime.

A.8 The Glasser–Greene–Johnson parameters

It is convenient to add to this appendix a generalized calculation of the magnetic curvature length, L_c, that appears in the Rutherford island width evolution equation, (12.15). Let $\hat{B} = |\mathbf{B}|/B_0$, $\hat{\nabla} = R_0\,\nabla$, and $d\psi_p/dr = f(r)/R_0$. Furthermore, let

$$J_1(r) = \oint \frac{1}{\hat{B}}\,\frac{d\Theta}{2\pi}, \tag{A.89}$$

$$J_2(r) = \oint \hat{B}\,\frac{d\Theta}{2\pi}, \tag{A.90}$$

$$J_3(r) = \oint \frac{1}{\hat{B}^{\,3}}\,\frac{d\Theta}{2\pi}, \tag{A.91}$$

$$J_4(r) = \oint \frac{1}{\hat{B}\,|\hat{\nabla}\psi_p|^2}\,\frac{d\Theta}{2\pi}, \tag{A.92}$$

$$J_5(r) = \oint \frac{\hat{B}}{|\hat{\nabla}\psi_p|^2}\,\frac{d\Theta}{2\pi}, \tag{A.93}$$

$$J_6(r) = \oint \frac{1}{\hat{B}^{\,3}\,|\hat{\nabla}\psi_p|^2}\,\frac{d\Theta}{2\pi}. \tag{A.94}$$

It follows that [6]

$$E(r) = -\frac{dP/d\psi_p}{(dq/d\psi_p)^2}\,\frac{1}{\gamma}\left[\frac{d}{d\psi_p}\!\left(\frac{J_1}{\gamma}\right) - g\,\frac{dq}{d\psi_p}\,\frac{J_1}{J_2}\right]J_5, \tag{A.95}$$

$$F(r) = \frac{(dP/d\psi_p)^2}{(dq/d\psi_p)^2}\,\frac{1}{\gamma^2}\left[g^2\left(J_5\,J_6 - J_4^{\,2}\right) + J_5\,J_3\right], \tag{A.96}$$

$$H(r) = \frac{dP/d\psi_{\rm p}}{dq/d\psi_{\rm p}}\,\frac{g}{\gamma}\left(J_4 - \frac{J_1\,J_5}{J_2}\right), \tag{A.97}$$

where $P(r) = \mu_0\,p(r)/B_0^{\,2}$, and $p(r)$ is the equilibrium plasma pressure. Finally,

$$D_{\rm R}(r) = E + F + H^2. \tag{A.98}$$

The value of the dimensionless parameter $D_{\rm R}(r_s)$ at a given rational magnetic flux surface is related to the magnetic curvature length, L_c, introduced in section 11.4, according to

$$D_{\rm R}(r) = -\frac{2\,c_\beta^{\,2}\,L_s^{\,2}}{L_c\,L_p}. \tag{A.99}$$

Here, $c_\beta^{\,2}$ is a dimensionless measure of the plasma pressure at the rational surface (see equations (4.65) and (4.66)), L_s the magnetic shear length at the rational surface (see equation (5.27)), and L_p the effective pressure gradient scale length at the rational surface (see equation (8.35)). The previous equation is a generalization of equation (11.57). The latter equation only holds in a large-aspect-ratio, low-β, tokamak plasma with magnetic flux surfaces of circular cross section.

References

[1] Fitzpatrick R 2014 *Plasma Physics: An Introduction* (Boca Raton, FL: CRC Press)

[2] Fitzpatrick R 2020 Modeling q_{95} windows for the suppression of edge localized modes by resonant magnetic perturbations in the DIII-D tokamak *Phys. Plasmas* **27** 102511

[3] Fitzpatrick R 2021 Further modeling of q_{95} windows for the suppression of edge localized modes by resonant magnetic perturbations in the DIII-D tokamak *Phys. Plasmas* **28** 022503

[4] Fitzpatrick R, Kim S and Lee J 2021 Modeling of q_{95} windows for the suppression of edge localized modes by resonant magnetic perturbations in the KSTAR tokamak *Phys. Plasmas* **28** 082511

[5] Fitzpatrick R and Nelson A O 2020 An improved theory of the response of DIII-D H-mode discharges to static resonant magnetic perturbations and its implications for the suppression of edge localized modes *Phys. Plasmas* **27** 072501

[6] Glasser A H, Greene J M and Johnson J L 1975 Resistive instabilities in general toroidal plasma configurations *Phys. Fluids* **18** 875

[7] Hirschman S P and Sigmar D J 1981 Neoclassical transport of impurities in tokamak plasmas *Nucl. Fusion* **21** 1079

[8] Galeev A A and Sagdeev R Z 1968 Transport phenomena in a collisionless plasma in a toroidal magnetic system *Sov. Phys. - JETP* **26** 233 http://jetp.ras.ru/cgi-bin/e/index/e/26/1/p233?a=list

[9] Kim Y B, Diamond P H and Groebner R J 1991 Neoclassical poloidal and toroidal rotation in tokamaks *Phys. Fluids* B **3** 2050

www.ingramcontent.com/pod-product-compliance
Ingram Content Group UK Ltd.
Pitfield, Milton Keynes, MK11 3LW, UK
UKHW051904100926
8556IPUK00005B/763

9 780750 353656